Streaming Media with Peer-to-Peer Networks:

Wireless Perspectives

Martin Fleury
University of Essex, UK

Nadia Qadri
COMSATS Institute of Information Technology, Pakistan

Managing Director:	Lindsay Johnston
Senior Editorial Director:	Heather A. Probst
Book Production Manager:	Sean Woznicki
Development Manager:	Joel Gamon
Acquisitions Editor:	Erika Gallagher
Typesetter:	Lisandro Gonzalez
Cover Design:	Nick Newcomer

Published in the United States of America by
Information Science Reference (an imprint of IGI Global)
701 E. Chocolate Avenue
Hershey PA 17033
Tel: 717-533-8845
Fax: 717-533-8661
E-mail: cust@igi-global.com
Web site: http://www.igi-global.com

Library of Congress Cataloging-in-Publication Data

Streaming media with peer-to-peer networks: wireless perspectives / Martin Fleury and Nadia Qadri, editors.
p. cm.
Includes bibliographical references and index.
Summary: "This book offers insights into current and future communication technologies for a converged Internet that promises soon to be dominated by multimedia applications, at least in terms of bandwidth consumption"--Provided by publisher.
ISBN 978-1-4666-1613-4 (hardcover) -- ISBN 978-1-4666-1614-1 (ebook) -- ISBN 978-1-4666-1615-8 (print & perpetual access) 1. Streaming technology (Telecommunications) 2. Internet. I. Fleury, Martin, 1951- II. Qadri, Nadia, 1980-
TK5105.386.S738 2012
004.67'8--dc23

2012002468

British Cataloguing in Publication Data
A Cataloguing in Publication record for this book is available from the British Library.

Editorial Advisory Board

List of Reviewers

Table of Contents

Detailed Table of Contents

Preface

This book presents a set of research-based investigations of a number of paradigms, which work together in various ways. From the book title we have: "peer-to-peer network," multimedia "streaming," and "wireless perspectives," in other words, mobile applications of peer-to-peer streaming. It turns out that there is or there is suspected to be a synergy between one type of wireless network and peer-to-peer networks. This is possible because peer-to-peer networks are overlay networks that are superimposed upon an existing physical network. The issue of the compatibility of the two paradigms is explored in Chapter 5. The wireless networks in question are ad hoc or infrastructureless, meaning that they can exist without the presence of a centralized access point to redirect traffic between access points over a wired infrastructure network. As this is a book about paradigms, it seemed fitting to include a final chapter which considered information-centric networks. These, too, are overlays, and may present the next step in locating multimedia streaming sources without the use of router-based addressing at the network layer of the protocol stack. In fact, service oriented architectures, XML routers, deep packet inspection, and content delivery networks (see later) together with peer-to-peer overlay technologies are a clear effort from the top to move towards a networking which is focused on information rather than router-based addressing. However, before considering the contents of the book's chapters further, it is worth asking the question "Where do these paradigms stand in the commercial world at the time of writing?"

Streaming is generally used for pre-encoded video streaming in order to avoid hefty storage requirements, particularly on mobile battery-powered devices. This requirement will grow as the era of high-definition laptops is entered. Streaming can also help protect content from copying, as there is no stored copy at the end-point. For live encoded video and two-way interactive video (such as video phone and video conferencing), streaming is certainly necessary. Chapter 4 has more on advances in adaptive streaming. Streaming seems to have been introduced commercially by Real Networks with a pioneering audio broadcast of a baseball event in 1995. The company went on to introduce video streaming in 1997 so that by 2000, about 85% of streaming content in the Internet was apparently in the Real format. However, the Real Player, which now streams using a proprietary codec rather than originally an H.263 standard codec, appears to have suffered a market decline, at least in respect to video streaming. One cause of this may have been the obtrusiveness of Real Player upgrades and auxiliary programs, or it could be the cost of the servers, or indeed competition with Adobe Flash Player (except on Apple devices). It is worth noting that though Real did prevent the writing of their streams to store, there now exists software to do this (de-streaming), thus allowing viewing times to be shifted. There are also filters to allow different players, other than the Real Player, to be used.

Commercial streaming tends to employ simulcast for pre-encoded content, in which different versions of the same pre-encoded stream are stored on a server, about five versions on the Akamai *content delivery network (CDN)*. Of course, this involves considerable storage and organization and management of that storage. The pixel/frame resolution is often varied resulting in different bitrates to suit the bandwidth bottleneck that commonly exists over an *Asymmetric Digital Subscriber Connection (ADSL* —marketed as "broadband"). Commercial streaming also tends to use pseudo-streaming, which is also known as progressive download. Pre-encoded bitstreams are broken into chunks, each of which is downloaded, while a previous chunk is played. Because browser plug-in compatibility is seen as important, connection-oriented *Transmission Control Protocol (TCP)* is used as the transport protocol for *HyperText Transfer Protocol (HTTP),* or in Adobe's case as the basis of *Real Time Messaging Protocol (RTMP),* which in turn may be encapsulated in HTTP. (RTMP is a proprietary protocol, which reportedly appears in one variant to use an anonymous Diffie-Hellman mechanism for its authentication, making content vulnerable to snooping.) The widespread adoption of the Adobe Flash Player (present on about 95% of browsers in July 2011) acts as a non-technical constraint on the form of streaming. Microsoft Silverlight has about 53% penetration of browsers. Both of these players use proprietary software.

It is sometimes said that TCP is necessary to penetrate firewalls and *network address translation (NAT)*. However, it is the negotiation of a connection that is needed for this purpose and a protocol such as *Datagram Congestion Control Protocol (DCCP)* can set up a connection before reverting to *User Datagram Protocol (UDP)* transport. In fact, Skype is thought to use the STUN protocol for NAT and firewall traversal by UDP streams. As is often observed, TCP, because of its built-in reliability, is unsuitable for a real-time service, as congestion can lead to frequent resends of the data and a slowing down of the transmission rate. The problem is worse in a wireless network as lost packets may be due to channel conditions, in which case, staunching the flow will not make the adverse channel conditions go away. Buffers can smooth out the delays. However, large buffers militate against *click-and-view* type streaming, as they lead to long start-up delays. This is only partially mitigated by placing the content in caches nearer the user. Apart from this, pseudo-streaming is unsuitable for the streaming of two-way video. However, it should be remarked that *peer-to-peer (P2P)* streaming also uses chunks rather than bitstreams, though the difficulty of lost chunks is now solved by means of having multiple sources for the same chunk. Chapter 8 considers the chunk method of streaming for wireless mobile ad hoc networks and their vehicle-bound related networks (VANETs).

In recognition of these problems with pseudo-streaming, commercial operators have introduced adaptive pseudo-download (one variant is known as HTTP Live Streaming) in which the target download rate is not fixed at start-up time but can be adapted according to measured or estimated conditions on the network. Chunk sizes tend to be between 2 and 15 s(econds) in duration. Unfortunately, this is unlikely to be suitable for wireless networks, as the channel can change over a time interval much smaller than a chunk's download time. On the other hand, if a chunk's size is reduced then compression coding gain is reduced as well. The complexity of managing selection of chunks in the context of stateless HTTP is also a considerable burden. Instead, there are a number of alternative ways to implement adaptation. If simulcast is to be used then an H.264/*AVC (Advanced Video Coding)* codec supports switching frames, which can remove the overhead of spatially coded I-frames as switching points and provide smoother transitions at those switching points. Bitrate transcoder banks can also serve to dynamically adapt to changes in congestion. However, transcoders represent a hardware investment that can be removed by scalable codecs. Such codecs as the *Scalable Video Coding (SVC)* extension to H.264 are said to have almost the same coding efficiency as H.264/AVC. The research perspective for scalable streaming is considered in Chapter 7. However, at the time of writing hardware H.264/SVC codecs do not appear to be available, which implies that mobile devices with comparatively limited processing ability are not able to decode SVC streams. A commercial solution by Vidyo for video conferencing appears to only send the base layer to 3G/4G devices, whereas other SVC layers are sent to static devices.

CDNs are another related area of commercial activity. In brief, a CDN distributes content from centrally placed servers to edge servers that act as caches for that content. The content may be load-balanced between these *point-of-presences (PoPs)*. When (say) a streamed video clip is selected from a central server, the request is diverted to an edge server or PoP and the originating requester application is also informed of the download source. Selection of a suitable edge server is by some criterion such as cost or quality of service. The problem of occasional high demand for (say) a popular TV program (flash crowds that can overwhelm client server systems) is reduced by a CDN but still presents a weakness.

A hybrid P2P CDN can help in this situation, as popular content is downloaded by the peers themselves to the network edge. Thus P2P assisted systems are more scalable than pure CDNs. Peers act as alternative content servers. In effect, this removes the cost of distributing data from a central server to the edge that a CDN provider must meet. However, there are some disadvantages of P2P CDNs. Due to the asymmetric nature of twisted-pair telephone line transmission under *Asymmetric Digital Subscriber Line (ADSL)*, there may not be enough bandwidth to conveniently support the uploading of content from a peer. A P2P CDN may also generate more traffic than a traditional CDN, as material is downloaded to multiple peers and then can be uploaded. It was probably the additional traffic generated and in particular that caused the original version of the BBC iPlayer (a UK catch-up TV service operated by a public broadcaster) to be converted from P2P delivery or rather a hybrid P2P delivery service from Kontiki. In doing that the BBC avoided the risk of *Internet Service Provider (ISP)* traffic blocking or traffic shaping. Another issue is the cost to the peer of acting as a P2P host in terms of reduction in compute power and occupation of upload bandwidth. The risk of downloading malware and the need to protect content providers' intellectual property must also be addressed in a commercial setting (as is promised for example by Kontiki, which operates a closed public-key infrastructure).

A clean-slate hybrid P2P solution to the delivery of streaming content is illustrated by China Telecom's *Media Telecom Network (MTN)*. This is intended to provide live and on-demand TV through cache servers and by exchange between the users' terminals. These can consist of PCs, set-top-boxes, and mobile phones. In contrast to research, which tends to concentrate on the P2P streaming mechanisms and architecture, a system such as MTN must manage subscribers, provide billing (including authentication, authorization, and accounting), and collect viewing statistics. In such a system terminal, churn is a constant issue for a P2P system but randomized selection of suitable peers is a way to counter that. Content exchange management is responsible for making the initial connection and the establishment of the initial streaming path, after which suitable peers are also identifies. There is also a need to maintain an electronic program guide. All content is subject to digital rights management, which requires authentication as part of the key exchange process. Thus, in such a large scale system, P2P exchange is one part of a much large system.

Streaming of TV programs is one of the commercial success stories of P2P (along, of course, with Skype, from the originators of the Kazaa file-sharing application — acquired by Microsoft in October 2011). The background to these streaming systems is considered in Chapter 3. These applications appear particularly popular in Asia and tend to concentrate on live programs rather than pre-recorded and stored TV programs. For example, QQLive is only available in Chinese but apparently attracts 3 million web site visitors per day. PPS.tv is available in a variety of Asian languages, with reportedly several hundred thousand users viewing at around 200-500 kbps per stream, i.e. low resolution TV, hosted by an originating server with a link no more bandwidth capacity than that of domestic ADSL. Other Asian P2PTV applications are TVUPlayer, PPLive, and PPStream. TVUplayer is one of the few that also carry U.S. channels as well. It should be noted that some re-broadcasting apparently comes without a license to do so.

CoolStreaming was a pioneering live P2PTV streaming solution from about 2000 onwards. At one point in time, CoolStreaming had as many as 80,000 concurrent viewers by 2008. CoolStreaming used a BitTorrent form of content distribution but targeting the chunk-based players of the Real Player, Windows Media Player variety (refer back to the earlier discussion). At the same time it carefully matched the resolution of streams to the download capability of access links. A gossip-based broadcast algorithm distributed TV program availability. The early version of CoolStream (according to its originators) suffered from a slow start-up time and possible reliance for download on a single source, which in turn could lead to rejection from the download of very popular content (called "flash crowds"). These problems were subsequently resolved firstly by servers acting as push agents (after the first pull operation by a client) rather than, as previously, clients pulling all video chunks from a server. Multipath and multi-stream delivery was deployed to make the service more reliable. However, notice that for live streaming it may be necessary to send redundant streams to ensure continuity, thus causing an increase in bandwidth usage, though probably no more than if multiple independent streams are distributed from the same server. Copyright issues seem to have closed CoolStreaming in 2005. However, Roxbeam Corporation, operating from Beijing, now has a P2P multicast TV solution, along with CDN support (a strengthened form of P2P using pre-positioned servers within China), VoD, and a service called RayCAST, which is a content production kit for P2P broadcasting. Roxbeam also now targets set-top boxes acting as P2P clients.

There is an issue over the quality of the viewing experience for such a service and Chapter 9 shows how mobile viewing experiences should best be assessed. While quality-of-service (QoS) assessment both on the network and in the quality of the video is sufficient for static networks, for mobile networks the variety of device displays, the expectations of viewers and the type of content are issues that need to be considered. These are generally grouped under a quality-of-experience (QoE) heading, which also takes into account the video viewing quality at the end device. Thus, QoE is not the same as QoS.

In Western Europe, a company like Zattoo has not apparently made as much progress as Asian P2PTV has. In part, this is due to licensing restrictions which are an obstacle to the re-broadcast of main-stream TV programs. There is also a problem with providing revenue, which in Switzerland at one point came from adding advertisements to re-broadcast German TV, which was then challenged in the courts by film content owners. Interestingly, from the point-of-view of this book is that there is now an iPad application for Zattoo, which implies that mobile streaming may provide a target for such P2P TV in Western Europe. Joost is another English-language P2P TV streaming service, with revenue clearly provided from advertisements, as these interrupt streams at regular intervals. Licensing issues restrict content viewing outside the USA.

Another area in which (hybrid) P2P TV may find a use is in narrowcasting, with corporate broadcasting being a case in point. The advantage of P2P in a corporate setting is that it reduces the cost of server deployment, while at the same time avoids the problems of ISP blocking or traffic shaping that occur in public networks. For example, Kontiki Inc. now offers such a service, after its previous foray into public Internet TV. Tele-education across university and college networks could also benefit from peer-assisted video streaming. Both corporate networks and educational establishments (at least in Anglo-

Saxon commercially minded universities) have a need for the delightfully-named ego-casting, in which the CEO or the principal of the university broadcast their views to the workers/academic and office staff.

Some of the chapters have been mentioned in the previous exploration of the commercial impact of P2P streaming. Chapter 1 is a short chapter that acts as an extension to this Preface, in that it considers the practicalities of extending the P2P paradigm to the commercial world. In particular, it considers Fonera wireless routers, which allow others to share some of their WLAN signal with other Fon users, thus creating a Fon spot (by analogy with wireless hotspot). One type of user exchanges free roaming for free access to others' Fon spots. It is also possible to receive revenue from the purchase of passes when these passes result in access to one's Fon spot. However, this business model is not widely implemented. In the UK, where this preface is being written, one of the largest ISP's had joined with Fon to create a network of Fon spots, and this editor was able to use such a hotspot when Internet access using ADSL was somehow discontinued by his erstwhile ISP. Thus, Fon spots can also act as backup networks. Chapter 1 also considers fixed-mobile convergence, which P2P can assist in. Chapter 2 explores this in more detail from the point-of-view of he convergence of the Internet with cellular phone networks, which requires an integration mechanism, namely the *IP Multimedia Subsystem (IMS)*. Chapter details a plan to integrate P2P with an overall system, which should be compared to the MTN of China Telecom.

Chapters 3 and 4 are related in that Chapter 3 reviews the development of P2P streaming from P2P file sharing while Chapter 4 continues the story by reviewing the need for streaming adaptation to varying network conditions. One place where such adaptation is certainly needed is in wireless networks, as the time granularity of channel change can be very rapid. In fact, as reported in Chapter 4, existing commercial solutions will be insufficient during periods of device mobility. Chapter 5 is the first of a number of chapters in this book that examine ad hoc networks. As previously mentioned, the chapter takes a hard look at the prospects of mapping the P2P paradigm onto ad hoc networks. Such networks have applications in emergencies and disasters when the existing network infrastructure has been removed. They are also attractive to security forces. In the cellular world, they have a role through multi-homing in providing backup capacity when cells are overcrowded or they can extend the coverage of a cellular system. However, the business model for capacity sharing for extensions of cellular systems remains unclear. In one area, that of wireless vehicular networks, there probably is a business model justifying the 'infotainment' as well as a case for public safety, reporting traffic congestion and accidents. Chapter 6 shows, particularly for vehicular networks, that more detailed modeling is required, particularly as it is not easy to set up wireless networks test-beds for automobiles or provide analytic solutions. Chapter 7 goes on to consider the feasibility of peer-to-peer video streaming in such environments.

This book is essentially about research, and from chapter 7 onwards, possible research directions are considered. Currently, as previously mentioned though scalable video coding has the potential to simplify video streaming within mobile networks, development of such systems is at the cutting edge of research, especially for peer-to-peer streaming. As Chapter 9 demonstrates, scalable video coding brings the prospect of adaptation to peer-to-peer networks. In that case, streaming should be aware both of the P2P network overlay and also the network underlay upon which the P2P network is mapped. If mobile P2P streaming is to become a reality then some way of judging the users' quality of experience is needed, if only to try to improve it in situations in which it is compromised. As previously mentioned Chapter 9 examines this issue as well as giving a guide to video streaming quality of experience assessment. Peer-to-peer streaming is a form of application layer multicast, occasioned by the failure of IP multicasting to be effectively and universally deployed. Chapter 10 places P2P multicast in the context of reliable application-layer multicasting and reviews the research that is still ongoing in this field.

Finally, the last section of the book considers related network development areas. Chapter 11 continues the theme of ad hoc networks but now instead of the familiar WLAN networks, the topic is extended into wireless mesh networks and in particular to broadband wireless technologies such as WiMAX. It is sometimes forgotten by European technologists, who believe that Long Term Evolution(-advanced) broadband wireless will predominate that IEEE 802.11m, is an at least if not better technology and that it is now widely deployed in much of the world, as WiMAX forum statistics confirm. Chapter 12 continues the theme of wireless network exploration with a look at multi-homed devices (ones that can communicate over more than one network type). This field presents a fertile research topic and represents a possible target of P2P streaming, once routing problems have been addressed. Chapter 12 represents a very detailed investigation of how to optimize routing in the mobile multi-homed environment. As was mentioned, at the beginning of this preface, Chapter 13 represents a future view of networks in which just as in P2P networking, an overlay is used. However, in this case routing is based on information rather than node virtual topology. In fact, looking back to chapter one, one can see that this chapter represents the latest turn in fixed-mobile convergence, of which P2P communication has a strategic part (see also Chapter 2). It is hoped that the books' content will be stimulating, giving a foundation in the first section, moving on to more specific topics in the second section, with research contributions on future networks in the final section.

In terms of the overall importance of understanding trends in video streaming, including P2P steaming, and the contribution of mobile video watching in particular, a number of salient facts can be pointed to. The number of users watching multimedia content over the Internet, both live and on-demand, has dramatically increased in recent years, and is now the leading source of Internet traffic, surpassing even peer-to-peer file sharing (before peer-to-peer streaming the leading contribution of the peer-to-peer paradigm). This trend has been reported in several surveys, and is confirmed by network operators. Over one third of the top 50 sites by volume are based on video distribution (Source: Cisco Visual Networking Index: Usage, 2010-2015). Netflix, a provider of on-demand Internet streaming video, at the moment accounts by itself for almost 30% of peak period downstream traffic (Source: Sandvine's Global Internet Phenomena Report - Spring 2011). The sum of all forms of video is expected to exceed 90% of global user traffic by 2014. The web is becoming one of the main broadcasting platforms; this is due to the recent availability of the so called *three screens* (digital TV, PC, and smartphones) capable of accessing it, and the increase of the number of faster connections to the network. The most recent sport events, such as the FIFA World Cup and the Vancouver 2010 Winter Games, have been watched through streaming by several million unique simultaneous viewers. Mobile Internet data traffic stands at 237 Peta Bytes (PB) per month in 2010 but is set to rise to 6,254 PB in 2015 and could exceed that of the wired Internet by 2015 according to Cisco's published estimates. Mobile Internet is by far the fastest growing new service with a forecast increase in data traffic over the next five years of 92%. In fact some markets such as South America according to Sandvine have largely replaced fixed or wired access with wireless or mobile access. Overall according to Cisco, unmanaged IP consumer traffic is predicted to be 53.3 Exabytes per month by the end of 2015, as opposed to 11.8 for managed networks, and 4.9 over mobile networks. This book is a modest contribution to understanding these developments.

The book will be of interest to industry managers, as it offers insights into current and future communication technologies for a converged Internet that promises soon to be dominated by multimedia applications (at least in terms of bandwidth consumption). Chapter 1, together with this preface, marks out that relevance, as well as indicating the challenges of selecting a viable business plan. Graduate researchers will find that individual chapters identify research areas and provide pointers to and a review of the research literature in the area. Some chapters in the final section also act as detailed research reports. Some of the chapters are also suitable as a basis of Masters' dissertation topics and likewise provide a kit to start such a project off. As such, a library copy will act as a valuable resource to such students and researchers.

Martin Fleury
University of Essex, UK

Section 1
Foundations

Chapter 1
Introduction:
Fixed-Mobile Convergence, Streaming Multimedia Services, and Peer-to-Peer Communication

Jason J. Yao
Graduate Institute of Communication Engineering, National Taiwan University, Taiwan

Homer H. Chen
Graduate Institute of Communication Engineering, National Taiwan University, Taiwan

ABSTRACT

Peer- to-peer technology has dramatically transformed the landscape of the Internet's traffic in recent years. In this introductory chapter, the authors highlight how the technology relates to the convergence of fixed and mobile networks with features that work irrespective of location, access technology, or user-interactive devices.

FIXED-MOBILE CONVERGENCE

Mobile telephony is regarded as the sweet spot of today's telecoms market. Its 1.7 billion user base still enjoys high growth rates of subscribers compared to other sectors of the telecommunications market. Unfortunately, mobile telephony is still based on traditional circuit-switching, which translates to higher costs and less flexibility to carry packet-based data traffic. Although serious efforts, both in publicity and in real work, have been made to migrate toward *Voice over Internet Protocol (VoIP)*, the transformation has barely begun. In developed countries, mobile phone operators face the challenge of finding new growth prospects. (For example, there are already more mobile phones than people in Sweden.) While battles for market share help consumers, they are costly and provide limited growth for operators. To expand or retain their customer base, operators constantly need to offer new services. One attractive selling point is to offer a single handset working as your mobile phone, home phone as well as office phone. Another is to offer a set of services accessible through multiple devices, such

DOI: 10.4018/978-1-4666-1613-4.ch001

as your office phone, your mobile handset, or even a speakerphone in a conference room. The key for these scenarios is the convergence of fixed and mobile networks that enables new features that work irrespective of location, access technology or user-interface device.

MAINSTREAM (INCUMBENT) APPROACHES

As a first step to achieve Fixed Mobile Convergence or FMC (Finnie, 2004), many equipment manufacturers now produce dual-mode mobile phones (SIP/WiFi & GSM). Motorola, Nokia and others have been offering such devices that connect to the Internet via a WiFi network whenever possible, or via a gateway of the traditional mobile network when the phone is beyond the range of WiFi networks. BT Fusion, a service rolled out by British Telecom based on *Unlicensed Mobile Access (UMA)* is an example of FMC on the service provider side. UMA, a 3GPP global standard, developed by major mobile and fixed/mobile operators and their primary vendors to create a cellular/Wi-Fi convergence solution to support all existing mobile voice and data services, can integrate into existing mobile networks, support all *Wireless Local Area Network (WLAN)* environments, and can easily fit in the future network evolution plans, including *IP Multimedia Subsystem (IMS)*.

IMS based on the protocols and principles of *Internet Protocol (IP)* telephony, is an evolving architecture for providing voice, video and other multimedia services to mobile and fixed phones. Compared to other IP-based protocols, IMS emphasizes central management and billing functions, thus allowing operators to offer centrally administered VoIP and other multimedia services on a managed IP network. This is a critical feature to the operators as they are accustomed to provide services to their customers in a "captive" environment. IMS standards come from the *3rd Generation Partnership Project (3GPP)*, a consortium focused on evolving GSM networks to 3G W-CDMA, and the detail specifications can be found at http://www.3gpp.org/specs/specs.htm.

Today mobile operators have a head start in the FMC race. Their VoIP service can run over any fixed broadband network and their mobile voice users rely on the mobile operators' existing infrastructure or their captive managed network. This advantage will disappear in time as fixed and mobile broadband Internet access (WiFi, WiMAX and 3.5G/4G) becomes widely available from multiple competing providers. Since 3rd party VoIP services, like Vonage or Skype, can run over anyone's broadband connection, mobile operators lose their lock on mobile voice services through market competition. Neither Skype nor Vonage uses or needs IMS. For now, however, mobile operators can launch FMC services more readily and attract more customers.

P2P COMMUNICATION: DISRUPTIVE TECHNOLOGY AGAIN FOR FMC?

Peer-to-Peer or P2P technology has drastically transformed the landscape of the Internet traffic in recent years, but so far we have not yet mentioned its potentials in FMC. P2P file-sharing services like Napster, eMule and BitTorrent, have been widely used and sparked legal controversies. P2P voice streaming created a tremendous sensation when eBay bought Skype for $2.6 billion in 2005. Meanwhile P2P video streaming is also becoming popular. P2P architecture offers a solution to the scalability problem often encountered by streaming networks. As a node joins a P2P network, it not only consumes resources but also contributes its bandwidth or computation power. By relaying data over P2P networks, users receiving data also help its distribution. In addition, the nature of overlay routing in P2P networks makes path diversity possible and this mechanism greatly relieves the load for streaming servers and facilitates traffic load balancing. But some difficulties exist in the current structure of P2P networks. More and more users who join the network are using Network Address Translation (NAT) or exist behind firewalls, and, thus, they cannot directly contribute their resources to relay data packets. This is especially true for WLAN users. Besides, routing traffic through end-users machines is not efficient considering the overhead involved.

Recent news about an investment deal to a startup company, FON, from Skype, Google and premier venture capital firms may reveal the thoughts of forefront proponents in the P2P camp about how P2P can cut into the FMC pie. As shown in Figure 1, the WLAN routers, while serving as a regular private wireless access point, are equipped with special software that would allow fellow users, called Foneros in FON's terminology, to share its bandwidth and gain wireless access to the Internet. Given enough bandwidth, the roaming Fonero can make voice calls, download data, and view streaming videos just like in their own home LAN. Likewise, the hosting Foneros enjoy the same benefit with other Foneros' services when they are away from home. The peers' resource sharing activities are pushed closer to the core network at the local wireless routers, and, thus, wireless routing does not increase the load of end user workstations. This is similar to the sharing idea of P2P networks, as the participants are offering and enjoying collectively Hotspot services. The traffic can also be routed in a P2P fashion while inside the Internet cloud shown in the Figure.

Despite the exciting prospect of such a scenario, which is referred to as P2P FMC for lack of a better term, many problems exist in the road ahead. Here is a list of them:

1. Pricing model: it would hardly seem fair if someone sets up a Hotspot in a remote area with little traffic and consumes a large amount of resources as he roams extensively. An accurate form of monitoring, account-

Figure 1. WLAN bandwidth sharing

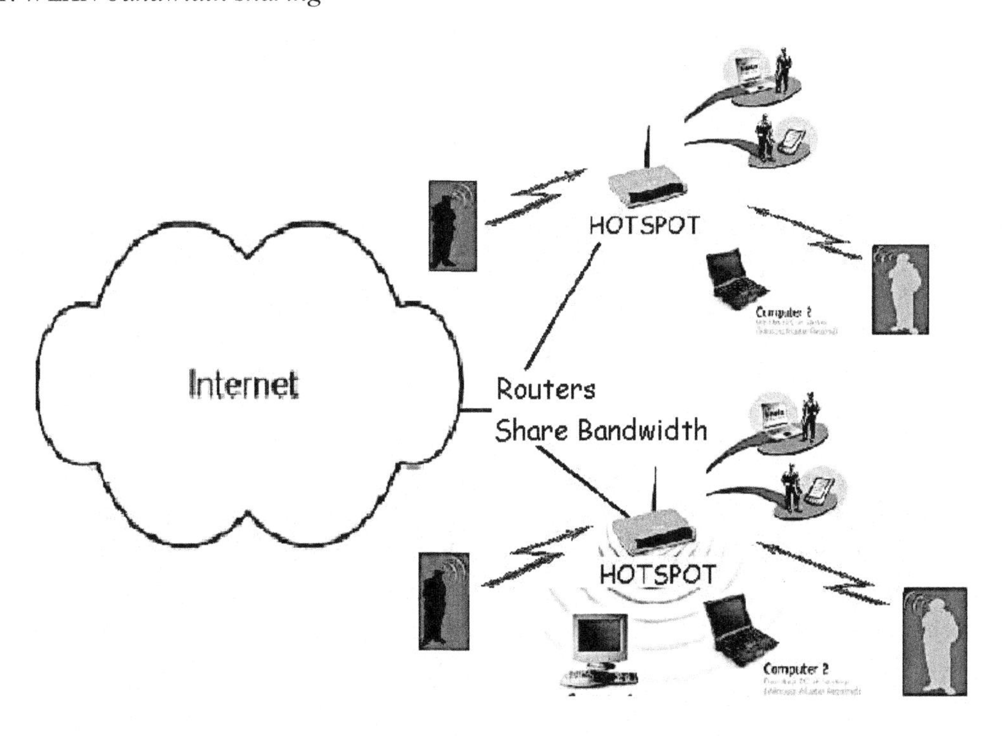

ing and pricing would be needed to make it attractive to participate. Without a sound business model, there will be limited capital investment for such operations to take off.

2. Range and mobility concerns: FON uses WiFi routers as its building blocks, taking advantage of its wide deployment. However, IEEE 802.11 was designed mainly for the range of homes and small offices, so it would be difficult for roaming Foneros to receive a good signal unless they are relatively close to the access point. Furthermore, the protocol does not support connection hand-over, so even when a Fonero is in range with multiple access points, a roaming device and the system cannot dynamically connect with the access point emitting the best signal when the receiving condition changes due to movement. This is a serious disadvantage because mobile users are accustomed to a seamless handover from cell phone operators. These issues may be partially solved when WiMAX matures and provides such capabilities, but then again, it may not deploy so widely as WiFi due to its initial cost.
3. Signal interference: Incumbent mobile phone operators purchased bandwidth from the government, so they can map the coverage of their base stations. In the disruptive scenario, private parties use unlicensed band to help each other. To attain better coverage, they can boost their signal strengths but interferences will occur. To mitigate the adverse effect, a negotiation protocol among peer hotspots is necessary to regulate their transmitting power. This has also being researched in the WiMAX License Exempt Standard Group.
4. Legal issues: At this time, some regulations are preventing a total VoIP takeover with issues such as emergency calls, government wiretapping and international tariffs. Fixed-line operators providing unlimited access with fixed monthly fee may oppose the extra traffic generated by bandwidth sharing because they do not profit from such activities. New broadband networks like Fiber-To-The-Home (FTTH) are being deployed as we write, so the task is upon advocates of FON ideas to convince established *Internet Service Providers (ISPs)* in working together with them.

NEW INFORMATION AGE

As explained above, P2P-FMC may totally disrupt the incumbents' game plan for FMC. Although the concept of resource sharing is not new, its implementation for telecommunications did not take off until recent years, thanks to innovative technologies and ideas. Indeed many obstacles still lie ahead, but none of them seems to be fundamentally insurmountable. Traditionally, a wireless operator first has to purchase a licensed bandwidth from regulators and pay for a very costly infrastructure which needs to satisfy many regulations and reach a wide coverage to achieve economies of scale. Only after that can it reap the profit. Naturally incumbent operators try to keep a lock on their customers to protect their investment interest, and since they own the whole network, they can probably implement some new features more quickly to satisfy their customer needs and charge a high premium. In P2P-FMC, every peer can establish its own Hotspot as a part of the whole infrastructure, sharing the startup cost of the operation. Although the underlying network may be heterogeneous, the peers can work together as long as they can communicate at the access level and above. With this unconventional model, anyone can join the effort of building a large infrastructure without a deep pocket, just sharing the proper amount of cost and benefit. In the past, the telecommunications' infrastructure was so costly that only very few entities could afford it, and oftentimes the government has to own and operate it. As always, it is hard to achieve efficiency in a large organization, especially in the public sector, but that seemed to be the necessary evil of the industry. P2P-FMC may finally break the old rules and change the whole landscape. It has happened before in the commercial world. Just a few decades ago, the capital requirement was huge to set up store fronts for retail business and even more so if it involved a global operation. Now average consumers can buy globally at eBay, and the cost of setting up an online store is only a tiny fraction compared to the cost before. Meanwhile, such online deals also enable average consumers to resell their unwanted goods which, in the past, would most likely just stay in a storage room. More people participate, more transactions take place, and the market is more efficient.

Welcome to the New Information Age.

REFERENCES

Finnie, G. (2004). Fixed-mobile convergence reality check. *Heavy Reading, 2*(26).

ADDITIONAL READING

Ahson, S. A., & Ilyas, M. (2010). *Fixed-mobile convergence handbook*. Boca Raton, FL: CRC Press. doi:10.1201/EBK1420091700

Shaneyderman, A., & Casati, A. (2008). *Fixed-mobile convergence*. Boston, MA: McGraw-Hill.

Watson, R. (2009). *Fixed-mobile convergence and beyond*. Amsterdam, The Netherlands: Elsevier (Newnes).

KEY TERMS AND DEFINITIONS

Broadband: A term that encompasses all forms of accessing the public Internet such as Asymmetric Digital Subscriber Line (ADSL) over twisted-pair telephone lines, fiber-to-the-home or fiber-to-the-kerb optical network access (usually through Passive Optical Networks), and wireless access such as IEEE 802.16 (WiMAX).

Fixed-Mobile Convergence: The realization of communications in which there is a seamless conjunction of fixed, wireline, mobile, wireless networks. This will provide a mobile environment for communications.

Hotspot: A WiFi (IEEE 802.11) hotspot is a service that offers access to the Internet to roaming users. A user's device is able to connect to the hotspot router, after which it gains access to the public Internet via an Internet Service Provider's (ISPs) network. Hostpots now exist not only in public spaces such as airports but in cafes, libraries and hotels throughout the world.

Peer-to-Peer: In a client server architecture all requests for service are mode by a client (or a server acting as a client) and all services are provided by a server. In a peer-to-peer system, individual peers act as clients and servers alike. Thus, a peer can download data and upload it to fellow peers. This potentially provides a distributed system that is robust to the failure of one entity (such as a server) and which can reduce the costs of running centralized servers.

Voice-over-IP: Audio streaming of speech conversations using IP framing together with call set-up and session control. The most prominent example of this is Skype which provides softphones across a variety of platforms, mobile and fixed.

Chapter 2
Peer-to-Peer Overlay for the IP Multimedia Subsystem

Ling Lin
Vodafone Group, UK

Antonio Liotta
Eindhoven University of Technology, The Netherlands

ABSTRACT

The growth of the Internet and its popular services are forcing telecom operators to provide advanced services to their subscribers, as traditional voice services are no longer enough to attract more customers. To enable more innovative and value-added IP services and take advantage of the services that the Internet provides, the IP Multimedia Subsystem (IMS) is introduced. The IMS provides a complete access-agnostic architecture and framework that facilitates the convergence of the mobile network, removing the gap between the two most successful communication networks: cellular and Internet network. The harmonized All-IP platform has the potential to provide all Internet services with a more cost-effective and more efficient architecture than the circuit-switched networks do. However, by merging two of the most successful networks, the integration of two network models with different concerns and motivations is not without its problems, among which, the scalability issue is the most essential when supporting content delivery services. The purpose of this chapter is to study and design a new content delivery network infrastructure, PeerMob, merging the Peer-to-Peer technology with the IMS framework, which benefits IMS with scalability, reliability, and efficiency features coming with decentralized P2P architecture. The chapter also puts this P2P IMS paradigm under realistic network conditions and strenuous simulation to evaluate the performance of the P2P IMS system.

INTRODUCTION

In recent years, the Internet has become the main medium to deliver contents to end users and the multimedia services in the Internet have experienced an explosive growth in many dimensions, including size, performance, and geographical span. The growth of the Internet and its popular multimedia services are forcing telecom operators to provide comparable services to their subscribers as traditional voice services are no longer enough to attract more customers. Therefore, to enable efficient and cost-effective value-added *Internet*

DOI: 10.4018/978-1-4666-1613-4.ch002

Protocol (IP)-based services, cellular networks are evolving from a traditional purely circuit-switched network to an All-IP packet switched network. Furthermore, to take the advantage of the multimedia services that the Internet provides, the telecom industry is as a whole is undergoing an evolutionary transformation to be an efficient content delivery platform.

With IP spreading throughout cellular networks, the challenge of integrating voice and data services in the fixed and mobile access networks becomes more formidable. Thus, a standard convergence networks platform is required to offer data, voice, and multimedia services. Then, the *IP Multimedia Subsystem (IMS)* has been introduced to deploy IP-based telephony and multimedia services on every access network, including both circuit-switched networks and packet-switched networks, to eventually replace the circuit-switched core network with an All-IP core network. The IMS is an attempt to provide a complete access agnostic architecture and framework removing the gap between the two most successful communication networks that are the cellular and the Internet network, while still integrating with legacy network for existing services. The IMS allows network operators to play a vital role in the network and that's why IMS has generated intense research and standardization efforts. (Camarillo et al., 2008).

PROBLEM STATEMENT

These emerging content delivery services have high demands for network resources and thus create new challenges in terms of network bandwidth, service management, configuration, and deployment. By merging two of the most successful networks, the cellular network and the Internet, the integration of two network models with different concerns and motivations is not without its problems. Among these, the scalability issues, inherited from the client-server architecture from the cellular network, are the most essential.

The IMS is realized through a collection of well specified logical nodes with clearly defined interfaces to achieve network functions, such as registration, subscription, session control etc. These nodes are designed according to the client-server paradigm as historically they have been located and operated in the fashion by network operators. In the IMS, both signal flows and media flows have to traverse a long list of nodes to reach their destinations. If one or more intermediate nodes in the communication link reach their capacity limit, congestion may occur and no more communication links via these nodes are possible. Thus, this will in turn limit the performance of the IMS in a large- scale telecom network, especially when supporting content delivery services.

MAIN CONTRIBUTIONS

Concerning the scalability issues in the IMS, the purpose of this chapter is to study and design a new content delivery network infrastructure, PeerMob, by merging the *Peer-to-Peer (P2P)* technology with the IMS framework, which benefits the IMS with scalability, reliability, and efficiency features coming with a decentralized P2P architecture.

Generally speaking, P2P is a technology that fosters resources self-deployment and self-organization, while still achieving optimized resource utilization for the deployed applications and services. P2P is designed for sharing computer resources, CPU power, storage and bandwidth, by direct exchange, rather than requiring the mediation or support of a centralized server. P2P architectures are characterized by their ability to adapt to failures and accommodate transient populations of nodes while maintaining acceptable connectivity and performance. The use of the P2P paradigm to provide content delivery services is gaining increasing attention, and has become a promising alternative to other legitimate approaches as the classical client-server model or *Content Delivery Network (CDN)*.

Another objective of the chapter is to put this P2P IMS paradigm under realistic network conditions and strenuous simulation to evaluate the performance of the P2P IMS system. For this purpose, a new P2P IMS proof-of-concept is carefully designed, implemented and deployed for demonstration. The P2P IMS proof-of-concept includes most of the IMS components, such as x-CSCF, HSS, IMS application servers and the IMS peer application. Also, a customized P2P IMS simulation has been built to demonstrate the performance of this P2P IMS approach.

IMS FRAMEWORK

Telecom. network architectures have evolved to several access networks coexisting both in the wireless and wired domains. The networks have also evolved from the original circuit-switched networks to the packet-switched networks. The intention of IMS is not to standardize applications within it but rather aid in building all services independently of access networks aiming to aid the deployment of All-IP networks, making most popular Internet services and application anywhere and everywhere.

3GPP release 7 states the complete solution to support IP multimedia applications consists of *User Equipment (UE)*, *IP-Connectivity Access Network (IP-CAN)*, and the specific functional elements of the *IP Multimedia Core network (IM CN)* subsystem. (Poikselkä & Mayer, 2009) (Oredope & Liotta, 2008a)

Figure 1 illustrates a simplified version of the IMS architecture. Dotted lines represent the

Figure 1. The IMS architecture

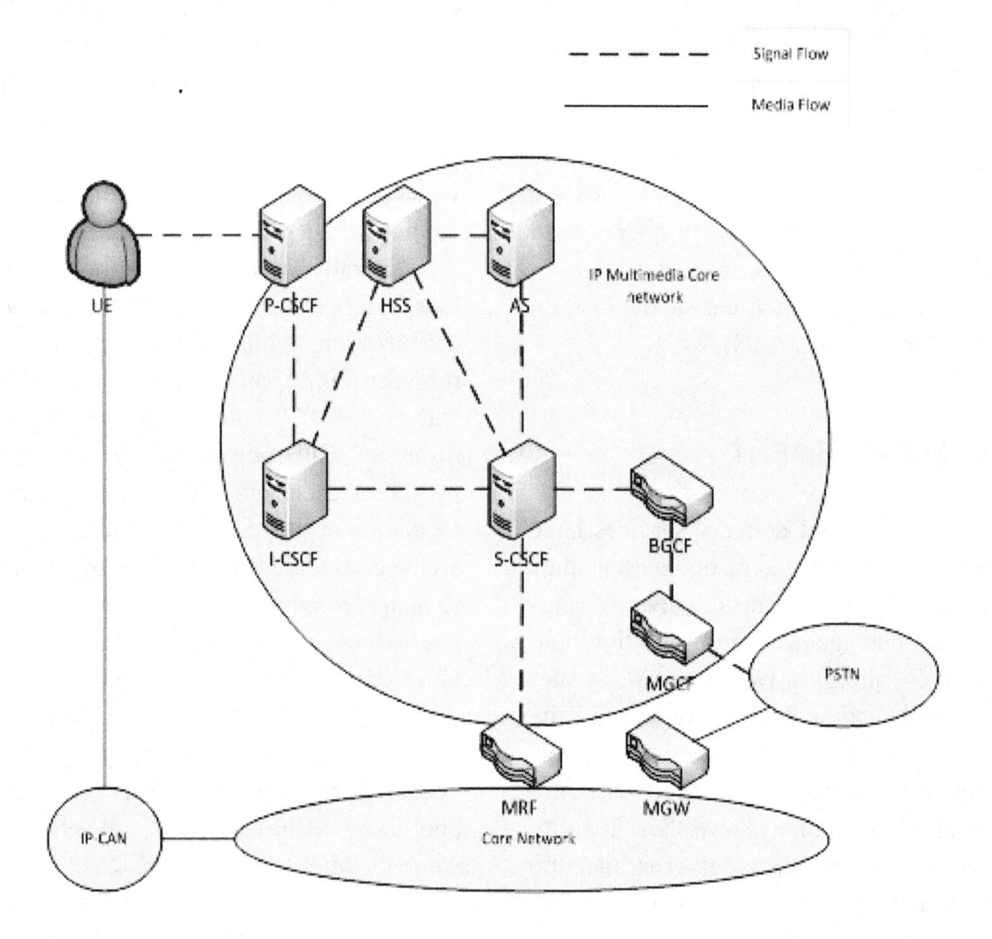

signal flows while the full lines give means for the media flows. It is important to notice that IMS deals just with the session signal and control while it does not tackle the actual transport of data or media flows of the sessions. The IMS architecture can be divided in three logical layers, the access layer, the session control layer and the application layer. (Poikselkä & Mayer, 2009) (Oredope & Liotta, 2008a)

The access layer provides a common access interface for UEs shielding the details of various access networks. UEs might connect to the IMS infrastructure through either a packet-switched network or a circuit-switched network that interfaces the IMS through the *Media Gateway Control Function (MGCF)*.The session control layer handles session setup and teardown, and manages the handover of sessions between service providers. The most important elements in this layer are the HSS and the three types of *Call Session Control Functions (CSCF)*. All of these nodes can be distributed over multiple nodes to increase redundancy and scalability.

The *Home Subscriber Server (HSS),* technically an evolution of the GSM *Home Location Register (HLR)*, provides a central repository for user-related information to handle multimedia sessions. The user-related information includes location information, security information, user profile information and the S-CSCF allocated to users. The HSS is always located in the home network. If there is more than one HSS in an IMS core, a *Subscription Locator Function (SLF)* is used to determine the HSS that a use's record is located.

The *Proxy Call Session Control Function (P-CSCF)* is a SIP proxy server, providing the first contact point in the signal plane between IMS UEs and the IMS core network. This means that all signal traffic from or to an IMS UE must traverse a P-CSCF, which will then forward SIP requests and responses in the appropriate direction. Here is how the function works. First, a P-CSCF terminates the IPSec security connection between an IMS UE and the P-CSCF. Then, the P-CSCF verifies the validity of SIP requests to or from the IMS UE. If the validation is successful, the P-CSCF forwards the SIP message to a suitable S-CSCF. In the early stage of IMS deployment, the P-CSCF needs to be placed in the home network as media traffic traverses always through GGSN and the P-CSCF has to be deployed close to GGSN. Currently, the P-CSCF may be located either in the home network or the visiting network.

The *Serving Call Session Control Function (S-CSCF)* is a SIP server with two key roles. One is a SIP proxy server routing SIP messages to appropriate SIP servers or application servers. The other is a SIP registrar server maintaining the binding between the current address of IMS UEs and user identities. S-CSCF has an interface to the HSS and the SLF to download authentication vectors from the HSS. The authentication vectors are required to authenticate IMS UEs to grant access. The S-CSCF also needs to inform the HSS which S-CSCF is allocated to handle a UE; subsequently the UE will send registration requests to the same S-CSCF and the P-CSCF. The S-CSCF is always located in the home network.

The Interrogating Call Session Control Function (I-CSCF) is also a SIP proxy server, placed at the edge of an IMS domain to communicate with other IMS domains, hiding the network capacity and topology from the outside. Because the I-CSCF is the first contact point to other IMS domains, the I-CSCF implements a functionality called *Topology Hiding Inter-network Gateway (THIG)* to encrypt sensitive information in SIP messages. The address of I-CSCF is listed in well-known domain DNS servers. When another IMS domain wants to find an entity within a destination domain associated with an I-CSCF, it will query the address of the I-CSCF from the DNS servers and forward the SIP message to the I-CSCF. Like S-CSCF, I-CSCF also has an interface to the HSS and the SLF. I-CSCF queries the HSS to obtain the address of an appropriate S-CSCF where a request should be forwarded, then assigns the

S-CSCF to handle the SIP request. The I-CSCF is usually located in the home network.

The Application Layer contains Application Servers, which are expansion slots for IMS networks where third party products and services are located. It consists of different application servers for the execution of various IMS services and the provision of end user service logic. The Application Layer offers a platform for value-added services which goes well beyond the integration of network and devices (Ilyas & Ahson, 2008).

Each function mentioned above is not necessarily implemented on a single node. The IMS architecture is a collection of functions exposed as standardized interfaces, while implementations are free to combine two or more functions into a single physical node. Similarly, a single function can be spread over two or more nodes for load balancing or availability purposes.

A P2P Architecture for the IMS

P2P is a distributed application architecture in which participants can act as clients and servers. These participants are able to work without a central server but are also ready to make use of a central server, if necessary, for the efficiency of the system. In P2P networks, the information is no longer concentrated on central servers but is provided by each participant, called a peer. The source of content is not the only one to upload the content and to share its resources, CPU power and bandwidth. Most participants download content but also upload it to other peers, sharing their resources to help other peers to get the content. As most participating peers compensate the additional workload on the system by providing their own resources, a P2P system typically scales well up to a large number of users.

Generally, P2P networks have the following characteristics: distribution, self-organization and robustness. Distribution means that the content is distributed to different peers, and the traffic is brought to the edges of the network. There is usually no central entity that is required for the functioning of the network. Self-organization means that the network coordinates itself, during which joins and leaves of peers and other forms of interruptions are handled automatically. Robustness means that the P2P system is already designed for the imperfect behavior of the network. The more content and resources are replicated among peers, the higher overall fault tolerance may grow. However, these advantages offered by P2P are coming with a certain cost, because in a P2P network peers become responsible for managing peers and network resources, including content management and peer management. Furthermore, the dynamic nature of peers poses presents challenges in the communication overlay (Oram, 2001) (Oredope et al., 2008b).

Topology Overview

The P2P topology of an overlay network is crucial to the operation of the system. The topological characteristics of P2P overlays need to be taken into account when designing P2P networks, as it is very cumbersome to change a topology module after deployment.

During its short history, P2P has already passed through several generations, which can be classified into two main groups: structured overlay network and unstructured overlay network, as shown in Figure 2. These depend upon the logical organization of the peers' topology. In structured architectures, peers are well organized and according to defined algorithms. By contrast, peers in unstructured overlay network have the flexibility to choose the number and destinations of their connections, and adapt them to network heterogeneity to improve network performance. The unstructured overlay topology can also be further divided into three different overlay topologies, according to the degree of centralization: centralized P2P architecture, pure P2P architecture and hybrid P2P architecture.

Figure 2. P2P overlay architectures

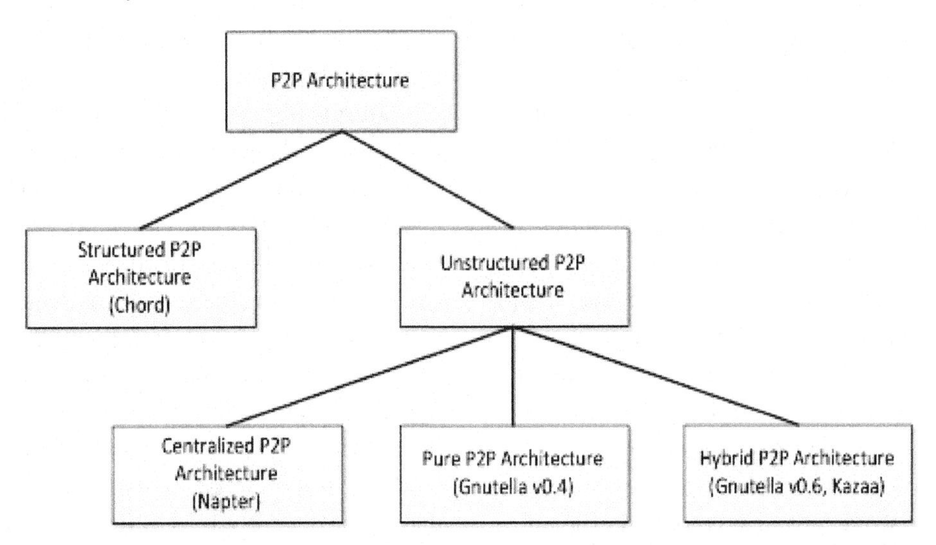

The fundamental problem of P2P is resource discovery, e.g. finding a particular peer, service or content. This is achieved differently in unstructured and structured networks. Resource discovery can be classified into three categories: mechanisms with specialized index nodes (centralized search), mechanisms without any index (flooding search), mechanism with indices at super peers (distributed search), and mechanism with a *Distributed Hash Table (DHT)* (Schollmeier & Schollmeier, 2002).

Centralized P2P Architecture

The centralized P2P architecture, often known as the Napster architecture, is an architecture where a centralized server holds all of the meta-data of peers, such as their address and content portfolio. When a peer searches for content, it first queries the central server. Then the central server replies with the address of the peers containing the requested content. Thus, the querying peer is able to retrieve content directly from the peers suggested by the central server. All further data exchanging is directly between these peers.

In this centralized P2P architecture, queries have to go through the centralized server to identify peers containing the content of interest, which is called a centralized search. Comparing to other P2P systems, a centralized P2P system is much easier to maintain and keep consistent. Moreover, the discovery is efficient and deterministic because only a single message is required to resolve a query. However, the central server represents a single point of failure. Relying on central servers leads to less scalable and, in particular, vulnerability to Denial-of-Service attacks.

Pure P2P Architecture

The pure P2P architecture has no centralized node and all nodes in this P2P network are truly equal. As there is no centralized node for resource discovery, searching is performed by forwarding queries from peers to peers, using a flooding based lookup algorithm (e.g. in Gnutella 0.4), which effectively broadcasts the search with limited scope. Another possibility is to use more intelligent routing methods (e.g. Freenet), whereby the query is iteratively routed towards the peer that is most likely to have the requested content (Jovanovic et al., 2001).

Flooding propagates the query messages to every node within a specified radius, based on the query's *Time-To-Live (TTL)* value. If some peers have the requested content, they reply to the original querying peer. Otherwise, queries are propagated from peer to peer, until matching resources are found or TTL has expired. In this way we avoid congesting the overlay network with search requests. Therefore, flooding can quickly retrieve any related documents within TTL hops of the initiator. However, it is possible that some peers, which are outside of the TTL limit, cannot be reached. The advantage of pure P2P systems is the fact that there is no single point of failure, which results in an enhanced reliability.

Although this mechanism is simple and robust, the resource discovery process is not very efficient, as the overlay structure is not deterministic. A node in a pure P2P network is in principle unaware of the resources its neighbor peers maintain. The flooding based discovery may lead to significant network usage by generating a large number of query messages. More importantly, the amount of traffic scales poorly as the number of peers in the network increases (Ripeanu et al., 2002).

Hybrid P2P Architecture

The hybrid P2P architecture is a combination of the centralized and pure P2P architectures. The hybrid architecture is used by many popular peer-to-peer applications, such as Gnutella 0.6, Kazaa, and JXTA.

Super peers hold meta-data about the content that single peers contains. Indices of single peers are maintained at super peers. Queries are forwarded form single peers to super peers. The super peers query each other if any of their single peers have the required content. If there is a match, peers exchange all further information directly, without super peer involvement. It is important to note that such super peers do not constitute single points of failure, since they are dynamically assigned and, if they fail, the network will automatically take action to replace them with others. Hybrid P2P networks allow a peer to function as either a single peer or a super peer, depending on the conditions, such as the available bandwidth and processing power.

In terms of performance, the hybrid architecture is located between the pure and the centralized P2P architecture. It scales quite well and has good resiliency. The hybrid architecture is also situated somewhere between the other two architectures when considering the search coverage in the network. When a peer sends a query, the request is forwarded to other super peers who then forward the request onwards. The search coverage is limited by the TTL of the request message, but the coverage is a lot larger in the hybrid architecture than in the decentralized P2P architecture, thanks to the backbone effect of super peers (Beverly & Garcia-Molina, 2003).

Structured P2P Architecture

The structured P2P architecture is also known as *Distributed Hash Table (DHT)* architecture. Each peer acquires an identifier based on a cryptographic hash of some unique attribute such as a peer's IP address or a peer's public key. The identifier for a content item is also obtained through the same hashing function. The hash table actually stores content items as values indexed by their corresponding keys. That is, node identifiers and key value pairs are both hashed to one identifier space. Then, peers are connected to each other in a certain predefined topology, for example, a circular space in Chord (Lua et al., 2005). Thanks to the structured topology, data lookup becomes a routing process with small routing table size. DHTs guarantee that the nodes distance and the number of connections per node is O(log(n)), where n is the number of nodes in the network. A structured P2P topology overcomes the limitations of an unstructured P2P topology by organizing the overlay with some structured content location mechanism such as DHT.

As a consequence of the deterministic organization of the structured P2P topology, maintenance efforts to keep the correct overlay are increased due to the churn. Churn is the rate by which the peers join and leave a network. If the churn handling of the algorithm is not efficient enough, the maintenance signaling would load the overlay unnecessarily and it may cause the entire algorithm to be inefficient.

Topology Summary

Different P2P architectures lead to different degrees of scalability, reliability, efficiency and the ability to control the network. The pure P2P architecture is the most resilient architecture against node failure. On the other hand, with the centralized P2P approach, the server may become a capacity bottleneck, causing the P2P system to stall. This type of problem has been evidenced with the Gnutella network, which was originally designed for relatively small growth rates. With the huge influx of users, Gnutella's performance degrades significantly and becomes poorly unusable.

The hybrid architecture is utilized to improve the discovery efficiency and advance the network scalability in P2P systems. The structured P2P architecture is developed in another direction to solve the weak discovery efficiency problem. The former shows its benefit in scalability and fuzzy search in a dynamic network environment, while the later has the advantage of a more search facility. Typically, structured P2P networks obtain better performance at steady state, but as the node churn increases they tend to show instability or congestion issues.

Neither the pure nor the centralized P2P networks can simultaneously offer both reliability and scalability in a single algorithm. In a highly dynamic, mobile environment, the hybrid P2P architecture is preferable. We have in fact chosen this approach to develop our P2P IMS system.

PeerMob P2P IMS

The aim of this study is to design a new content delivery network infrastructure for the IMS (the PeerMob P2P IMS platform), incorporating a P2P framework into the IMS. Our system is realized through a collection of well-defined PeerMob P2P IMS components and protocols. Since the IMS components and interfaces have already been standardized by 3GPP, our strategy is to extend the IMS using the existing recommendations for the incorporation of new mechanisms and applications. Thus, we don't require any modifications to the existing IMS platform and suggest an approach to implement P2P networking in the form of a software upgrade.

The PeerMob P2P also adopts a group-based strategy within a hybrid architecture to support and maintain a low-diameter discovery network. Each peer must be assigned to one of two mutually-exclusive roles (a single peer or a super peer), based on a super-peer election algorithm. Single peers do not join any peer groups and are not responsible for peer group management. Super peers must join a peer group, based on a cryptographic hash of its identity, to provide peer group services. Super peers trace the single peers that contain the content which might be relevant to the group, and maintain content chunk bitmaps for the peer group.

Super peers know all other super peers within the same peer group. The assignment of roles is not permanent: a peer may start as a single peers, and later become a super peer if more super peers are required for a particular peer groups. Alternatively, a super peer may decide to move all of its single peers to the other super peers and become a single peer by itself, to reduce the number of super peers and thus reduce the traffic generated by communication between super peers.

Both single peers and super peers are able to not only download content but also upload content to other peers, sharing material to help other peers to retrieve relevant information. Each set of content belongs to a particular peer group based on a cryptographic hash of its content ID. Super peers and content are both hashed to one identifier space. Each peer group is responsible for managing content within the peer group only.

We describe below the key entities of P2P IMS:

Peer: a peer refers to a participant in a P2P system, which not only consumes content, but also shares content to other participants. Peers can be User Entities (UEs) or cached servers. A UE must be either a super peer or a single peer. A cached server is always a super peer located in the IMS network, which aims to improve the performance of the PeerMob P2P IMS.

Peer Group: peers self-organized into peer groups, which are responsible for distributing specific category of content based on a cryptographic hash of content IDs. Each super peer must join one but only one peer group based on cryptographic hash of its identity. Single peers do not belong to any peer groups.

Cached Server: the cached servers refer to network entities that are usually deployed at the network edge and act as initial super peers. A cached server also belongs to a peer group based on cryptographic hash of its identity. A cache server keeps all copies of the content belonging to its peer group.

Tracker Server: the tracker server is a logical entity being a super peer management server and a key store server.

Source Server: the source server is an application server providing user interface for content provision.

Chunk: a chunk is a basic unit of partitioned streaming content, which is used by a peer for the purpose of storage, and exchange among peers.

By introducing the concept of group-based super peers, the topology is now organized through a two-level peer hierarchy: the super peers, included in different peer groups, and the single peers. Super peers are nodes that usually have more bandwidth, CPU power and reliable links than single peers. Because of that they can take server-like responsibilities, providing indexing services to a set of single peers. The design based on super peers has two main advantages. Firstly, certain system tasks, such as the indexing service, can be assigned to super peers, to improve the overall system reliability and performance. Secondly, the use of super peers allows the P2P system to limit the number of participants in certain distributed algorithms, such as search, which do not scale well and become too expensive when the system size is large. In the evaluation section below, we show how the super peer design can improve the system scalability.

However, the use of super peers introduces a number of new challenges to be addressed. To manage super peers, the P2P system needs to maintain a super peer list and the peer group structure. Furthermore, the P2P system has to continuously adjust the super peer list in response to peers' arrivals and departures. Also, the load between super peers must be kept balanced to ensure the system's scalability and fault-tolerance.

Figure 3 depicts a typical content delivery user case for PeerMob P2P IMS, illustrating the interaction between the PeerMob P2P IMS and a user, Alice, who wants to view content.

The sequence of operations can be described as follows:

1. User Alice completes the IMS registration and authentication processes. Then, Alice updates her content list from a tracker server.
2. Alice wants to watch content and sends a watch query to the tracker server (if no super peer cached for the peer group that the content belongs to). The tracker server computes the peer group for the content, depending on a pre-defined hash function, and responds to Alice with both a cached server list and a super peer list of the peer group. The tracker server returns also a license key to Alice for content decryption. Alice caches the super peer list for future discovery.

Figure 3. P2P IMS use case

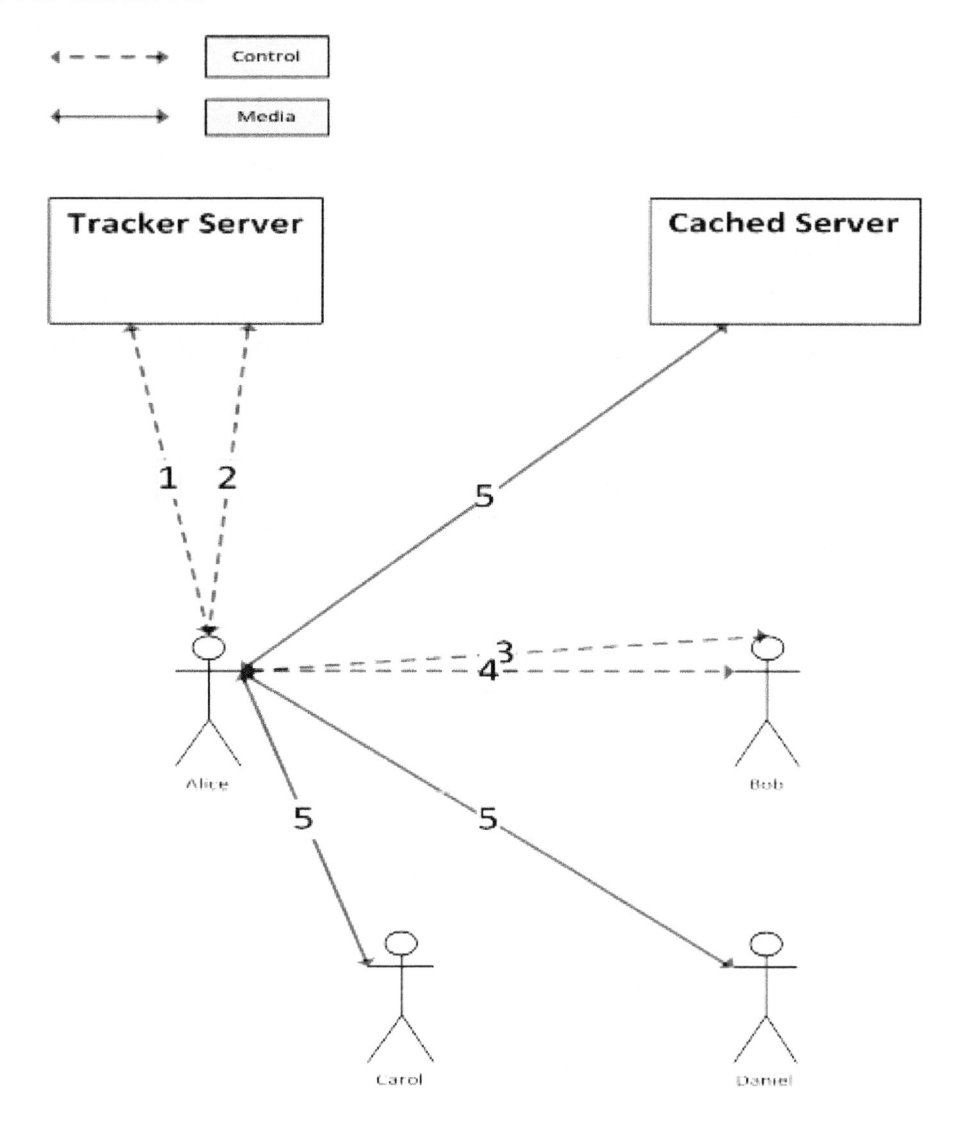

3. Alice is now able to query either a cache server in the cache server list or a super peer (Bob in this case) in the super peer list, in order to find peers who actually contain the required content.
4. Bob responds to Alice that both Carol and Daniel contain the requested content.
5. To speed up downloading, user Alice downloads some chunks of the content from cached server A (cache servers always keep the content in the peer group) and others from Carol and Daniel. Alice decrypts these chunks in real time with the key provided in step 1.

System Architecture

Figure 4 depicts the high-level architecture of the proposed PeerMob P2P IMS. The IMS is mostly used for signal control while P2P is mainly to optimize the media traffic in the media plane. Apart from the standard IMS network nodes, such as core IMS network elements (including P-CSCF, I-CSCF and S-CSCF) and HSS, three new application servers are introduced: the source server, the cache server, and the tracker server.

In IMS terms, the source server is an application server providing a new user interface for content provision. The content contained in the

Figure 4. The P2P IMS architecture

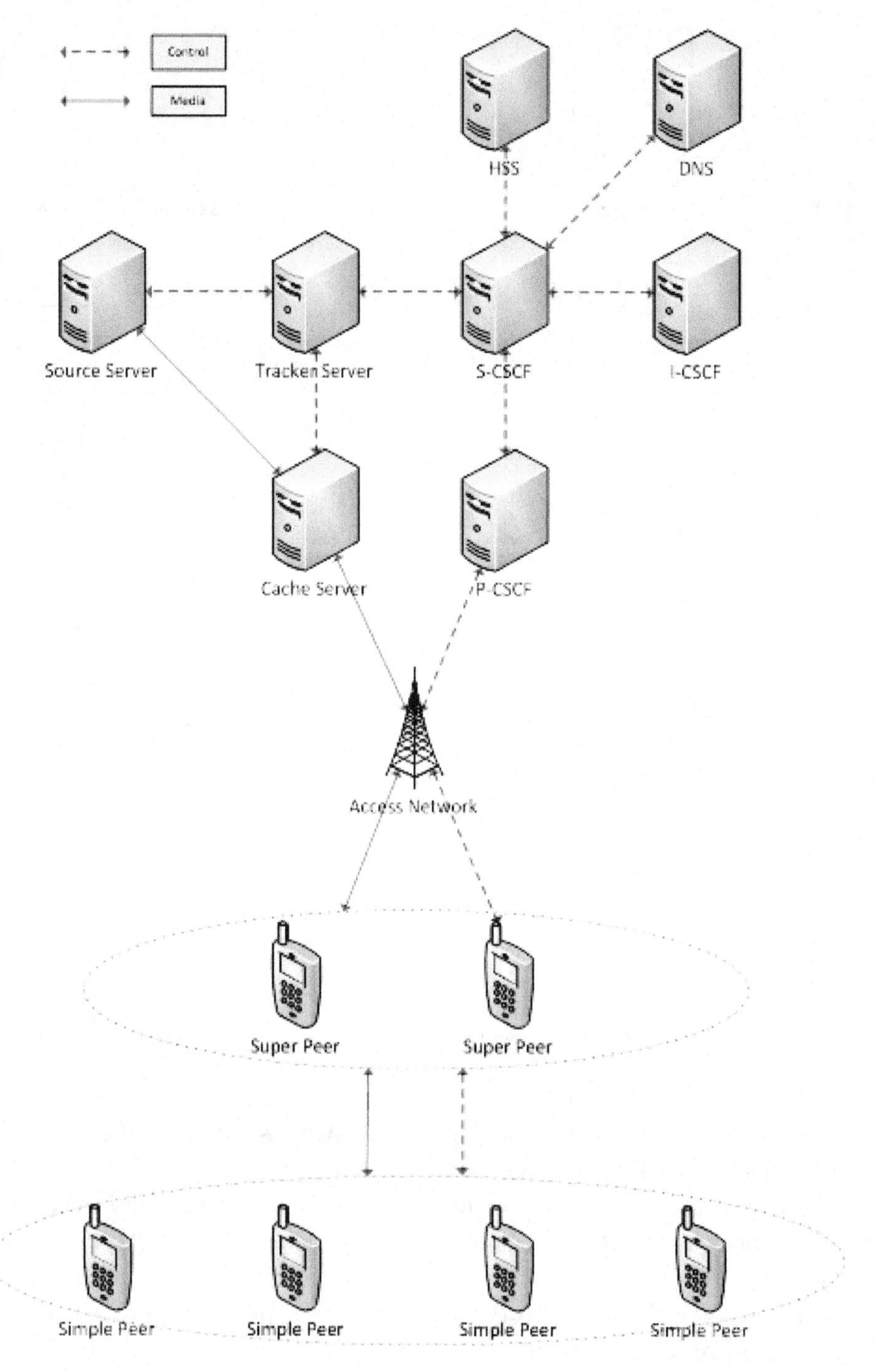

source server will be automatically distributed to a cached server belonging to the same group as the content itself. The cached server caches content from the source server, acting as the initial super peer. To save the bandwidth in the core network, the cached server should be deployed in the access network, sitting as close to the UEs as possible.

The tracker server plays an essential role in PeerMob P2P IMS. It has two main roles: a super peer management server and a key store server. To be a super peer management server, the main focus is to construct super peer lists for different peer groups and to respond with relevant information to the UE's queries. To be a key store server, the tracker server is required to maintain the pairs of content and license keys and to issue proper keys to a user, based on the user's subscription.

Tracker Protocol

To deploy a content delivery service on our group-based, hybrid P2P IMS framework, two key protocols are required. One is the signaling protocol between tracker servers and peers, namely the tracker protocol. The tracker protocol is responsible for the discovery of the correct cache servers, super peers and peer groups. The latter manage the single peers that contain the actual content.

The second protocol, namely the peer protocol, handles the communication signals among the peers. The peer protocol is responsible for the efficient transmission of data within peer groups. For instance, it locates the relevant peers (both single peers and super peers), which actually contain the content.

1. The tracker protocol handles the initial and periodic exchange of metadata between trackers and peers, such as peer lists and content information, as shown in Figure 5.
2. The peer protocol controls the advertising and exchange of media data availability between peers. Both tracker protocols and peer protocols can be carried over TCP or UDP, when delivery requirements cannot be met by TCP.

The sequence of operations of the Tracker Protocol is depicted in Figure 5.

3. A cache server registers at the tracker server.
4. The tracker server computes the group ID which the cache server belongs to, based on the cache server ID.
5. The tracker server records the cache server's address together with its group ID.
6. The tracker server responds to the cache server with the group ID and cache server list in the peer group. From the returned cache server list, the cache server knows the entire current cache servers of the peer group.
7. The cache server downloads content in the peer group.
8. A peer is selected to be a super peer, based either on its capacity or randomly.
9. The super peer registers to the IMS core.
10. The super peer registers to the tracker server.
11. The tracker server computes the group ID which the super peer belongs to, based on the super peer ID.
12. The tracker server records the super peer's address together with its group ID.
13. The tracker server selects the super peers of the peer group.
14. The tracker server responds to the super peer with group ID, cache server list and super peer list of the peer group.
15. The tracker server responds to the super peer with group ID, cache server list and super peer list of the peer group.

Peer Protocol

To prevent isolation between super peers in the same peer group, a mechanism based on the gossip protocol (Allavena et al., 2005) is adopted in the peer protocol. The peer protocol behavior periodically performs the following actions: first it selects a super peer from the super peer list of its peer group and exchanges information. Then, both participants update their actual information according to the received one. The peers involved in the exchange send to each other information about their current super peer lists and content

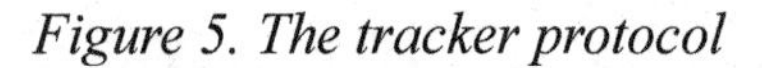

Figure 5. The tracker protocol

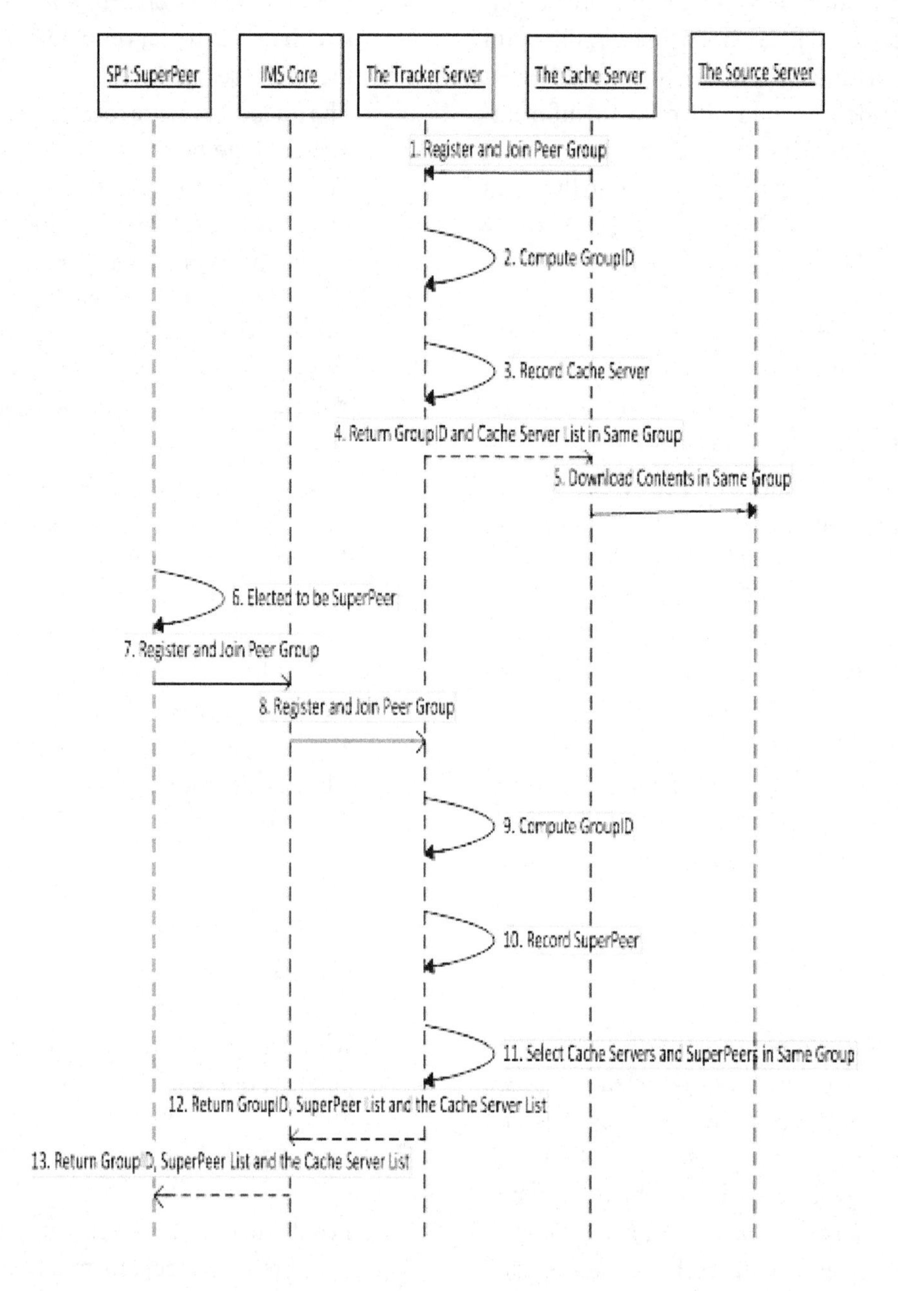

chunk availability. Based on the received information, super peers update their super peer lists in order to obtain a better approximation of the super peer topology.

The continuous gossiping of topology information captures the dynamic nature of P2P systems. In this way super peers learn about new super peers in the peer group by receiving their identifier in an exchange, while crashed super peers are progressively forgotten and then removed from the super peer overlay. Thus, super peers are provided with continuously fresh samples of the entire peer group. The gossip protocol has been used successfully to implement several P2P protocols, including newscast, broadcast and aggregation protocol.

1. The sequence of operations of the Peer Protocol is depicted in Figure 6.
2. Super peer 1 is elected to be a super peer of its peer group.
3. Super peer 1 exchanges information with other super peers and cached servers in same peer group, in this case, cached server 1.
4. The cached server updates its super peer list and cache server list.
5. The cached server response to the super peer 1 with its super peer list and cache server list.
6. Super peer 1 updates its super peer list and cache server list.

System Implementation

In order to validate the protocols proposed in PeerMob P2P IMS, we realized a prototype which runs over a standard IMS platform. Figure 7 depicts the deployment diagram of the PeerMob P2P IMS proof-of-concept testbed.

The PeerMob implementation follows the module design approach, in which functionalities are grouped into modules and each module depends on the services provided by other modules. This ensures maximum module reuse. The PeerMob P2P IMS proof-of-concept has been successfully demonstrated in the University of Essex, UK, using HP IPAQ 5500 hand-held devices.

The server side is implemented in Java; the peer implementation is coded in J2ME to make it runnable in hand-held devices. The PeerMob proof-of-concept has been tested successfully on the J2ME mobile phone simulator (running J2ME CLDC1.1 and MIDP2.0) and HP IPAQ 5500 with familiar Linux. The server implementation is based on Jain SIP (JainSipApiv1_1.jar) and NIST SIP (Nist-sip-1.2.jar). Java media framework has been integrated for media player.

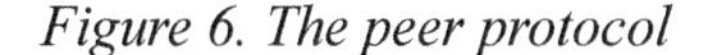
Figure 6. The peer protocol

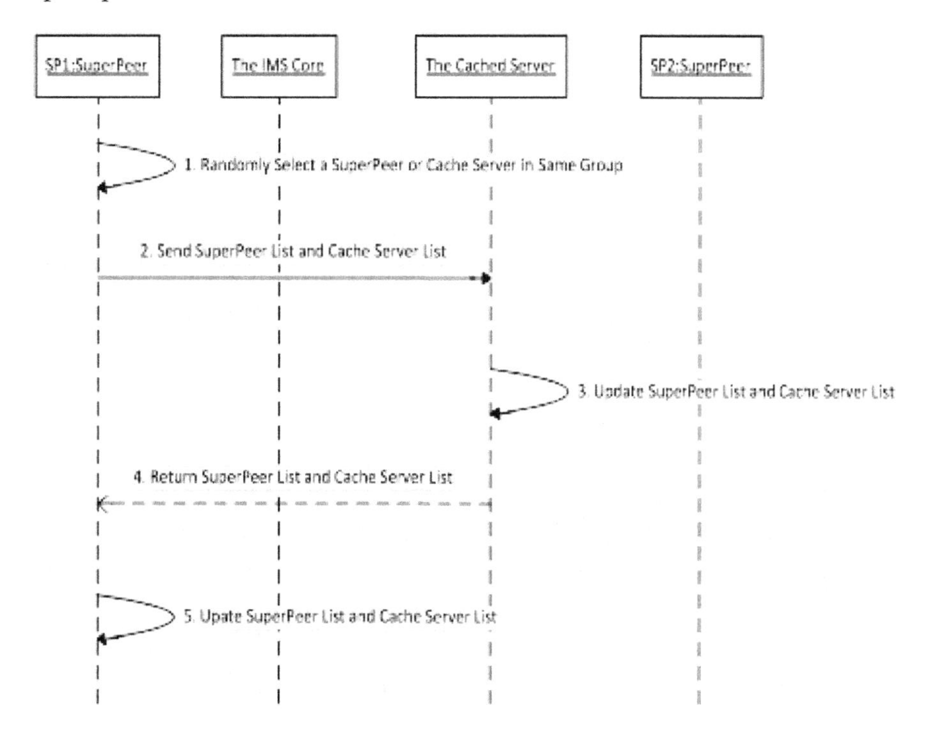

Figure 7. Deployment diagram

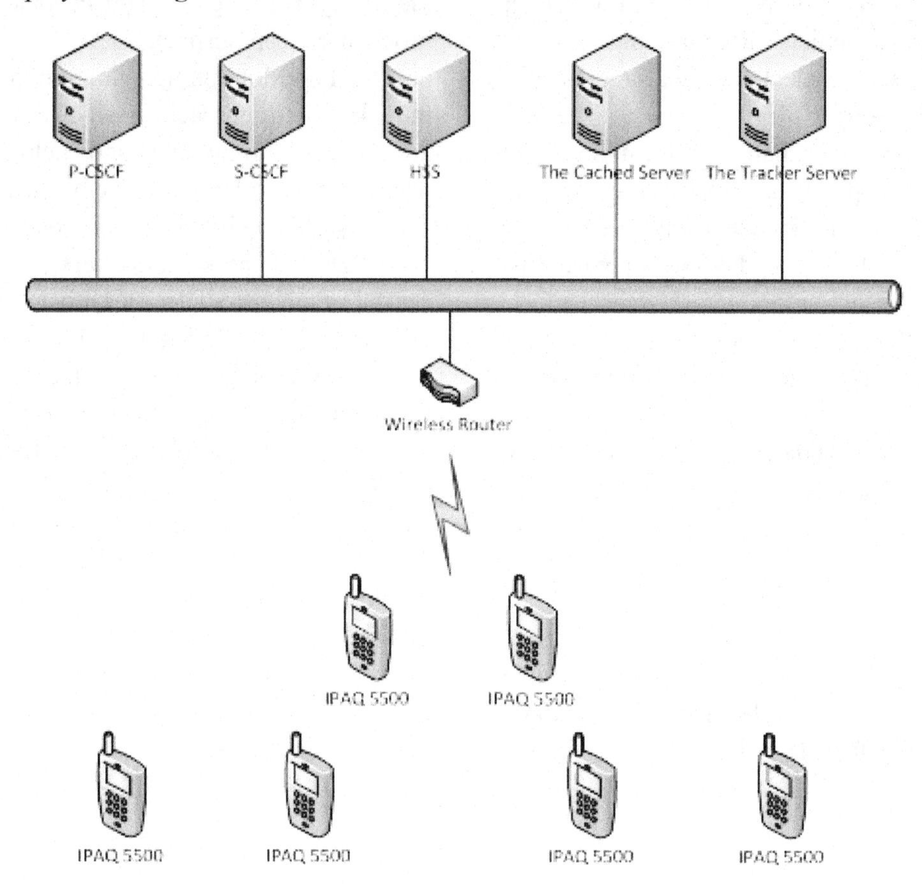

PeerMob Simulation

Evaluation is vital in designing P2P systems by comparing a range of conditions which is broader than those tested on the test-bed. A customized simulator has been developed to analyze the behavior of PeerMob P2P IMS. In order to avoid the tight scalability limits imposed by packet-based simulators and model networks dynamic with acceptable accuracy level, the PeerMob simulator follows discrete-event and flow-based approach, with a simplified network layer, aimed at modeling PeerMob P2P IMS protocols accurately and efficiently in a dynamic P2P environment.

The PeerMob simulation follows the procedure for the standard discrete-event model. The simulator is initialized by a configuration entity targeted to read a configuration file, which is a plain xml file used to specify the simulation scenarios. After the initialization of the system variables and clock, the simulator calculates next clock time and processes subsequent events in the scheduling queue. The ending condition of the simulator happens once the maximum simulation time has been reached or when the scheduling queue is empty. Instead of advancing the simulation time at fixed increments and processing events synchronously at each clock tick, in the PeerMob simulator the processing and time advancement is triggered by the occurrence of events.

Inspired by the modular design of the P2P simulator OMNet++, everything in the PeerMob simulator is a module. These modules can be further divided into simple modules and compound modules. Simple modules include node, event and resource class; compound modules contain process and engine classes. Compound modules may consist of one or more other modules. Each module is defined by a time base, input, states, output and functions to compute the next states and outputs.

For the purpose of portability, ease of development and extensibility, the PeerMob simulator is implemented in Java. In the PeerMob experiments, the PeerMob simulator consumes around 1 Gigabyte of memory when simulating 100,000 peers. We could have simulated a larger system if we adopted a more powerful computer.

The simulator's core class diagram is depicted in Figure 8. The engine class represents the simulation engine. First, the engine parses the configuration file and loads nodes, events and processes from the configuration file into memory. Each event is assigned a time-stamp which indicates the logical time at which the event should be processed. Then, the engine initializes the scheduling queue, which is in control of the execution driven

Figure 8. The simulator core class diagram

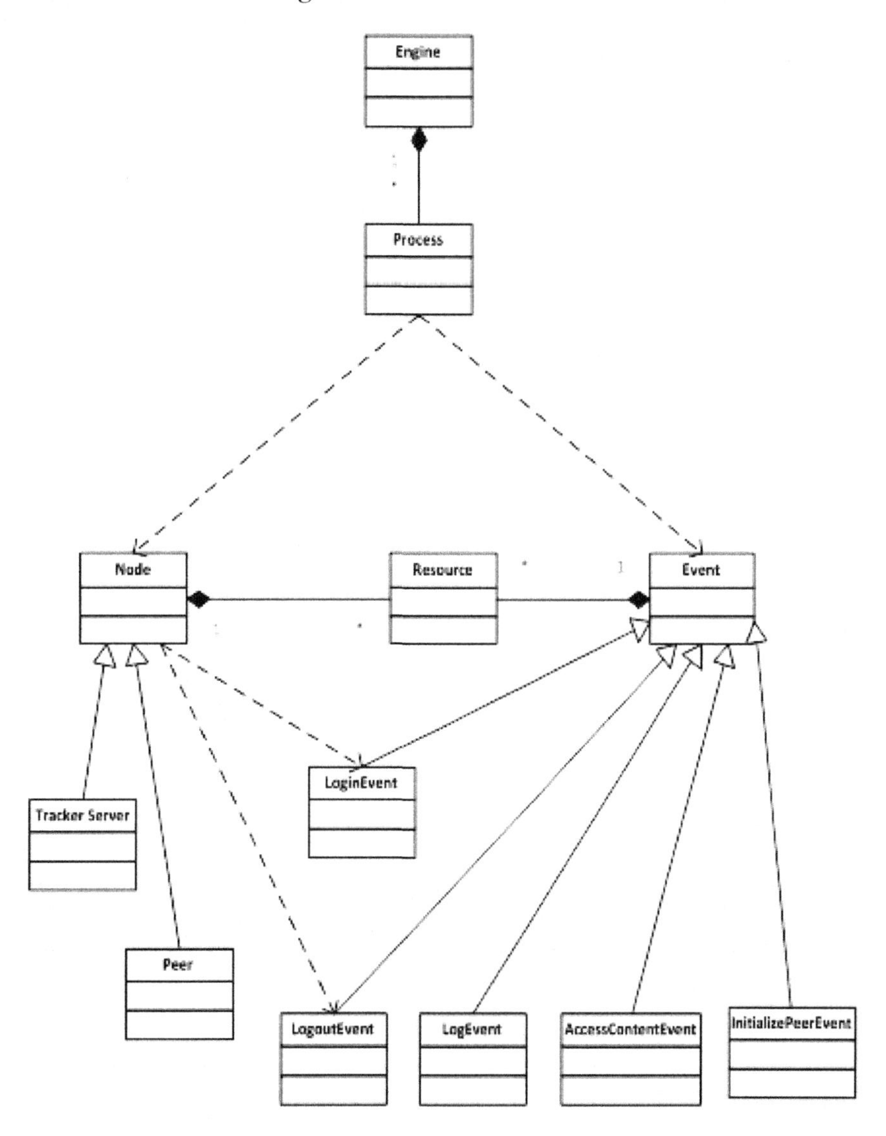

by events. After initialization, the engine is ready to run. The engine starts by inserting events into the scheduling queue and processes these events individually one after the other. The events are stored in the scheduling queue in increasing time-stamp order. The engine processes each event in the scheduling queue until a maximum virtual time is reached or the queue is empty. After being executed, each event is removed from the top of the scheduling queue and the virtual time of the simulation is updated.

SIMULATION PARAMETERS

The parameters of the PeerMob simulator can be divided into four categories: network parameters, cached server parameters, peer parameters and super peer parameters. Table 1 lists the network parameters and their default values in the PeerMob simulator. These default values are used as the initial setting for the experiments, as further detailed in the next section. In each experiment, only one parameter at a time is varied, while other ones are kept fixed. The total default number of peers, including single peers and super peers, in the simulation network is 100,000; the default number for peer groups is 10; and 10,000 data files are available in the simulation network.

Table 2 illustrates the parameters required for the cached server. Cached Servers are usually deployed at the edge of the network before the content delivery service is started. The Number of Cached Servers indicates the servers deployed at service start-up time. The Cached Server Refresh Rate defines the delay that every cached server uses to randomly selects another cached server or super peer in its peer group, to exchange information (according to the gossip protocol). The Max Super Peer in Cached Server represents the length of the super peer list of the cached servers, which defines the maximum number of super peers that could be possible cached in a cached server. Cache servers will remove older cached super peers from the super peer list as soon as the length of the super peer list reaches its limit.

In Table 3, the parameters required to simulate the peers are listed. Peer Join Rate signifies the churn of a simulation network, which indicates the average number of peer joining the simulation network at every second. Content Access Rate here refers to the content access rate of each peer (including single peer and super peer).

Table 4 includes the simulation parameters of the super peers. Super Peer Percentage refers to the probability of a joining peer is elected super peer. The Super Peer Refresh Rate represents the period after which super peers may randomly

Table 1. Network parameters

Item	Value
Number of Peers	100,000
Number of Peer Groups	10
Number of Data Files	10,000

Table 2. Cache server parameters

Item	Value
Number of Cached Server	10
Cached Server Refresh Rate	100 seconds
Max Super Peer in Cached Server	1,000

Table 3. Peer parameters

Item	Value
Peer Join Rate	10 peer per second
Content Access Rate	0 content per second

Table 4. Super peer parameters

Item	Value
Super Peer Percentage	10%
Super Peer Refresh Rate	100 seconds
Max Cached Super Peer in Super Peer	100
Max Cached Single Peer in Super Peer	20

select some other cached server or super peer for the purpose of information exchange. Max Cached Super Peer in Super Peer specifies the maximum number of super peers that can be cached in a super peer. Max Cached Single Peer in Super Peer represents the number of single peer that a super peer can afford to handle.

Scalability Experiment

This section provides an evaluation of the proposed PeerMob P2P overlay network, based on simulation results and cross-validations with the prototype implementation. Our scalability experiments look at how the communication costs increase as the network grows. We select the results obtained at steady state, i.e. not during peer joining/leaving transients. Each peer selects content randomly across the network, switching content every 100 seconds.

The communication cost is measured by the average signal messages that traverse super peers, including outgoing requests and incoming responses. The super peers are the most limiting entities in the group-based hybrid P2P overlay network and that is why we focus on their signalling overheads.

Figure 9 depicts the results of the scalability experiment. The x-axis represents the total number of peers (including single peers and super peers) in the simulation network, while the y- axis refers to the average amount of signalling messages, including the outgoing requests and the incoming responses that each super peer sends and receives every 200 seconds. The average amount of signalling messages that traverses the super peer remains stable as the number of peers grows, showing that we have a scalable architecture.

Reliability Experiment

As mentioned previously, an important characteristic of P2P networks is that peers join and leave the overlay on a continuous basis (this is referred to as churn). The churn rate refers to the average fraction of peers joining and leaving the system per time unit. The reliability of a P2P network is evaluated by assessing the P2P network under different churn rates.

The first reliability experiment tries to assess the performance of the PeerMob network at fixed churn. In this experiment, initially there is no peer. When the simulation starts, new peers join the network at the rate of 10 peers per second. Also, as soon as a peer (single or super) joins the network, it starts accessing randomly selected content, switching content every 100 seconds. All other parameters are set to the default values.

Figure 10 illustrates how the number of single peer and super peer grows in this fixed churn reli-

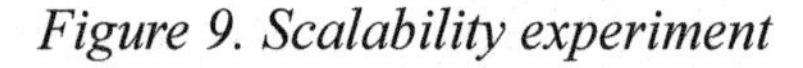

Figure 9. Scalability experiment

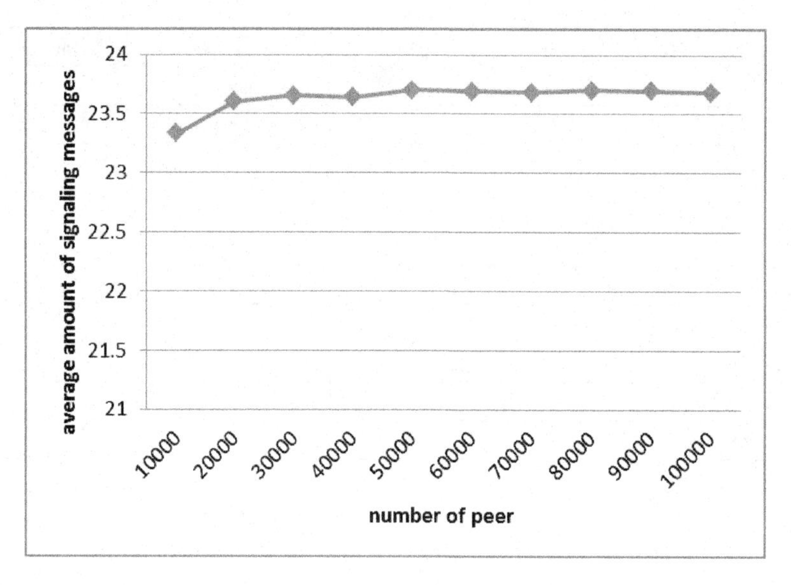

ability experiment. In this figure, X-axis represents the time eclipses in seconds while the y-axis refers to the number of peers in the simulation network.

Figure 11 depicts the load of super peers in this fixed churn reliability experiment. The x-axis represents the elapsed time slots (in seconds), while the y-axis refers to the average amount of signaling messages (including outgoing requests and incoming responses) that each super peer sends and receives in every 200 seconds. The chart shows that the average signaling increases sharply in the first 600 seconds and remains stable at around 45 signals for every 200 seconds thereafter.

The second reliability experiment aims at evaluating the load of super peers at different churn rates. We repeat the first reliability experiment but churn rates are varied. Only the rate of

Figure 10. Fixed churn reliability experiment

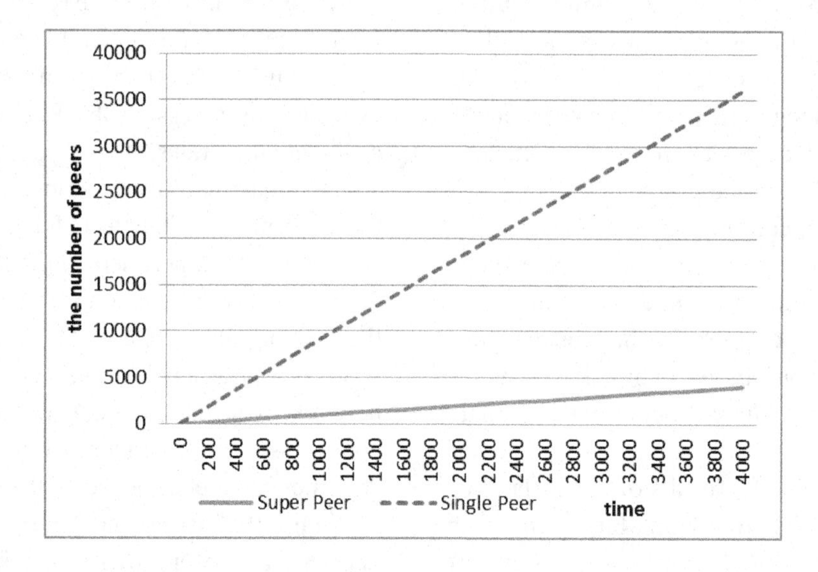

Figure 11. Fixed churn reliability experiment

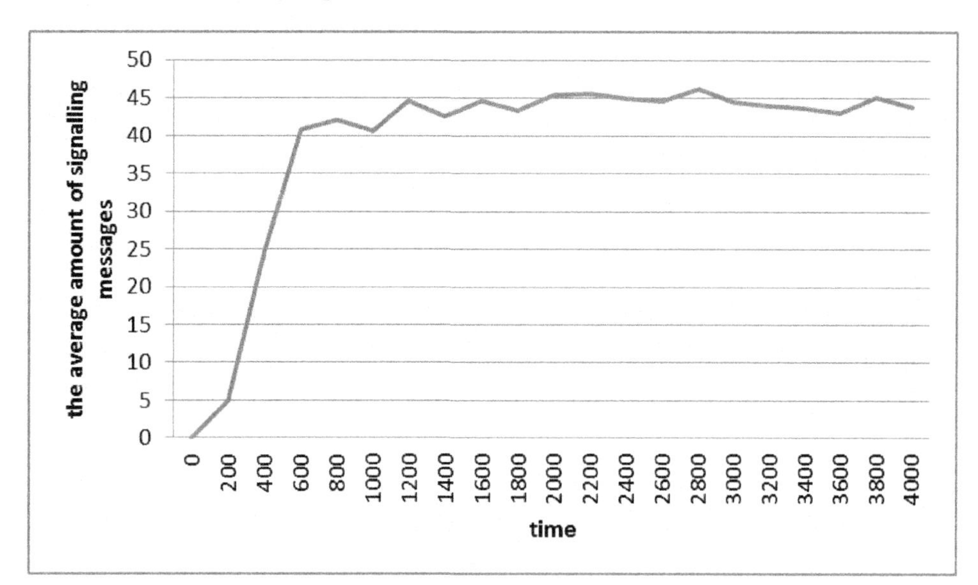

joining peers is variable, while we keep the other parameters fixed. The x-axis represents the churn rate, which is the number of peers per second joining the PeerMob simulation network. On the other hand, the y-axis refers to the average signal messages that traverse the super peers in 200 seconds at steady state. As shown in Figure 12, the average signalling messages per super peer raises slightly with churn rate. This result indicates that the average message per super peer is not determined by churn rate.

Other Experiments

Figure 13 illustrates how the load of super peers is affected by their churn rate, which is the per-

Figure 12. Variable churn reliability experiment

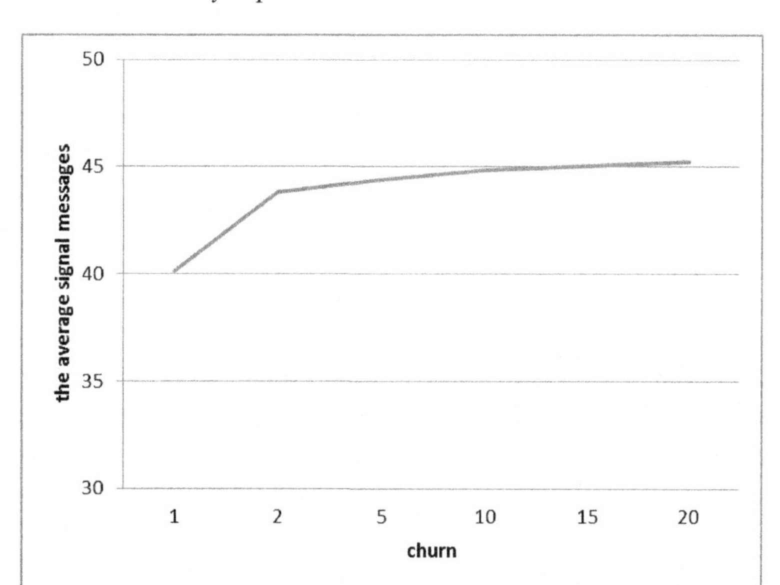

Figure 13. Super peer rate experiment

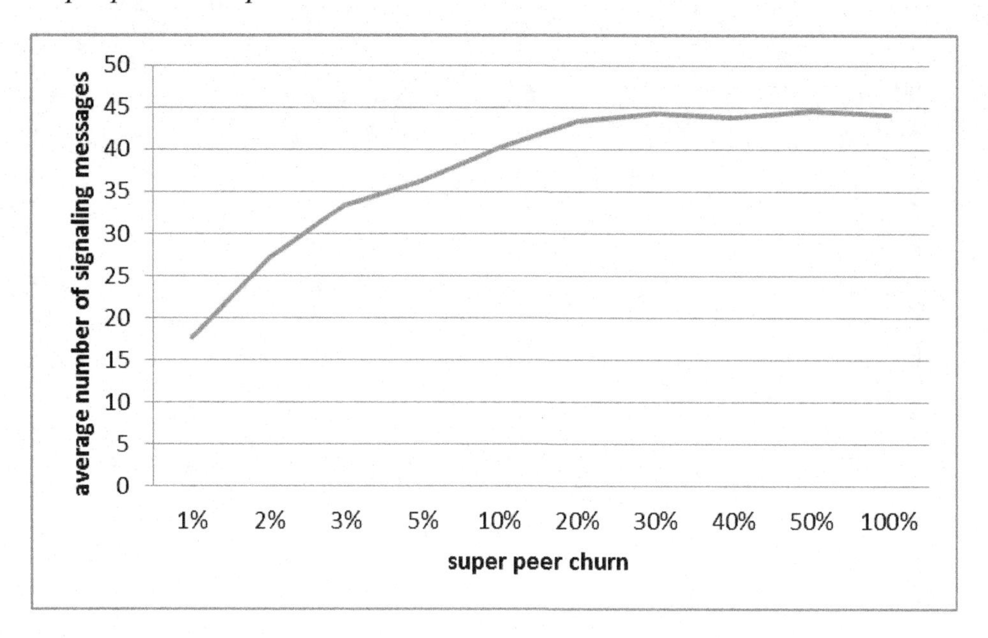

centage of super peers in the PeerMob simulation network. The super peer rate experiment follows the method used in the fixed churn reliability experiment (above). The churn rate is kept constant at 1 peer per second while the super peer rate is increased. The x-axis represents the super peer rate while the y-axis refers to the average number of signaling messages that traverses the super peers in 200 seconds (at steady state). We can see that the signaling load increases gradually with the percentage of super peers.

Figure 14 illustrates how the load of super peer is affected by the number of peer groups. The group experiment follows the method of the fixed churn reliability experiment. The churn rate is fixed at 1 peer per second, while the number of groups increases. The x-axis represents the number of groups, while the y-axis refers to the average number of signaling messages that traverses the super peers in 200 seconds (at steady state). We observe a sustained load decrease as more groups are introduced in the system.

FUTURE RESEARCH DIRECTIONS

The use of the P2P paradigm to provide content delivery services in the IMS is gaining increasing attention recently. China Mobile Research Institute started IMS based P2P Streaming project in 2009. China Mobile is going to trail this IMS based P2P Streaming service in a few provinces in China in 2011. Furthermore, China Mobile extended this P2P idea and proposed a P2P media streaming standardization, *Peer-to-Peer Streaming Protocol (PPSP),* as an IETF draft in 2011. The key distinction between PPSP and PeerMob proposed in this Chapter is that PPSP is based on Napster P2P topology while PeerMob is relying on group-based Hybrid P2P topology.

A large scale content delivery platform over telecommunication networks is a complex task. There are some challenges that remain for possible future work. First, P2P content delivery services tend to increase the traffic of network providers, as peers first download content and then upload the content to other peers. The P2P approach approximately doubles the amount of traffic in the network compared to client-server

Figure 14. Group experiment

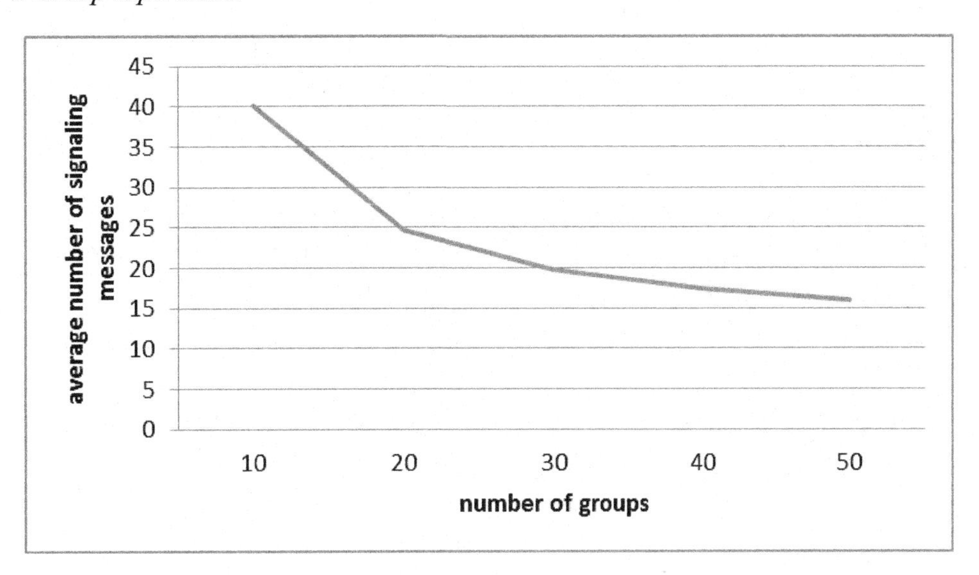

architecture, where most of the traffic flows one direction. This motivates research in locality-aware P2P optimization solutions that exploit network proximity between peers to reduce the traffic in the core networks. Adept PeerMob P2P IMS with locality-aware P2P protocols remains open for future research.

CONCLUSION

The current trend is going more and more towards IP-based delivery for all kind of digital content. At the same time, the increasing quality of the digital content is simultaneously increasing the size of the content, which will also increase the requirements for the capacity of the content delivery platforms. At the time of writing this chapter, *Content Delivery Networks (CDNs)*, such as Akamai, have been widely deployed to provide content delivery services. However, CDNs can only cover a small portion of the open Internet, leaving a large chunk of traffic to ordinary web servers (Client-Server delivery). The same applies to IPTV solutions which employ multicast within the IPTV network only (global IP Multicast has not been deployed yet). Therefore, more scalable solutions are required for content delivery services and P2P has been seen as a potential mechanism to complement the other media delivery approaches.

P2P fosters resource self-deployment and self-organization, reducing operational costs, while still achieving optimized resource utilization for the deployed applications and services. P2P architectures are characterized by the ability to adapt to failures and accommodate transient populations of nodes, while maintaining acceptable connectivity and performance. Although P2P technologies do not directly reduce the overall network traffic in comparison with the traditional client-server approach, the network load is distributed more evenly to the whole network. On the other hand, the IMS, as a control plane technology, addresses primarily issues of heterogeneous technologies for access, addressing schemes, *Authentication, Authorization and Accounting (AAA)*, security and mobility management. These functions are missing in ordinary P2P systems. So our construction of a P2P IMS system brings the benefits of P2P and IMS into a single platform for network operators, service providers and content distributors.

We have prototyped PeerMob P2P IMS, assessing its performance through a purpose-made simulator. Our work highlights the benefits of extending the conventional IMS platform (which is designed according to the client-server paradigm) with the P2P paradigm.

REFERENCES

Allavena, A., Demers, A., & Hopcroft, J. E. (2005). Correctness of a gossip based membership protocol. *Twenty-Fourth Annual ACM Symposium on Principles of Distributed Computing*, (pp. 292-301).

Camarillo, G., & Garcia-Martin, M. A. (2008). *The 3G IP multimedia subsystem (IMS): Merging the Internet and the cellular worlds*. Chichester, UK: Wiley. doi:10.1002/9780470695135

Ilyas, M., & Ahson, S. A. (2008). *IP multimedia subsystem (IMS) handbook*. Boca Raton, FL: CRC Press.

Jovanovic, M. A., Annexstein, F. S., & Berman, K. A. (2001). *Scalability issues in large peer-to-peer networks-A case study of Gnutella*. University of Cincinnati Technical Report.

Lua, E. K., Crowcroft, J., Pias, M., Sharma, R., & Lim, S. (2005). A survey and comparison of peer-to-peer overlay network schemes. *IEEE Communications Surveys and Tutorials*, *7*(2), 72–93. doi:10.1109/COMST.2005.1610546

Oram, A. (2001). *Peer-to-peer: Harnessing the benefits of a disruptive technology*. Sebastopol, CA: O'Reilly Media.

Oredope, A., & Liotta, A. (2008a). Service provisioning in the IP multimedia subsystem . In Mahbubur, R. S. (Ed.), *Multimedia technologies: Concepts, methodologies, tools, and applications* (pp. 491–500). Hershey, PA: Information Science Reference. doi:10.4018/978-1-59904-953-3.ch036

Oredope, A., Liotta, A., Morphett, J., & Roper, I. (2008b). P2P-SIP in highly volatile networks. *IEEE International Conference on Next Generation Mobile Applications, Services and Technologies* (pp. 76-82).

Poikselkä, M., & Mayer, G. (2009). *The IMS: IP multimedia concepts and services*. Chichester, UK: John Wiley & Sons Inc.

Ripeanu, M., Foster, I., & Iamnitchi, A. (2002). Mapping the Gnutella network: Properties of large-scale peer-to-peer systems and implications for system design. *IEEE Internet Computing Journal*, *6*(1), 50–57.

Schollmeier, R., & Schollmeier, G. (2002). *Why peer-to-peer (P2P) does scale: An analysis of P2P traffic patterns*. Second International Conference on Peer-to-Peer Computing.

Stoica, I., Morris, R., Karger, D., Kaashoek, M. F., & Balakrishnan, H. (2001). Chord: A scalable peer-to-peer lookup service for internet applications. *ACM SIGCOMM Computer Communication Review*, *31*(4), 149–160. doi:10.1145/964723.383071

Yang, B., & Garcia-Molina, H. (2003). *Designing a super-peer network*. 19th International Conference on Data Engineering.

ADDITIONAL READING

Bertrand, G. (2007). The IP multimedia subsystem in next generation networks. *Architecture (Washington, D.C.)*, *7*(3).

Borosa, T., Marsic, B., & Pocuca, S. (2003). QoS support in IP multimedia subsystem using DiffServ. *7th International Conference on Telecommunications*, (pp. 669-672)

Camarillo, G., Kauppinen, T., Kuparinen, M., & Ivars, I. M. (2007). Towards an innovation oriented IP multimedia subsystem. *IEEE Communications Magazine*, *45*(3), 130–136. doi:10.1109/MCOM.2007.344594

Dabek, F., Li, J., Sit, E., Robertson, J., Kaashoek, M. F., & Morris, R. (2004). Designing a DHT for low latency and high throughput. *1st Conference on Networked Systems Design and Implementation*, (pp. 669-672).

Ding, G., & Bhargava, B. (2004). Peer-to-peer file-sharing over mobile ad hoc networks. *2nd IEEE Annual Conference on Pervasive Computing and Communications*, (pp. 104-108).

El-Ansary, S., Alima, L., Brand, P., & Haridi, S. (2003). Efficient broadcast in structured P2P networks. *2nd International Workshop on Peer-to-Peer Systems*, (pp. 304-314).

Faccin, S. M., Lalwaney, P., & Patil, B. (2004). IP multimedia services: Analysis of mobile IP and SIP interactions in 3G networks. *IEEE Communications Magazine*, *42*(1), 113–120. doi:10.1109/MCOM.2004.1262170

Gagliardi, J. D., & Munger, T. S. (2008). *Content delivery network: US patent app. 20,090/157,850.*

Galán, F., García, E., Chávarri, C., Gómez, M., & Fernández, D. (2006). *Design and implementation of an IP Multimedia Subsystem (IMS) emulator using virtualization techniques. Technical Report*. Madrid Polytechnic University.

Gupta, A., Liskov, B., & Rodrigues, R. (2004). *Efficient routing for peer-to-peer overlays*. 1st Conference on Networked Systems Design and Implementation.

Gupta, I., Birman, K., Linga, P., Demers, A., & Van Renesse, R. (2003). Kelips: Building an efficient and stable P2P DHT through increased memory and background overhead. *2nd International Workshop on Peer-To-Peer Systems*, (pp. 160-169).

Harren, M., Hellerstein, J., Huebsch, R., Loo, B., Shenker, S., & Stoica, I. (2002). *Complex queries in DHT-based peer-to-peer networks* (pp. 242–250). Peer-to-Peer Systems.

Huang, C. M., & Li, J. W. (2007). One-pass authentication and key agreement procedure in IP multimedia subsystem for UMTS. *International Conference on Advanced Information Networking and Communications*, (pp. 482-489).

Koukal, M., & Bestak, R. (2006). Architecture of IP multimedia subsystem. *48th International Symposium on Multimedia Signal Processing and Communications*, (pp. 323-326).

Larsen, K. L., Matthiesen, E. V., Schwefel, H. P., & Kuhn, G. (2006). Optimized macro mobility within the 3GPP IP multimedia subsystem. *International Conference on Wireless and Mobile Communications.*

Maymounkov, P., & Mazieres, D. (2002). *Kademlia: A peer to peer information system based on the xor metric* (pp. 53–65). Peer-to-Peer Systems.

Melnyk, M. A., Jukan, A., & Polychronopoulos, C. D. (2007). A cross-layer analysis of session setup delay in IP Multimedia Subsystem (IMS) with EV-DO wireless transmission. *IEEE Transactions on Multimedia*, *9*(4), 869–881. doi:10.1109/TMM.2007.895680

Pierre, G., & Van Steen, M. (2006). Globule: A collaborative content delivery network. *IEEE Communications Magazine*, *44*(8), 127–133. doi:10.1109/MCOM.2006.1678120

Rao, A., Lakshminarayanan, K., Surana, S., Karp, R., & Stoica, I. (2003). Load balancing in structured p2p systems. *Peer-to-Peer Systems*, *II*, 68–79. doi:10.1007/978-3-540-45172-3_6

Rebahi, Y., Sher, M., & Magedanz, T. (2008). Detecting flooding attacks against IP multimedia subsystem (IMS) networks. *IEEE/ACS International Conference on Computer Systems and Applications*, (pp. 848-851).

Rhea, S., Geels, D., Roscoe, T., & Kubiatowicz, J. (2004). *Handling churn in a DHT*. USENIX Annual Technical Conference.

Singh, K., & Schulzrinne, H. (2005). *Peer-to-peer internet telephony using SIP*.

Stutzbach, D., & Rejaie, R. (2006). Understanding churn in peer-to-peer networks. *International Workshop on Network and Operating Systems Support for Digital Audio and Video*, (pp. 63-68).

Zhang, Z., Shi, S. M., & Zhu, J. (2003). SOMO: Self-organized metadata overlay for resource management in P2P DHT. *Peer-to-Peer Systems, II*, 170–182. doi:10.1007/978-3-540-45172-3_16

Zhu, B. (2003). *Analysis of SIP in UMTS IP multimedia subsystem*. MSc Dissertation, North Caraolina State University.

Zhuang, W., Gan, Y. S., Loh, K. J., & Chua, K. C. (2003). Policy-based QoS architecture in the IP multimedia subsystem of UMTS. *IEEE Network Magazine*, *17*(3), 51–57. doi:10.1109/MNET.2003.1201477

KEY TERMS AND DEFINITIONS

Content Delivery Network: *CDN* is a system of networked computers containing replicas of the data to be transmitted at various points in the network.. When properly designed and implemented, a CDN can significantly improve network utilization and latency.

Distributed Hash Table: DHT is a type of decentralized distributed system that provides a lookup service similar to a hash table. Key and value pairs are stored in a DHT and any participating node can efficiently retrieve the value associated with a given key.

Interrogating-Call Session Control Function: *I-CSCF* is a SIP proxy server placed at the edge of an IMS domain to communicate with other IMS domains, hiding the network capacity and topology from the outside.

IP Multimedia Subsystem: *IMS* is an architectural framework for delivering multimedia services over converged all-IP networks. It was originally designed by the wireless standards body known as 3rd Generation Partnership Project (3GPP), as a part of the vision for evolving mobile networks beyond GSM.

Peer to Peer: P2P is a distributed application architecture in which participants can act both as clients and servers to pursue efficient resource utilization. These participants are able to work without a central server but, in same implementations, make use of servers for purposes such as registration, authentication and indexing.

Proxy-Call Session Control Function: *P-CSCF* is a SIP proxy, which is the first point of contact for the IMS terminal.

Serving-Call Session Control Function: *S-CSCF* is a SIP server acting as both a SIP proxy server and a SIP registrar server.

Session Initiation Protocol: *SIP* is an IETF-defined signaling protocol, widely used for controlling communication sessions such as voice and video calls over the Internet Protocol (IP).

Chapter 3
Pervasive Streaming via Peer-to-Peer Networks

Majed Alhaisoni
University of Ha'il, Saudi Arabia

Antonio Liotta
Eindhoven University of Technology, The Netherlands

ABSTRACT

Media streaming is an essential element of many applications, including the emerging area of mobile systems and services. Internet broadcasting, conferencing, video-on-demand, online gaming, and a variety of other time-constrained applications are gaining significant momentum. Yet, streaming in a pervasive environment is not mature enough to address challenges such as scalability, heterogeneity, and latency. In a client-server system, streaming servers introduce computational and network bottlenecks affecting the scalability of the system and mobile client exhibit intermittent behavior and high-latency connections. This chapter explores ways that several proposed peer-to-peer (P2P) streaming systems deploy to address some of these challenges. An initial introduction on P2P network fundamentals and classifications provides the necessary background information to focus on and assimilate the different mechanisms that enable scalable and resilient streaming in a pervasive environment. The most interesting developments are presented in an accessible way by revisiting the features of common P2P streaming applications. This approach helps in identifying a range of burning research issues that are still undergoing investigation.

INTRODUCTION

In recent years, *Peer-to-Peer (P2P)* networking has gained much consideration after its successful achievement from its file sharing ability such as BitTorrent (Liu et al., 2008) and, more recently, in multimedia streaming. The concept of P2P networking is realized by encouraging the end users to contribute to the network resources and act as client and server simultaneously. That is, each user can upload and download directly from/to other users avoiding central entities. The motive behind is the cooperation among them to overcome various limitations of the more conventional *Client-Server*

DOI: 10.4018/978-1-4666-1613-4.ch003

(CS) paradigm to attain user and bandwidth scalabilities. At its early times, P2P Computing was considered as a file sharing platform across different networks and environments. Over the time, P2P TV has gained both academic and industrial attention as the next potential killer technology for streaming multimedia to the public. Overlay networks play an important role as the deployment infrastructure of different P2P applications. The underlying mechanism is based on the distribution of streams or files through an application-level overlay, including the user terminals in the role of peers, i.e. content distribution relays.

Following this introduction, the chapter is organised into four main sections. The first provides a quick overview of general knowledge on video streaming delivery platforms. The following two give more emphasis on Unstructured and Structured P2P systems respectively presented from the video streaming viewpoint. The fourth section provides an analysis of recent advances on P2P streaming. The focus of that section is on the identification of those features that enable efficient streaming even on highly dynamic networks like P2P. The chapter concludes with the authors' point of view and an outlook for the use of those systems to pervasive streaming.

VIDEO DELIVERY PLATFORMS

The Client–Server Model is the opposite extreme of Peer-to-Peer Computing. However, the latter can be viewed as the evolution of the former. In a CS-based streaming setting, the client initiates a connection with the video source address and the server replies back by directly delivering the content. Though simplicity and manageability are two major advantages of this scheme, its weaknesses pushed its evolution to P2P Networks. As all the content is located and provided by a single central entity, any failure of that entity may deactivate the whole streaming service to any client. There is a plethora of unpredictable either accidental (e.g. power cut-offs) or deliberate (e.g. security attacks) reasons for these failures. Traffic bursts are also conditions that these architectures are not designed to handle. Sudden increases of content requests from clients can quickly consume all the resources of the server and force it to drop any excessive load. In an effort to tackle those problems, system administrators increased the initial investment and maintenance costs by building very powerful, highly secured infrastructures that only very specialized personnel could operate. They quickly became unscalable systems that few could afford (Androutsellis-Theotokis & Spinellis 2004; Liu et al. 2008) and without even providing guarantees for service quality and reliability in case of flush crowds. The need for a more distributed architecture drove the development of *Content Delivery Networks (CDNs)* (Pallis & Vakali 2006).

The main concept behind CDNs is the use of many strategically placed video content delivery servers on the edges of the internet. Any video is delivered by the closest server to the requesting client. Initially, the server pushes all the content to its counterparts. Clients access a content server and if that is not the closest one, its requests are redirected to the nearest one. From the server perspective, this mechanism is a load sharing approach and from the client perspective it reduces the start-up delay. As a locality-aware mechanism, it also saves the network from traffic as streams tend to have the shortest possible paths. One of the real-world video streaming applications over the internet that exploits this mechanism are YouTube (Hossfeld & Leibnitz 2008), MySpace (myspace), Veoh (VEOH), and Akamai (akami). Due to its higher resource distribution level, compared to the client-server model, companies can save some of their operational costs and handle more requests. However, maintaining multiple servers across the planet introduces higher costs in designing and maintaining the system. For instance, the decision to place the servers at the most appropriate location includes incoming traffic monitoring, distributed maintenance units or even system management outsourcing. These actions and decisions are to be taken by a central entity; even though a more distributed model compared to CS, it still has weak scalability properties as the centralized management is still a requirement. Finally, the servers' operation and the whole CDN content need to comply with the diversification of national regulations they are installed at. All these restrictions may pose high bootstrap costs and delays to companies.

From the paragraphs above, a tendency to decentralization becomes more evident with respect to the streaming services. A considerable increase to the number of servers of a CDN would make the system more reliable but also an unrealistic solution due to its management costs. A way around, can be computational resource and management sharing. P2P Networks come to close exactly this gap as they represent an economical, robust, and scalable alternative to the more conventional CS approach (Liu et al. 2008). The number of providers can, in theory and practice, be as many as the requestors. Moreover, the content, if replicable, can live in many copies around the network and simultaneously be streamed to as many or more other nodes. Both these two features improve the scalability of the system as every new node not only does consume bandwidth but adds to it, too. This high decentralization of content, management, resource ownership, maintenance cost and resource scalability render more robust and reliable delivery mechanisms. It increases the complexity but offers resistance to several kinds of attacks prevents bottlenecks and increases the service availability.

Finally, theoretical P2P approaches have demonstrated their benefits in comparison to their client-server counterparts, in terms of scalability. However, in practice they pose huge management hurdles to network operators as P2P systems do not consider important network operational and management requirements. Therefore, many operators choose to throttle P2P traffic as their scalability is countered by severe network inefficiency (Alhaisoni et al. 2010). Therefore, network awareness is a P2P feature that gets more and more attention for improving their applicability (Kovacevic et al. 2008).

Peer-to-Peer systems are networks built on top of and decoupled from the underlying network. This is known in most P2P applications as the 'overlay' network. An important classification factor of these systems is their resources' distribution over the nodes. If there is a common distributed algorithm that maps resources to indexing nodes (not necessarily to their owners), then these networks (structured) build a structured formation. That is, both the links between nodes and resource index location are controlled by that algorithm. At any moment any resource can be located through that lookup algorithm. On the other hand, unstructured networks do not deploy any well-defined mapping algorithm to place resource indexes into nodes. In these cases, they may randomly exist within the network; symmetrically, links between nodes usually can be between any two nodes thus forming random networks.

The topology structure and degree of centralization of the overlay network are two important classification factors of those systems as they both may influence the fault tolerance, self-manageability, performance, and scalability (Milojicic, Kalogeraki et al. ; Androutsellis-Theotokis & Spinellis 2004). In general, although P2P networks are made-up to be fully decentralized and unstructured, many existing P2P systems are designed with different characteristics. The following *Figure* and paragraphs present this centralization level- and structure- based classification.

There are several factors also affecting the structure of P2P networks. Peers are free to join or leave at any time without any prior notification or change their connectivity by disconnecting and/or connecting to different subset of peers. This results in the dynamicity nature of these systems and the heterogeneity among the contributing peers. In structured networks, these actions

Figure 1. Distribution of P2P networks

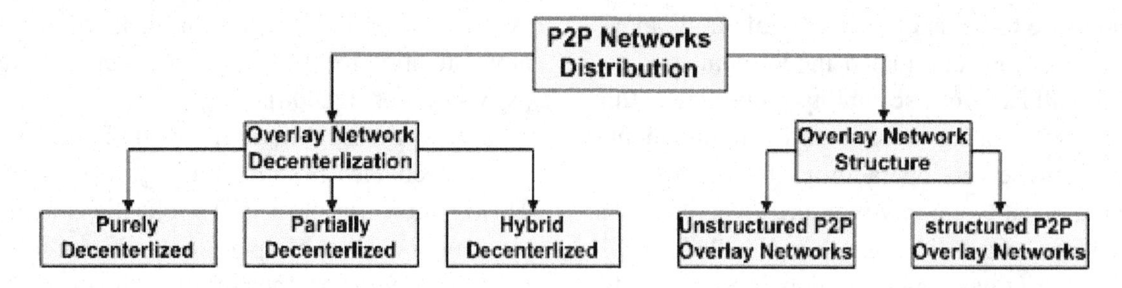

are driven by specific algorithms whereas in unstructured ones can happen randomly.

UNSTRUCTURED P2P OVERLAYS

In this type of overlay, peers are organized in a random graph. There are a variety of searching mechanisms deployed on these networks; the basic principle behind them is the recursive query forwarding from a peer to a random subset of its neighbours. This query propagation within the overlay builds parallel searching paths from the requesting peer to its vicinity. For bandwidth saving purposes, this propagation stops after a number of steps based on some criteria. These propagation stopping and targeted neighbor selection criteria are may differ between unstructured search mechanisms. A popular and well-documented such algorithm is flooding, which was also used in the first P2P network of this kind: Gnutella (gnutella). To name a few, modified-Breadth-First-Search (Kalogeraki et al. 2002), Random Walkers (Lua et al. 2005) and Iterative Deepening (Yang & Garcia-Molina 2002) are widely known algorithms. A more comprehensive analysis of then can be found in Li & Wu (2006).

A widely used propagation stopping criterion is the *Time-to-Live (TTL)* parameter of each query. It is a technique to avoid network overloading by infinite cyclic forwarding paths. It prevents queries from exploring the whole network and/or getting trapped within loops. Setting a horizon to queries is a limiting factor of the discovery mechanism efficiency in the case of rare files (Lua et al., 2005). In addition to that, each peer on the overlay needs to evaluate every incoming query in its path, thus, introducing extra computational overhead in the system. Despite their occasional inefficiency, unstructured P2P look-up mechanisms help networks cope with the problem of single point of failure. In this regard, there are common unstructured P2P overlay approaches like BitTorrent (BitTorrent), FastTrack (FastTrack), Freenet (Freenet) and Gnutella.

Most of the P2P file sharing applications are adopting the FastTrack architecture such as KaZaA(Kazaa) and iMesh (iMesh). Moreover, *Voice over IP (VoIP)* and multimedia applications have also used this architecture such as Skype (Skype) and Joost (Joost). The advantages of FastTrack is that it supports meta-data searching and is based on a hierarchical approach using nodes with high bandwidth, disk space and processing power as supernodes. These nodes index metadata sent by their leaf ordinary ones. Queries from ordinary nodes are first examined at their Supernode and in case of no matches; they are forwarded to the overlay of supernodes. Freenet (Clarke et al. 2001) is designed for providing anonymity and can adapt to various network conditions. Each peer in the overlay maintains a dynamic routing table and also uses *hop-to-live (HTL)* query propagation stopping criterion to prevent infinite chains. Freenet is mostly deployed for serving file sharing applications.

Gnutella (gnutella) is based on a flat topology in which the peers in the overlay can assume the role of either a server or a client. New peers join the overlay using one or more a-priori known bootstrap nodes which reply back with a list of present neighboring nodes and sets connections with a subset of them. Once its neighbors are known, it periodically sends PING messages to discover other participating nodes. Any subsequent query is flooded within the overlay and stops propagation after TTL steps. Although this approach is not scalable and generates a lot of traffic in the network, it makes the overlay highly resilient, as the continuous random joins and leaves of nodes hardly affect the lookup efficiency. The latest version of Gnutella uses the notion of ultra-peers which are synonymous to the supernodes in FastTrack discussed earlier. The use of Gnutella is quite common in file-sharing applications such as LimeWire (Limewire) and is the base platform for PeerCast (PeerCast), a P2P multimedia streaming application.

From the above, it becomes clear that some P2P architectures use different peer roles depending on their discovery mechanism. Super-peers and ultra-peers are a small subset of nodes in these networks with some increased resource indexing role. In a way, this scheme introduces a level of centralization in the system and improves the discovery efficiency. Though P2P systems have emerged to overcome the centralization of the traditional CS paradigm, certain unstructured P2P systems may use exactly this limitation for their benefit. Based on the decentralization degree, Unstructured P2P systems can be classified into the following architectures: Purely, Partially and Hybrid Decentralised P2P Networks.

Purely Decentralized Architecture

This type of architecture reflects the pure P2P systems without any centralization in the design of their architecture. All peers are equal and are freely connected and they communicate with each other in the network performing their tasks and acting as client and server simultaneously, as shown in *Figure 2*. The peers (nodes) of such architectures are termed servents (server+client). There are different applications adopting this architecture such as Gnutella (gnutella) and Freenet (Freenet). Pure decentralization gives robustness and reliability. However, discovery mechanisms deployed on those networks can bring high costs in messages wasting bandwidth and time without providing any discovery guarantees. For instance, flooding can be considered as the most expensive mechanism in this architecture.

Figure 2. Purely decentralized

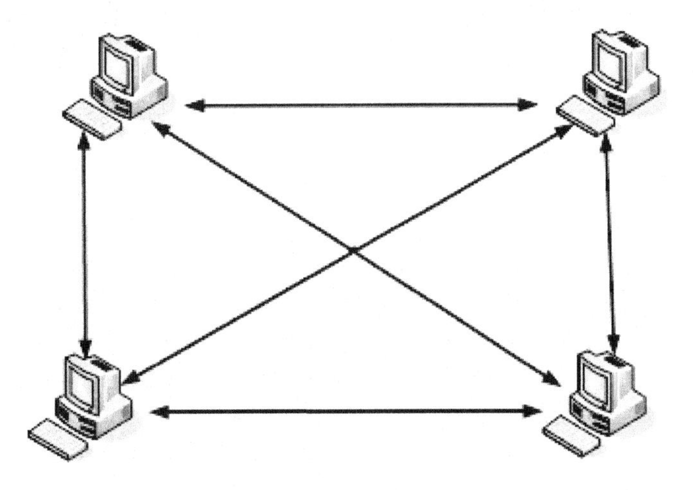

Gnutella is described as a representative system of the purely decentralized P2P architectures. It has its own routing principles in contrast to other P2P applications. In terms of the interconnections among peers, Gnutella end-users connect to each other freely through software which acts as client and server. In the Gnutella framework, at the applications level, there are four types of messages which form the main communications methods between the participants. These messages include:

- Ping: any peer uses this message flooded within the overlay to discover new peers and potentially append in its neighbour list. It is initiated by a new peer and periodically by all existing ones.
- Pong: reply to an incoming Ping message containing the IP and port of the responding host. It follows the same path backwards as the Ping that reached the responder.
- Query: This initiates the search request to all the peers. It floods the overlay though multiple propagations from peers to their neighbours. Besides requestors details (IP, Port) and search terms, it carries a TTL and Hop counter to avoid its infinite forwarding on the overlay.
- Query Hits: is the reply to the Query message including IP, port, and speed of the responding host as well as the number of matching files found and their indexing results.

As previously mentioned, the time-to-live field (TTL) was introduced to save the network from unnecessary messages. It works as a safety guard from infinite propagation of not only queries but ping messages, too; lack of that field would overwhelm the network with messages. However, it divides the network into sub-networks to ensure that the message does not go beyond certain sub-area (Jovanovic 2001). These issues were investigated by many researchers such as (Lv et al., 2002). Finally, Gnutella uses TCP connections between servents which remain open throughout the construction of the ping/query paths. This is a way to drive the pong/query hit messages back to the requestor avoiding some firewall restrictions. However, there are a maximum number of TCP connections each servent can open simultaneously, thus limiting overlay's scalability.

Partially Centralized Architecture

The partially decentralized architecture is evolved to improve the content search of the purely decentralized architectures. Per contra, partially decentralized architectures introduced the concept of "supernodes" connected to set of peers (leaf peers). The supernodes build an extra interconnection layer on top between each other. Their main role is to index the files located in their leaf nodes and facilitate their discovery by introducing a low degree of centralisation. All the queries from leaf peers go through these supernodes; thus, they act as search proxies for the leaf peers. The topology they form can be any unstructured one; a widely used one is the full-mesh since they are few and fully-aware of each others' locations. *Figure 3* illustrates a sketch diagram of the partially decentralized architecture where each set of peers are connected to a Supernode and the Supernodes are connected with each other.

There are several applications based on this concept of partially decentralized architectures including KaZaA (Kazaa) and Morpheus (Morpheus), and iMesh (iMesh). As previously mentioned, these applications are mainly deployed on FastTrack which can be considered one of the most popular file sharing networks. Many such architectures deploy autonomic leaf node promotion/demotion mechanisms to supernodes for keeping a good balance of indexing peers to ensure efficient discovery. However, in the case of a Supernode failure, its leaf peers may even get disconnected if they are not aware of another Supernode. Client-Server characteristics may start

Figure 3. Partially Decentralized Architecture

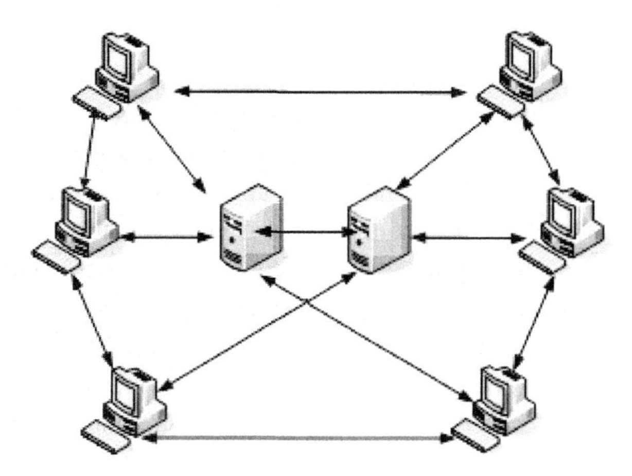

appearing in the behavior of these systems if that is massive Supernode failures take place simultaneously. Every designed system implements a certain technique to overcome this issue. One well known solution is rapid dynamic switching to other supernodes no matter how many supernodes go down. Supernode determination is still under investigation but different applications deploy Supernode election mechanisms are according to specific criteria. For instance, in KaZaA, supernodes are selected dynamically based on the availability of the bandwidth and processing power.

Summing up the advantages of the partially over purely decentralized architectures, it is important to briefly analyse a few.

- **Discovery time**: It takes less time to find the desired content. Issues with potential single point of failure situations are prevented by switching over to other supernodes. Gnutella2 (Gnutella2) is an example of the applications that have implemented the Supernode switching mechanism to overcome the single point of failure problem.
- **P2P Heterogeneity**: In purely decentralized systems, all peers are treated similarly irrespective of their heterogeneity. CPU power and bandwidth limitations are not considered as all nodes are equal. On the contrary, in partially centralized systems, nodes promotion to supernodes happens when these peers are capable of carrying out the heavy load posed by leaf nodes (typical peers). More comparisons can be found in Lv et al. (2002) and Xu et al. (2003).

Hybrid Decentralized Architecture

In this architecture, there is a central server (directory) to keep up-to-date information on all the peers and their resources. This server takes over only the discovery part of the P2P system. It maintains a central index of all the resources within the network. All the search requests go through this central entity; thus, it is responsible for maintaining the interconnectivity among the peers since. Any new node contacts the server to submit a list of the resources it brings into the network; this is the bootstrapping process. A requesting node has to access that central point to collect a list of potential peers that host the requested files. The requesting peer, then, directly contacts a subset of this list to fetch the resource. In this way, the server acts as a searching tool for the peers but the actual content delivery is done between the peers. This architecture has a higher scalability than a traditional Client-Server model as the central entity hosts only an index and not the resources, too. It basically saves server's bandwidth, thus, allowing more simultaneous connections to the server. However, the central index is prone to inconsistencies when peers fail or change their resources; frequent updates are necessary. As long as a requesting node has a list of provider peers, there is no need of communication with the central index. However, as in all centralised architectures, hybrid P2P networks also suffer from bottleneck issues and failures of the server. *Figure 4* depicts a sketch of a hybrid decentralized peer-to-peer architecture.

The BitTorrent (BitTorrent)and Napster (Napster) P2P systems use the hybrid architecture by centralizing a server which acts as a directory to all the resources in the network. While Napster has a well-known dedicated server for that purpose, BitTorrent groups all the available peers into swarms with one such indexing server. The former is vulnerable to attacks but the latter is more robust. In general, the two main advantages of this architecture is its implementation simplicity and search efficiency. On the other hand, there are some issues with this architecture such as the vulnerability to censorship and malicious attacks. It is easy to trace the host of the server and take actions against its normal operation. If, for instance, it indexes illegal content the host might be accused by law of sharing illegal content at his consent. Finally, despite its improved scalability, the server remains a limitation of the system based on the users it can concurrently serve.

The most common P2P system of this type is BitTorrent protocol. This system allows all the end users to download and share their files over the network (Pouwelse et al., 2005). It involves trackers, peers, and seeds (Pouwelse et al., 2005) gathered in swarms. BitTorrent was designed mainly to reduce the free riders in the overlay network. It is based on a tit-for-tat approach allowing peers with high upload speed to gain access to high download rates. The main advantage of BitTorrent is that it breaks large files into smaller pieces allowing peers to download various bits from various peers simultaneously allowing for faster downloads.

Another system that gained popularity until its shut-down by legal authorities is Napster. It supported music exchange and was released on the latest of the nineties. It was based on a central index of different music files.

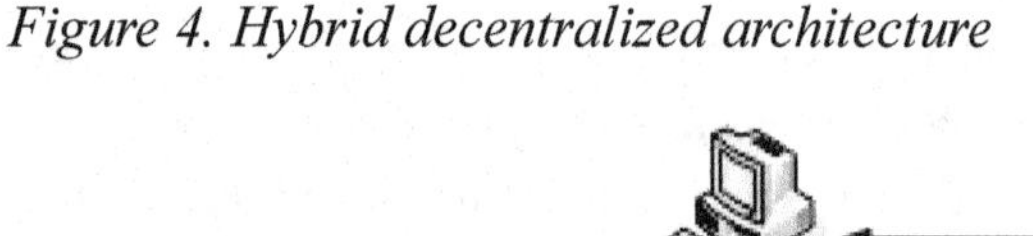

Figure 4. Hybrid decentralized architecture

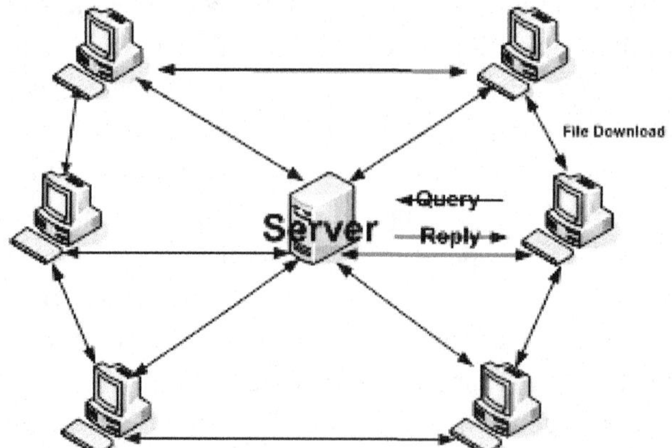

STRUCTURED P2P OVERLAY NETWORKS

Due to the limitations of searching and scalability of unstructured overlay networks, various applications and services are deployed over the Internet based on structured P2P overlays. The most common approaches are based on Chord (Stoica et al., 2003), *Content Addressable Network (CAN)* (Ratnasamy et al. 2001), Kademlia (Maymounkov & Mazieres 2002), Tapestry (Zhao et al., 2004) and Pastry (Rowstron & Druschel, 2001). These systems are known as the second generation of P2P systems. Generally, structured P2P overlay networks deploy a *Distributed Hash Table (DHT)* to increase their scalability and provide discovery guarantees. In this approach, the P2P overlay network is built using keys generated from various hash algorithms. The data and nodes are usually hashed to provide unique keys which are the searching key and node ID, respectively. Each data in the system is mapped to a key; the key is then mapped to a node ID. Their lookup algorithms are well-founded and studied through mathematical models. Different implementations use different graph theories for their lookup mechanisms (Lua et al., 2005).

The main disadvantage of this approach is that exact only matches of queries can be retrieved due to the resource hashes. For instance, partial file names or metadata cannot be hashed to the same hash key as the corresponding file names. Therefore, complex queries are difficult perform; there is high research interest in this area to develop algorithms that allow complex queries (Triantafillou & Pitoura, 2004) (Liu et al., 2005). The following paragraphs give a brief description of the most common structured P2P systems.

Chord

Chord (Stoica et al., 2003) is a decentralized P2P system which improves the routing and searching mechanism. It uses consistent hashing based on the SHA-1 hash function to provide its unique keys to nodes and resources. These keys are based on an m-bit identifier, long enough to prevent duplication. It also provides each peer with equal amount of keys to balance the load in the system and allow only a little movement of keys when nodes join and leave the overlay. All the identifiers are arranged in an identifier circle of modulo $2m$ which is known as the chord ring. This allows the look up cost in the overlay of N peers to be $O(\log N)$ allowing chord to be a highly scalable platform. Chord has been used in several applications such as cooperative mirroring (cooperative file system) (Dabek et al. 2001), which allows data to be spread across various content providers balancing the system. It is also used in the Chord-DNS (Cox et al., 2002) which is similar to the Internet based DNS for looking up hostnames from IP address.

Content Addressable Network (CAN)

Another approach is the *Content Addressable Network (CAN)* (Ratnasamy et al., 2001) which is designed to be highly scalable, fault tolerant and self-organizing. It is based on a virtual multidimensional Cartesian coordinate space on a multi-torus in which each peer is given a unique and individual zone in the overall space. The peer's point P on the coordinate space is made up of the key k and value v, (i.e. k,v) pair, whereby k is mapped on to P using a uniform hash function. Each peer also stores the IP address and coordinates values of its neighbors and then uses a greedy algorithm to route messages to them. CAN also has its own *Domain Name System (DNS)* service (Ratnasamy et al., 2001) which allows peers joining the overlay to locate the bootstrap nodes closest to them. In order to implement redundancy and fault tolerance into CAN, each peer is mapped to different unique coordinates (Lua et al., 2005). CAN has been used in the implementation of large scale storage management systems such as Oceanstore (Kubiatowicz et al., 2000), Farsite (Bolosky et al., 2000) and Pubilius (Waldman et al., 2000) which require efficient insertion and retrieval of large content and a scalable mechanism.

Table 1. Classification of P2P applications

	Overlay Decentralization		
Overlay Structure	Purely	Partially	Hybrid
Unstructured	Gnutella, Freenet	KaZaA, iMesh, Morpheus	BitTorrent, Napster
Structured	CAN, Chord, Pastry, Tapestry		

Pastry

Pastry (Rowstron & Druschel 2001) is very similar to Tapestry (Zhao et al., 2004) as both are built on the Plaxton's Mesh but it is based on a 128-bit peer identifier in which it provides node IDs to each node which is placed in a circular node ID space. Each node also maintains a leaf set in addition to the neighborhood and routing set which is maintained in Tapestry. The leaf set contains an entry of a set of nodes with comparable larger and smaller node IDs which are contacted as a form of last resort in times of node failures. Pastry has been implemented in Scribe (Castro et al., 2002) for application level multicasting, SplitScream (Castro et al., 2002) for multicast cooperative environments, PAST (Emule) for distributed storage and Pastiches for inexpensive and distribute backup.

Lastly, the following table summarizes the above-mentioned classification of P2P networks with an example of every used technique.

P2P VIDEO STREAMING

Due to a number of advantages with regards to load balancing and robustness, P2P architectures have attracted both academic and industrial attention as potential content delivery platforms. This section tries to depict the potentials of such systems for video streaming and provides a brief description of the key characteristics of well-known P2P applications used for that purpose.

Streaming Modes

After the successful launch of P2P file sharing applications, the academic and industrial attention gradually shifts towards streaming services. Within the context of P2P video sharing, there are two distribution modes over P2P Networks: *offline* and *streaming* mode as shown in Figure 5. The former is a view-after-download and the latter a view-while-downloading approach.

Offline mode is the classic way of video sharing over P2P networks similar to any file downloading process. The receiving peer requests the desired file from other contributing peers on the network. FTP and HTTP protocols are well known protocols used in this mode of downloading. P2P sharing applications play a vital role in this mode. As mentioned earlier, BitTorrent(BitTorrent), KaZaA (Kazaa), Gnutella (gnutella), and iMesh (iMesh) are the most popular applications for the video sharing. Recent studies show that BitTorrent

Figure 5. Video modes over P2P

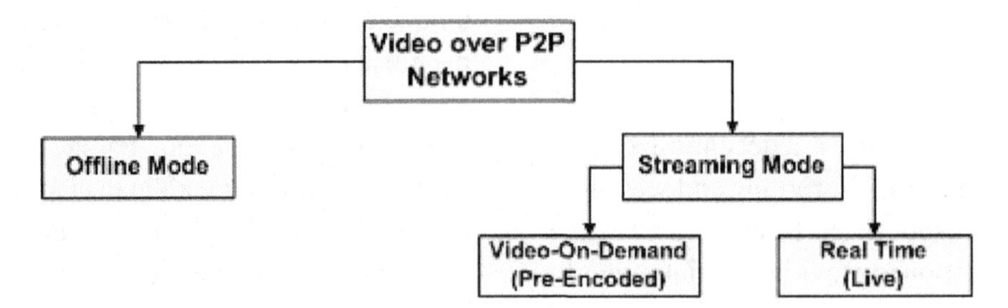

is the most popular among others for offline mode downloading (Qiu & Srikant 2004). There are several approaches to the way a video is fetched. Requesting peers tend to choose the best provider among all the discovered ones. They may download the whole file from one peer or fetch different chunks of the same video from different peers (Pouwelse et al., 2005).

Considering the source of a P2P video stream, we can identify two subclasses of P2P Streaming field: *Video on Demand (VoD)* and *Real Time streaming (RT)* services. Real time videos demand tighter time constraints as the production and provision of the stream to other peers happen simultaneously. In this regard, the receiver is playing-while-downloading which is sensitive to continuity-indexing. Live streaming is time sensitive, as it has stringent playback deadlines and the end-user has no control over the stream. Sopcast (Sopcast) is an application that provides both VoD and RT services. Zattoo(Zattoo) offers only real time streaming. If the video is pre-recorded, VoD, the receiver can playback any part of it over different channels. It allows random access at any point in the stream including actions such as pause, fast forward, fast rewind, slow forward, slow rewind, and jump to previous/future frame etc. There are different applications supporting this kind of service such as Joost (Joost) and Babelgum (Babelgum).

Recent interest in P2P video streaming is also triggered by its potentials as an alternative to conventional IPTV for distributing video content over the internet (Liu et al., 2008). The underlying mechanism is based on the distribution of the stream through an application-level overlay including the user terminals in the role of peers, i.e. content distribution relays. This is in contrast to other IPTV approaches, which are based on content distribution networks. These require a dedicated multicasting infrastructure whose cost increases dramatically with the scale and dynamics of the system (Liu et al., 2008). On the other hand, in a P2P environment, the number of distribution points increases with connected users. Consequently, the system not only improves *Quality of Service (QoS)* but also scales much better than CS counterparts (Hossfeld & Leibnitz 2008).

Video Delivery Mechanisms

In a P2P environment, every peer might be in one of three phases: content provider to any other peer, content delivery re-transmitter and content receiver. While the first phase refers to peers that are the video source, the second is about those that have already received part of the video and retransmit it to another one. This video retransmission can create a number of parallel deliver paths of the same video with the same or different origin. Apart from the video sources, all the other peers involved in these paths are or used to be content receivers. These paths build a virtual topology of the content delivery. The main two well-documented ones are the Tree- (Figure 6) and Mesh-based ones (Figure 7).

Figure 6. Tree architecture

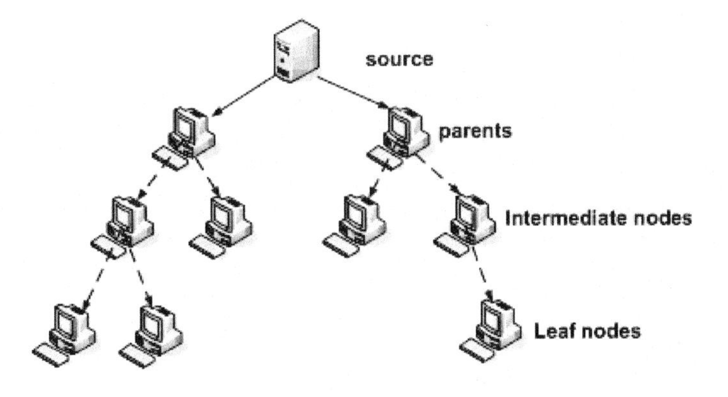

Figure 7. Mesh architecture

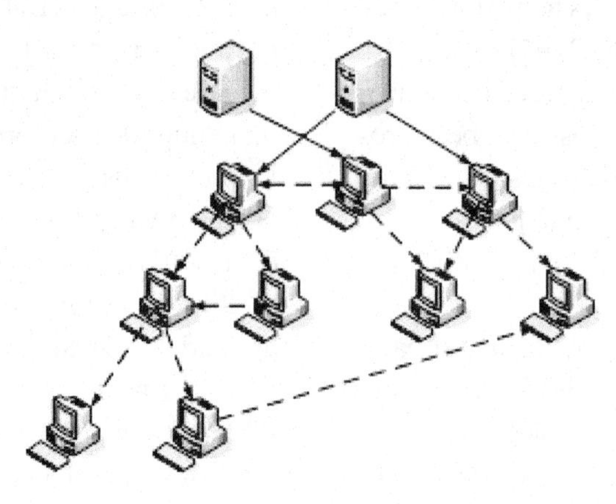

In the tree-based architectures, the peers are ordered hierarchically by the source, known as the parent. A parent node, in turn, sends data packets to intermediate nodes, and these nodes relay them iteratively until leaf nodes are reached. An example of a tree-based streaming application is Peercast (Marfia et al. 2007), open-source software for streaming both audio and video. A peculiarity of Peercast is that any node can specify the maximum number of incoming connections allowed.

Despite introducing a good level of parallelism and distribution, this approach suffers from a number of limitations. The root, or data source, is a single point of failure, which limits the robustness of the system. As another problem, if peers join and leave frequently the tree has to be rebuilt too often, which has a negative impact on signaling overheads, latency, and stability. This is because the receiving peer needs to first find a new source when rebuilding the tree, before continuing with the streaming session from where it was interrupted. This lack of continuity in the streaming, often leads to a repeat or a skip during playback of the multimedia content. Rebuilding of a tree generates a lot of control traffic, which can overload the network (Diot et al., 2000).

Another issue arising in such an organization is that the peers are heterogeneous in uplink and bandwidth capacities. This requires extra measures for efficient content delivery. Leaf peers are highly dependent on the intermediate peers, if any intermediate peer is unable to provide the required bandwidth then the child peers suffer from the high delay and in some cases the content delivery is not guaranteed. Furthermore, the leaf peers at the tree do not have any child peer and thus they do not contribute for their uplink bandwidth. This maximizes the chance of the free riders not to contribute to the network, which in return will degrade the peer bandwidth utilization.

To overcome some of the limitations of the single tree-based system such as load balancing and vulnerability posed by the heterogeneity and dynamicity of peers, the multi-tree architecture is introduced. This architecture is used to overcome the potential single point of failure, by using multiple trees to split the stream on network paths. The multi-tree architecture leads to a balanced tree that makes the system self-organizing and more resilient than the single-tree architecture. Each peer acts as an intermediate peer, so it can contribute its uplink bandwidth capacity by serving other peers. In this architecture, churn of peers during the media transmission do not effect much as each peer only contributes to a part of media content. Example of multi tree-based streaming application are SplitStream (Castro et al., 2003) and CoopNet (Jurca et al., 2007) (Padmanabhan et al., 2003). The multi-tree model overcomes the high delay problem of the single tree model and provides highly resilience, when every peer shares their resources with others. On the other hand, multi-tree model also suffers from some disadvantages. The complexity in organizing peers in multiple trees is much higher than that of single-tree models. The main limitation of the multi-tree architecture is that peers can only receive data packets from a single source even though the data packets go along different paths.

In the mesh-based architecture, the overlay network supporting the stream distribution is a mesh. Data is divided into chunks in such a way which allows a peer to receive portions of the stream from different peers and assemble them locally. This approach is more robust than the tree-based architecture, since when a stream comes from various sources communication does not break when only a subset of peers disconnect. It addresses the single-point of failure problem and provides a robust solution for streaming applications. Another benefit is that this transport method reflects well the asynchronous nature of many access network technologies (e.g. *Asymmetric Digital Subscriber Line (ADSL)*). In fact, peers can download a stream at full quality whilst uploading only a fraction of it. The same feature is exploited in P2P file sharing application BitTorrent (Qiu & Srikant 2004).

The selection among the available sources, which a peer will receive media streams from, forms the main challenge in this approach. There is the additional stringent requirement of the re-ordering data packets at the receiving peers since media packets are transmitted and received from multiple sources. Common streaming applications in the mesh-based category are GNuStream (Jiang et al., 2003), PPlive (Hei et al., 2007), Coolstreaming (Zhang et al., 2005), and Joost (Lei et al.). This method is becoming more widespread than the tree-based one.

P2P Streaming Applications

The P2P streaming concept has now led to a number of trials of P2P IPTV systems. There is now clear commercial interest in these new technologies which are revolutionizing the online broadcasting arena. Some of these systems are providing live streaming such as Zattoo (Zattoo) and others are V-oD only such as Joost (Joost). In this section, some of the popular P2P streaming applications are given with some details about each.

Joost is a P2P streaming application that delivers television-quality VoD and RT streaming[1] services via a P2P network overlay. It was created by the founders of Skype and KaZaA and is currently in Beta version (v. 1.1.4 at the time of writing)(Joost). Joost supports more than 15,000 TV programs through more than 200 channels (Joost). At system start-up, Joost performs a number of operations:

- **Port Selection**: During the initial bootstrapping, Joost selects a specific port to connect to and communicate with other peers via UDP. The Internet Assigned Numbers Authority has recently assigned port 4166 for this purpose.
- **Local Cache**: Each Joost Client stores all the media data as "anthill cache" in the C drive, so to get into there, using the following link: C:\Documents and settings\Application Data\Joost\"anthill cache". Anthill (Babaoglu et al., 2002) is agent based supporting media distribution services. The size of the local cache depends on how and for how long the program has been launched, so it increases with the size of the programs. However, in the case of watching the whole channel, all the media data are stored in the local cache, so in the second watching, the channel will be running from the local cache instead of connecting to the server except for some codecs. In our experiment, it caches more

than 2 GB. However, this is will affect the user resources when watching more channels.

- **Installation**: One of the functions of Joost is the Installation, so in this phase the Joost client connects to the server sending an HTTP request to retrieve the available channel list and downloads the SQLite file (Hipp) which gives the initial available channel lists. Moreover, SQLite is used for managing the database of Joost channels.
- **Bootstrapping**: In Joost, there are three servers and two super nodes. Initially, the *JC (Joost Client)* connects to the server lux-www-lo-2.joost.net over HTTP. Then, the JC will receive some available super nodes. After that, an HTTP request will be sent to lux-www-lo4.joost.net to get the updated version of the software. Lastly, the J-C will connect to the super nodes such as lid-snode-1-eth0.joost.net to get the available list of channels and the available peers that they watching the channels. Before that, the JC has already started communicating with other servers and peers. A schematic view of Joost's architecture is depicted in Figure 8.

Zattoo is a peer-to-peer application which delivers a live streaming service via a P2P network overlay (Zattoo). This means that the data is not streamed from one central server to all users that are watching a certain program, but is relayed among the users. Zattoo was founded in 2005 by Sugih Jamin, Beat Knecht, and Wenjie Wang (Zattoo) and, as of today, it offers 237 channels across Europe in different languages (Zattoo ; Hossfeld & Leibnitz, 2008), distributing those

Figure 8. Joost architecture

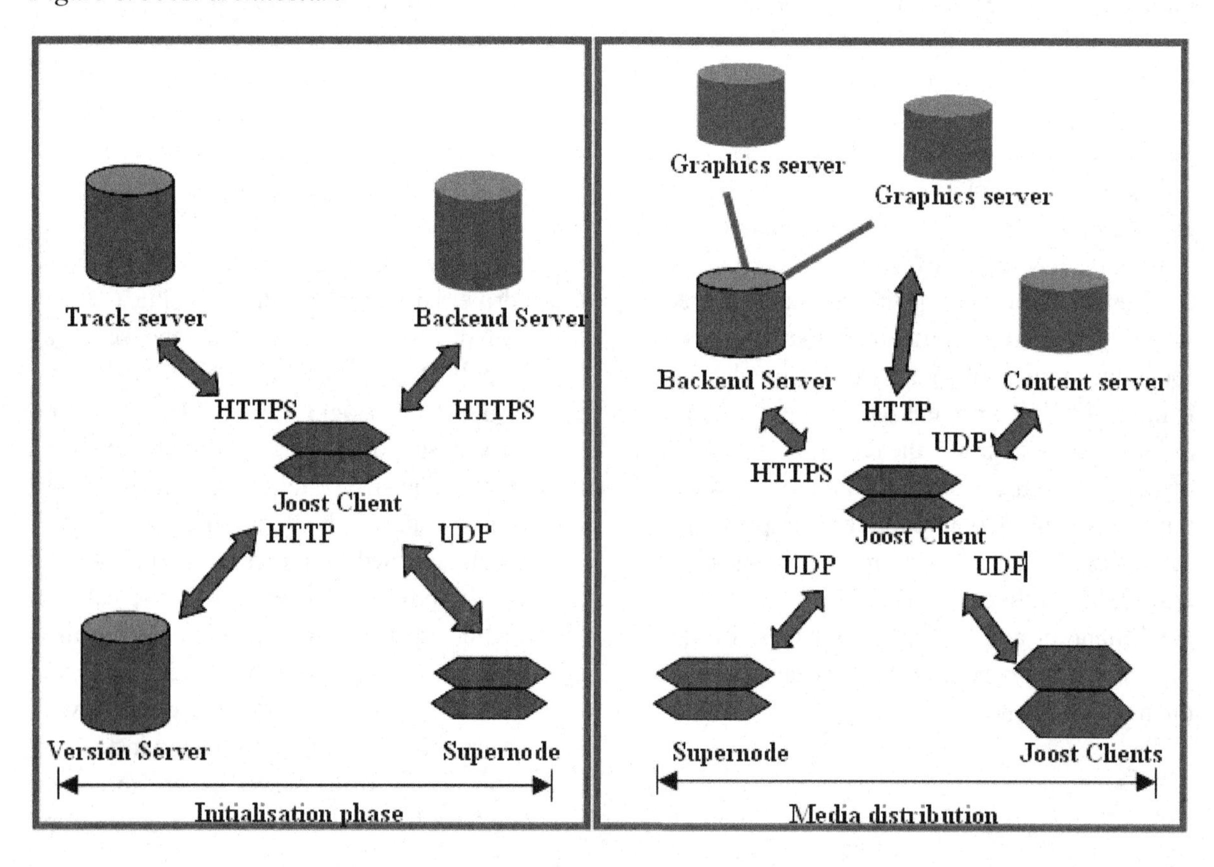

channels between countries based upon their IP. Zattoo maps each client (or peer) to its country, regulating in this way the channels in a per-country way (Hossfeld & Leibnitz, 2008). European countries that are at the time of writing in the Zattoo network include Belgium, Denmark, France, Germany, Norway, Spain, Switzerland, and the UK (Zattoo).

Zattoo adopts a video coding and player based on the H.264 video codec and it is compatible with all major operating systems. From the *DRM (Digital Rights Management)* viewpoint, Zattoo seeks the right to retransmit the channels from different countries.

Sopcast is a freeware application for Windows and Linux that permits broadcast of video or audio over the network. It makes use of the peer-to-peer technology, so that every user can become a broadcaster, building their own TV stations, creating interest groups, and permitting in this way to save bandwidth and to avoid the costs of servers. Every single peer can serve 10,000 online users with the only help of a personal computer and a home broadband connection. The name Sopcast comes as the abbreviation of *Streaming over P2P* (Sop) Broadcasting (cast). System and its core is the communication protocol produced by Sopcast Team, named sop: // or Sop technology. Sopcast provides both the *Real Time (RT)* and the VoD services. It supports streaming protocols like mms http for the RT streams, and formats like asf, wmv, rm, rmvb, mp3 for the streaming of media files. It also provides the loop file playing function, and, in addition, the possibility to record the clips while they are playing. The stream can be played with Windows Media Player, RealPlayer, VLC, or others. The first two are recommended, otherwise the stream has to be opened and played manually. This can be done specifying the URL: http://localhost:8912/ to the player URL field. For a rm (rmvb) stream, in RealPlayer, it has to be specified the url: http://localhost:8912/tv.rm.

As an additional feature, every viewer or broadcaster can control his channels through the authentication service. Every time a general user enters their username and password, in fact, they can edit the list of the favourite channels, subscribe a new private or public channel (upon registration), edit their preferences and view the subscribed 48 Architecture and functionalities channels. After having registered a channel, the user becomes a "channel user", which means that they can edit the subscribed channel's information and join a broadcasting group. In addition, the channel user can become a group administrator by applying for a broadcasting group. The system administrator has in this case to give them the permission to manage their broadcasting group. Once the user has been allowed, they can manage the permissions for the interest group, and, as a consequence, permit or deny the joining of some channels by certain users. Another feature is the possibility to run up to 10 different channels on the same computer. To do so, the "Allow only one instance of Sopcast" checkbox has to be ticked. It is also possible to choose a different language. English, Simplified Chinese and Traditional Chinese are the only languages available. Lastly, the Sop Player can be embedded into a webpage, or to any software applications, like Windows Media Player.

Finally, an overview of our experimental results based on their performance and characteristics is shown in Table 2.

FUTURE RESEARCH DEVELOPMENTS

Despite of the advantages that P2P streaming has brought to the end consumers, there are still some challenges concerning the end users and the network operators. These issues include:

- **Network Efficiency:** Every single application can make use of different policies to provide the P2P video streaming service. However, there are some features that a P2P system should provide, to exploit all

Table 2. Peer-to-peer streaming application taxonomy

P2P Applications	Sopcast	Zattoo	Joost	Babelgum
Service	Live / VOD	Live	VOD	VOD
Start-up delay	1-5 mins	6.2 sec	25 sec	10 sec
Data rate	350 kbps	560 kbps	500 kbps	450 kbps
Architecture	Mesh	Tree / Forest	Mesh	Forest
Protocol	UDP	TCP / UDP	UDP	TCP

the advantages of the resources distribution and network usage. First of all, for efficient network utilisation, locality is one of the major goals (challenges). Network locality (efficiency) is the ability to maintain the P2P overlay in such a way as to create logical connections among the peers that are physically close to each other. The ideal condition occurs when the most intensive data exchanges happen among the nearby peers with significant spreading and swapping the utilised resources.

- **Computational Efficiency:** A second challenge that needs to be considered from a structural point of view is the computational efficiency. This concerns how peers are selected for the download of a certain video file; how often they are substituted by the new peers; how many peers are involved in the whole process of downloading. A P2P application is computationally efficient if it makes use of the largest possible number of peers and, at the same time, provides reliability by the means of frequently renewing the involved peers, which will in turn increase the bandwidth as well as the QoS without any negative effect on network locality. Another characteristic that has to be considered is the fairness policy, which indicates the percentage of downloaded traffic with respect to the uploaded one. Most of the applications give priority to the download rather than the upload link. Figure 9 shows the performance of four popular P2P streaming applications with respect to these parameters (Alhaisoni et al., 2010).
- **Reliability:** Reliability concerns with the ability of a P2P systems to be recovered after any failure. The factors which play vital

Figure 9. Network and computational efficiency

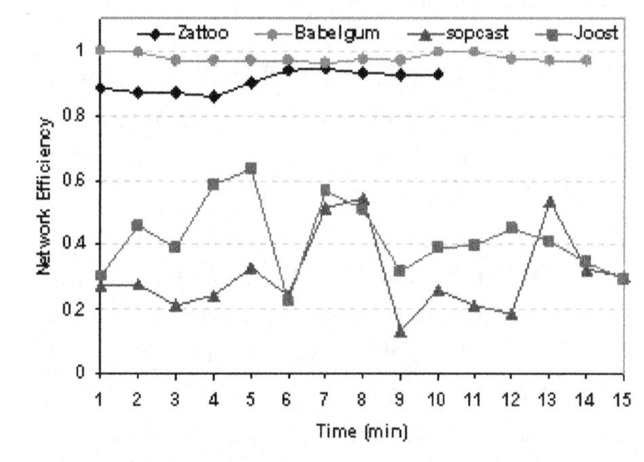

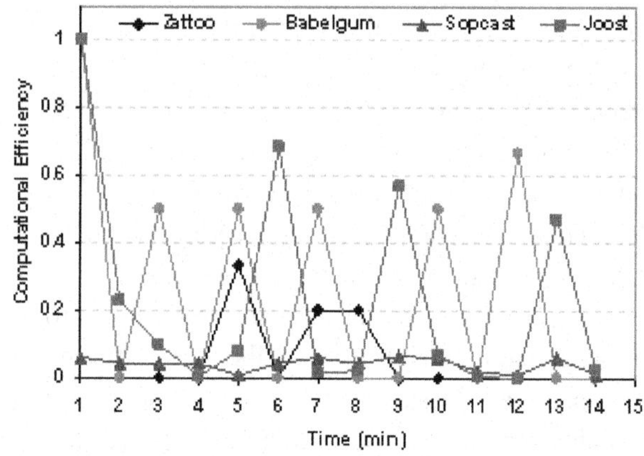

role in reliability is the data availability, the detection of any nodes failures or departures to avoid a single point of failure and the availability of various paths to data to be transmitted. Data replication increases reliability by increasing redundancy and locality. Replication can be implemented either on the client side or over the paths (Lv et al., 2002).

- **Security:** Distributed systems face additional challenges for security as compared to the C-S architecture. Due to the dynamics of nodes in P2P applications, peers do not trust each other, so achieving a high level of security in peer-to-peer systems is more difficult than in non-peer-to-peer systems. The difficulty in securing peers results from the global connectivity among the intercommunicating peers. Therefore, normal security mechanisms can not be applied easily over P2P data and systems to shield from attacks such as firewalls can not protect peer-to-peer systems and also these mechanisms can inhibit peer-to-peer communication. For that reason much attention should be given to P2P security (Wallach, 2003).
- **Copyright and DRM:** Recently, DRM has attracted much consideration in P2P file sharing and more recently in media streaming. Hence, DRM in P2P systems is of vast significance and research is ongoing in this area. However, there is still a critical problem within most of the P2P based IPTV systems such as PPlive (PPLIVE) as it does not take into account content management. Digital content can be copied or redistributed without any restriction, which of course is undesired. Therefore, a content management system is required for the practical deployment of P2P based IPTV.

CONCLUSION

This chapter is a concise overview of video content streaming platforms over highly decentralised and dynamic environments. Principal candidate platforms for these cases are peer-to-peer networks as they have been designed for content sharing over large scale heterogeneous environments. However, the focus of the chapter is not the review of well-known P2P principles and practices. It tries to depict the features and functionalities based on which a new P2P platform, which can be designed to facilitate pervasive streaming applications. The sections above provide a critical review of traditional and mature technologies such as Client-Server models and Content Delivery Networks (CDN). After the identification of their main weaknesses, the focus shifts to P2P networks which are able to cope with the dynamics of video delivery environments for the masses.

The client-server model is compliant with most of the current content delivery business models according to which companies want to be fully in control of the content. This facilitates their resource updates, service personalisation, billing strategies and helps them control their content sharing, protecting them from legal actions. However, the viability of this model can be at risk, as P2P Networks already operate in parallel restricting their clientele. In an effort to reach high QoS to bring customers back, these companies need high investments to serve the masses. Moving one step further, Content Delivery Networks try to lift the bottleneck and single-point-of-failure problems of the expensive Client-Server models by practically introducing more servers and content replicas dispersed around the globe. The scalability improvements, however, are still limited.

P2P networks are a technology that has not reached the maturity of the other two. They can support massive simultaneous users with intermittent behavior but there is a need to cope with a number of challenges. Therefore, this chapter presents a variety of P2P architectures with sample real applications and their classification based on their structure and decentralisation degree. Finally, through an analysis of the main streaming modes (i.e. offline-playback, VoD, live), streaming overlay structures (i.e. tree or mesh-like) and existing application features (e.g. Zattoo, Joost), the chapter finishes with the identification of the main challenges yet to be addressed so that P2P Networks can be an applicable and preferable content delivery paradigm for companies and users. It concludes with the factors that challenge the deployment of P2P streaming systems such as network and computational efficiency, DRM, and security. These are the parameters that real-world and viable implementations of pervasive streaming platforms for massive content delivery should consider.

REFERENCES

Akamai. (n.d.). *Akamai home page*. Retrieved from www.akamai.com.

Alhaisoni, M., Ghanbari, M., & Liotta, A. (2010). Localized multi-streams for P2P streaming. *International Journal of Digital Multimedia Broadcasting, 2010*. doi:10.1155/2010/843574

Alhaisoni, M., Liotta, A., & Ghanbari, M. (2010). Resource-awareness and trade-off optimisation in P2P video streaming. *International Journal of Advanced Media and Communication*, *4*(1), 59–77. doi:10.1504/IJAMC.2010.030006

Androutsellis-Theotokis, S., & Spinellis, D. (2004). A survey of peer-to-peer content distribution technologies. *ACM Computing Surveys*, *36*(4), 335–371. doi:10.1145/1041680.1041681

Babaoglu, O., Meling, H., & Montresor, A. (2002). Anthill: A framework for the development of agent-based peer-to-peer systems. *22nd IEEE International Conference on Distributed Computing Systems,* (pp. 15-22).

Babelgum. (n.d.). *Babelgum home page*. Retrieved from www.babelgum.com

BitTorrent. (n.d.). *BitTorrent home page*. Retrieved from www.bittorrent.com

Bolosky, W., Douceur, J., Ely, D., & Theimer, M. (2000). Feasibility of a serverless distributed file system deployed on an existing set of desktop PCs. *ACM SIGMETRICS Performance Evaluation Review*, *28*(1), 34–43. doi:10.1145/345063.339345

Castro, M., Druschel, Kermarrac, A. M., & Rowstron, A. I. T. (2002). SCRIBE: A large-scale and decentralized application-level multicast infrastructure. *IEEE Journal on Selected Areas in Communications*, *20*(8), 1489–1499. doi:10.1109/JSAC.2002.803069

Castro, M., Druschel, P., et al. (2003). SplitStream: High-bandwidth multicast in cooperative environments. *Proceedings of the Nineteenth ACM Symposium on Operating Systems Principles*, (pp. 298-313).

Clarke, I. Sandberg, Wiley, B., & Hong, T. W. (2001). Freenet: A distributed anonymous information storage and retrieval system. *International Workshop on Designing Privacy Enhancing Technologies*, (pp. 46-66).

Cox, R., Muthitacharoen, A., & Morris, R. (2002). Serving DNS using a peer-to-peer lookup service. *First International Workshop on Peer-to-Peer Systems*, (pp. 155-165).

Dabek, F., & Kaashoek, M. (2001). Wide-area cooperative storage with CFS. *ACM SIGOPS Operating Systems Review*, *35*(5), 202–215. doi:10.1145/502059.502054

Diot, C., Levine, B. N., Lyles, B., Kassem, H., & Balansiefen, D. (2000). Deployment issues for the IP multicast service and architecture. *IEEE Network*, *14*(1), 78–88. doi:10.1109/65.819174

Emule. (n.d.). *Emule home page*. Retrieved from http://www.emule.com/

FastTrack. (n.d.). *FastTrack home page*. Retrieved from http://developer.berlios.de/projects/gift-fasttrack/

Freenet. (n.d.). *Freenet Project home page*. Retrived from http://freenetproject.org/

Gnutella2. (n.d.). *Gnutella2 home page*. Retrieved from www.Gnutella2.com

Gnutella. (n.d.). *Gnutella home page*. Retrieved from www.Gnutella.com

Hei, X., Liang, C., Liang, J., Liu, Y., & Ross, K. W. (2007). A measurement study of a large-scale P2P IPTV system. *IEEE Transactions on Multimedia*, *9*(8), 1672–1687. doi:10.1109/TMM.2007.907451

Hipp, R. (n.d.). *SQLite*. Retrieved from http://www.sqlite.org/

Hossfeld, T., & Leibnitz, K. (2008). A qualitative measurement survey of popular Internet-based IPTV systems. *Second International Conference on Communications and Electronics*, (pp. 156-161).

iMesh. (n.d.). *iMesh home page*. Retrieved from www.imesh.com

Jiang, X., Dong, Y., Xu, D., & Bhargava, B. (2003). GnuStream: A P2P media streaming system prototype. *International Conference on Multimedia and Expo*, (pp. 325-328).

Joost. (n.d.). *Joost home page*. Retrieved from http://www.joost.com/

Jovanovic, M. (2001). *Modeling large-scale peer-to-peer networks and a case study of Gnutella*. Master's thesis, University of Cincinnati.

Jurca, D., Chakareski, J., & Wagner, J.-P. (2007). Enabling adaptive video streaming in P2P systems. *IEEE Communications Magazine*, *45*(6), 108–114. doi:10.1109/MCOM.2007.374427

Kalogeraki, V., Gunopulos, D., & Zeinalipour-Yatzi, D. (2002). A local search mechanism for peer-to-peer networks. *Proceedings of the Eleventh International Conference on Information and Knowledge Management* (pp. 300-307).

Kazaa. (n.d.). *Home page of Kazaa*. Retrieved from www.kazaa.com

Kovacevic, A., Heckmann, O., Liebau, N. C., & Steinmetz, R. (2008). Location awareness—Improving distributed multimedia communication. *Proceedings of the IEEE*, *96*(1), 131–142. doi:10.1109/JPROC.2007.909913

Kubiatowicz, J., Bindel, D., Chen, Y., Czerwinski, S., & Eaton, P. (2000). Oceanstore: An architecture for global-scale persistent storage. *ACM SIGARCH Computer Architecture News*, *28*(5), 190–201. doi:10.1145/378995.379239

Lei, J., Shi, L., & Fu, X. (2009). An experimental analysis of Joost peer-to-peer VoD service. *Peer to Peer Networking and Applications*, *3*(4), 351–362. doi:10.1007/s12083-009-0063-5

Li, X., & Wu, J. (2006). Searching techniques in peer-to-peer networks. In Wu, J. (Ed.), *Handbook of theoretical and algorithmic aspects of ad hoc, sensor, and peer-to-peer networks* (pp. 613–642). Boston, MA: Auerbach Publications.

Limewire. (n.d.). *Limewire home page*. Retrieved from http://www.limewire.com/

Liu, B., Lee, W. C., & Lee, D. L. (2005). Supporting complex multi-dimensional queries in P2P systems. *The 25th IEEE International Conference on Distributed Computing Systems*, (pp. 155-164).

Liu, Y., Guo, Y., & Liang, C. (2008). A survey on peer-to-peer video streaming systems. *Peer-to-Peer Networking and Applications*, *1*(1), 18–28. doi:10.1007/s12083-007-0006-y

Lua, K., Crowcroft, J., Pias, M., Sharma, R., & Lim, S. (2005). A survey and comparison of peer-to-peer overlay network schemes. *IEEE Communications Surveys & Tutorials*, *7*(2), 72–93. doi:10.1109/COMST.2005.1610546

Lv, Q., Cao, P., Cohen, E., Li, K., & Schenker, S. (2002). Search and replication in unstructured peer-to-peer networks. *Proceedings of the 16th International Conference on Supercomputing*, (pp. 84-95).

Lv, Q., Ratnasamy, S., & Schenker, S. (2002). Can heterogeneity make Gnutella scalable? *1st International Workshop on Peer-to-Peer Systems*, (pp. 94-103).

Marfia, G., Pau, G., Di Rico, P., & Gerla, M. (2007). P2P streaming systems: A survey and experiments. *ST Journal of Research*.

Maymounkov, P., & Mazieres, D. (2002). Kademlia: A peer-to-peer information system based on the xor metric. *First International Workshop on Peer-to-Peer Systems*, (pp. 53-65).

Milojicic, D., Kalogeraki, V., et al. (2003). *Peer-to-peer computing.* HP Labs. Technical Report.

Morpheus. (n.d.). *Morpheus home page*. Retrieved from www.morpheus.com

MySpace. (n.d.). *Myspace home page*. Retrieved from www.myspace.com

Napster. (n.d.). *Napster home page*. Retrieved from http://free.napster.com/

Padmanabhan, V., Wang, H., & Chou, P. A. (2003). Resilient peer-to-peer streaming. *IEEE Interntional Conference on Network Protocols*, (pp. 16-27).

Pallis, G., & Vakali, A. (2006). Content delivery networks. *Communications of the ACM*, *49*(1), 101–106. doi:10.1145/1107458.1107462

PeerCast. (n.d.). *PeerCast home page*. Retrieved from http://www.peercast.org/

Pouwelse, J., Grabacki, M., Epema, D., & Sips, H. (2005). The BitTorrent P2P file-sharing system: Measurements and analysis. *Peer-to-Peer Systems*, *IV*, 205–216. doi:10.1007/11558989_19

PPLIVE. (n.d.). PPLIVE. Retrieved from www.pplive.com

Qiu, D., & Srikant, R. (2004). Modeling and performance analysis of BitTorrent-like peer-to-peer networks. *ACM SIGCOMM Computer Communication Review*, *34*(4), 367–378. doi:10.1145/1030194.1015508

Ratnasamy, S., Francis, P., Handley, M., Karp, R., & Schenker, S. (2001). A scalable content-addressable network. *ACM Conference on Applications, Technologies, Architectures, and Protocols for Computer Communications*, (pp. 161-172).

Rowstron, A., & Druschel, P. (2001). Pastry: Scalable, distributed object location and routing for large-scale peer-to-peer systems. *IFIP/ACM International Conference on Distributed Systems Platforms*, (pp. 329-350).

Skype. (n.d.). *Skype home page*. Retrieved from http://www.skype.com

Sopcast. (n.d.). *Sopcast home page*. Retrieved from www.sopcast.com

Stoica, I., & Morris, R. (2003). Chord: A scalable peer-to-peer lookup service for Internet applications. *IEEE/ACM Transactions on Networking*, *11*(1), 17–32. doi:10.1109/TNET.2002.808407

Triantafillou, P., & Pitoura, T. (2004). Towards a unifying framework for complex query processing over structured peer-to-peer data networks. *First International Workshop on Databases, Information Systems, and Peer-to-Peer Computing*, (pp. 169-183).

VEOH. (n.d.). *VEOH home page*. Retrieved from www.veoh.com

Waldman, M., Rubin, A., & Cranor, L. (2000). Publius: A robust, tamper-evident, censorship-resistant web publishing system. *9th Conference on USENIX Security Symposium*, (pp. 59-72).

Wallach, D. (2003). A survey of peer-to-peer security issues. *International Conference on Software Security*, (pp. 42-57).

Xu, Z., Mahalingam, M., & Karlsson, M. (2003). Turning heterogeneity into an advantage in overlay routing. *Proceedings of the IEEE INFOCOM*, (pp. 1499-1509).

Yang, B., & Garcia-Molina, H. (2002). *Efficient search in peer-to-peer networks*. 22nd International Conference on Distributed Computing Systems.

Zattoo. Zattoo home page from www.zattoo.com

Zhang, X., Liu, J., Li, B., & Yum, Y.-S. P. (2005). CoolStreaming/DONet: A data-driven overlay network for efficient live media streaming. *Proceedings of the IEEE INFOCOM*, (pp. 2102-2111).

Zhao, B., & Huang, L. (2004). Tapestry: A resilient global-scale overlay for service deployment. *IEEE Journal on Selected Areas in Communications*, *22*(1), 41–53. doi:10.1109/JSAC.2003.818784

ADDITIONAL READING

Antonopoulos, N., Exarchakos, G., & Liotta, A. (2009). *Handbook of research on P2P and Grid systems for service-oriented computing: Models, methodologies and applications.* Hershey, PA: Information Science Reference.

Shen, X., Yu, H., Buford, J., & Akon, M. (Eds.). (2010). *Handbook of peer-to-peer networking*. New York, NY: Springer Verlag. doi:10.1007/978-0-387-09751-0

Wu, J. (Ed.). (2005). *Handbook on theoretical and algorithmic aspects of sensor, ad hoc wireless, and peer-to-peer networks*. Boca Rato, FL: Auerbach Publications. doi:10.1201/9780203323687

KEY TERMS AND DEFINITIONS

Client-Server: Platform for information sharing between any two parties whose communication is driven through and controlled by a central entity.

Overlay Network: is a computer network which is built virtually or logically on top of another network.

P2P Networks: Virtual Networks build on the OSI application layer that enable the direct communication and resource sharing among any two users.

P2P Streaming: On demand or live content (audio or video) delivery via a P2P Network.

Pervasive Streaming: Content (audio or video) delivery on highly dynamic (e.g. churn) networks.

Structured P2P overlays: Structured P2P network employ a globally consistent protocol to ensure that any node can efficiently route a search to some peer that has the desired file

Unstructured P2P Overlays: is formed when the overlay links are established arbitrarily. Such networks can be easily constructed as a new peer that wants to join the network can copy existing links of another node and then form its own links over time.

ENDNOTE

[1] RT service was pending for development purposes at the time of writing.

Chapter 4
Adaptive P2P Streaming

Fabrizio Bertone
Politecnico di Torino, Italy

Vlado Menkovski
Eindhoven University of Technology, The Netherlands

Antonio Liotta
Eindhoven University of Technology, The Netherlands

ABSTRACT

Widespread adoption of broadband Internet and the introduction of multimedia-capable mobile devices enable the proliferation of many streaming video services. However, best-effort networks are not natively designed to such a purpose. They do not provide any guarantees for delivering the content on time or offer constant service quality. Furthermore, video streaming presents a heavy load for servers. This is especially the case for special events that bring an enormous amount of users causing so called "flash crowds," which overload unsuitable systems. Peer-to-peer (P2P) techniques can be successfully exploited to build scalable streaming systems using the distributed resources of users themselves. In this chapter, the authors explore the different techniques proposed in the context of adaptive streaming, both live and on demand. Each covered approach addresses the video streaming problem from a different perspective, and so, brings specific advantages and disadvantages in its solution.

INTRODUCTION

The number of users watching multimedia contents over the Internet, both live and on demand, has dramatically increased in recent years, and is now the first source of traffic, surpassing even *Peer-to-Peer (P2P)* file sharing. This trend has been reported by several surveys, and is confirmed even by network operators. Over one third of the top 50 sites by volume are based on video distribution. (*Cisco Visual Networking Index: Usage*, 2010) Netflix, a provider of on-demand internet streaming video, at the moment accounts by itself for almost 30% of peak period downstream traffic (*Global Internet Phenomena Report - Spring 2011*, 2011). The sum of all forms of video is expected to exceed 90% of global user traffic by 2014 (*Cisco Visual Networking Index: Forecast and Methodology, 2009 – 2014*, 2010). The web is becoming one of the main broadcasting

DOI: 10.4018/978-1-4666-1613-4.ch004

platforms, this is due to the recent availability of the so-called "three screens" (digital TV, PC and smartphones) capable of accessing it, and the increase of the number of faster connections to the network. The most recent sport events like FIFA World Cup, Olympic games etc. have been watched in streamed form by several million of unique simultaneous viewers (Begen et al., 2011).

The common solution used to stream over the Internet nowadays is the classic client/server architecture, In which a connection is opened between the client and the server, and the content is transferred directly. The major issue for this technology is its lack of scalability. One single server is not able to handle a big number of connections, both due to its limited resources and the network bandwidth. One extension of the client/server approach that tries to overcome the scalability problem introduces clusters of servers scattered throughout the network. Instead of downloading directly from the source server, clients are redirected to the nearest server, reducing latency and exploiting more efficiently network resources. This approach is called *Content Delivery Network (CDN)*. CDNs are costly and complex to maintain, so that a small number of specialized enterprises (e.g., Akamai, CDNetworks and EdgeCast) share all the market. Besides that, even if CDNs are dimensioned for normal traffic conditions particular events can still attract a great number of users and overload the system.

Peer-to-peer networking emerged as a new paradigm to build distributed applications. In P2P systems, each client contributes with its own resources to the operation of the network. A peer is distinguished from a normal client because not only does it download data, but also uploads to other users. This way the load of the source server is considerably reduced, allowing for great degree of scalability. P2P systems can be classified in two main classes, according to the structure in which peers are organized: trees or meshes (Y. Liu et al., 2008). Figure 1 shows the different link topology between tree (a) and mesh (b) structures.

In a tree topology, the source server is placed at the root of the distribution paths and actively *pushes* the stream to peers directly attached to it. Those peers will then forward received data to their child nodes in a cascade. To reduce the delays, a short and wide tree is preferable. The main drawback of tree structures is the complexity of its maintenance. In P2P systems, users typically join and leave the network quite frequently. When a node abandons the tree, it stops forwarding data to its descendant peers, which necessitates restructuring of that part of the tree. The alternative to trees is using meshes (Hei et al., 2008). In a mesh there is not an explicit parent-child configuration; peers can freely create links between them. This enables peers to close slower connections and connect to faster nodes, in the attempt to improve performances. Mesh-based systems

Figure 1. Tree (a) and mesh (b) based structures.

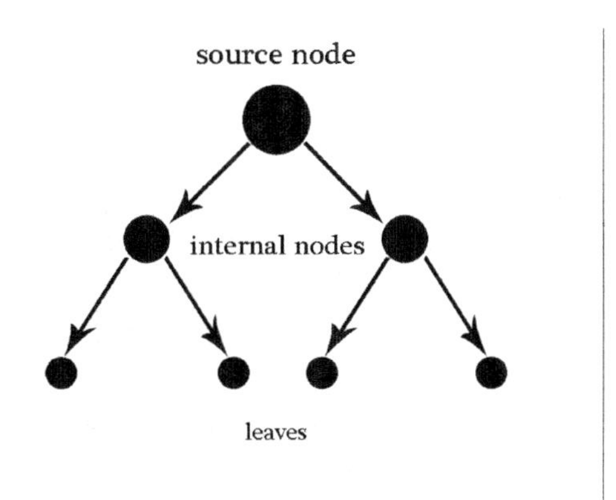

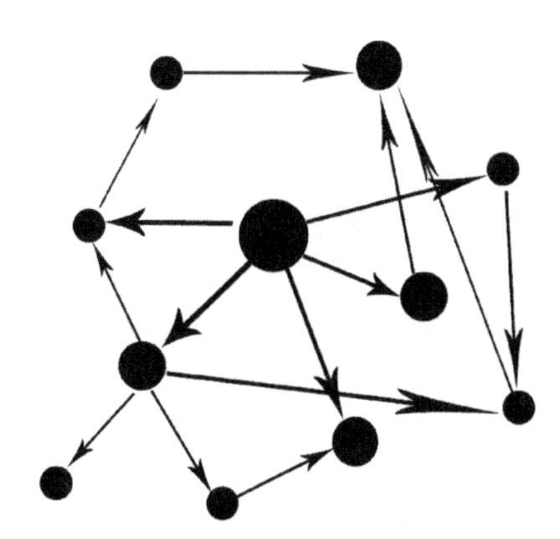

can use both "push" and "pull" strategies, unlike tree-based systems that are forced to use push transfers. In pull protocols, each node is responsible for actively retrieving required content from its neighbours. While tree structures provide lower delays and overhead in static networks, experimental simulations show that meshes are perform better in highly dynamic systems like P2P, and appear to be more stable over a long time period (Magharei et al., 2007).

Scalability is not the only limiting factor in current streaming systems. Best-effort networks do not natively provide ways to ensure Quality-of-Service and on-time delivery of data. Receivers usually deal with delays and retransmission due to errors by using buffers. Bigger buffers handle bigger delays. However, buffer size also adds delay in the playback, which is not desirable particularly in live streaming. Dealing with lack of bandwidth or low quality transmission channels can be implemented by using a lower quality version of the video. This way the video bit rate is adapted to the transport conditions or device capabilities. In this chapter we present various techniques available to do adaptive streaming, especially applicable to P2P networks.

Transcoding allows adapting a video stream on-the-fly by taking a previously encoded video and changing its format as required. Typical parameters that can be changed are bit rate, frame rate and frame size. Transcoding used with lossy compression schemes is a destructive process, this mean that lost information cannot be recovered later to obtain the original quality level. This effect is called digital generation loss, and should always be considered when dealing with multiple transcoding conversions. Transcoding can also be used to change the codec when the original one is not supported by a client's player.

Layered encoding is a technique that offers the possibility to stream at different bitrates to the child nodes. It does so by subdividing a single stream in a set of substreams that can be transmitted separately. A so-called "base layer" contains the minimum quality level, and is always needed to correctly decode the video. In addition to a base layer, optional enhancement layers can be used to increase the quality of the video. Various layered encoding have been designed. In this Chapter we will focus specifically on one of them, called *Scalable Video Coding (SVC)*. SVC is standardized is an extension of the successful H.264 codec and provides three types of enhancing layers: temporal, spatial and fidelity. The advantages of SVC are the simplicity of its scalability process, and the possibility of combining layers received through different paths to obtain again the original quality.

Multiple Description Coding (MDC) is a technique very similar to layered encoding, but is more oriented to resiliency than scalability. The original video is converted in a way to create multiple independent sub-streams, called descriptions. Each description can be decoded by itself, differently from layered encoding in which a base stream must always be present in order to decode the others. Combining multiple descriptions leads to better quality compared to a single description. A higher redundancy is usually introduced, leading to a less efficient delivery system in the event of no losses of data experienced during transmission.

Finally we consider the technique of stream switching, that is recently gaining popularity in web-based streaming systems. The video is encoded in multiple streams of different rates, suitable for different network conditions or target devices. Each stream is then divided, physically or virtually in sections of fixed length. Clients can then dynamically choose the appropriate quality level considering the instantaneous characteristics of network. The advantage of this technique is its outstanding simplicity of implementation, while the drawback is the need for multiple versions of the video at the source, increasing storage cost. This approach shows promise because it relies on well established web-based pull techniques combined with single layer typical video encoding technologies. Furthermore, it shows potential for implementation into a distributed P2P environment.

In the remainder of this chapter we discuss various optimized transcoding techniques and their application in adaptive p2p streaming systems. Layered Encoding is considered in the next section, that focuses in particular on the Scalable Video Coding implementation. In the following section, Multiple Description Coding is presented, and some streaming systems that exploit its characteristics are proposed. The last presented adaptive technique is the stream switching for web streaming.

TRANSCODING

The purpose of transcoding, or re-encoding, is to convert a previously encoded media signal (video and/or audio) into another one with different format. When a video is digitized for the first time, the original signal acquired from the recorder is converted in a particular format, so that it can be stored on a file system. Often this signal is saved with a high bit-rate, which is not very suitable for transmission directly over a network. For this purpose, the acquired file is converted using more complex compression schemes that reduce the file size and consequently the transmission duration. Websites or other services that allow users to upload and stream their own contents must support various input formats, from videos recorded using mobile phones to others recorded using professional equipment. All those devices support different file formats, which need to be converted to other homogeneous standard formats suitable for streaming. Furthermore, new better performing compression algorithms are constantly developed, which offer better compression rates. The opposite is also possible, when an unsupported format needs to be converted to serve an obsolete player. Watermarking, editing in general and content protection are also reasons for the use of re-encoding.

Transcoding is usually a key step in building an adaptive streaming workflow. The starting point of media transcoding should be a stream of the highest quality possible, sometimes being directly the original signal captured from the recorder, like in live broadcasting. The parameters of the re-encoding process, such as frame size, frame rate and bit rate can be varied in order to adapt the stream to various network conditions, so that clients connected through a broadband connection can be served with the best quality possible, while clients using lower speed connections can still have access to a lower quality versions of the video in a reasonable amount of time. Since network conditions can differ widely from one user to another, in a multi-stream approach a lot of different versions of the video should be created in order to offer the best experience to each viewer. For example, at the moment of writing YouTube supports seven different resolutions, ranging from 144p[1] to the most recently introduced 4K[2]. Usually the technique of transcoding is used on-the-fly, while the media stream is sent from a source node to another one. In this way, the need is avoided of storing multiple copies of the same content in different formats. The cost of this operation is anyhow an increase in the latency and complexity of the system.

Each conversion from one format to another lossy format causes a degradation of the media quality. Compression artefacts are cumulative and it is then necessary to pay attention to how the transcoding process is implemented, and the total number of times in which transcoding is used from the source to the receiver. For this reason, when possible it is better to retain a copy of the original content in a lossless format, and then re-encode starting from that when necessary. Converting from a higher quality level to a lower bitrate, some information is lost and it is no more possible to reverse the process to reconstruct a better quality level. For this reason, transcoding is a destructive process.

Transcoding is often implemented directly on the source side, but can also be used in some of the transit network nodes, in order to adapt the stream to the local network conditions. For instance, one could decide to furnish different bitrates for local users connected via fast Ethernet cables and slower wireless accesses.

Since our purpose is to review adaptive techniques, we'll mainly focus on bitrate reduction transcoding. Ideally the quality of a reduced bitrate stream created using a transcoding process should be the same as a stream encoded for the first time, from an uncompressed source, directly at the target bitrate. The complete process of re-encoding would require the use of a pixel-domain approach, in which the original signal is fully decoded, processed as necessary and then re-encoded according to the constraints of the newly wanted stream. While this is always possible to do and sometimes necessary, this process is usually very costly and difficult to be implemented efficiently in a real-time environment. The actual challenge in transcoding is to find more approximate and efficient techniques while still maintaining an acceptable quality level.

Two types of transcoding systems can be defined: open-loop and closed-loop (Assuncao & Ghanbari, 1998). Open-loop systems are the simplest one, and require a very limited elaboration of the input stream because only the encoded *Discrete Cosine Transform (DCT)* coefficients (Watson, 1994) are modified. The roughest technique consists of discarding all the coefficients below a certain threshold level, or above a certain frequency. The number of coefficients removed can be adjusted to obtain the required final bit rate. A slightly more sophisticated method consists in the re-quantisation of the coefficients. They are decoded and subsequently re-quantised using a larger step size. The resulting stream will be improved in comparison to the simple discarding of some coefficients, because in this case the rate reduction is spread all over the frequency domain. An high level introduction to modern video compression and encoding techniques can be found in Wu & Rao (2005).

However, the open-loop systems introduce a major source of distortion called drift, which leads to a visual blurring. This is due to incoherence between the pictures used for prediction by the encoder and the decoder. As a matter of fact, the new predictively-encoded P-frames are directly derived from the old P-frames, without considering the inaccuracies introduced by the re-quantisation. This way, a continuous drop of quality is accumulated over predicted frames. For each following P-frame, the introduced distortion is not only caused by the current frame, but is also a residue of the transcoding of previous frames. Since the intra-coded I-frames are not decoded using prediction but are fully described, whenever such a type of frame is encountered, the drift error is recovered to zero. Likewise, bi-predictively-encoded B-frames do not propagate the drift to others frames because they are not used for prediction, but are still affected by their own distortion.

The closed-loop approach tries to solve this problem by approximating the complete decoder-encoder architecture. The stream is still not fully re-encoded, but a feedback loop containing a frame buffer is used in order to compensate the transcoding distortion, so that it does not propagate into the successive frame. The most power-consuming part of the encoding process is the motion estimation. Suppressing this step from the transcoding process, we obtain two great benefits. First of all, the computing cost is greatly reduced, being one of the best sources of optimisation. Secondly, we preserve a more precise version of the motion vector, because they are calculated from a more accurate and defined signal. Comparing the outputs of the implementation of this transcoder, and the encoding at the same bitrate of the original stream, the reduction in *Peak Signal-to-Noise Ratio (PSNR)* appears to be less than 1 dB, and even better than the complete re-encoding.

A problem similar to bit-rate reduction is spatial resolution reduction. The introduction of smartphones and other hand-held devices in the media market greatly increased the need for converting videos originally encoded for home video to lower resolutions. Reuse of motion parameters and macroblock information can help reduce the encoding complexity and consequently transcoding efficiency. A simple way to implement spatial reduction is to filter the input signal in order to retain only low-frequency coefficients, and then reconstruct new reduced macroblocks with the retained ones. Pixel averaging is a technique in which a square of pixels is compressed to a single pixel having as a value the average of all of them. An issue of this technique is that the resulting pictures can become blurred.

In the case of temporal transcoding, some of the frames originally present in the video are dropped. The most prominent task in this kind of transcoding is the re-estimation of motion vector. Bilinear interpolation techniques can be repeated iteratively to approximate all skipped frames. Other alternatives are: Forward Dominant Vector Selection, Telescopic Vector Composition and Activity-Dominant Vector Selection (Ahmad et al., 2005). Additional refinements of the resulting motion vector are suggested anyway, for better coding efficiency.

When the video is transmitted through a channel with high error rates like a wireless network, the transcoding process can be optimized using stronger protection mechanisms in order to compensate for these losses and supply users with a better quality of experience while maintaining the same required rate. Spatial resilience for instance can be increased using preferably intra-blocks. By adding more redundancy, the total video size is augmented, so it's necessary to reduce the encoding quality in order to retain the same bitrate. A trade-off between redundancy and SNR can be computed in order to improve perceived quality in the presence of transmission errors.

In a peer-to-peer system, the nodes could be organized according to their networking capabilities, so that the fastest connections can keep the original video stream at the best quality level, while transcoding and serving streams of lower rates to less performing peers. A P2P system that performs an adaptive transcoding service, independently from the type of overlay used, is proposed in Chen & Repantis (2011). Both local and distributed adaptation mechanisms are used in order to respond to changes in infrastructure and user's QoS requirements. Three roles can be taken by each node: Media Source, Media Receiver and Media Transcoder. A streaming session can make use of various sources and transcoders, organised in multiple streaming paths. A requesting node starts searching for a media object, and the returned results contain information about the available streams. A number of source nodes are selected by the requesting peer. Eventually some transcoding nodes can also be selected as needed to match a receiver's constraints. Each transcoding node performs only a limited number of transformations on independent fragmented media units. The modified media units coming from different paths are then merged by the receiver.

The proposed coordinated media streaming service (Chen & Repartis, 2011) is composed of four techniques: streaming graph composition, local quality adaption, feedback-based coordination and graph reconstruction. A comparison between different adaptation schemes showed that the proposed one is able to ensure an average good quality level with smooth degradation in the case of a very loaded network, maintaining a buffer of just four seconds. This system shows a good scalability because of the coordination between the various nodes. The end-to-end delay slightly decreases with an increasing number of available streams and system load, because the transcoders are forced to reduce their output quality level in order to elaborate quickly their task queue.

In Liu et al. (2007), a peer-assisted transcoding service is proposed. The network is composed of two overlays: the first overlay is used for normal streaming; the second one is used to share information and metadata useful for transcoding, in order to reduce the global complexity of this operation.

The transcoding process can be decomposed in various steps, and some information resulting from these intermediate steps can be successfully used by other peers engaged in the same task. Simulation shows that this system can reduce the average load of the transcoding system by 58% for bitrate scaling and 39% for frame rate reduction, if compared with the same streaming architecture in the absence of metadata sharing. Sharing of more information is a source of more network traffic; however this increment seems to be limited to a negligible 6% of the total traffic.

In Sun et al. (2005), users requesting different bitrates are sorted in a unique cascading transcoding tree. The tree root is the original source of the video; users that need higher quality are placed near the root, while users needing lower quality levels are located near the leaves. Available system's resources and transmission bandwidth are taken into account while constructing the delivery tree and positioning the nodes. Peers are grouped in layers composed of a controlled number of similar participants. The quality level that the whole group will receive is computed as the average requirement of the peers that belong to it. The size of each group, therefore, has to be chosen wisely in order to meet both system performances and a single user's requirements. The transcoder tree is decided in a centralized way, so that each user willing to be part of the streaming system has to send an initial request to the video source, including its resources and requirements. The tree is periodically reconstructed in order to compensate imbalances caused by peers joining and leaving the streaming system. During reconstruction time, streaming is stopped for all peers, and data contained in buffers is used to fill this period. Experimental evaluation of MTcast system was done in Shibata & Mori (2007). Results shows that an average PC can calculate a tree composed of 100 000 nodes in less than 2 seconds, and the maximum latency of such a tree between the root and the most external leaves is usually under 20 seconds, considering various configurations in terms of number of layers and the size of them. Rejoining time in the case of parent's disconnection is under 2 seconds.

Transcoding plays an important role in adaptive streaming. It is the technique that allows the greatest level of flexibility. This is because each parameter of the video can be varied in order to obtain the most appropriate format. However, time constraints introduced by live broadcastings can force some restrictions in the transcoding process. It should also be noticed that transcoding can only be used to downgrade the original quality of a video, and the dropped information cannot be recovered afterwards. The major issue of transcoding is the introduction of a certain quantity of artifacts that can occur during the recompression, or can occur as a result of approximations used to speed up the process.

SCALABLE VIDEO CODING

Scalable Video Coding (SVC) is an extension (Schwarz et al., 2007) of the H.264/MPEG-4 Advanced Video Coding standard (Wiegand et al., 2003), which introduces a special layered encoding, similar to the progressive JPEG used for image compression and transmission over the Web. A video encoded in SVC format is created composing multiple sub-streams derived from the original video signal. Each sub-stream can be transmitted independently from the others. However, in order to reconstruct the video the client needs to start from the base layer and then sequentially use the available improvement or enhancement layers. This way, from a single encoded video stream it is possible to achieve a multi-bitrate streaming service. In contrast other currently used technologies require separate encoding of each individual stream for multiple streams at different bit rate. This operation needs to be done once, if the streams are stored as independent files, or each time the content is required if done on the fly. In the first case, for a typical streaming service, at least three, and perhaps more than 10 copies of each video are created and stored on the transmitting source, introducing a great redundancy. While the storage costs are becoming less of a factor, this approach greatly increases the initial efforts and the ongoing maintenance complexity of the system. In the second case, we remove the storage requirements, but the encoding process has significantly higher computing cost. SVC tries to reduce those costs while keeping encoding efficiency and sufficient granularity of scalability.

SVC also brings a penalty in increasing the amount of data by 10 to 20% in order to reach the same quality of a traditional single-layer H.264 encoding. However, the advantage of layered coding becomes evident in streaming systems that use a significant number of different bitrates for each available media content. For example, consider a single video provided at 6 different quality levels of approximately doubling size, ranging from 0.15 to 3 Mbps. Using single-layer coding would lead to 6 different files for a total of 6.25 Mbps (0.15 + 0.3 + 0.5 + 0.8 + 1.5 + 3 Mbps). SVC uses instead just a single file 20% bigger than the higher bitrate, requiring approximately 3.6 Mbps (-42% of storing space). Using more streams would increase this saving even more. For backward compatibility reasons, the base layer is encoded using the standard H.264/AVC scheme, so that even the devices not supporting SVC can still accept and reproduce at least a low quality version of the video. Nowadays H.264 is more and more widespread, so even this extension has potential for the distribution of content.

After the encoding of the video, a descriptor indicating the position of the various bit-stream entities is created. Each layer is composed of its own header and texture data. A motion vector can also be included, but not all layers need to include it. The bit stream is organised in portions called *Network Abstraction Layer (NAL)* units, so that the truncation can be executed in points with low elaboration complexity. Each NAL unit is a packet of data that consists of a simple one-byte header, identifying the type of information contained, followed by the video payload. Multiple NAL units with specific properties are grouped together in Access Units. One access unit represent the encoding of a frame. A set of consecutive access units starting with an *Instantaneous Decoding Refresh (IDR)* frame represents the smallest independent decodable video sequence, also identified as a *Group of Pictures (GOP)*.

There are three main classes of scalability implemented in the SVC definition (Schwarzet al., 2007):

- Temporal: reducing the temporal resolution (number of video frames per second)
- Spatial: reducing the spatial resolution (number of pixels per spatial region, frame size)
- Fidelity: also called *Signal-to-Noise Ratio (SNR)* or quality scalability, reducing the fidelity of the video (coarsely quantised pixels)

A fourth class of scalability is obtained by a combination of the other three.

Figure 2 shows an example of video encoded using multiple layers of enhancing quality, from bottom (base layer; lowest quality) to top (highest quality). Each layer uses a different type of scalability.

Temporal scalability is the result of the introduction of hierarchical B-frames. In the classical prediction structure, B-frames are derived only from preceding and subsequent I/P frames. In the new scalable model, B-frames are used to predict other B-frames of enhancement layers in a cascading way. For instance, the base layer could be composed of I and P frames, which are used to predict the B frames of the first temporal enhancement layer. In turn, B-frames of the first layer are used to predict B-frames in the second layer (together with key frames of the base layer), and so on. As a result, the total number of B-frames between two key frames (of type I or P) is pro-

Figure 2. Example of layered structure of a SVC encoded video

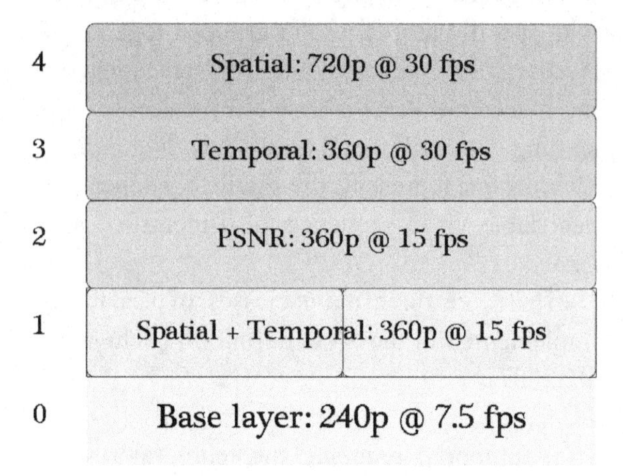

portional to 2^k -1, where k is the number of temporal layers. Hierarchical B-frames can be organised in different orders. A simple sequence is to iteratively put one B-frame of a higher layer between two predicting frames of the lower layer. A consequence of this kind of hierarchy of B-frames is that the first frame to be predicted and decoded is the central one, while in the classical approach is usually the first one of the group. In general, the coding efficiency improves by increasing the GOP size, where the best performances are achieved using groups of between 8 and 32 pictures. It is nevertheless necessary to notice that increasing the GOP size also increases decoding latency, because later frames are needed to predict B-frames shown earlier. Different prediction sequences can be used to solve this issue while retaining the temporal scalability property. Another benefit of using hierarchical prediction structures is that usually an improvement in coding efficiency is also introduced if compared to other classic frame structures used in single-layer encoding.

In spatial scalability, intra and inter-layer prediction mechanisms are exploited. Each following layer corresponds to one of the supported layers increasing spatial resolutions. Again, the base layer is used to encode the following enhancement layers in a bottom-up fashion. An intuitive constraint is that the resolution cannot decrease in enhancement layers. An important property of SVC is that each spatial layer is decodable using a single motion compensation loop, keeping a low complexity level. Spatial scalability generally performs better when high resolution material is used as input.

Quality scalability is based on an improvement of the concept of *coarse-grain quality scalable coding (CGS)*, called *medium-grain quality scalability (MGS)*. The improving signal contained in each successive enhancment layer is re-quantized using a finer quantization step compared to the previous layer. Motion compensation is done using only the key frames composing the base layer. For that reason, any enhancement NAL unit can be dropped by the decoder without introducing significant drift. Both spatial and quality scalability are sources of loss of encoding efficiency compared to a single-layer system. Several analyses of SVC performances have been conducted. In Van der Auwera et al. (2008), traffic characteristics of the various scalability modes proposed by SVC are compared with MPEG-4 Part 2.

Temporal enhancement layers have significantly lower bit rates compared to the base layer. That's because the base layer contains the key I-frames, which not being predicted but fully encoded are clearly less compressible. The quality of the unique base layer results in more variation than when enhancement layers are also available to the decoder. Compared to MPEG-4 Part 2, SVC enhancement layers present a more variable and difficult to manage traffic. This is probably due to a better compression algorithm of the SVC encoder that exploits redundancies more efficiently and as consequence is more sensitive to variations in frame content. Spatial scalability causes a more variable traffic. Contrary to temporal scalability, the enhancement layers are more than twice larger than the base layer for the highest quality. This means that the same final bit rate can be obtained by combining different types of scalability layers.

Not surprisingly, also the fidelity scalability is characterized by a more variable traffic than the MPEG-4 Part 2 standard.

Even though SVC is still a little more costly in terms of encoding efficiency or overhead, the gap with H.264/AVC single layer coding is small enough for SVC to be considered a good alternative in most cases. Furthermore, rate-distortion comparison shows that SVC outperforms almost all other video encoding technologies currently available (Wien et al., 2007). Another positive property of SVC is the strong error resilience, which results in a graceful degradation in quality when the transmission is impaired by packet losses (T. Wiegand et al., 2009). Some layers are more important than others for the final perceived quality of the video displayed. For this assumption, a stronger protection (redundancy) is used for these layers. This leads to the keeping of a better quality level while reducing the general overhead of protection mechanisms.

For its characteristics, SVC has the potential to be exploited for efficient adaptive P2P streaming services. The most important and indispensable layer is the base one. It must be received by all peers, so it should be transferred using reliable links. This can be achieved using multiple sources, which in turn helps avoid possible delays caused by peers disconnections or inadequate throughput. On the other hand, enhancement layers are optional and independent from each other. This way, each peer can request and store just the layers that she needs. If a desired layer is not received in time for display, it's still not a big issue; the quality will be a little inferior but the stream can continue without interruptions. These additional layers can therefore be delivered using less reliable or efficient links.

In Mushtaq & Ahmed (2007), a P2P system using hierarchical hybrid overlay networks is proposed. Peers offering the same video quality layers are placed in the same group of the overlay network. A locality-aware organization is also performed based on end-to-end *round trip time (RTT)* measurements. The streaming system is receiver-centric, so that each receiver node actively requests portions of the video from the peers that it chooses. The peers are organized in a tree structure for each overlay network, when the peers offering lower RTT are placed near the root, in order to be preferably selected as senders. When a streaming link is no more efficient enough, a different peer is selected for switching. If no peers with lower RTT are available, a stream switching is performed instead of peer switching. Using this system, simulations showed an improvement in throughput, smoother video delivery, lower delay and lower loss in comparison to a random selection scheme.

An implementation of the Stanford *P2P Multicast protocol (SPPM)* is proposed in Baccichet et al. (2007). Peers are organized in multiple complementary multicast trees. Source peers are roots of the trees and schedule packet distribution between themselves, to ensure load balancing and error resilience. Each peer keeps track of its parents by sending probing messages. In case of packet loss, retransmission requests are used to retrieve from other distribution trees. Packets are scheduled and transmitted in a prioritized order, where most important packets for decoding are sent first. Enhancement data are sent only when network conditions allow that to take place without impairments. When the uplink capacity of peers is close to the source bitrate, significant gains in average quality are noticed over all the peers.

Several streaming systems based on BitTorrent protocol have been conceived and implemented. In BitTorrent, file chunks to be downloaded are chosen in a rarest-first fashion. While this technique is good for a traditional file sharing system to increase the average availability of chunks, this is not efficient for streaming. Using a technique of this kind will surely lead to frequent interruption in video playback, since chunks are not received sequentially. For this reason, various enhancements are proposed.

In Abboud et al. (2009) and Abboud et al. (2011), a mesh-based streaming architecture with two stages of adaptation is proposed. In the first stage, called *Initial Quality Adaptation (IQA)*, parameters like screen size, processing power and network bandwidth are evaluated in order to choose the best initial quality level supportable. During peer discovery, the proper streaming nodes are selected and added to the list of elected senders. The idea is to have neighbour peers with the same or higher quality level, in order to have more probability of finding any of the necessary blocks. After the IQA phase, periodically *Progressive Quality Adaptation (PQA)* mechanisms are executed as a control loop in order to keep a good streaming experience with smooth adaptation to changing network conditions, processing power and block availability. At this stage, only temporal and quality scalability are supported, to avoid annoying artefacts derived from changing spatial resolution. Stream complexity is always kept under control to assure that less powerful devices are not overloaded. Priorities of the chunks are calculated in order to run a smart block selection algorithm that chooses the next block to request. A tracker is used to manage the list of all peers participating in a certain video streaming. Each node periodically renews its registration to the tracker, updating the list of available nodes. Peers linked together inform each other of the supported layers; when a node remains with an inadequate number of peers of the same level, it asks the tracker for an updated list. Seeding nodes manage requests in a prioritized order. When a higher priority request arrives, this takes the place of a lower priority one, if already present in queue. If not enough free slots are available, then the lower priority connection is dropped gracefully. Simulations show that the PQA process tends to decrease the average layer level to avoid stall events during playback. This effect is more evident the more frequently the algorithm is invoked. The interval between each execution of PQA is an important parameter to choose wisely in order to compensate streaming session quality (in terms of stalls, delays...) and average video quality.

In Abbasi & Ahmed (2010), a streaming system based on Small World overlays (Hui et al., 2006) and hybrid push/pull mechanisms is compared to CoolStreaming/DONet. A peer receives the video from multiple senders with a PUSH mechanism. The missing parts are then actively requested by the receiver using a PULL mechanism. Received packets are stored in buffers constantly controlled in relation to playback status, in order to determinate when to start the pull phase. Nodes are divided into four classes: sources, trackers, super peers and ordinary peers. Participating peers are grouped in a small world organization, forming different local clusters. When a packet enters the cluster, it can be easily retrieved by other peers in the same cluster. The overlay is organized in two hierarchical layers. The first layer is composed by linked super peers. Ordinary peers form the second layer, organized in several compact clusters, linked to the rest of the network by a super peer. Trackers are in charge of updating the peers list and dynamically elect super peers in accordance with network conditions and resources available. Peers mostly contributing to the overlay network (in terms of upload capacity) are selected as super peers and are able to connect directly to source peers, resulting in a lower latency. Periodical keep-alive messages are exchanged between trackers and super peers, and between super peers and ordinary peers in order to keep an up-to date awareness of the network topology. The purpose of the push mechanism is to quickly spread a substantial part of the video into the network. Maps of chunk availability are held in various nodes, and are used to determine from which peers the required blocks are downloaded during the pull phase. If a particular block is not available in the current cluster, a request is forwarded to others clusters through the leading super peer. Initially source peers distribute base layer to its overlay of super peers. Enhancement layers are then exchanged with clusters that need them. The packet delivery ratio increases with an increasing number of peers, because more clusters are available to exchange data.

Lee et al. (2008) proposed an architecture called GaiaSharp, especially designed for live streaming. Nodes are subdivided into Channel Servers and Client Engines. Channel server gets videos from multiple sources, and split each stream into Segments and Crumbs. Torrents are then prepared and announced to other peers. A crumb is an atomic packet composed of header and a base layer segment. Each crumb is tracked with an independent torrent. Streams offer a limited temporal window of packets available to download. Client Engines download torrents related to crumbs, exchanging information with trackers. A single segment at a time is selected to be downloaded. Priority is given to base layer segments; when the temporal window is big enough, enhancements packets are successively requested. The proposed architecture was simulated using a base layer of 128 kbps and two additional enhancement layers of 512 and 2048 kbps performed well with a minimum channel size of 144 kbps.

A similar approach is proposed in Asioli et al. (2010) and Ramzan & Izquierdo (2011). GOPs and layers are mapped into torrent segments using an index file, which must be downloaded firstly by the client. After that, the pre-buffering phase begins. In this stage, a large size of W * t(GOP) seconds is available to download the first part of the video. Higher priority is assured to base layer segments, which must be completely downloaded before enhancement layers. If enough time remains, a client can decide to download further segments belonging to enhancement layers. A different strategy is used in the two cases: for the base layer, segments are downloaded in playing time order, while enhancement layers are downloaded using the classic torrent "rarest first" approach. After pre-buffering time is passed, the sliding window is shifted after each GOP. Yet, at least the base layer has to be downloaded; otherwise the system will stay in a waiting state. All pending requests that expired outside the window are dropped. A variant using Multiple Description Coding is proposed in Ramzan & Izquierdo (2011). Videos are not diffused as a single scalable stream, but multiple compensating streams are created. Each description can be completely decoded by itself, but several descriptions can be joined together obtaining an improved quality. A slight downside of this technique is that a certain quantity of redundancy is added to the stream, because each independent description has to contain a base layer.

Scalable Video Coding represents the state of the art in layered encoding. Its standardization as an extension of H.264 codec ensures top performances in encoding efficiency and resulting quality. Since its introduction is quite recent, not a lot of streaming systems are using it so far. However, the great flexibility allowed by its particular coding strategy makes it an interesting alternative when peers can be organized in a cascade of decreasing required quality level. The drawback of this approach is that enhancement layers rely on lower layers to be correctly decoded, so that when a layer is not received, or is received too corrupted, all following layers will be wasted. While SVC is a very interesting approach to the solution of adaptive streaming, transcoding is still a necessary step when we need to convert from one format to another for compatibility reasons, or in particular applications like multipoint videoconferencing, or for any other application that needs to edit somehow the original video.

MULTIPLE DESCRIPTION CODING

The idea behind *Multiple Description Coding (MDC)* was originally investigated in the late 1970s. The purpose was to increase the robustness and reliability of the circuit-switched telephone network, without using physically separated "standby links" activated only in case of necessity (Goyal, 2001). The same technique can be adapted and transferred to the more modern packet switched networks, when high packet losses are expected and retransmission is not a tolerable option. Media streaming over wireless networks is a good example of applicability. The idea is similar to Scalable Coding, where a single signal is encoded in multiple streams. The more of these streams are received by the final user, the better the quality of the decoded signal is. However, to decode a scalable coding stream the client needs to start with the base layer and improve the quality by the consecutive enhancement layers. In MDC, each single stream is called a *description* and can be ideally self decoded. Variants of MDC involving additional enhancement layers (not self-sufficient) that increase the flexibility of the system are also possible. Since the purpose of MDC is to counter packet losses and sudden link failures increasing redundancy, each description should be sent through different paths, even if this can reduce routing efficiency. Using multiple paths can however have also additional positive impacts in traffic dispersion and load balancing, which is useful to prevent or mitigate congestions. The benefits of MDC come obviously at the price of using more bits to achieve the same quality than using a "single description" encoding (Reibman, 2005). The various descriptions are not necessarily required to have the same bitrate, allowing a certain degree of adaptability to specific transmission channels.

Different ways exist to create *MDs (multiple descriptions)*. The simplest one is to divide somehow the source data into several subsets and compress them separately to produce different descriptions. Interpolation is then used to decode any received combination of descriptions. A classic example is to separate odd and even samples, for example alternate frames, obtaining two subsets at half the rate of the original signal. Three decoders can then be used by the receiver: two for each separate description and one in case both the streams are properly received. The general case with N descriptions would then require $2^{(N-1)}$ decoders. A different approach is to use successive refinements to reduce the total number of decoders. In this case, one decoder is used when a single stream is received and a different one is used to decode two descriptions and so on. Another possibility is to repeat some part of the data in every description created. Not all information has the same usefulness, so the most important parts are replicated more than the others. This approach is called *Unequal Error Protection (UEP)*.

Modern compression systems use prediction and decorrelating transforms to reduce redundancy and increase efficiency. If the encoder uses a predictor that relates on information not received by the decoder, there will be a mismatch. This is a major concern in modern MD systems. Using more redundancy to protect earlier frames of the GOP helps reduce eventual mismatch in subsequent frames, because later frames are encoded in cascade using information contained in previous ones.

A hybrid streaming system called CoopNet, which takes advantage of a P2P network to support normal client-server architecture, is proposed in Padmanabhan et al. (2002). Even when robust CDNs are used to stream videos, exceptional events of global interest can cause "flash crowds" that saturate their resources. To address the problem of sudden changes in the distribution network due to peers joining and leaving, Multiple Description Coding is exploited. The original content is divided in multiple sub-streams and each one of them is delivered to the final user through a different peer. When the streaming service is of the type "on demand," clients are required to store in a cache contents they recently received. When the server is overloaded, it responds to new requests with a list of other clients that have the requested content. The client then contacts some of the peers in the list to begin the streaming process. If the client cannot retrieve content from other peers, it is forced to use directly the server as source.

When "live streaming" is used, clients are arranged in multiple trees having roots in the server. To provide the best performances, the distributing trees should be as short as possible. A short tree means less intermediaries peers between the source and the viewer, reducing latency and the probability that an ancestor node will leave the network interrupting the flow of data. The management of the trees is completely delegated to the central server. In general this can be risky because if it fails, the tree will also break, but since the server is also the source of data it doesn't really matter because even the stream will stop. When a new client wants to join the net, it contacts the server and receives from it a list of designated parents, one for each different description available. Each node is required to determine its "network position" by measuring the latency between itself and a set of fixed hosts. This information will be used to optimize the distributing trees grouping together neighboring nodes. The server itself is the direct parent of some peers, especially in the beginning phases when the trees must be initially constructed. When a node decides to leave the net, it should signal this to the server in order to allow a quick reconstruction of the tree. Unavoidably some peers will leave the net inadvertently, so each node constantly monitors packet losses and delays for every parent. When a loss ratio became unacceptable, the node experiencing the losses asks its parent if it is aware of the problem. If so, some ancestor node is responsible for resolving the problem, and no actions are taken. Otherwise the current node requests the server to have a new parent assigned. Experiments were executed to evaluate the performances of the system. As the number of descriptors increases, the percentage of clients that is able to receive all of them decreases, but decreases also the percentage of clients that receive only a small portion of the descriptions. In case of link failure it is important to quickly recover the distribution path, to avoid buffer draining. With a repair time of 1 second, involved clients are able to receive 90% of the descriptions, while when 10 seconds are required the percentage drops to 30%.

In Akyol et al. (2007), an interesting way to create MDC for a P2P system is proposed. *Firstly, Motion Compensated Temporal Filtering (MCTF)* is used to decorrelate frames within a GOP. Subsequently, JPEG2000 codec is used to separately encode each code-block of frames, which can then be treated independently. JPEG2000 natively supports multiple quality layers; the obtained codeblocks can then be truncated at different points: one at high rates and the other one at lower rates. The remaining information can also be used to create multiple additional enhancement descriptions (that cannot be decoded by themselves but relay on information from other layers). Base descriptions are then created by mixing in various combinations of codeblocks at high and low rates. With this approach, descriptions of different bitrates can be easily constructed, by changing the proportions of different quality blocks. Motion vectors and other necessary data are inserted losslessy in each description. When a decoder receives multiple descriptions it retains only the highest version available of every block, discarding unnecessary data.

In Taal et al. (2004), another P2P system using multiple multicast trees is proposed. The original video is encoded in multiple layers, protected differently in order of importance of the contained information, using erasure coding. The resulting data are then split and mixed to create multiple descriptions. Information from the base layer is inserted in all descriptions, while data of superior layers are used in a decreasing number of layers. By using a small base layer and larger enhancement layers, the redundancy can be significantly reduced. This approach tends to create descriptions of different sizes, according to the number and type of layers inserted. A metric that takes into account perceived quality in relation to used and unused bandwidth is used to evaluate the efficiency of the system and optimize parameters like the number of descriptions and their rates. Results show that the number of different quality levels perceived by the clients is not the same as the number of descriptions, but significantly lower. This means that increasing the number of received descriptions doesn't always lead to a better quality of experience. For instance, using eight descriptions led to only four distinguishable levels of quality. This is due to the high redundancy introduced by the proposed process. This is good from a resilience point of view, but is not very efficient in terms of flexibility and adaptability.

MDC is based on multiple streams as are layered encoding methods, but it does not have the hierarchy of base and improvement layers. This way, all streams can be treated equally, without the need to assure the delivery of one of them in particular. This is a great characteristic that can be exploited when high packet losses may be experienced in the considered transport network. The downside of the MDC solution is a relatively higher redundancy, which translates into a potential waste of network resources. Various modification of the pure approach can be introduced to increase the flexibility of the system, as using asymmetric descriptions or supplementary enhancement layers. In general, MDC should be used in conjunction with *Multiple Path Transport (MPT)*, where each description reaches the receiver through (physically) different routes. The idea is that some link may over time collapse and stop serving contents, while the probability that all links fails at the same moment is very low. If at least one description can be correctly transferred, then the service is still assured, even if the quality will be low. This is particularly useful when retransmission of lost packets is not a suitable option due to the time constraints of the system. In videoconferencing, late-comer packets are useless and it is pointless to retransmit them; so it's preferable to receive a minimum amount of data in time than the whole but late.

Even though MDC is not designed to be the most resource efficient solution, when implemented wisely it outperforms traditional streaming architectures. Compared to a Content Delivery Network with the same total serving capacity, a P2P network using MDC shows better performances with smaller playback delay and packet losses (Khan et al., 2004).

STREAM SWITCHING IN WEB ADAPTIVE STREAMING

Stream switching is a paradigm for adaptive streaming that is recently gaining a lot of interest within web based services. Its success is due to many factors. First of all, it can be simply built on top of existing and widespread open technologies like HTTP. Dedicated and optimized servers exist for adaptive streaming; however it can be completely implemented using normal web servers already deployed. One of the main advantages of using plain HTTP is that it is usually permitted even in the most restricted environments, passing through firewalls and proxy servers. Another advantage derives from the cacheability of web contents. Since videos are seen as normal web content, intermediate caches placed between the original server and end user allows great reduction of network load, bringing contents closer to the consumer.

Historically, the so called "progressive download," also known as pseudo-streaming, has been used for online video viewing, due to its simplicity. Content is transferred as a single file, and playback can start as soon as enough data is buffered, while still downloading the remaining part of the video. Multiple versions of a single video may be available to users, diversified according to bitrate, and as consequence, quality level. At the beginning of a streaming session, a user is required to select the most suitable version that meets its requirements and available resources. In case of considerable variation in bandwidth, the viewer

Figure 3. Example of an adaptive streaming session

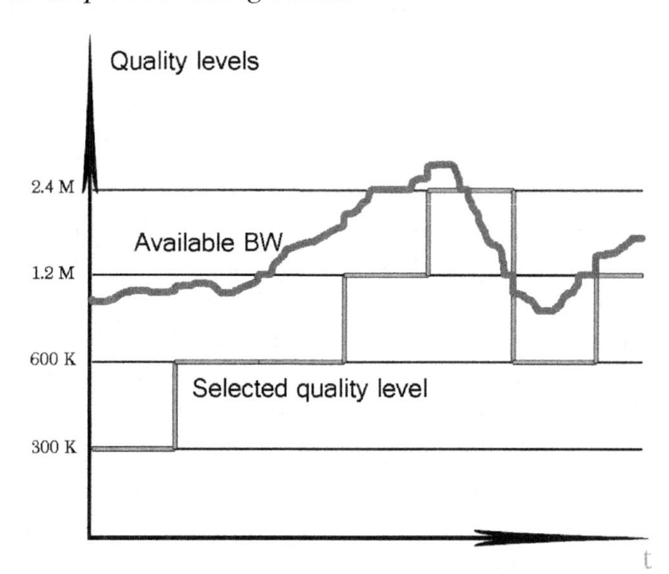

may experience frequent freezes due to buffer draining. Moreover, in progressive streaming usually the client tries to greedily get and cache in the buffer as much content as possible, regardless of the current position in the video playback. This attitude can waste a lot of network resources, because many videos are not watched entirely by the user. To overcome those issues, while still retaining the simplicity of pseudo-streaming, the HTTP Adaptive Streaming approach has been proposed, and is currently deployed by several media content providers.

The idea is to encode and store multiple versions of each single video at different bitrates, and let the client choose instant by instant the best suitable version to download. To do so, every stream is fragmented in chunks of fixed length, typically in the range between 1 and 10 seconds. The reference frame at the beginning of each chunk is synchronized, so that switching between different streams does not create glitches. The client keeps a limited buffer and does not try to get more chunks than necessary to fill it. Available bandwidth can be evaluated by the download time required for previous chunks. When changes in network conditions are detected, a different version of the video can be selected for the next chunk. Frequent changes in video quality should be avoided because they are annoying to the user, and can result in a poor experience. Figure 3 shows an example of an adaptive streaming session. The bold line represents the instantaneous available bandwidth, while the piecewise linear curve represents the quality levels selected to download.

Since HTTP is a pull-based protocol, each client is required to autonomously manage its streaming session by actively requesting each chunk. This means that servers are relieved from keeping track of the status of each client. Moreover, each client can potentially implement its own adaptive algorithm. The list of available streams and chunks must be known by the client. To do so, some kind of manifest file is published by the server, and is the first file that a client requests and parses before starting its streaming session.

Two different strategies can be used to store the videos used in HTTP adaptive streaming. The first one uses physically separated chunks. This is the most straightforward and simple to adopt, because any existing web server can be used without modifications. The downside is represented by the high number of files that are created. A one-hour long video encoded in five different quality levels and subdivided in chunks of five seconds each, requires 3600 files to be stored in the system. The maintenance of such a high number of files can be a complex task for a service that offers hundreds or thousands of different videos. The alternative is to use virtually segmented files. In this case each stream is stored on the server side as a single file. An index is necessary to map each chunk to a precise location of the file. For this reason, an additional manifest file is kept for exclusive use of the server. When a chunk is requested, the server application looks it up in the index file and access the stream at the right location, responding with the needed portion of video. This technique requires the deployment of a specialized server application.

Three main commercial implementations for HTTP adaptive streaming exist nowadays: Microsoft's Smooth Streaming (Zambelli, 2009), Apple's HTTP Live Streaming (Pantos & May, 2011) and Adobe's dynamic streaming (Adobe, 2010). A new open protocol called *Dynamic Adaptive Streaming over HTTP (DASH)* is in the process of standardization by MPEG (Stockhammer, 2010).

Akamai is one of the most important Content Delivery Network service available. The performances of its adaptive streaming implementation were analyzed in Cicco & Mascolo (2010). Each video appears to be encoded in five different bitrate versions. The client communicates with the server passing variables and commands through POST messages. The frame rate is never modified, so the adaptation mechanism relies only on the quality level of the video. The control algorithm is executed on average each two seconds. Despite that, when a sudden increase of the available bandwidth occurs, the time required by the transition to fully match the new bandwidth is roughly 150 seconds. This large actuation delay can also be noticed in the case of a drop in the bandwidth, leading to short interruptions in video playback.

A study of client-side intelligence for adaptive streaming was conducted in Jarnikov & Ozcelebi (2010). A system using reward and penalty parameters is used in order to personalize the streaming strategy. Experiments showed that considering the past two minutes of network conditions is enough to have a sufficient feedback for the adaptation algorithm. Increasing this period leads to a more conservative strategy that tends to keep a lower quality level. Comparisons with Apple's implementation shows that the proposed solution can deliver video at a quality level at least as good as the Apple solution, and usually exploit greater the available bandwidth at a cost of more changes of quality levels. It is also pointed out that the Apple implementation sometimes wastes bandwidth requesting more than once the same temporal chunk at various quality levels.

A system that uses bitstream switching and a customized congestion control for adaptive streaming over RTP/UDP is proposed in Schierl & Wiegand (2004). The pre-buffering time is fixed at 1 second, and the maximum signaling overhead due to feedback mechanisms is kept below 5% of the total data sent. Experimental results shows that bitstream switching, combined with temporal scalability (drops of frames), provide the best performances, compared to only temporal scalability.

Since the success of stream switching is quite recent and mainly bound to web streaming services, there is still a lack of research in P2P networks exploiting this technique. However, its simplicity of implementation and its similarity to other classic file sharing systems based on chunks (e.g. BitTorrent) could lead stream switch to be an interesting field worthy of further researches. Unlike other techniques previously reviewed, stream switching by itself does not provide resilience or error protection mechanisms. If used with normal encoding like H.264, retransmission of lost packets may still be needed. Protocols that use stream switching usually are built on top of TCP connections, so that retransmission of lost or corrupted packets is transparent to the application.

This should be taken in account when deploying a streaming system that requires real-time delivery or that uses highly unreliable networks.

FUTURE RESEARCH DEVELOPMENTS

Since video streaming over the Internet is more and more popular, the need for optimized and reliable services is increasing. Adaptive streaming techniques are subject of active research in the last years, with the goal of improving the Quality of experience perceived by the users. While in the past PUSH-oriented strategies were more used, the current trend is to move the download logic on the client side, that becomes an active part of the streaming process. Implementation of smarter selection strategies for the bitstream to request is an important topic of future developments. Another interesting field of research for optimizing P2P streaming, not covered in this chapter, is the network locality awareness, that permits a better selection of peers to connect to, limiting data transmission over long distances and reducing network congestion.

CONCLUSION

In this chapter we have reviewed a set of popular techniques used for adaptive streaming, with a particular attention to applicability to P2P. The purpose of these techniques is to allow a good Quality of Experience of streaming services even in presence of adverse network conditions characterized by sudden bandwidth variations and packet losses.

Transcoding allows changing on the fly some parameters of the original video, converting it in a different format. Frame rate and quantization step can be easily reduced in order to obtain a video with lower bitrate. Frame size and encoding scheme can be changed as well, but require some additinal computation. Since transcoding processes are usually optimized with approximations, a certain quantity of drift may be introduced. Transcoding can also be used to increase the resilience of a stream by introducing more redundancy.

Scalable Video Coding is the most promising implementation of layered encoding techniques. A single video is divided in multiple layers that can be transmitted independently. Starting from the base layer, that is necessary to decode the video, each additional layer brings improvement to the delivered quality. Layered encoded streams are also more resilient to packet losses than the single layer streams, because errors in one layer will affect only that layer and the upper ones, while lower layers can still be decoded successfully. The drawback is that if a layer is not correctly received, all higher layers transmitted are wasted.

Multiple Description Encoding streams on the other hand can be decoded independently from the each other. The more descriptions are received the better will be the final quality of the decoded video. This approach is particularly useful when different paths are used for the different descriptions, increasing the overall robustness of the transmission. While transcoding and even more so layered encoding are oriented towards scalability, MDC is more optimized for resilience purposes. The obvious drawback of MDC is the higher redundancy, which is acceptable when dealing with unreliable networks.

Stream switching is gaining popularity in web adaptive streaming mostly due to its simplicity and use of legacy technology. It is the best candidate at this time to substitute for progressive download. Furthermore it can be adapted for P2P use to deal with the scalability issue. Each video is encoded in multiple bitrates, suitable to different network conditions or target device. The client can then monitor its bandwidth and change the bit rate when necessary by requesting a different stream for the next chunk to download. Stream switching does not increase the resilience of the streaming session by itself, so retransmission mechanisms are still needed. This technology is also suitable for P2P implementation, which can address the resilience and scalability. The servers or primary nodes would contain all the different stream bit rates while the other nodes only receive and redistribute one selected bit rate. A new joining node can then choose a bit rate level based on the availability of nodes with free resources that stream at that bit rate and its own resource restrictions. Additionally, any node that desires to change to another bit rate needs to take the same considerations. The effectiveness of this approach in dealing with scalability issues needs to be examined in the context of tree and mesh P2P overlays.

REFERENCES

Abbasi, U., & Ahmed, T. (2010). SWOR: An architecture for P2P scalable video streaming using small world overlay. *Proceedings of the 7th IEEE Conference on Consumer Communications and Networking Conference*, (pp. 1-5).

Abboud, O., Pussep, K., Kovacevic, A., & Steinmetz, R. (2009). *Quality adaptive peer-to-peer streaming using scalable video coding* (pp. 41–54). Wired-Wireless Multimedia Networks and Services Management.

Abboud, O., Zinner, T., Pussep, K., Al-Sabea, S., Steinmetz, R., & Darmstadt, T. U. (2011). On the impact of quality adaptation in SVC-based P2P video-on-demand systems categories and subject descriptors. *Proceedings of the Second Annual ACM Conference on Multimedia Systems*.

Adobe (2010). *HTTP dynamic streaming on the Adobe Flash platform.* Retrieved from http://www.adobe.com/products/httpdynamicstreaming/pdfs/httpdynamicstreaming_wp_ue.pdf

Ahmad, I., Wei, X., Sun, Y., & Zhang, Y. Q. (2005). Video transcoding: an overview of various techniques and research issues. *IEEE Transactions on Multimedia*, *7*(5), 793–804. doi:10.1109/TMM.2005.854472

Akyol, E., Tekalp, A. M., & Civanlar, M. R. (2007). A flexible multiple description coding framework for adaptive peer-to-peer video streaming. *IEEE Journal of Selected Topics in Signal Processing*, *1*(2), 231–245. doi:10.1109/JSTSP.2007.901527

Asioli, S., Ramzan, N., & Izquierdo, E. (2010). Efficient scalable video streaming over P2P network. *User Centric Media*, (pp. 153-160).

Assuncao, P., & Ghanbari, M. (1997). Transcoding of single-layer MPEG video into lower rates. *IEEE Proceedings: Vision . Image and Signal Processing*, *144*(6), 377–383. doi:10.1049/ip-vis:19971558

Assuncao, P. A. A., & Ghanbari, M. (1996). Post-processing of MPEG2 coded video for transmission at lower bit rates. *Proceedings of the IEEE International Conference on Acoustics, Speech, and Signal Processing*, (pp. 1998-2001).

Baccichet, P., & Schierl, T. Wiegand, Thomas, & Girod, B. (2007). *Low-delay peer-to-peer streaming using scalable video coding*. Presented at Packet Video Workshop.

Begen, A., Akgul, T., & Baugher, M. (2011). Watching video over the web: Part 1: Streaming protocols. *IEEE Internet Computing*, *15*(2), 54–63. doi:10.1109/MIC.2010.155

Chen, F., & Repantis, T. (2005). Coordinated media streaming and transcoding in peer-to-peer systems. *Proceedings of the 19th IEEE International Parallel and Distributed Processing Symposium*.

Cicco, L. D., Mascolo, S., Bari, P., & Orabona, V. (2010). An experimental investigation of the Akamai adaptive video streaming. *Proceedings of the 6th International Conference on HCI in Work and Learning, Life and Leisure: Workgroup Human-Computer Interaction and Usability Engineering*, (pp. 447-464).

Cisco Systems. (2010). Cisco visual networking index: Forecast and methodology, 2009 – 2014. Retrieved from http://www.cisco.com/en/US/solutions/collateral/ns341/ns525/ns537/ns705/ns827/white_paper_c11-481360_ns827_Networking_Solutions_White_Paper.html

Cisco Systems. (2010b). *Cisco visual networking index: Usage*. Retrieved from http://www.cisco.com/en/US/solutions/collateral/ns341/ns525/ns537/ns705/Cisco_VNI_Usage_WP.html

Goyal, V. K. (2001). Multiple description coding: Compression meets the network. *IEEE Signal Processing Magazine*, *18*(5), 74–93. doi:10.1109/79.952806

Hei, X., Liu, Y., & Ross, K. (2008). IPTV over P2P streaming networks: the mesh-pull approach. *IEEE Communications Magazine*, *46*(2), 86–92. doi:10.1109/MCOM.2008.4473088

Hui, K. Y. K., Lui, J. C. S., & Yau, D. K. Y. (2006). Small-world overlay P2P networks: construction, management and handling of dynamic flash crowds. *Computer Networks: The International Journal of Computer and Telecommunications Networking*, *50*(15), 2727–2746.

Jarnikov, D., & Ozcelebi, T. (2010). Client intelligence for adaptive streaming solutions. *IEEE International Conference on Multimedia and Expo*, (pp. 1499-1504).

Khan, S., Schollmeier, R., & Steinbach, E. (2004). A performance comparison of multiple description video streaming in peer-to-peer and content delivery networks. *IEEE International Conference on Multimedia and Expo*, (pp. 503-506).

Lee, T.-C., Liu, P.-C., Shyu, W.-L., & Wu, C.-Y. (2008). Live video streaming using P2P and SVC. *Proceedings of the 11th IFIP/IEEE International Conference on Management of Multimedia and Mobile Networks and Services: Management of Converged Multimedia Networks and Services*, (pp. 104-113).

Liu, D., Setton, E., Shen, B., & Chen, S. (2007). PAT: Peer-assisted transcoding for overlay streaming to heterogeneous devices. *Proceedings of 17th International Workshop on Network and Operating Systems Support for Digital Audio and Video*.

Liu, Y., Guo, Y., & Liang, C. (2008). A survey on peer-to-peer video streaming systems. *Peer-to-Peer Networking and Applications*, *1*(1), 18–28. doi:10.1007/s12083-007-0006-y

Magharei, N., Rejaie, R., & Guo, Y. (2007). Mesh or multiple-tree: A comparative study of live P2P streaming approaches. *IEEE INFOCOM*, (pp. 1424-1432).

Mushtaq, M., & Ahmed, T. (2007). Hybrid overlay networks management for real-time multimedia streaming over P2P networks. *Proceedings of the 10th IFIP/IEEE International Conference on Management of Multimedia and Mobile Networks and Services: Real-Time Mobile Multimedia Services*, (pp. 1-13).

Padmanabhan, V. N., Wang, H. J., Chou, P. A., & Sripanidkulchai, K. (2002). Distributing streaming media content using cooperative networking. *Proceedings of the 12th International Workshop on Network and Operating Systems Support for Digital Audio and Video*.

Pantos, R., & May, W. (2011). *HTTP live streaming*. IETF Draft. Retrieved from http://tools.ietf.org/html/draft-pantos-http-live-streaming-06

Ramzan, N., & Izquierdo, E. (2011). Scalable and adaptable media coding techniques for future Internet. In Domingue, J. (Eds.), *The future internet* (pp. 381–389). Berlin, Germany: Springer Verlag. doi:10.1007/978-3-642-20898-0_27

Sandvine. (2011). *Global internet phenomena report - Spring 2011*. Retrieved from http://www.wired.com/images_blogs/epicenter/2011/05/SandvineGlobalInternetSpringReport2011.pdf

Schierl, T., & Wiegand, T. (2004). H. 264/AVC rate adaptation for internet streaming. *14th International Packet Video Workshop*.

Schwarz, H., Marpe, D., & Wiegand, T. (2007). Overview of the scalable video coding extension of the H. 264/AVC standard. *IEEE Transactions on Circuits and Systems for Video Technology*, *17*(9), 1103–1120. doi:10.1109/TCSVT.2007.905532

Shibata, N., Mori, M., & Yasamuto, K. (2007). P2P video broadcast based on per-peer transcoding and its evaluation on PlanetLab. *Proceedings of the 19th IASTED International Conference on Parallel and Distributed Computing and Systems*.

Stockhammer, T. (2011). Dynamic adaptive streaming over HTTP–design principles and standards. *Proceedings of the Second Annual ACM Conference on Multimedia Systems*, (pp. 133-144).

Sun, T., Tamai, M., Yasumoto, K., & Shibata, N. (2005). MTcast: Robust and efficient P2P-based video delivery for heterogeneous users. *Proceedings of the 9th International Conference on Principles of Distributed Systems*.

Taal, J., Pouwelse, J., & Lagendijk, R. (2004). Scalable multiple description coding for video distribution in P2P networks. *24th Picture Coding Symposium*.

Van der Auwera, G., David, P. T., Reisslein, M., & Karam, L. J. (2008). Traffic and quality characterization of the H.264/AVC scalable video coding extension. *Advances in Multimedia*, *2*, 1–27. doi:10.1155/2008/164027

Wang, Y., Reibman, A. R., & Lin, S. (2005). Multiple description coding for video delivery. *Proceedings of the IEEE*, *93*(1), 57–70. doi:10.1109/JPROC.2004.839618

Watson, A. B. (1994). Image compression using the discrete cosine transform. *Mathematica Journal*, *4*(1), 81–88.

Wiegand, T., Noblet, L., & Rovati, F. (2009). Scalable video coding for IPTV services. *IEEE Transactions on Broadcasting*, *55*(2), 527–538. doi:10.1109/TBC.2009.2020954

Wiegand, T., Sullivan, G. J., Bjontegaard, G., & Luthra, A. (2003). Overview of the H.264/AVC video coding standard. *IEEE Transactions on Circuits and Systems for Video Technology*, *13*(7), 560–576. doi:10.1109/TCSVT.2003.815165

Wien, M., Schwarz, H., & Oelbaum, T. (2007). Performance analysis of SVC. *IEEE Transactions on Circuits and Systems for Video Technology*, *17*(9), 1194–1203. doi:10.1109/TCSVT.2007.905530

Wu, H. R., & Rao, K. R. (2005). *Digital video image quality and perceptual coding*. Boca Raton, FL: CRC Press.

Zambelli, A. (2009). *IIS smooth streaming technical overview*. Microsoft Corporation. Retrieved from http://users.atw.hu/dvb-crew/applications/documents/IIS_Smooth_Streaming_Technical_Overview.pdf

ADDITIONAL READING

Setton, E., & Girod, B. (2010). *Peer-to-peer video streaming*. New York, NY: Springer Verlag.

Simpson, W. (2008). *Video over IP*. Amsterdam, Holland: Focal Press.

KEY TERMS AND DEFINITIONS

Adaptive Streaming: A kind of streaming service that automatically adapts the bitrate of the sent stream according to the momentary state of the transmission channel (mainly the available bandwidth) and the client state (screen resolution, CPU load, battery level…)

Content Delivery Network: A cluster of servers used to deliver media contents over the network to a wide number of users. The servers are usually located in different geographical locations in order to be closer to users.

Layered Encoding: A class of media encoding techniques that permits to divide the data in hierarchical "layers" of information. Starting from a base layer, that provides the lowest level of quality, each subsequent layer improves the final quality of the reconstructed media file. It is important to notice that higher levels are dependent on the lower ones to be correctly decoded.

Multiple Description Coding: A class of media encoding techniques that separates the original data in multiple representations (*descriptions*) of lower quality. When several descriptions are combined together, the resulting quality is enhanced compared to the quality of a single description.

Scalable Video Coding: A particular implementation of layered encoding that is an extension of the popular H.264/MPEG-4 AVC video compression standard.

Stream Switching: A technique used to change the streaming bitrate. Multiple versions of the same video are available at different bitrates, and at fixed time intervals is possible to choose a different bitrate in order to adapt the stream to the network conditions.

Transcoding: the process of converting a media stream to a format different from the original. Transcoding can cause a decreasing of the media quality.

Unequal Error Protection: when some parts of the information that needs to be transmitted are more important than others, stronger protection should be applied to these parts, in order to reduce the impact of errors during the transmission process. Unequal Error Protection codes do so by adding more redundancy on the most important parts.

ENDNOTES

1. 144p is the shorthand name for a video having resolution of 176x144 pixels. The p stands for *progressive scan*.
2. 4K represents a class of emerging video standards having an horizontal resolution of approximately 4000 pixels.

Chapter 5
P2P Video Streaming over MANET

Nobal Bikram Niraula
University of Memphis, USA

Anis Laouiti
Telecom SudParis, France

ABSTRACT

Video streaming in Mobile Ad hoc NETwork (MANET) is a real challenge due to frequent changes in network topology, and sensitiveness of radio links. Recent approaches make use of Peer-to-Peer (P2P) technologies to combat these challenges because the technologies have been already found to be effective for content delivery on the Internet. However, as the Internet and MANET operate differently, the P2P technologies used in Internet need modifications before employing to MANET. In this chapter, the authors discuss the recent P2P approaches, the adaptations to be made, and the major challenges to be faced while using P2P approaches in MANETs.

INTRODUCTION

With the proliferation of wireless and pervasive communication technologies, the number of mobile devices such as mobile phones, laptops, PDAs etc. is growing by each day. These devices can be used to form *Mobile Ad Hoc Networks (MANETs)*. As a result research communities saw the potential use of MANETs. MANETs are special as they don't need any infrastructure for communication and thus have potential use in emergency scenarios such as military and disaster relief operations. The recent research (Kristiansen et al., 2010) trend in MANET is towards streaming that allows playing multimedia while it is downloading. Unlike file sharing systems, streaming requires continuous data delivery. Each data packet should be delivered within a strict deadline to guarantee a smooth playback. This is very challenging in a MANET because of frequent changes in topology that take place due to node mobility, high transmission errors due to channel fading and interference, and network partitioning.

P2P networks are developed by different communities from MANETs. They are highly successful in the Internet due to their special

DOI: 10.4018/978-1-4666-1613-4.ch005

characteristics such as decentralization, distributed design etc. These make them very popular for bandwidth intensive tasks such as video streaming. Thousands of users can simultaneously watch live broadcasts of popular TV programs using *Peer-to-Peer (P2P)* technologies (Hei et al., 2006). Despite the fact that they originated from different communities, MANET and P2P share many similarities. By seeing this, the obvious research question is whether we can use a P2P approach for streaming applications in MANET. Applying off-the-shelf P2P approaches to MANET is not efficient as Internet operate differently from MANET. The physical topology must be taken into consideration while using P2P in MANET. There are many other challenges and issues to be dealt with before employing P2P approaches to MANET. Video coding, network coding and cross-layers are the main issues to be considered.

The chapter is organized as follows: we start by giving a short overview of the P2P networks and we remind the reader the basic concepts of MANET in the next section. In the following Section, we introduce video streaming, its types and requirements. We also discuss several approaches used for video streaming over MANET with examples. We then talk about P2P architectures used for video streaming and explain some research works for P2P based video streaming in MANET. Following that, we talk about the challenges of P2P video streaming over MANET. Adaptations and issues are then discussed. We provide future research directions before concluding remarks. References and additional reading sections are given at the end of this chapter.

MANET AND P2P

In this section we review the basic concepts of P2P networks like their organization and functioning, then we present the MANETs and their characteristics, before discussing the convergence issues between these two technologies.

Peer to Peer (P2P) Networks

The birth of Napster paved the way for exponential growth of the Peer-to-Peer (P2P) technology (Brosnan et al., 2011). Consequently, Internet users have benefitted from many decentralized P2P file-sharing programs like Kazaa, Limewire, iMesh and so on. Its popularity and repercussions motivated researchers to conduct a lot of research not only in file sharing but also in telephony applications. For example, Skype is a well known Internet telephony application which uses P2P network to deliver telephony services (Baset & Schulzrinne, 2006). These P2P applications generate the majority of IP traffic in the Internet (Eberspächer et al., 2004).

Unlike traditional server/client based network where some nodes (servers) are responsible for serving others (clients), P2P consists of a network of nodes in which each node has equal responsibilities. It means each node in a P2P network can start request like a client as well as serve to other peers like a server. Thus, peer node is often called "servent", meaning server plus client. Due to these peculiar characteristics, P2P applications do not limit only to file-sharing applications. Content distribution and streaming applications use P2P network. One such example is P2PTV. It is an application designed to redistribute video streams or files on a P2P network. Similarly, BitTorrent [1] based applications such as QTorrent, μTorrent, FlashGet etc. are very popular for distributing contents among a large number of users.

P2P networks can be classified as unstructured and structured (Eberspächer et al., 2004). In structured P2P networks, the topology of the network and location of the content is determined by the employed P2P protocol. In other words, the P2P protocols determine how a peer is connected to other(s) in overlay network and distribution of content among the peers. In unstructured P2P networks, nodes and contents are randomly distributed. Searching is conducted using flooding. Thus, searching for nodes and contents is difficult

Figure 1. A hybrid P2P network

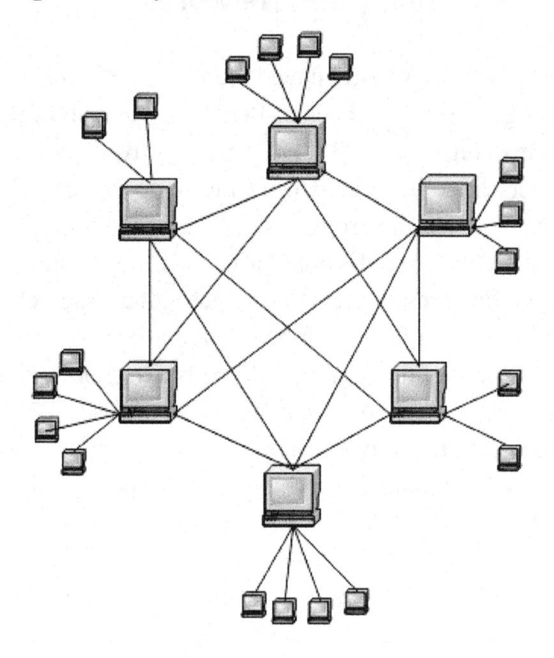

in the case of unstructured P2P network. In both unstructured and structured P2P systems, each of the participating nodes (i.e. peer nodes) points one or more nodes which form an overlay network. *Distributed Hash Table (DHT)* is used to make a structured overlay network. Some popular DHTs are Content Addressable Network (Ratnasamy et al., 2001), Chord (Stoica et al., 2003) and Pastry (Rowstron & Druschel, 2001). DHT provides efficient "content" based routing and searching. A typical P2P network can be seen in Figure 1 where connections between peers are represented by lines.

P2P networks are very popular for video streaming. IPTV applications such as PPLive[2] use P2P technologies. More than 100, 000 simultaneously online users are observed on PPLive for a live broadcast of a popular TV program (Hei et al., 2006).

Emergence of MANET

Like P2P applications, wireless technology is another popular technology. Use of small portable devices (like laptops, PDAs) is increasing due to the drop in prices of the hardware technologies. Since wireless technologies are useful in mobile environments, they are the best solutions to provide communication in many situations. Both short-range and long-range communications are possible using wireless networks. Many hand-held devices now can support very popular wireless technologies like Bluetooth and Wi-Fi. Therefore, the popularity of wireless technology has been increased further.

Two types of wireless networks exist: network with and without infrastructure. Cellular network is infrastructure network since it uses the pre-established network (towers, satellite) for communication. On the other hand, a MANET does not need the infrastructure for the communication. It is either inconvenient to make an infrastructure for the scenario like connecting MP3 players with laptops or impossible to make such infrastructures as in a battlefield. Also, these types of networks can be formed anywhere and anytime. MANET can be deployed quickly, easily and at a minimal cost. These are the reasons behind the popularity of MANET. A typical MANET with five nodes can be seen in Figure 2.

Figure 2. A mobile ad hoc network

In other words, a MANET can be seen as a collection of mobile autonomous nodes that use radio medium to communicate with each other. The nodes use the same radio channel and usually need to organize the channel access to avoid collisions (e.g. the Carrier Sense Multiple Access (CSM) channel technique (Tanenbaum, 2003)). They are free to move without any constraints and there should be no central entity or control over the nodes in the network. These assumptions of mobility and the autonomy of the nodes imply that the resulting network topology may change without any prior indication as the nodes move, leave and join the network. Topology change could be also the result of an immediate environment change or radio interferences of other distant transmissions. Remember that radio transmissions are very sensitive, and subject to fading with distance.

The broadcast nature of a radio transmission means that all the nodes within its reach will receive it. It means also, that when two distant nodes want to communicate, they will need that some intermediate nodes to play the role of a router and relay the data from the source to the destination. Routing protocols will be in charge of this task, and will insure the multi-hop communication by transmitting the data hop by hop through the intermediate nodes.

Basically routing protocols can be classified into reactive protocol family, and proactive protocol family (Giordano, 2002). In the first class of protocols, nodes will look for routes on demand only when they have data to send. This process consists of, first, the flooding of route request message to the entire network by the source node, and second, route reply messages generation (sent back to the source) by the nodes which have a valid route to the destination. The propagation of route reply messages will indicate or establish (depending on protocol) the route to reach the destination. In the second protocol, a family nodes will exchange topological information periodically. The collected information allows all the nodes to calculate and provide all the routes toward all the destinations. Routes are available immediately for the application layer. The price is a continuous consumption of a constant portion of the available bandwidth even when there is no application data to exchange between the participating nodes.

Notice that to insure the connectivity between distant nodes, the routing protocol reactivity (i.e. topology update diffusion within the network) must be equal or higher than the speed of the topology changes; otherwise the routing protocol will use stale information, and consequently fails to calculate valid routes.

The MANET IETF working group was created to develop routing protocols for mobile ad hoc networks. Different informational RFC documents for some routing protocols have been already published such as the OLSR and AODV routing protocols. The standardization process is still ongoing; the working group will develop two standards track routing protocol specifications:

- Reactive MANET Protocol
- Proactive MANET Protocol

The interested reader can find a comprehensive overview and discussion on MANET routing protocols in Giordano (2002).

MANET & P2P CONVERGENCE

Both P2P and MANET systems share many common characteristics (Shollmeier et al., 2002). For example, both systems have self-organization and decentralization properties. These systems have differences too. In this section we review the differences and the similarities between P2P and MANET networks.

Differences between P2P and MANET

Motivation in creating network: P2P is data driven, meaning that peer joins in P2P networks to search data and doesn't necessary want to communicate to a particular node. Usually the same data could be available in many nodes, and the downloading target is decided using a particular policy such as the available bandwidth of the node. All the effective data exchange is guaranteed by an underlying infrastructure. However, the first motivation for a MANET is to provide wireless communication for a mobile group of nodes. Routing protocols enable multi-hop communication and data exchange.

Focus on different Layers: P2P refers to application layer where as MANET focuses on network and lower layers in protocol stacks. P2P network can be seen as an overlay network that uses the network layer information to build up its connectivity among the participants. The resulting topology is made of a set of tunnels that link different nodes.

Different Constraints: MANET peers are mobile, using unreliable radio links, constrained by limited energy and computation power. In P2P networks, we do not consider these issues, and we suppose that we have unlimited computation power, unlimited energy, and using a reliable communication medium.

Different Broadcast Mechanisms: Since a P2P overlay is a single cast network, execution of broadcast in P2P is always virtual. It means that broadcast in P2P overlay consists of a number of single cast messages via tunnels. However, MANET always performs a physical broadcast. The transmission reaches all the neighbor nodes located within the radio range of the originator. Moreover, the MANET networks suffer from the well-known problem of broadcast storm, which imply high bandwidth usage and a large number of collisions when a packet is flooded within the network (Ni et al., 1999).

Similarities between P2P and MANET

Despite these differences, MANET and P2P share many things in common. They are given below:

No central server: Both systems do not have peers which act explicitly as a central server. To work by the whole system, each node should collaborate with other peers.

Common Major Problem: Enable communication and data exchange between distant nodes. For P2P nodes have to locate the requested data and to download it. For a MANET network, nodes have to calculate routes efficiently which is actually the common major problem of both systems.

Dynamic Topology: Nodes movement changes topology. Nodes can join, or leave the network, hence new adjacencies or neighbors are created or can disappear that give a dynamic topology.

Basic Routing Principle: Broadcasting (Virtual in P2P and Physical in MANET), flooding is the basic routing principles in both systems.

Connection Establishment: Hop by hop. In P2P uses TCP links and MANET uses via radio links.

Limit on scalability: P2P scalability is limited by flooding of bandwidth consuming signaling traffic and in MANET it is also limited by flooding of signaling traffic and additional physical constraints.

These similarities motivate researchers to enable P2P network over MANET. P2P mechanisms are complementary to MANET routing techniques in order to distribute and exchange data contents, and enable P2P video streaming.

MANET is the best solution for many group communications such as occurs in the battlefield and disaster relief areas. In the case of disaster relief operations, many teams are formed for searching for survivors, distributing foods and supplies, providing medical services, constructing buildings, providing security services and so on. In all these scenarios no infrastructure will be available for immediate use; MANET networks, thus, offer a communication platform for group communication of great help to coordinate the work of different stakeholders in the field.

However, sharing information among nodes in resource-limited MANET is very important. Consider a scenario where a commander in a battlefield wants to see a live video stream of the area surrounding his soldiers, who have the portable devices with camera support. To get the video contents, he uses the P2P application running on his device to search the solders' devices that are enabled for streaming. Based on his interest, the commander selects one of the soldier's cameras to watch the surrounding environment. If any other person is also interested in watching the same video, he also follows the same procedure and joins the P2P network. Interested nodes share information with each other and, thereby, reduce the server's load. Since nodes in MANET are constrained by energy consumption and bandwidth capacity, efficient streaming protocols are always demanding. This chapter focuses on P2P streaming technologies for MANET in order to share resources and loads among MANET nodes.

The next section is dedicated to explain the video streaming techniques used in the Internet and the different approaches used in MANET to provide such a service.

VIDEO STREAMING

Streaming is a technique for transferring data in such a way that it can be processed at a steady and continuous stream. Users don't need to wait for the complete media file (audio, video, animation) to be downloaded before playing. For instance, streaming allows user to watch a movie while they are downloading its later portion, which could take a long time. Also, there are many applications such as distance learning, telemedicine, and video-on-demand that use multimedia streaming.

Types of Streaming

Generally, we can divide streaming applications into three classes (Kurose & Ross, 2005):

1. Streaming Stored
2. Streaming Live
3. Real-time Interactive

In the first class of streaming, the multimedia is stored at source and transmitted to client when it requests to the server. It begins playing before all data arrives. Transmitted data from the server should arrive before the scheduled time; hence, time constraint exists while playing the stored multimedia. Watching a movie on the Internet, watching some stored videos are examples of this category. Client can pause, rewind, fast forward, push slide bar and so on.

In streaming stored video, users can fast forward the media. However, this is not possible with streaming live multimedia applications. Internet radio talk shows and live sporting events are typical examples of streaming live multimedia. In such examples, user cannot fast forward but rewind and pause are possible. Similar to stored type of streaming, live streaming still has timing constraints. Client uses a playback buffer to improve continuous playout of the media. The playback can lag tens of seconds after the transmission.

The third type of streaming multimedia is real-time interactive multimedia. IP telephony and video conference are two well-known examples of this category. In the first two classes, the data is sent from a server to a client. In contrast to both classes of streaming, it allows both sender and receiver to transmit data.

In on-demand streaming (stored streaming), multimedia content of a known size is available at the source, however, in the case of the second and third classes of streaming, the data stream has unpredictable length and the data are available for a small period of time. Also, start-up delay is the most important challenge for the first class of streaming and playout delay is the most important challenge for the second and third type of streaming.

Delay of a packet is the difference between the received time and the corresponding send time. It is caused by many factors, for example, packetization, propagation, queuing in the network etc. Different packets, thus, have different delays.

Jitter is another quality measurement for streaming. Jitter is caused by the variation of the delays and plays a significant role in streaming. Perfect streaming has jitter value equal to 0 meaning that all packets reach the destination with a constant delay. A high value of jitter is not good for streaming because it reduces the smoothness of the playback.

Packet delay and jitter can be seen in the Figure 3. Clearly, if a client starts playing the video as soon as the first packet arrives, then it has to suffer from a stoppage problem because the next packet might not arrive after a constant delay. Thus, jitter should be reduced in order to provide a better streaming quality.

Generally, a jitter buffer is used to reduce the jitter effect. As shown in Figure 3, if a client buffers the video for the playout time and starts playing it, it gets a constant bit rate output although there exists a jitter in the network. The size of a jitter buffer depends upon the type of the streaming technique. Stored playback applications can allow for larger jitter buffers while live streaming and live interactive streaming cannot afford it because this increases the end-to-end latency.

VIDEO STREAMING IN MANET

Common Approaches

As discussed previously, video transmission in an ad-hoc network is challenging due to the high bit error rates ranging from single bit errors to burst errors or frequent loss of the connection. These errors are due to multi-path fading and loss of connection is due to the mobility of nodes in the networks.

To provide continuous video transmission over such networks, server diversity and path diversity schemes are used.

Server Diversity

Server diversity is also called multipoint-to-point communication and resembles to multi-sources architecture. In this approach the same video content is available in many peers. Thus, it can

Figure 3. Role of jitter buffer

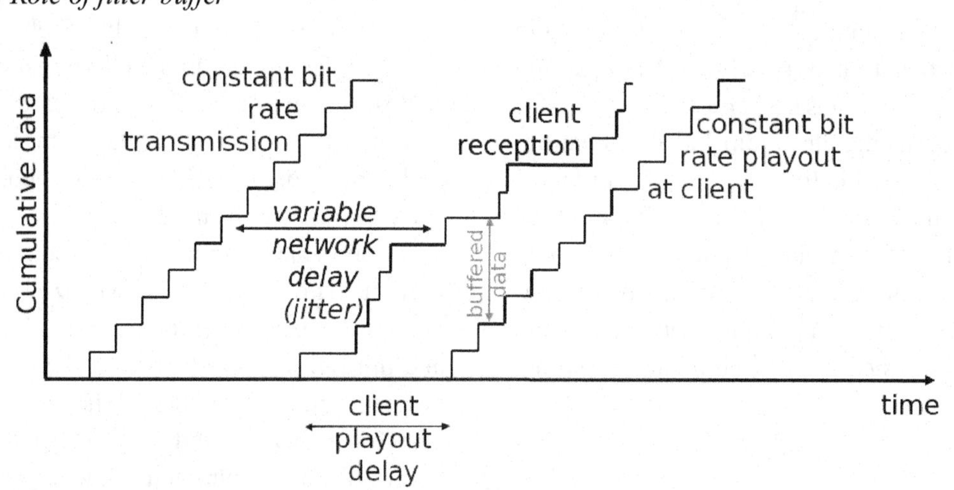

Figure 4. Server diversity scheme

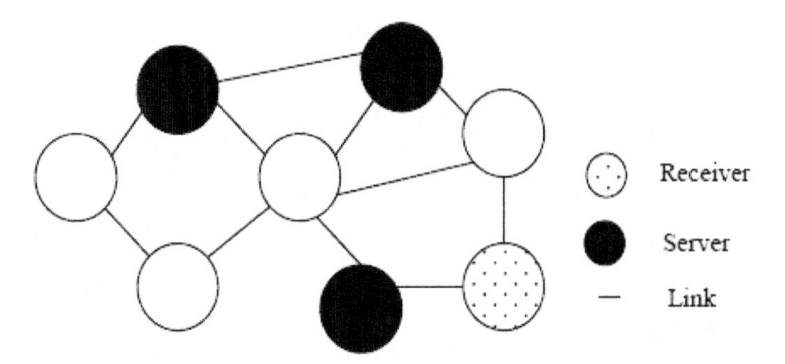

be sent to the receiver(s) using one of them. If any one of the servers fails, another will provide the content. The scheme is shown Figure 4. There are three MANET nodes, represented by filled circles, which have the same video content. There is a single receiver, represented by partially filled circle.

Example of Server Diversity

Chow and Ishii (2007) enhanced real-time video streaming using multipoint-to-point (server diversity) communication for MANET. If the same video is contained in two nodes, then the nodes are called source nodes. Each source node sends only a portion of the video coded using Multiple Description Coding (MDC) (described later) scheme. There are three important stages involved:

i. *Source Searching Stage*: In this state, a receiver searches for video sources. In the proposed model, the receiver is responsible for selecting the best video sources if many sources are available. The receiver informs the selected source about the description it needs to provide. Searching is accomplished using PUSH and PULL methods.
ii. *Route Establishment Stage*: The authors modified the *Dynamic Source Routing (DSR)* (Johnson & Maltz, 1996) protocol (they called it "extended DSR"), which is used to find the routes between source and receiver, to get optimally disjoint routes for each description of MDC coding schemes. Disjoint routes are necessary to cope with the problem caused by link breakage.
iii. *Video Streaming Stage*: This stage starts as soon as a valid route is obtained. The sender uses MDC encoder to generate descriptions (number of descriptions to be generated is sent by requester with the feedback information during source searching stage). Then, the traffic selector sends one description as assigned by the receiver. The receiver collects the descriptions and puts in a playout / resequencing buffer to reduce the frame jitter and reorganize the frames into the proper sequence before sending them to the MDC decoder, which also performs the frame compensation too.

By simulation, the authors show that the number of transmission points should be limited because too many transmission points increase the network overhead. They propose that two or three transmission nodes are practical numbers for nodes to get maximum streaming efficiency with acceptable control overhead.

Path Diversity

Since link failures are very common in MANET, the use of multi-path to send streaming data improves the quality of video streaming. The simplest

Figure 5. Path diversity scheme

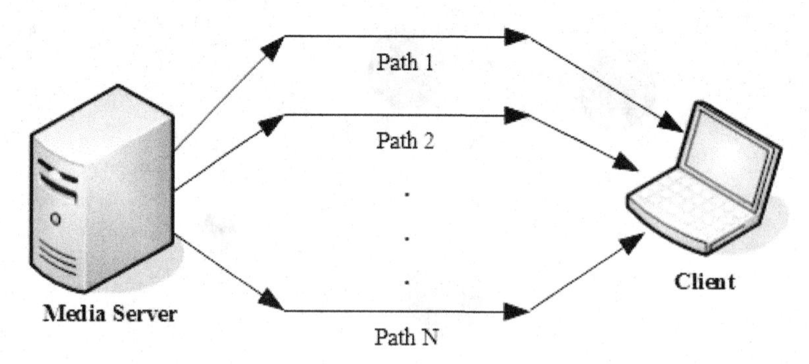

model includes sending a stream through multiple paths. If any of the streams reaches the destination, the video can be played. This simplest solution is not efficient because the same stream is sent via different routes. However, the use of MDC together with path diversity solves the problem. Each sub-stream from MDC is sent through a different path. If any of the sub-streams is lost in the path, it won't much affect the video playing; it only degrades the video quality. A simplified diagram of a path diversity scheme is shown in Figure 5.

Example of Path Diversity

Xu and Cai (2007) claim that existing routing protocols (either reactive or proactive) do not support live streaming applications properly. Routing overheads make proactive protocols inefficient for streaming applications. Furthermore, delay in finding routes causes a problem in reactive protocols. To solve these problems they proposed a *proactive link protection (PLP)* and *receiver-oriented adaptation (ROA)* based routing protocol. PLP is used to □nd an alternative link in a streaming path before its current one becomes broken. Thus, PLP reduces the probability of having to find a new path to urgently repair a broken link. ROA is used to improve streaming efficiency by reducing the hop number of a streaming path.

P2P APPROACH

Use of P2P for video streaming in MANET is an interesting and challenging research area to be explored. In this section we discuss more about this issue, and we start by presenting a classification of P2P streaming architectures. According to Liu et al. (2008), P2P streaming systems can be classified into two categories: *tree-based* and *mesh-based.*

Tree-Based Architecture

The tree-based architectures have a well-organized overlay structure. Peer nodes are ordered hierarchically with the root being the source node of the P2P system. The source node distributes video to its children. The child nodes distribute video to their children and so on. A tree-based architecture can be either single-tree or multi-tree.

Single-Tree Streaming

In a single-tree streaming architecture, user nodes participating to a video streaming session form a tree at the application layer by creating an overlay tree. Each participating node joins the overlay tree at a certain level. A typical example of single-tree streaming formed by ten nodes can be seen in Figure 6. There are two peers at level 1, 4 in level 2 and 4 in level 3. Peers at level 1 receive the video directly from the source whereas the peers at level 2 and level 3 receive from their parents respectively.

Figure 6. Single-tree architecture

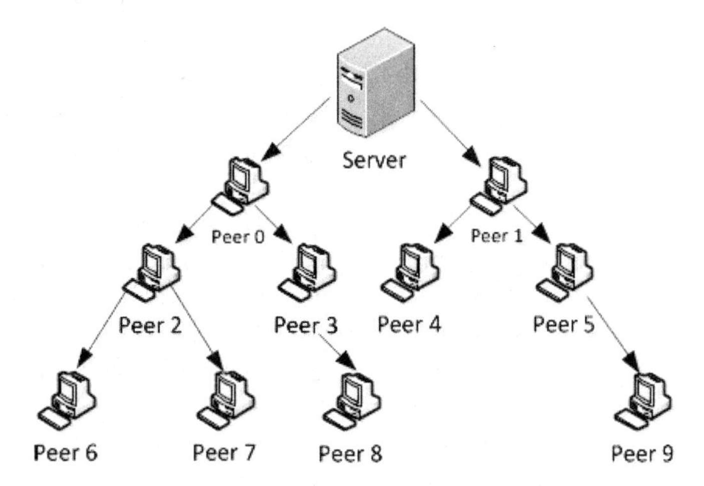

The single-tree architecture suffers from the tree construction problem and the tree-maintenance problem. Another major drawback is that leaf nodes are pure receivers i.e. they do not contribute their uploading bandwidth. This degrades the peer bandwidth utilization efficiency as leaf nodes account for a large portion of the peers in the single-tree architecture.

Multi-Tree Streaming

The multi-tree streaming architecture is proposed to overcome the problem associated with leaf nodes in single-tree streaming. Source node divides the video into a number of sub-streams. Peers form different trees for each sub-stream. A peer that appears as a leaf of a tree could appear as a non-leaf peer in another tree. Thus, utilization of peer bandwidth is improved. Figure 7 shows

Figure 7. Multi -tree architecture

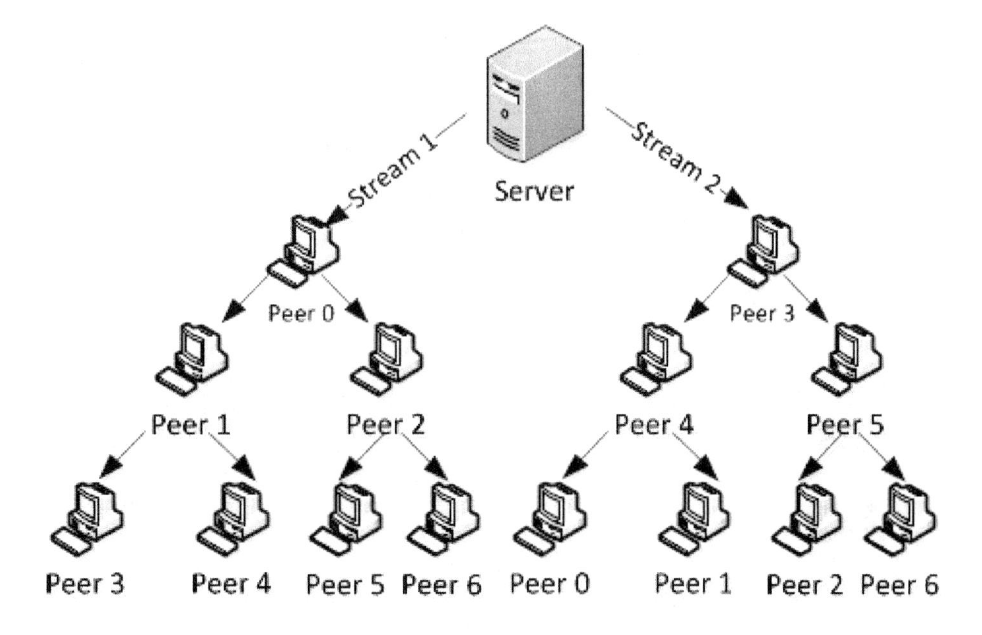

seven peers participating to two sub-stream trees for stream1 and stream2 respectively. Peer0 always receives stream2 but contributes stream1 to peer1 and peer2.

Mesh-Based Architecture

The tree based architecture has a number of drawbacks. The source node is a single point of failure and limits the robustness of the system. Moreover, if a peer's parent leaves, it not only affects the immediate peer but to all its descendants because a peer node receives a video only from its parent. Peer dynamicity causes a tree to be rebuilt too often which impacts negatively on signaling overheads, latency and stability.

In mesh-based architecture, peers don't have a static streaming topology. They establish and terminate relationships dynamically. Video to be shared is divided into chunks and distributed among the peers. Peers cooperate to assemble the original data. Thus they can upload/download a video from multiple neighbors simultaneously. This provides robustness to the system against peer churn. Research study has shown that mesh-based streaming performs better than tree-based streaming (Magharei et al., 2007).

In a P2P system, peers join and leave frequently. However, the volatility is much higher in a MANET because of mobile nodes. Frequent construction of trees due to dynamicity increases overheads on signaling, latency and stability. Thus tree-based P2P systems are inefficient for a MANET. As a result, mesh-based P2P streaming is of interest for MANET researchers. By simulation study Quadri et al. (2008) found that highly dense MANETs are able to sustain P2P streams.

Niraula et al. (2009) proposed a *Cross-Layer and P2P based Solution (CLAPS)*, a mixture of a tree- and mesh-based streaming architecture. In their design, a number of interested nodes want to watch a video from a source node as shown in Figure 8. There are 18 mobile nodes (nodes 1 to 17 and a source node A) out of which 7 peer nodes

Figure 8. CLAPS based video streaming over MANET

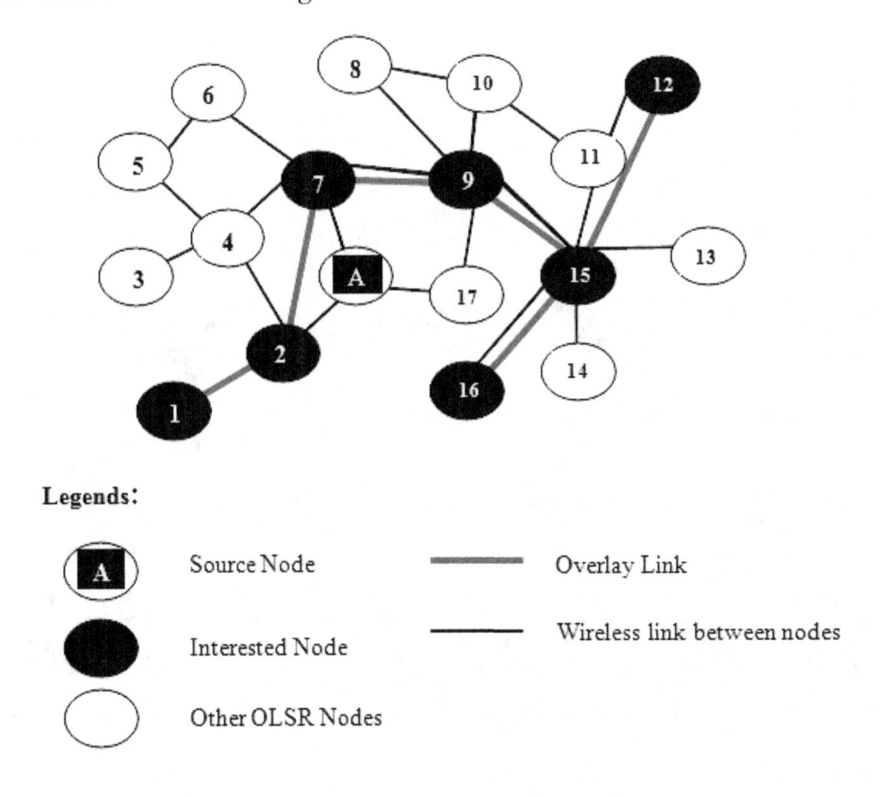

(1, 2, 7, 9, 12, 15, and 16) are interested to watch the video at the source node. The interested peers form an overlay (the red line), a minimum spanning tree, using the Multicast Overlay Spanning Tree Protocol for Ad Hoc Networks (Rodolakis et al., 2007).

The server node divides the video into a number of streams using multiple description coding. The server then divides each video stream into chunks. A peer node is randomly selected for each chunk and the chunk is sent to it. Once a chunk is received by a peer, it distributes the chunk among other peers using the overlay tree. Even if a chunk is missing in a peer, the same chunk of another stream could already be there to play. They found by simulation study that the continuity index, an index to measure the quality of streaming applications, of the video stream was significantly better than in a simple tree based system.

In this section we have described some proposed approaches from the literature for video streaming in MANET. We started by giving a classification of the common approaches, then we give an overview of some P2P based system for video streaming. In the next section, we discuss the challenges that must be taken into account and should be overcome in order to build an efficient P2P video streaming system over a MANET.

Challenges of P2P Video Streaming over MANET

A P2P solution over MANET is not as straightforward as in wired network because of its unique working principle. Any P2P streaming solution over MANET faces three important challenges: inefficient usage of link layer connection, high routing overhead and frequent loss of application layer connection. In this section, we explain these challenges with examples.

Inefficient Usage of Link Layer Connection

The first challenge is clearly the lack of effective topology knowledge at the P2P application level. The absence of such information leads to an inefficient usage of link layer topology, and, hence, to bandwidth wastage. Remember that contrary to wired networks, bandwidth is a scarce resource in a MANET so that the usage should be optimized as much as possible. To understand the impact of not using topology information in P2P application level, let's consider a P2P network formed in a MANET as shown in Figure 9. The physical topology is shown on the right side and the overlay network they formed is shown on the left side. Let's say node N6 wants to send a

Figure 9. Internet P2P approach is inefficient in MANET

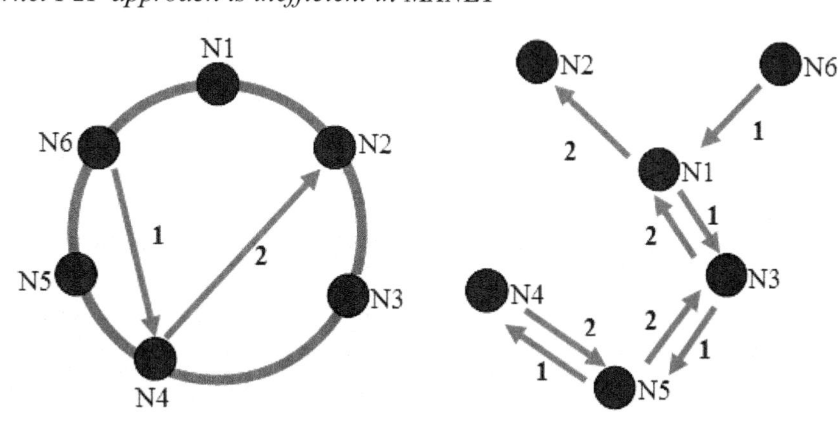

message to N2 and computes N6-N4-N2 path in the overlay network. Thus, the message is first sent from N6 to N4 and then N4 forwards it to N2. The corresponding path is a zigzag one in the physical topology. An important point here is that the data to be sent from N6 to N2 are sent multiple times over physical links between N1 and N4, which is not good. To be efficient, it should have been routed via N6-N1-N4 in the physical topology so that it would have avoided the unnecessary communication between N1 and N4. Thus, considering only the overlay network is not sufficient for a MANET.

High Routing Overhead

Nodes are mobile in MANET. This mobility creates extra work in the network layer to maintain the routing. To see this effect, consider a MANET as shown in Figure 10 where thick lines represent application layer connections and thin lines represent link layer connections. Further, consider that originally node E is in the lighter circle. In that situation, node D and node F are connected by D-E and E-F link. When a mobile node E goes to its current position, a route discovery should be started to maintain the connectivity between node D and F. This route discovery causes a lot of routing overhead. Routing overhead is a serious problem to maintain static overlay network (Klemm et al., 2004).

Figure 10. Application layer VS link layer connections

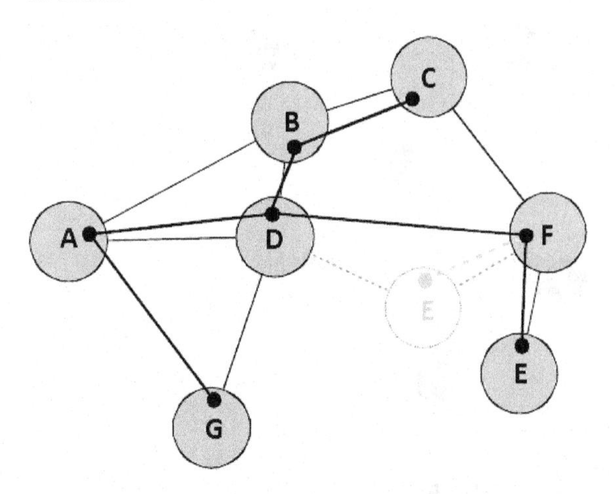

Frequent Loss of Application Layer Connections

Applications are vulnerable to the link layer connection failure. Loss of link layer connection may cause application failure. To see this effect, consider a case, in Figure 10, when link layer A-D and A-B is lost. The link failure disconnects node A from the network. Node G cannot remain in overlay network although it has a connection to node D because G is connected to A in the overlay network. To join in overlay network, it has to restart peer searching.

In order to build an efficient P2P video streaming over MANET, these issues must be addressed. This means that off-the-shelf P2P systems cannot be implemented efficiently in MANET. P2P applications over MANET therefore must utilize the underlying topology information. We'll discuss more about this in the following section.

ADAPTATIONS AND ISSUES

In this section, we present the adaptations and issues related to P2P video streaming in MANET. Specifically we discuss on cross-layer design, video coding and network coding.

Cross-Layering

Eberspächer et al. (2004) provides reasons to support why strict layering of the structured overlay approach on top of wireless routing protocols is unlikely to work. Firstly, the strict layering would require proactively maintaining of the overlay links that would be multi-hop in the physical network. Secondly, keeping the network connected would require a significant amount of traffic to detect and repair failed overlay links. The same

Figure 11. Cross-layer design framework for media streaming over ad hoc wireless network

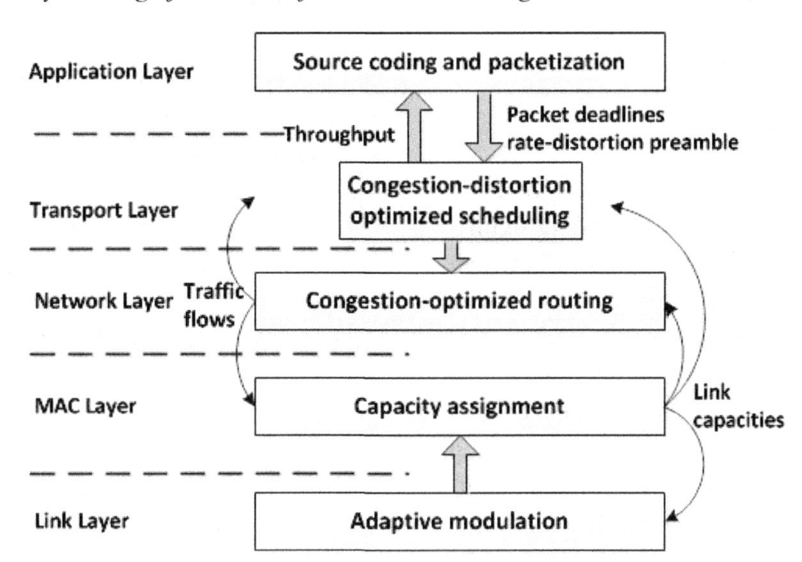

reasons are applicable to unstructured P2P overlays too. Information sharing among layers may improve the efficiency.

As there is no established infrastructure in MANET, it suffers from unprecedented challenges caused by channel dynamics and nature of the wireless medium. As a result, picture quality of video streaming suffers greatly. The problem could be handled by cross-layer design of protocol stack which helps by optimally adopting the underlying channel conditions and specific requirements.

Although traditional network design, which is based on protocol layering, provides important abstraction, it is not well-suited to wireless networks as the nature of the medium makes it difficult to decouple the layers (Setton et al., 2005). Interaction between protocol layers is necessary to meet the end-to-end performance requirements of video-streaming applications. The information exchange in cross-layer design increase robustness. For example, application layer can adapt the transmission rate based on the underlying network throughput and latency. Routing protocol can avoid links having deep fades. The information to be exchanged between the layers is a crucial thing to be considered while designing cross-layer architecture.

A cross-layer design for low-latency media streaming for ad hoc wireless network is proposed by Setton et al. (2005) and can be seen in Figure 11. In this design, adaptive techniques are used at the link layers to maximize the link rates under varying channel conditions. The MAC layer uses link-state information to select one point of the capacity region. The network layer and MAC layer works together to minimize congestion. The transport layer runs congestion-distortion optimized scheduling to control the transmission and retransmission of video packets. The application layer identifies the most efficient encoding rate.

The authors revealed that live video streaming was better in a cross-layer architecture than in strict layering of protocol stacks. An interesting use of network layer information by application layer for streaming in MANET is described by Niraula et al. (2009). In their design, source node uses routing table information while distributing the video pieces to the peers.

Although cross-layering architecture improves live video streaming quality, it is more complex to implement than a layered architecture. Moreover, design advantages provided by layering should not be eliminated by cross-layering.

Network Coding

One of the problems to be addressed in a MANET is the packet loss. Use of network coding significantly improves the packet delivery ratio which in turn improves the video quality.

The principle behind the network coding is that instead of sending packets separately and independently, a sender transmits a linear combination of packets (Chen et al., 2006). Receiving such packets with all linearly independent coefficients helps a node to recover or decode the content of the original packet. The idea can be understood by looking at the Figure 12. The line between each pair of node means that they are in the transmission range of each other. CPR refers to co-operative peer-to-peer where peers repair the packets by sharing their packets. In NC-CPR, packets are repaired using network coding. In the figure m1, m3, m3, m2 packets are missing in nodes n1, n2, n3 and n4 respectively. Using CPR, it needs 4 rounds to repair the packets where as NC-CPR needs just two rounds.

Chen et al. (2006) are motivated with this idea and proposed Codecast, a network-coding-based ad hoc multicast protocol. They found that CodeCast significantly improved the packet delivery ratio. It is also found that network coding is especially well-suited for multimedia applications with low-loss, low-latency constraints such as audio/video streaming.

Liu et al. (2008) extended the use of network coding in cooperative peer-to-peer repair in wireless ad-hoc networks and found it to be effective. As multiple path schemes are popular for video streaming in a MANET, use of network coding is very useful in such an environment.

Video Coding

As mentioned earlier, video applications have strict requirements on bandwidth, delay and jitter. Packet loss and error propagation greatly reduces the quality of video at the receiver end. Video coding is another technique used to handle this issue. In a conventional method, video is encoded into a single stream of video frames in order to make it suitable for storage and transmission. Generally, it uses motion-compensation prediction between frames, block *discrete cosine transform (DCT)* for error prediction and entropy coding (Chow & Ishii, 2007). If a single frame is lost in

Figure 12. Advantage of network coding for cooperative P2P recover

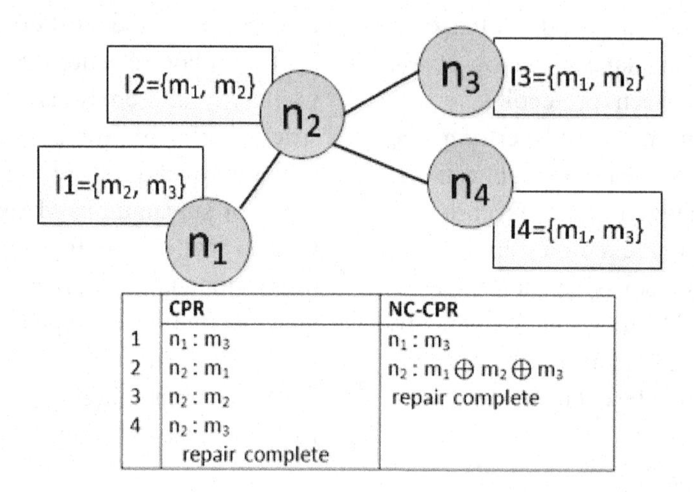

	CPR	NC-CPR
1	$n_1 : m_3$	$n_1 : m_3$
2	$n_2 : m_1$	$n_2 : m_1 \oplus m_2 \oplus m_3$
3	$n_2 : m_2$	repair complete
4	$n_2 : m_3$ repair complete	

Figure 13. Layered coding scheme for video encoding

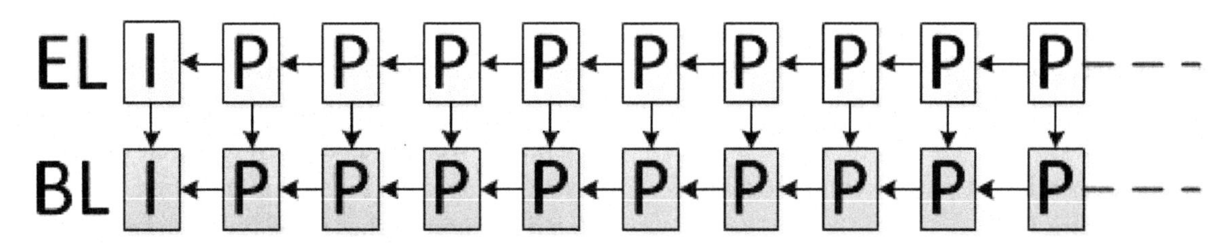

this scheme, it affects not only that frame but its subsequent frames too.

To solve this problem, many other encoding techniques are developed. Commonly used techniques are *Multiple Description Coding (MDC)* and *Layered Coding (LC)* (Chow & Ishii, 2007).

Layered Coding

Layered coding divides the video stream into two or more layers: the base layer and the enhancement layers. The logic behind (say) two layers is that a video can be played if a user gets only the base layer but with lower quality. Enhancement layers are used to increase the quality of the video. In the normal case, both layers are sent to a client. In case there exists some constraints such as bandwidth, only the base layer is sent.

Fig. 13 shows a layered scheme for video coding. Here, only two layers are generated. BL and EL stand for base layer and enhancement layer respectively. Although this method is robust, it has the same problem as the conventional one. For example, if the base layer is lost, there is no use of enhancement layers. MDC is used to solve this problem.

Multiple Description Coding

MDC splits the multimedia stream into many descriptions. A raw video stream is divided into several sub-streams before it is given to the video encoder. A simplified approach is shown in Figure 14. Then, the sub-streams are encoded independently to obtain several video descriptions. The encoder may be H.264, MEPG-4 or any one that has better error resilience.

At the receiver side, each stream is decoded separately since they are encoded as independent streams. The decoding process is shown in Figure 15. If some frames in a stream are lost, the compensation scheme is used to compensate the loss. Suppose a video is encoded into two streams say Description 1 and Description 2 and consider

Figure 14. Encoding using multiple description coding

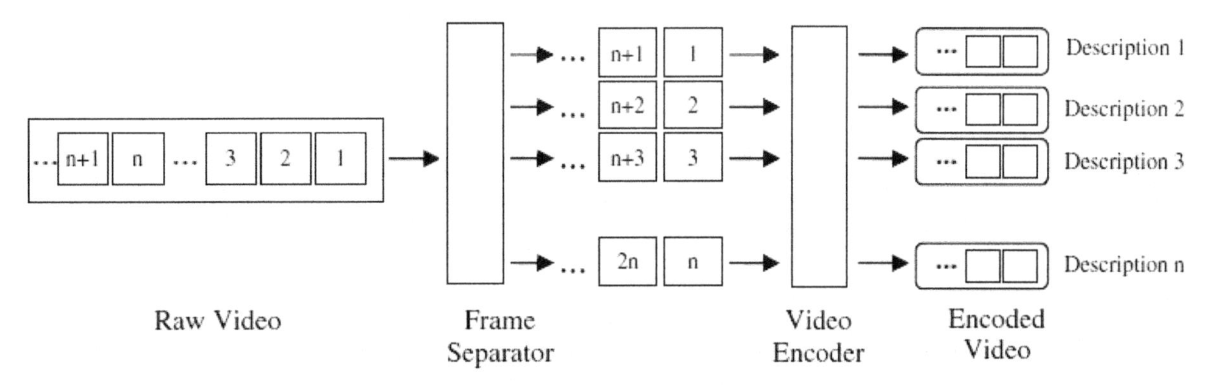

frame P5 is lost. Therefore, we cannot decode P7 and P9 as well since they depend on P5. However, it can be decoded after I11 since I-frame can be decoded independently. Lost frames P5 can be replaced by P4, P7 by P6 and P9 by P8 so that the video streaming is played continuously, as shown in Figure 16.

Besides the benefits of MDC, it has some limitations, too. Consider Table 1. We can see that the average frame size increases as we increase the number of descriptions for a given video. For example, if we use two descriptions instead of one, the average frame size increases by scaling factor of 1.124 i.e. by 12.4%. Furthermore, the increase in size is 20.1% and 26.00% for a number of descriptions equal to 3 and 4 respectively.

Both MDC and LC improve the video quality as the peer "connect probability" increases. If the probability is small, the performance of an MD system is better than the layered systems (Meddour et al., 2006).

FUTURE RESEARCH DIRECTIONS

Video streaming over a MANET is demanding and is still a challenging research problem. In the literature, we found very few research works that use P2P approach for streaming over a MANET. Available approaches are simulation based. They suggest that P2P technologies are useful for video streaming. Real implementation based analysis of P2P video streaming for MANET is still lacking. Thus, P2P based streaming over MANET has to be explored more and real implementation and analysis of such techniques are remaining as open research problems.

Although a significant number of published research papers focused on cross-layer issue in MANET, it is still an open research domain, as there is no complete solution which gives complete satisfaction. Network coding in MANET is also an important point to look at and could be an interesting research direction, because it offers redundancy in all transmitted packets no matter what is carried inside.

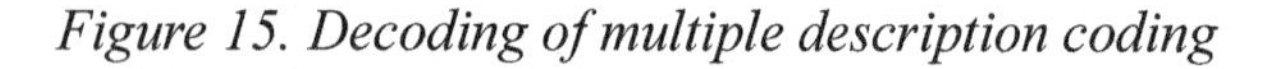

Figure 15. Decoding of multiple description coding

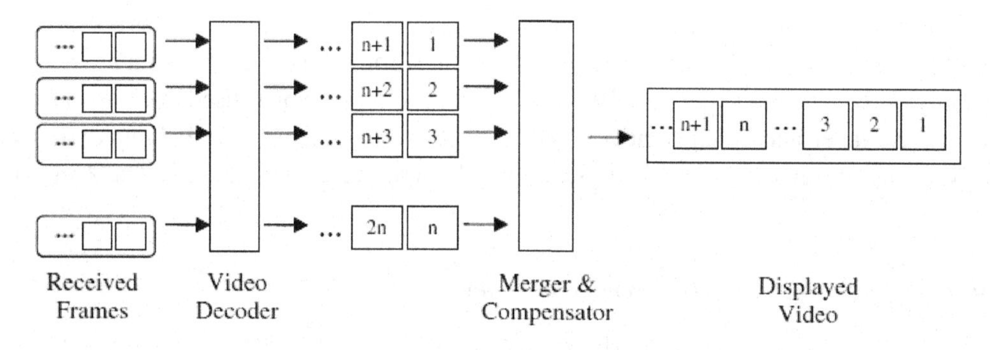

Figure 16. Frame compensation scheme with presence of frame lost

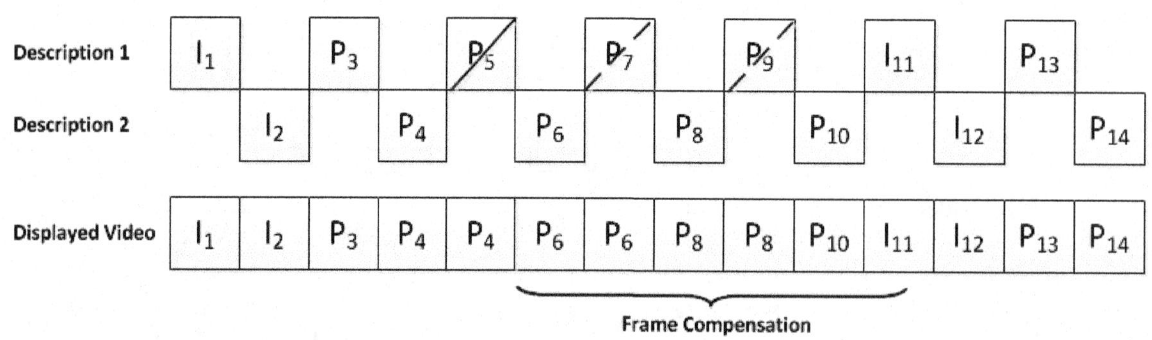

Table 1. Average frame size versus number of video descriptions

No. of Video descriptions (n)	1	2	3	4
Average frame size (bytes)	4142.93	4656.52	4977.62	5220.21
Scaling Factor	1	1.124	1.201	1.2600

CONCLUSION

Both P2P and MANET evolved independently. P2P is very successful in the Internet especially for file sharing and bandwidth intensive tasks such as video streaming. MANET evolved by its own but shares similarities with P2P. The key motivation of this chapter is to provide an in depth study of P2P over MANET for video streaming applications, the approaches, the challenges to be faced and the mitigating schemes.

We started by introducing the P2P and MANET concept, motivations behind them, their differences and commonalities, and the motivations behind using P2P streaming protocols over MANET. We presented by an example that off-the-shelf P2P applications are not suitable for MANET. They have to deal with the underlying network conditions to be efficient. We also introduced the video streaming concepts, their requirements and types.

Next, we presented the video streaming approaches in a MANET. Server diversity and path diversity schemes are the most common approaches researchers are using for this task. We explained by an example of each type.

We also described the P2P architectures. Tree-based and mesh-based architectures are two popular architectures for P2P. The tree-based architecture has limitations because it suffers from a single point of failure and limits the robustness of the system. Peer dynamicity is a big problem for such an architecture as the tree management (construction and maintenance) is inefficient. On the other hand, a mesh-based architecture utilizes the peer bandwidth efficiently. Moreover, it provides robustness to the system against peer churn. A mesh-based architecture is a suitable choice for a MANET when peer dynamicity is high.

Finally, we presented the challenges to be faced by P2P applications over MANET. They are: inefficient usage of link layer connection; high routing overhead; and frequent loss of application layer connection. Then, we discussed three important issues to be considered to address the problems of using P2P in MANET. The first one is the use of cross-layer design instead of strict layering. The second one is the use of network coding to improve the packet delivery ratio. The third one is to use the appropriate video coding scheme, especially MDC. We provided examples to illustrate the issues.

REFERENCES

Baset, S. A., & Schulzrinne, H. G. (2006). An analysis of the Skype peer-to-peer internet telephony protocol. *Proceedings - IEEE INFOCOM*, 1–11. doi:10.1109/INFOCOM.2006.312

Brosnan, A., Maitrat, T., Colhoun, A., & MacArdle, B. (2011, May 11). *Historical development*. Retrieved from http://ntrg.cs.tcd.ie/undergrad/4ba2.02-03/p1.html

Chakareski, J., Han, S., & Girod, B. (2003). Layered coding vs. multiple descriptions for video streaming over multiple paths. *Proceedings of Internation Conference on Multimedia.*

Chen, C.-C., Lien, C.-N., Lee, U., Oh, S. Y., & Gerla, M. (2006, October). Codecast: A network-coding-based ad hoc multicast protocol. *IEEE Wireless Communications*, *13*(5), 76–81. doi:10.1109/WC-M.2006.250362

Chow, C., & Ishii, H. (2007). Enhancing real-time video streaming over mobile ad hoc networks using multipoint-to-point communication. *Computer Communications*, *30*(8), 1754–1764. doi:10.1016/j.comcom.2007.02.004

Eberspächer, J., Schollmeier, R., Zöls, S., Kunzmann, G., & Für, L. (July 2004). Structured P2P networks in mobile and fixed environments. *Proceedings of the International Working Conference on Performance Modeling and Evaluation of Heterogeneous Networks.*

Giordano, S. (2002). Mobile ad-hoc networks . In Stojmenovic, I. (Ed.), *Handbook of wireless networks and mobile*. New York, NY: Wiley. doi:10.1002/0471224561.ch15

Hei, X., Liang, C., Liang, J., Liu, Y., & Ross, K. W. (2006). Insights into PPLive: A measurement study of a large scale P2P IPTV system. *Proceedings of the Workshop on Internet Protocol TV (IPTV) Services over the World Wide Web in conjunction with WWW2006.*

Klemm, A., Lindemann, C., & Waldhorst, O. (2004). Peer-to-peer computing in mobile ad hoc network . In Calzarossa, M., & Gelenbe, E. (Eds.), *Performance tools*. Berlin, Germany: Springer Verlag. doi:10.1007/978-3-540-24663-3_9

Kristiansen, S., Lindeberg, M., Rodríguez-Fernández, D., & Plagemann, T. (2010). *On the forwarding capability of mobile handhelds for video streaming over MANETs.* MobiHeld 2010 — The Second ACM SIGCOMM Workshop on Networking, Systems, and Applications on Mobile Handhelds.

Kurose, J., & Ross, K. (2005). *Computer networking: A top down approach featuring the internet* (3rd ed.). Boston, MA: Addison-Wesley.

Liu, X., Raza, S., Chuah, C.-N., & Cheung, G. (2008). Network coding based cooperative peer-to-peer repair in wireless ad-hoc networks. *IEEE International Conference on Communications*, (pp. 2153 - 2158).

Liu, Y., Guo, Y., & Liang, C. (2008). A survey on peer-to-peer video streaming systems. *Peer-to-Peer Networking and Applications*, *1*(1), 18–28. doi:10.1007/s12083-007-0006-y

Magharei, N., Rejaie, R., & Guo, Y. (2007). Mesh or multiple-tree: Acomparative study of live P2P streaming approaches. *Proceedings of the IEEE INFOCOM*, (pp. 1424-1432).

Meddour, D., Mushtaq, M., & Ahmed, T. (2006). Open issues in P2P multimedia. *Proceedings of, MULTICOMM2006*, 43–48.

Ni, S.-Y., Tseng, Y.-C., Chen, Y.-S., & Sheu, J.-P. (1999). The broadcast storm problem in a mobile ad hoc network. *Proceedings of the Fifth Annual ACM/IEEE International Conference on Mobile Computing and Networking (MOBICOM 99).*

Niraula, N. B., Kanchanasut, K., & Laouiti, A. (2009). Peer-to-peer live video streaming over mobile ad hoc network. *Proceedings of the 2009 International Conference on Wireless Communications and Mobile Computing: Connecting the World Wirelessly,* (pp. 1045-1050).

Qadri, N. N., Alhaisoni, M., & Liotta, A. (2008). Mesh based P2P streaming over MANET. *Proceedings of the 6th International Conference on Advances in Mobile Computing and Multimedia,* (pp. 29-34).

Ratnasamy, S., Francis, P., Handley, M., Karp, R., & Shenker, S. (2001). A scalable content-addressable network. *Proceedings of ACM SIGCOMM.*

Rodolakis, G., Laouiti, A., & Merarihi, N. A. (2007). *Multicast overlay spanning tree protocol for ad hoc networks*. International Conferences on Wireless/Wired Internet Communications.

Rowstron, A., & Druschel, P. (2001). *Pastry: Scalable, distributed object location and routing for large-scale peer-to-peer systems*. International Conference on Distributed Systems Platforms (Middleware).

Setton, E., Yoo, T., Zhu, X., Goldsmith, A., & Girod, B. (2005). Cross-layer design of ad hoc networks for real-time video streaming. *IEEE Wireless Communications Magazine, 12*(4), 59–65. doi:10.1109/MWC.2005.1497859

Shollmeier, R., Gruber, I., & Finkenzeller, M. (2002). *Routing in mobile ad hoc and peer-to-peer networks, a comparison*. International Workshop on Peer-to-Peer Computing.

Stoica, I., Morris, R., Nowell, D. L., Karger, D. R., Kaashock, M. F., & Dabek, F. (2003). Chord: A scalable peer-to-peer lookup protocol for internet applications. *IEEE/ACM Transactions on Networking, 11*(1), 17–32. doi:10.1109/TNET.2002.808407

Tanenbaum, A. (2003). *Computer networks*. Upper Saddle River, NJ: Prentice Hall.

Xu, T., & Cai, Y. (2007). Streaming in MANET: Proactive link protection and receiver-oriented adaptation. *IEEE International Conference on Performance, Computing, and Communications.*

ADDITIONAL READING

Camp, T., Boleng, J., Davies, V., & Golden, C. (2002). *A survey of mobility models for ad hoc network research* (pp. 483–582). Wireless Communication and Mobile Computing.

Ding, G., & Bhargava, B. (2004). *Peer-to-peer file-sharing over mobile ad hoc networks*. Second IEEE Annual Conference on Pervasive Computing and Communications Workshops.

Fonseca, R., Ratnasamy, S., Zhao, J., Ee, C. T., Culler, D., Shenker, S., et al. (2005). *Beacon vector routing: Scalable point-to-point routing in wireless sensor nets*. Second Symposium on Networked Systems Design and Implementation.

Iliofotou, M., Vlavianos, A., & Faloutsos, M. (2006). *Bitos: Enhancing BitTorrent for supporting streaming applications.* 9th IEEE Global Internet Symposium.

Jiang, X., Dong, Y., Xu, D., & Bhargava, B. (2003). GnuStream: A P2P media streaming system prototype. *Proceedings of the 2003 International Conference on Multimedia and Expo.*

Johnson, D. B., & Maltz, D. A. (1996). Dynamic source routing in ad hoc wireless networks. *Mobile Computing, 353*, 153 181. doi:10.1007/978-0-585-29603-6_5

Klemm, A., Lindemann, C., & Waldhorst, O. P. (2004). A special-purpose peer-to-peer file sharing system for mobile ad hoc networks. *Wireless Communications and Networking Conference*, (pp. 114-119).

Kortuem, G., & Schneider, J. (2001). An application platform for mobile ad-hoc networks. *Proc. of the Workshop on Application Models and Programming Tools for Ubiquitous Computing.*

Kulkarni, S. (June 2006). Video streaming on the internet using split and merge multicast. *Proceedings of the Sixth IEEE International Conference on Peer-to-Peer Computing*, (pp. 221-222).

Li, Z., Yin, X., Yao, P., & Huang, J. (2006). Implementation of P2P computing in design of MANET routing protocol. *Proceedings of the First International Multi-Symposiums on Computer and Computational Sciences (IMSCCS'06)*, (pp. 594 - 602).

Liao, B., Jin, H., Liu, Y., Ni, L., & Deng, D. (2006). Analysis: P2P live streaming. *Proceedings of the IEEE INFOCOM.*

Mao, S., Lin, S., Panwar, S., Wang, Y., & Celebi, E. (2003). Video transport over ad hoc networks: Multistream coding with multipath transport. *IEEE Journal on Selected Areas in Communications, 21*(10), 1721–1737. doi:10.1109/JSAC.2003.815965

Meddour, D., Mushtaq, M., & Ahmed, T. (2006). Open issues in p2p multimedia streaming. In *MULTICOMM2006*, (pp. 43--48).

Motta, R., & Pasquale, J. (2010). Wireless P2P: Problem or opportunity? *AP2PS 2010: The Second International Conference on Advances in P2P System.*

Oliveira, L., Siqueira, I., & Loureiro, A. (2003). Evaluation of ad-hoc routing protocols under a peer-to-peer application. *Proceedings of the IEEE Wireless Communications and Networking* (WCNC 2003), (pp. 16-20).

Pucha, H., Das, S., & Hu, Y. (2004). Ekta: An efficient DHT substrate for distributed applications in mobile ad hoc networks. *Proceedings of the 6th IEEE Workshop on Mobile Computing Systems and Applications,* (pp. 163- 173).

Rajagopalan, S., & Shen, C.-C. (2006). Cross-layer decentralized BitTorrent for mobile ad hoc networks mobile and ubiquitous systems workshops. *3rd Annual International Conference*, (pp. 1-10).

Rodolakis, G., Naimi, A. M., & Laouiti, A. (2007). Multicast overlay spanning tree protocol for ad hoc networks. *Proceedings of the 5th International Conference on Wired/Wireless Internet Communications.*

Schollmeier, R., Gruber, I., & Niethammer, F. (2003). Protocol for peer-to-peer networking in mobile environments. *Proceedings of the IEEE 12th International Conference on Computer Communications and Networks.*

Shah, P., & Paris, J. (2007). Peer-to-peer multimedia streaming using bittorrent. *IEEE International Conference on Performance, Computing, and Communications.*

Yan, L., Sere, K., & Zhou, X. (2004). Towards an integrated architecture for peer-to- peer and ad hoc overlay network applications. *Proceedings of the 10th IEEE International Workshop on Future Trends of Distributed Computing Systems,* (pp. 312–318).

Zhang, X., Liu, J., Li, B., & Yum, T. (2005). Coolstreaming/donet: A data-driven overlay network for live media streaming. *Proceedings of the IEEE INFOCOM, (pp. 2102-2111).*

Zhou, L., & Haas, Z. J. (1999). Securing ad hoc networks. *IEEE Network Magazine,* 24-30.

KEY TERMS AND DEFINITIONS

Cross-Layer Design: A design of protocol stack in which helpful information is exchanged between the protocol layers rather than enforce strict protocol layering.

Mobile Ad Hoc Network: A wireless network in which there is no infrastructure or access point. Instead the wireless nodes themselves route packet between each other, usually by multi-hop routing.

Multiple Description Coding: A multimedia stream is split into two or more sub-streams in order to exploit path diversity. Each stream if correctly received is capable of providing a reasonable quality representation of the original stream but if both substreams are combined then a higher quality stream is produced at the destination.

Network Coding: A technique to protect and repair packets as they pass across the network rather than at the end point.

Peer-to-Peer Network: An overlay network that exists above the physical network. Packets are routed between the peers of such a network by the peers themselves. A peer can act as both a source of packets and a sink of packets.

Video Streaming: Rather than download a complete video file, in video streaming a compressed video bitstream is broken into packets prior to continuous dispatch of the packets across the network.

ENDNOTES

1 (http://www.bittorrent.com)

2 (http://www.pplive.com)

Chapter 6
Overview of Mobile Ad Hoc Networks and their Modeling

Nadia N. Qadri
COMSATS Institute of Information Technology, Pakistan

Martin Fleury
University of Essex, UK

ABSTRACT

Infrastructureless or ad hoc wireless networks have long been a target of research, because of their flexibility, which is matched by the difficulty of managing them. As this research approaches maturity, it behoves researchers to be more responsible in modeling such networks. In particular, as this chapter discusses, the range of ad hoc networks has been extended to vehicular networks, for which it is no longer possible to loosely define their topology. The chapter then discusses how to improve the modeling of such networks in terms of more representative wireless channel models and more realistic mobility models for vehicular networks. The chapter also contains a review of a topic that has been the subject of intensive research: selection of suitable multi-hop routing. The chapter serves as a prelude to the study of applications on such networks, including multimedia streaming.

INTRODUCTION

This chapter gives an overview of Mobile Ad Hoc Networks (MANETs) and Vehicular Ad Hoc Networks (VANETs). In particular, the technologies used at each layer of the standard protocol stack are discussed. There follows an overview of key studies carried out so far on the performance of mobile ad hoc routing protocols. This leads to taxonomy of a wide variety of different protocols, based on mechanisms including route construction, maintenance, and update, topology formation, network configuration, and exploitation of specific resources. Then well-known ad-hoc routing protocols with their advantages and disadvantages are discussed. Furthermore VANET's and IEEE 802.11p standard for vehicular networks is explained. In addition to this, an overview of ad hoc and vehicular mobility models are given. Finally the GloMoSim network simulator, which is specialized for the study of ad hoc networks is introduced.

DOI: 10.4018/978-1-4666-1613-4.ch006

MOBILE AD HOC NETWORKS

Mobile Ad hoc Networks (MANETs) are a type of pervasive network that truly support pervasive computing, pervasive networks allow users to communicate anywhere, anytime, and on-the-fly. Future advances in pervasive computing rely on advancements in mobile communication, which includes both infrastructure-based wireless networks and non infrastructure-based MANETs. The traditional infrastructure-based communication model is not adequate for today's user requirements. In many situations, communication between mobile hosts cannot rely on any fixed infrastructure. The cost and delay associated with installation of infrastructure-based communication model may not be acceptable in dynamic environments such as disaster scenes, battle field, and inter-vehicular communications. MANET would be an effective solution in these scenarios.

MANET technology draws great attention of worldwide researchers and scientists. Since the first appearance of wireless ad-hoc networks in the DARPA packet radio networks in the 1970 (Jubin and Tornow, 1987), it became an interesting research object in the computer industry. In the 1990s, the concept of commercial ad-hoc networks arrived with notebook computers and other viable wireless communications equipment. At the same time, the idea of a collection of mobile nodes was proposed at several research conferences. The IEEE 802.11 subcommittee had adopted the term "ad-hoc networks" and the research community had started to look into the possibility of deploying ad-hoc networks in other areas of application. During the last couple of years tremendous improvements have been made in the research of ad hoc networks. Due to their ability to create and organize a network without any central management, MANETs are characterized as the art of networking without a network (Jiang et al, 1984).

A MANET can be defined as a self-organizing and autonomous system of mobile nodes that communicate over wireless links. Since the nodes are mobile, the network topology may change rapidly and unpredictably over time. The network is decentralized, where all network activity including discovering the topology, routing functionality and message delivering is executed by the nodes themselves. MANETs introduce a new communication paradigm, which does not require a fixed infrastructure - they rely on wireless terminals for routing and transport services. Therefore MANET can be flexibly and rapidly deployed. The special features of a MANET bring about great opportunities together with severe challenges. Due to their highly dynamic topology, the absence of an established infrastructure for centralized administration, bandwidth constrained wireless links, and limited resources, MANETs are hard to design in terms of efficient and reliable networks (Goldsmith & Wicker, 2002).. Figure 1 shows a typical MANET.

MANET Protocol Layers

In most MANET studies simulation models are used for evaluation of higher -layer protocols and applications, as it is otherwise difficult to formulate solutions, due to the complex computation required by the many variables involved. Similarly testing on a real testbed is not easily implementable as it requires movement of peoples or vehicles with handheld devices in different patterns. Typically for such testing the focus is only on higher layers ignoring the details of models at other layers. There are many factors (Takai et al., 2001; Xu et al., 2002; Kekmat & Mieghem, 2004; Yang & Vaidya, 2005; Gupta & Kumar, 2000; Weber et al., 2007) such as physical layer modelling, signal reception method, path-loss model, fading models, mobility models, MAC models, interference and noise calculations which have a huge effect on performance of higher layer protocols and applications.

In this Section, brief explanations of models for the lower layers are presented.. Network layer routing protocols are discussed in detail in the Section after that.

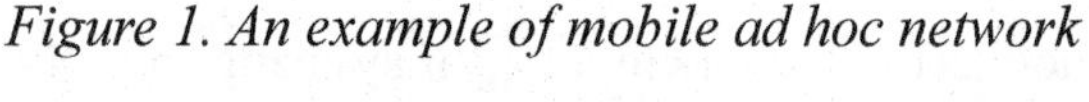

Figure 1. An example of mobile ad hoc network

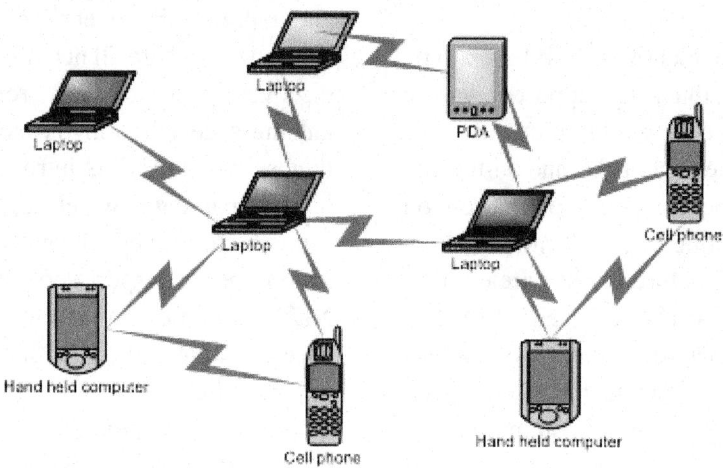

Physical Layer Modeling

The IEEE 802.11 standard specifies two wireless physical layer types for vehicular networks: one is based on Direct Sequence Spread Spectrum (DSSS) and the other is based on Frequency Hopped Spread Spectrum (FHSS) (O'Hara & Petrick, 2004)). The DSSS physical layer uses an 11-chip Pseudo-Noise (PN) Barker code (ISO/IEC, 1999) combined with Differential Binary Phase Shift Keying (DBPSK) or Differential Quadrature Phase Shift Keying (DQPSK) to achieve 1 and 2 Mbps data rate.

Another important factor to model the signal propagation is path loss, which defines the average signal power loss of a path over the terrain. The simplest propagation models are: free space and the ground reflection (or two-ray) models. The free space model is considered to be an idealized propagation model. The model predicts received signal strength when the transmitter and receiver have a clear and unobstructed line-of-sight between them. The transmission power is predicted to be attenuated in proportion to the square of the distance. Accordingly, the free space equation for a non-isotropic antennas is the following:

$$P_r = P_t \left(\frac{\lambda}{4\pi d} \right)^n G_t G_r \qquad (1)$$

where P_r is the received power, P_t is the transmitted power (in Watts or milli-Watts), λ is the carrier wavelength (in meters), d is the distance between transmitter and receiver (in meters), n is the path loss coefficient, G_t is the antenna gain at the transmitter and G_r is the antenna gain at the receiver.

The antenna gain is a measure of the directionality of an antenna. Antenna gain is defined as the power output, in a particular direction, compared to that produced in any direction by a perfect omni-directional antenna or isotropic antenna. For example, if an antenna has a gain of 3 dB, that antenna improves upon the isotropic antenna in that direction by 3 dB, or a factor of 2. The increased power radiated in a given direction is at the expense of other directions (Stallings, 2000). Therefore for isotropic or omni-directional antenna the gain will be 0. The decibel is a measure of the ratio between two signal levels. The decibel gain is given by the following equation:

Figure 2. Two-ray model geometry

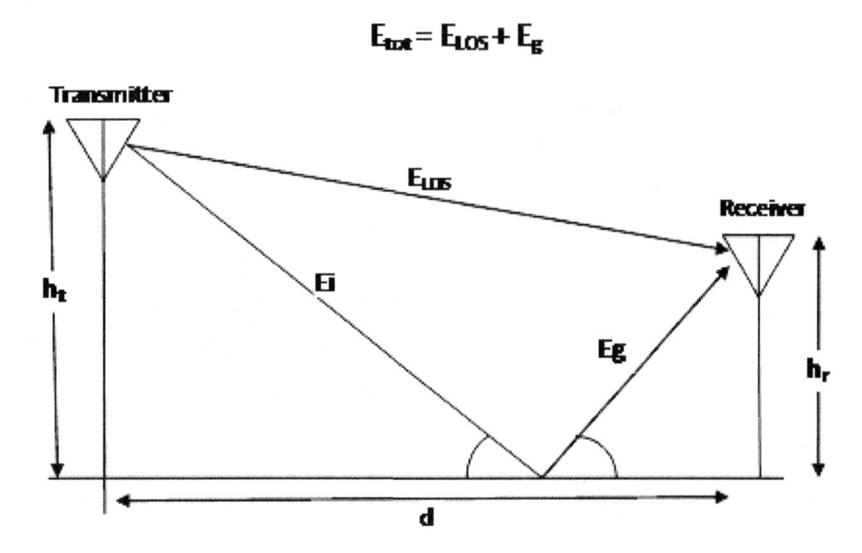

$$G_{db} = 10log_{10} \frac{P_{out}}{P_{in}} \quad (2)$$

For the ideal omni-directional antenna, the free space loss equation is:

$$\frac{P_t}{P_r} = \frac{(4\pi d)^2}{\lambda^2} = \frac{(4\pi fd)^2}{c^2} \quad (3)$$

where c is the speed of light (3 x 10^8 m/s) and is the frequency (in hertz or 1cycle/sec).

With this path loss model, even nodes far from the transmitter can receive packets in MANETs, which can result in fewer hops to reach the final destination. Therefore, simulation results with the free space path loss model tend to be more optimistic than with other path loss models but unfortunately they are unrealistic. However, as signal propagation with little power loss may cause stronger interference for concurrent transmissions, it does not necessarily yield the best performance under all scenarios.

The two-ray path-loss model is suited to Line-Of-Sight (LOS) channels in urban environments (Rappaport, 2001). Its use for MANETs can be justified by the environmental similarities (low transmit power and low antenna height) (Takai et al., 2001) The two-ray model considers both the direct path and a ground-reflected propagation path between transmitter and receiver. The model predicts that received power decays with distance raised to the fourth power, or at a rate of 40 dB/decade. This is a much more rapid path loss than is experienced in free space (Rappaport, 2001) but it is more realistic. This model is represented in Figure 2.

The following equation is used to calculate the receiver power and radio range distance for two-ray model.

$$P_r = P_t \left(\frac{h_t^2 h_r^2}{d^4} \right)^n G_t G_r \quad (4)$$

$$d = \sqrt[4]{\frac{P_t h_t^2 h_r^2}{P_r}} \quad (5)$$

where h_t and h_r is the height of the transmitter and receiver antennas.

General-purpose MANET simulations tend to restrict modeling of the wireless channel to either free space or two-ray, line-of-sight (LOS) modeling and do not account for wireless fading through multi-path, as can obviously occur, especially in a urban environment. In the next Section fading models are discussed in detail.

Fading Models

Propagation models such as fading, shadowing and path loss are not part of the radio physical models, but they control the input given to the physical models and have great impact on their performance, and including these models is relevant to the emphasis of multiple-layer interactions.

An additive propagation model (Takai et al., 2001), which sums path loss attenuation and multi-path interference or fading, is modeled by Rician or Rayleigh probability density functions (Goldsmith, 2005). Fading is a variation of signal power at receivers, caused by the node mobility that creates varying path conditions from transmitters. In a Rician model, there is one LOS signal. In Rayleigh fading, there is no LOS component and, consequently, Rayleigh fading represents a worst-case scenario1. Specifically, the Rayleigh distribution of the signal envelope is given by equation (6):

$$f\left(z\right) = \frac{z}{\sigma^2} exp\left[-\frac{z^2}{2\sigma^2}\right], z \geq 0 \qquad (6)$$

where $\tilde{A}^2$ is the variance of the zero-mean Gaussian distributed in-phase, r_I, and quadrature, r_Q, components of the signal envelope:

$$z\left(t\right) = \sqrt{r_I^2\left(t\right) + r_Q^2\left(t\right)} \qquad (7)$$

The Rician distribution is given by

$$f\left(z\right) = \frac{z}{\sigma^2} exp\left[-\frac{\left(z^2 + s^2\right)}{2\sigma^2}\right] I_0\left(\frac{zs}{\sigma^2}\right), z \geq 0 \qquad (8)$$

where the in-phase and quadrature components are no longer Gaussian, as there is a fixed LOS component. I_0 is the modified Bessel function of zeroth order. The average power in the non-LOS components is given by $2\tilde{A}^2$ and s^2 is the power of the LOS component. The fading parameter $K = s^2 / 2\tilde{A}^2$ is the ratio of these powers. In fact, the right-hand-side of equation (8) can be rewritten in terms of K and the average received power (Goldsmith, 2005). If $K = 0$ then this equation reverts to a Rayleigh distribution and with $K = \infty$ there is only a LOS component.

Interference Computation and Signal Reception

Computation of interference and noise at each receiver is a critical factor in wireless communication modeling, as this computation becomes the basis of *Signal-to-Interference-Noise Ratio (SINR)* or *Signal-to-Noise-Ratio (SNR)* that has a strong correlation with *Packet Error Rate (PER)* or *Frame Error rate (FER)* on the channel. The power of interference and noise is calculated as the sum of all signals on the channel other than the one being received by the radio plus the thermal (receiver) noise. The resulting power is used as the basis of SNR, which determines the probability of successful signal reception for a given frame. For a given SNR value, two signal reception models are commonly used in wireless network simulators: *SNR threshold (SNRT)* based and *Bit Error Rate (BER)* based models.

The SNR threshold based model uses the SNR value directly by comparing it with an SNRT, and accepts only signals whose SNR values are above the SNRT at any time during the reception.

The BER-based model probabilistically decides whether or not each packet is received successfully based on the packet length and the BER deduced by SNR and modulation scheme used at the transceiver. As the model evaluates each segment of a packet with a BER value every time the interference power changes, it is considered to be more realistic and accurate than the SNRT-based model. However, the SNRT-based model requires less computation cost and can be a good abstraction if each packet length is long. However, the latter is not normally the case for video communication. Both of these models are used for simulations depending on type of scenario and application (Takai et al., 2001).

MAC (Medium Access Control) Layer Modelling

The MAC layer technology used in MANETs is IEEE 802.11 *Distributed Coordination Function (DCF)* with *RTS/CTS (Request To Send/ Clear To Send)* support. The WLAN working group of (IEEE 802) defined IEEE 802.11 (Wi-Fi) (ISO/IEC 802.11, 1999). There are different variations of the 802.11 such as 802.11a or 802.11n as each of them uses different modulation techniques and modifications. For simplicity of modeling in MANET's the simpler IEEE 802.11b is sometimes assumed for MANETs. There are two different types of service for the MAC layer of IEEE 802.11b. The DCF, which is based on CSMA/CA, is an access contention-based service, and the *Point Coordination Function (PCF)*, based on a poll-and-response mechanism, is a contention-free service. The DCF service is used for the ad hoc mode communications, as the PCF needs a central coordination point.

The 802.11b standard uses the unlicensed 2.4-GHz band. This frequency range is unregulated and the devices working in this range can interfere with each other. The maximum bit rate for 802.11b can be 11 Mbps at most, but considering the overhead share of the bandwidth, the maximum data rate (excluding overheads and packetization) is not more than 5.9 Mbps over *Transmission Control Protocol (TCP)* and 7.1 Mbps over the *User Datagram Protocol (UDP)* transport networks, because of the relative size of the packet headers.

In order to avoid the hidden node problem, which is caused by the limited range of the antennas, a collision avoidance mechanism is employed, which uses RTS/CTS packets to assure no interruption happens by the listening nodes during the communication. The extra overhead of the CSMA/CA limits the available bandwidth of 802.11b. However, this method makes communication more reliable.

Transport Layer Modeling

IP framing is employed with UDP transport protocol (Postel, 1980), as the TCP (Stevens, 1993) can introduce unbounded retransmission delays, which are not suitable for delay-intolerant video streaming applications (Zheng & Boyd, 2001).

MOBILE AD HOC ROUTING PROTOCOLS

Due to the highly dynamic nature of MANETs, designing suitable ad hoc routing protocols is a challenging issue. A robust and flexible routing approach is required to efficiently use the limited resources available, while at the same time being adaptable to the changing network conditions such as network size (scalability), traffic density, and mobility. The routing protocol should be able to provide efficient route establishment with minimum overhead, delay, and bandwidth consumption, along with a stable throughput. Furthermore, the possibility of asymmetric links, caused by different power levels among mobile hosts and other factors such as terrain conditions, make routing protocols more complicated than in other networks.

For this purpose, various protocols has been introduced and authors of each proposed protocol claim that the algorithm proposed by them brings in enhancements and improvements over a number of different strategies, under different scenarios and network conditions. However only few protocols have actually been implemented (beyond the simulation stage) and not all of these have been assessed in depth. Many articles such as Broch et al., (1998) Johansson et al. (1999) Das et al. (2000) Perkins et al., (2000) Jiang & Garcia-Luna-Aceves (2001) Lee et al., (2002) Boukerche (2004) Trung et al. (2007) have provided a protocol assessment which is specific and often do not allow drawing general conclusions. Therefore, it is difficult to determine which protocols may perform better under different network scenarios.

To achieve the required efficiency, routing protocols for MANETs must satisfy special characteristics. Important characteristics are identified by the Internet Engineering Task Force (IETF) MANET Charter in RFC 2501 (Corson & Macker, 1999). The fundamental characteristics required by ideal mobile ad hoc routing protocol are exemplified below.

- Distributed routing: Routing protocols must be fully distributed, as this approach is more fault tolerant than centralized routing.
- Adaptive to topology changes: Routing must adapt to frequent topological and traffic changes that result from node mobility and link failure.
- Proactive/Reactive operation: The routing algorithm may intelligently discover the routes on demand. This approach will be useful to efficiently utilize the bandwidth and energy resources but comes at the cost of additional delay. However, in certain conditions the delay incurred by on-demand operation may be unacceptable.
- Loop free routing: Routes free from loops and stale paths are desirable. Perhaps to increase robustness, multiple routes should be available between each pair of nodes.
- Robust route computation and maintenance: The smallest possible number of nodes must be involved in the route computation and maintenance process, so as to achieve minimum overhead and bandwidth consumption.
- Localized state maintenance: To avoid propagation of overheads, localized state maintenance is desirable.
- Optimal usage of resources: The efficient utilization and conservation of resources such as battery power, bandwidth, computing power and memory is required.
- Sleep mode operations: To reduce energy consumption, the routing protocol should be able to employ some form of sleep mode operation. Nodes that are inactive should switch to 'sleep mode' for arbitrary periods.
- Quality of Service: Routing algorithms are required to provide certain levels of QoS in order to meet specific application requirements.
- Security: Some form of security protection is desirable to prevent disruption due to malicious modifications of protocol operations.

Taxonomy of Mobile Ad Hoc Routing Protocols

Mobile ad hoc routing protocols can be classified in many ways depending upon their route construction and maintenance mechanisms, route selection strategy, topology formation, update mechanism, utilization of specific resources, type of cast and so on (Murthy & Manoj, 2004). Here they are classified using several characteristics - the basis of the classification is discussed below. The taxonomy of routing protocols is shown in Figure 3. In this Section we focus on those protocols that will be discussed in the remainder of this Chapter.

Figure 3. Taxonomy of mobile ad hoc routing protocols (for abbreviations please refer to the list of abbreviations)

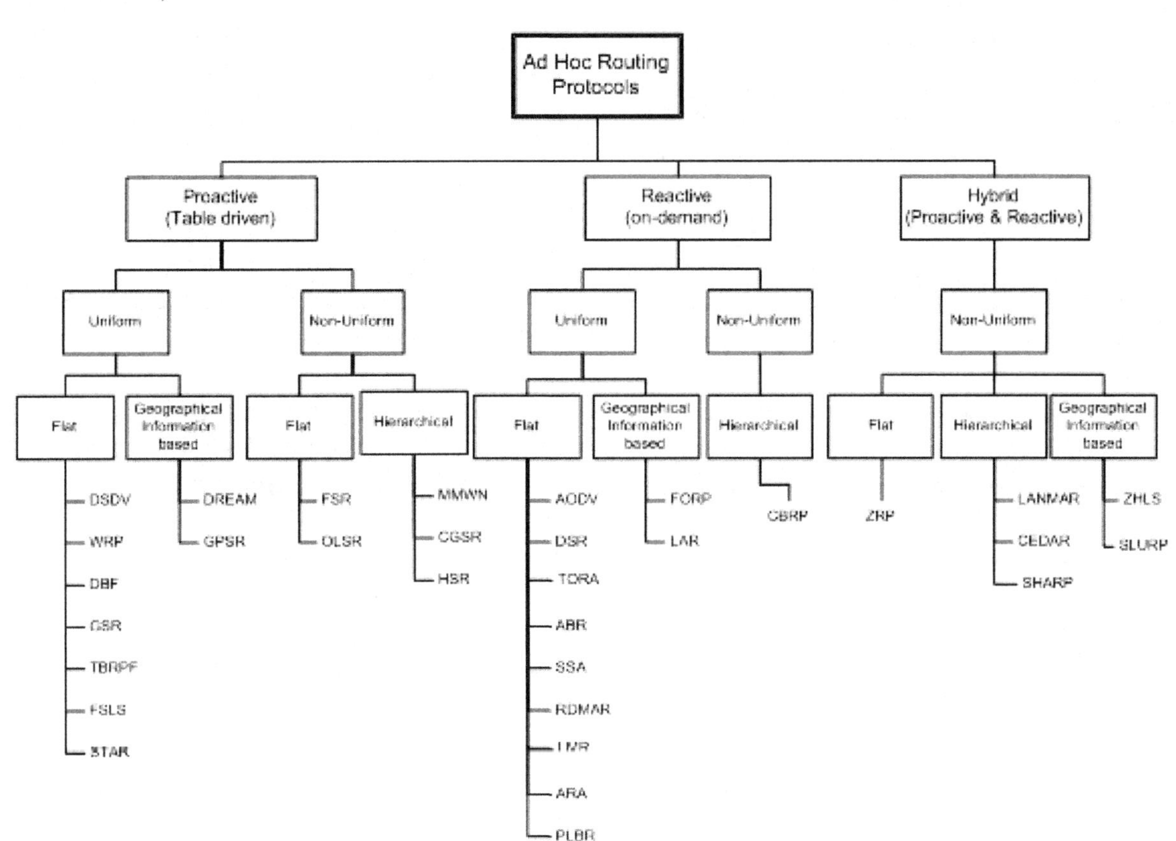

Approaches Based on Route Construction, Maintenance, and Update Mechanisms

These protocols can be described as the way the route is constructed, updated, and maintained, and route information is obtained at each node and exchanged between the nodes. Based upon these characteristics routing protocols can be divided broadly into three categories.

Proactive (Table-Driven) Routing

The first category is proactive or table-driven routing protocols in which each node consistently maintains up-to-date routing information for all known destinations. These types of protocols keep routing information in one or more tables and maintain routes at each node by periodically distributing routing tables throughout the network or when the topology changes. Each node keeps information on all the routes, regardless of whether or not these routes are needed. Therefore, control overhead in these protocols can be significantly high especially for large networks or in a network where nodes are highly mobile. However, the main advantage of these protocols is that the routes are readily available when required and end-to-end delay is reduced during data transmission in comparison to the case in which routes are determined reactively, which introduces a latency in discovering a route to the destination. The most popular proactive protocols are Destination Sequenced Distance Vector (DSDV) (Perkins & Bhagwat,

1994), Optimized Link State Routing (OLSR) (Clausen & Jacquet, 2003), Wireless Routing Protocol (WRP) (Murthy & Garcia-Luna-Aceves, 1996), Fisheye State Routing (FSR) (Pei et al., 2000; Pei et al., 2002), and Distance Routing Effect Algorithm (DREAM) (Basagni et al., 1998).

Reactive (On-Demand) Routing

The second category is reactive or on-demand routing protocols the routes are discovered only when they are actually needed. These protocols consist of route discovery and route maintenance processes. The route discovery process is initiated when a node wants to send data to a particular destination. Route discovery usually continues by flooding the network with route-request packets. When a destination node or node holding a route to a destination is reached, a route-reply is sent back to the source node by instantiating routing information at the appropriate intermediate nodes. Once the route reply reaches the source, data can be sent to the destination. The route maintenance process deletes failed routes and re-initiates route discovery in case of topology change. The advantage of this approach is that overall overhead is likely to be reduced compared to proactive approaches. However, as the number of sessions increases the overhead generated by route discovery becomes high and may exceed that of proactive protocols. Some example reactive routing protocols are Ad-hoc On-demand Distance Vector (AODV) (Perkins & Royer, 1999) (Perkins et al., 2003), Dynamic Source Routing (DSR) (Broch et al., 2003) (Johnson et al.,, 2001), Location Aided Routing (LAR) (Ko & Vaidya, 1998) (Ko & Vaidya, 2000), and Temporally Ordered Routing Algorithm (TORA) (Park & Corson, 1997).

Hybrid Routing

The third category is hybrid routing protocols combine the advantages of both proactive and reactive routing. These protocols usually divide the network into zones such that each node sees the network as a number of zones. The routes to nodes close to each other or within a particular zone are proactively maintained and the routes to far-away nodes are determined reactively using a route-discovery strategy. The well-known hybrid routing protocols are Zone Routing Protocol (ZRP) (Zygmunt & Marc, 2001) Sharp Hybrid Adaptive Routing Protocol (SHARP) (Ramabramanian et el., 2002), and Zone-based Hierarchical Link State (ZHLS) (Joa-NG & Lu, 1999).

Approaches Based on Logical Organization, Network Configuration, and Utilization of Specific Resources

These protocols are based on the way nodes organize themselves logically, the way nodes participate into route computations, nodes location or route is obtained based on geographical information. Based on these characteristics routing protocols are divided into two main categories. a) Uniform Routing b) Non-Uniform Routing.

Uniform Routing

The first category is uniform routing. In uniform routing, all nodes are equal and each node participates in route computations. Each node generates routing control messages and replies to routing control requests in the same way. Thus, every node supports exactly the same functionality as the other. Uniform protocols can be sub-divided into Flat and Geographical Information (GI) based routing protocols. The geographical information based protocols proposed to date are mostly uniform except for the ZHLS and Scalable Location Update Routing Protocol (SLURP) (Woo et al. 2001) routing protocols, which are non-uniform routing protocols, that is discussed in next category.

Flat Uniform Routing

In flat-routing, nodes do not form a specific structure or hierarchy. Each node has similar roles. Nodes that are within the transmission range of each other form a connection, where the only limitations are determined by connectivity conditions or security constraints. The major advantage of this routing structure is that there are multiple paths between source and destination, which reduces traffic congestion and traffic bottlenecks in the network. Single points of failure in the case of a hierarchical routing protocol where there is a cluster head (a special node within each cluster that coordinates the traffic in and out of the cluster) could lead to larger control overheads arising from network reconfiguration. Nodes in flat routing require significantly lower power for transmission in comparison to cluster heads (Haas & Tabrizi, 1998).

Geographical Information Based Uniform Routing

In these types of protocols the location of the nodes can be obtained by utilizing the *Global Positioning System (GPS)*; alternatively, the relative coordinates of nodes can be obtained by calculating the distance between the nodes and exchanging this information with neighbouring nodes. The distance between nodes can be estimated on the basis of incoming signal or time delays in direct communications (Stojmenovic, 2002). The main advantage of this approach is that the protocols can improve routing performance and reduce control overheads by effectively utilizing location information. All the protocols in this category assume that all nodes know their positions and the network topology of nodes corresponds well with the geographical distance between them. The drawback of this approach is that its above mentioned assumptions are often not acceptable and location information may not be accurate at all times (Jiang & Garcia-Luna-Aceves, 2001).

Non-Uniform Routing

In non-uniform routing, the way of generating and/or replying to routing control messages may be different for different groups of nodes. In these protocols, only a few nodes are involved in route computation. For instance, some nodes broadcast received routing requests, others do not. Non-uniform protocols attempt to reduce routing overhead by reducing the number of nodes involved in route computation. Moreover, they have a cost introduced for maintaining a high-level structure and the use of complex algorithms. Non-uniform protocols can be logically sub-divided into flat (based on neighbour selection) and hierarchical routing.

Flat Non-Uniform Routing

In this routing approach, each node selects some subset of its neighbours to take a distinguished role in route computation and/or traffic forwarding. Each node makes its selection independently and there is no negotiation between nodes to attain nodes consensus. The node's selection is also not affected by non-local topology changes (Feeney, 1999).

Hierarchical Non-Uniform Routing

In hierarchical routing protocols the nodes organize themselves into groups, called clusters. Within each cluster a cluster head or gateway node is selected which coordinates all the traffic in and out of their clusters. Routing between two nodes from different clusters is usually performed by their cluster heads. The depth of the network can vary from single to multiple levels, depending upon the number of hierarchies. The advantage of this approach is that each node maintains route to its cluster head only, which means that routing overheads are much lower compared to flooding routing information through the network. However, these protocols require complex algorithms for the creation and reconfiguration of clusters when cluster heads fails. Along with this there are significant overheads associated with maintaining clusters (Haas & Tabrizi, 1998), such as: instability in the system due to a high rate of change of cluster heads at high mobility, the need for additional interfaces to avoid cluster head conflicts, and the fact that power consumption at the cluster head is higher than for normal nodes, which results in frequent changes of cluster heads and may lead to frequent multiple path breaks and overhead involved in exchanging packets for the sake of cluster head selection process (Chaing et al., 1997).

Overview of Well-Known Routing Protocols

Dynamic Source Routing (DSR)

DSR is an on-demand routing protocol based on the concept of source routing (Broch et al., 2001) (Johnson & Maltz, 1996). Mobile nodes are required to maintain route caches that contain the source routes of which the mobile is aware. The route cache entries are continually updated as new routes are learned. The protocol consists of two main phases: 1) route discovery and 2) route maintenance. When a node wants to send a packet to a destination it first checks its route cache to determine whether it already has a valid route to the destination. If it has a valid route to the destination, it will use that route to send the packet. Otherwise, it initiates a route discovery process by broadcasting a route request packet (Johnson et al., 2001).

Maintaining a route cache is very beneficial for networks with low mobility as in this way routes will be valid for a longer period. In addition, the route cache information can also be utilized by intermediate nodes to efficiently reduce control overheads. However the broken links are not locally repaired by a route maintenance mechanism. Therefore, this is a disadvantage of this protocol. Along with that, stale route cache information could also result in variations during the route reconstruction phase. The connection setup delay is higher than in table-driven protocols. The protocol performs better with static nodes and slow-moving nodes but its performance degrades rapidly with an increase in mobility (Abolhasan et al., 2003).

Ad hoc On-demand Distance Vector (AODV)

The AODV routing protocol is a type of on-demand (reactive) protocol (perkins & Royer, 1998). AODV share the same on-demand characteristics of DSR and uses the same discovery process to find routes when required. There are two major differences between AODV and DSR. AODV uses a traditional routing table with one entry per destination, whereas DSR maintains multiple route cache entries for each destination. Another difference is that AODV relies on routing table entries to propagate route replies back to the source and subsequently to route data packets to their destination. AODV employs the destination sequence numbers procedure to identify the recent route. All routing packets are tagged with sequence number assigned by the destination in order to indicate the freshness of route and avoid the formation of routing loops. Sequence numbers are incremented each time a node sends an update. A route is considered to be more favorable if its sequence number is higher. A node updates its route only if the sequence number of the last stored packet is greater than the sequence number of the current packet received. AODV uses the periodic beaconing (periodic broadcast of routing path updates) and sequence numbering procedure of DSDV but minimizes the number of required broadcasts by creating routes on demand, as opposed to maintaining a complete list of routes as in DSDV (Perkins et al., 2003).

One of the disadvantages of AODV is that intermediate nodes can lead to inconsistent routes if the source's sequence number is very old and the intermediate nodes have a higher (but not the latest) destination sequence number, thereby having stale entries. Also multiple RouteReply packets in response to a single RouteRequest packet can lead to heavy control overheads, thereby introducing extra delays as the size of network increases. Another shortcoming is that periodic beaconing leads to unnecessary bandwidth consumption (Abolhasan et al., 2003).

Fisheye State Routing (FSR)

FSR is a proactive, non uniform routing protocol, employing a link-state routing algorithm (Pei et al., 2000). To reduce the overhead incurred by periodic link-state packets, FSR modifies link-state routing in the following three ways:

1. Link-state packets are no longer flooded instead, only neighboring nodes exchange the topology table information;
2. The link-state exchange is solely time-triggered and not event triggered;
3. Instead of periodically transmitting the entire link-state information, FSR uses different exchange intervals for different types of entries in the topology table.

Link-state entries corresponding to nodes within a predefined distance (scope) are propagated to neighbors more frequently (intra updates) than entries of nodes outside the scope (inter updates). FSR is suitable for large and highly mobile network environments, as it triggers no control messages on link failures. Broken links will not be included in the next link state message exchange. This means that a change on a far away link does not necessarily cause a change in the routing table. However, scalability comes with a price of reduced accuracy because as mobility increases the route to remote destinations becomes less accurate (Murthy & Manoj, 2004) (Pei et al., 2002) (Abolhasan et al., 2003).

There are four configuration parameters for FSR, the value of which depends on factors such as mobility, node density and transmission range:

1. Size of the scope: This parameter specifies the scope radius of a node in number of hops.
2. Time-out for the neighbouring nodes: If a node does not hear from a neighbour specified by this value, the neighbour node will be deleted from the neighbour list.
3. Intra scope update interval: Update interval of sending the updates of the nodes within the scope radius.
4. Inter scope update interval: Update interval of sending the updates of the nodes outside the scope radius.

Location Aided Routing (LAR)

LAR is analogous to on-demand routing protocols such as DSR but it uses location information to reduce routing overheads (KO & Vaidya, 1998). LAR assumes that each node knows its physically location by using the GPS. GPS information is used to restrict the flooded area of route request packets. In Ko & Vaidya (2000) two different schemes are proposed. In scheme 1, the source defines a circular area in which the destination may expected to be present is the ExpectedZone. The position and size of the ExpectedZone is decided based on the past location and speed information of the destination. In case of non-availability of past information of destination, the entire network area is to be considered as the ExpectedZone. The smallest rectangular area that includes this circle and the source is the RequestZone and is determined by the source. This information is attached to a route request by the source and only nodes inside the RequestZone propagate the packet. In scheme 2, the source calculates the distance between the destination and itself. The source includes the distance and location of destination in route request and sent it to neighbors. When neighboring nodes receives this packet, they compute their distance to the destination, and relay the packet only if their distance to destination is less than or equal to the distance indicated by the packet. When forwarding the packet, the node updates the distance field with its distance to destination. In both schemes, if no Route Reply is received within the timeout period, the source retransmits a route request via pure flooding (Lee et al., 2002). The major advantages of LAR are an efficient use of geographical position information, reduced control overhead, and increased utilization of bandwidth. The disadvantage is that each node must support GPS.

VEHICULAR AD HOC NETWORK (VANET)

A Vehicular Ad-Hoc Network, or VANET, is a form of MANET. VANETs are a cornerstone of the envisioned *Intelligent Transportation Systems (ITS)*. By enabling vehicles to communicate with each other via *Inter-Vehicle Communication (IVC)* as well as with roadside base stations via *Roadside-to-Vehicle Communication (RVC)*, vehicular networks will contribute to safer and more efficient roads by providing timely information to drivers and concerned authorities. The interesting research area of Vehicular Networks is where ad hoc networks can be brought to their full potential (Biswas et al., 2006).

The main goal of VANET is providing safety and comfort for passengers. To this end a special electronic device will be placed inside each vehicle which will provide ad-hoc network connectivity for the passengers. This network tends to operate without any infra-structure or legacy client and server communication. Each vehicle equipped with VANET device will be a node in the ad-hoc network and can receive and relay others messages through the wireless network. Collision warning, road sign alarms and in-place traffic view will give the driver essential tools to decide the best path along the way.

There are also multimedia and internet connectivity facilities for passengers, all provided within the wireless coverage of each car. Automatic payment for parking lots and toll collection are other examples of possibilities inside VANET. VANETs can further take part in the exchange or sharing of personal video sequences, as may occur in a social network. They may also have a role in reporting traffic conditions as seen by roadside cameras and likewise reporting conditions in an emergency. An example scenario is shown in the Figure 4.

Most of the concerns of interest to MANETs are of interest in VANETs but the details differ. Rather than moving at random, vehicles tend to move in an organized fashion. The interactions with roadside equipment can likewise be characterized fairly accurately. And finally, most vehicles are restricted in their range of motion, for example by being constrained to follow a paved highway. In general, wireless VANETs, just like general ad hoc networks can: relieve congested cells; extend coverage; and service dead-spots within cellular networks, assuming dual cellular and ad hoc interfaces in-vehicle transceivers or relay vehicles for those vehicles not so equipped.

VANET integrates on multiple ad-hoc networking technologies such as WiFi IEEE 802.11a/b/g/n, WiMAX IEEE 802.16 (IEEE 802.16e-2005, 2005), Bluetooth IEEE 802.15.1 (Bluetooth Special Interest Group, 2001), ZigBee IEEE 8-2.15.4 (Zigbee Alliance, 2003) for easy, accurate, effective and simple communication between vehicles on dynamic mobility. Effective measures such as media communication between vehicles can be enabled as a method to track the automotive vehicles is also preferred. VANET helps in defining safety measures in vehicles, streaming communication between vehicles, infotainment and telematics.

VANETs are implementing variety of wireless technologies such as *Dedicated Short Range Communications (DSRC)*, which is a type of WiFi. Other candidate wireless technologies are cellular, satellite, and WiMAX. IEEE 802.11p (Jiang &

Figure 4. VANET with captured video relayed by cars to first responder emergency vehicle

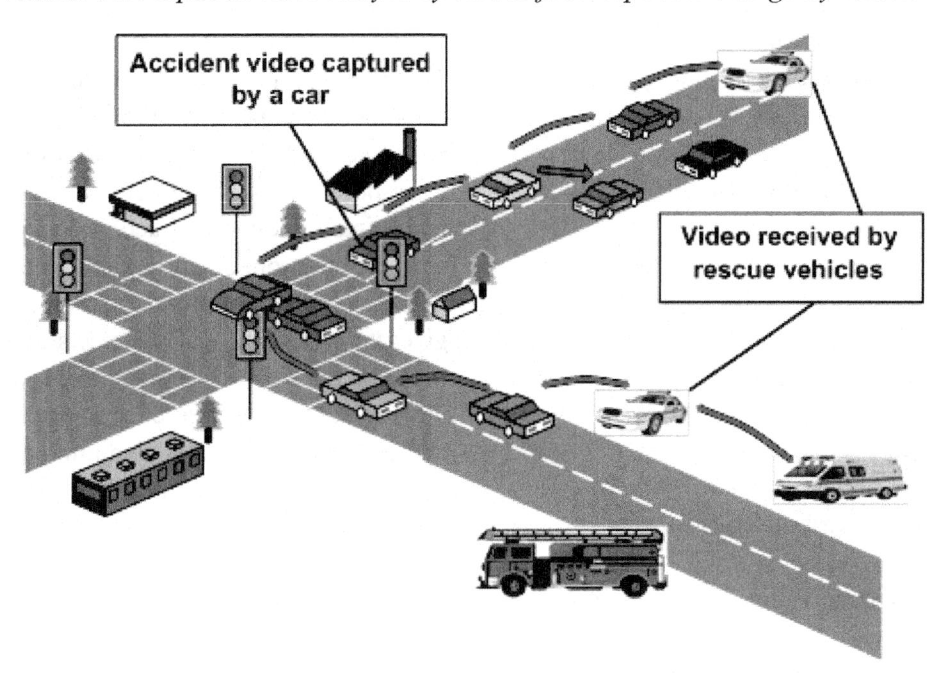

Delgrossi,) is a basis for a DSRC standard (Jiang et al., 2006), as is explained in the next section.

IEEE 802.11p Standard

IEEE 802.11p is an amendment to the IEEE 802.11 standard that adds *Wireless Access in Vehicular Environments (WAVE)* (Eichler, 2007). It defines enhancements to 802.11 required to support *Intelligent Transportation Systems (ITS)* applications. This includes data exchange between high-speed vehicles and between the vehicles and the roadside infrastructure in the licensed ITS band of 5.9 GHz (5.85-5.925 GHz).

The IEEE 802.11p standard (Xu et al., 2004). operating in ad hoc mode will facilitate VANET development, taking advantage of spectrum allocated both in Europe (30 MHz) and the USA (75 MHz) in the 5.9 GHz range using 10 Mbps channels[2] operating at up to 27 Mbps depending on modulation mode. The increased safety (Biswas et al., 2006) that may arise from wireless provision is under active investigation. As well as safety alerts through wayside access points, the possibility of advertising localized services provides an additional commercial incentive to wireless take-up. Car manufacturers are moving towards equipping cars with WLAN capability and they may already have done so at the high end of their range. It is thought that early adoption will result in around 20% of WLAN-enabled cars (Blum et al., 2004) in the near term. Therefore, at least 20% of the available cars in a city are likely to be available as relays to aid in video communication in an emergency.

Mobility Models

Mobility Model Generators

For general mobile ad hoc simulations the Random Way Point (RWP) mobility model (Broch et al., 1998) is used. For vehicular network simulations BonnMotion and VanetMobiSim mobility generator were can be used. Except for the Manhattan Grid Mobility model many other vehicular mobility models can be generated using VanetMobiSim (Fiore et al., 2007). The Manhattan Grid Mobility model is conveniently generated using the BonnMotion mobility generator.

BonnMotion Mobility Generator

BonnMotion (http://iv.cs.uni-onn.de/wg/cs/applications/bonnmotion/) is Java software which creates and analyses mobility scenarios. It is developed within the Communication Systems group at the Institute of Computer Science IV of the University of Bonn, Germany.

VanetMobiSim Mobility Generator

VanetMobiSim is an extension for the CANU Mobility Simulation Environment (CanuMobiSim) (Stepanov et al., 2003), a flexible framework for user mobility modeling. CanuMobiSim is JAVA-based and can generate movement traces in different formats, supporting different simulation tools for mobile networks. The VanetMobiSim extension focuses on vehicular mobility, and features new realistic automotive motion models at both macroscopic and microscopic levels. In VanetMobiSim, a variety of urban road layouts can be generated of increasing density by means of the random backbone mode using Voronoi tessellations. Also, it adds support for multi-lane roads, separate directional flows, differentiated speed constraints and traffic signs at intersections. The real advantage of VanetMobiSim is the variety of driver behavior models. The micromobility models presented by VanetMobiSim are of increased sophistication in driver behaviour. Notice though that when driver behaviour is introduced into simulations it is no longer possible to easily examine node speed dependencies, as the vehicles will have a range of speeds depending on local conditions, even though the minimum and maximum speeds are not exceeded.

MANET Mobility Model

The RWP mobility model (Broch et al., 1998) is commonly used as a synthetic model for mobility in ad hoc networks. It is an elementary model which describes the movement pattern of independent nodes in simple terms. Each node moves along a zigzag line from one waypoint to the next. The way points are uniformly distributed over the given terrain area. A node selects a destination from the physical terrain area and moves in the direction of that destination at a speed uniformly chosen between the defined minimum and maximum speed. Upon reaching its destination, the node stays there for some time, i.e., the pause time, and then repeats the whole process until the simulation ends.

Vehicular Mobility Models

VANETs are self-organized networks built up from moving vehicles, and are part of the broader class of MANETs. Because of their peculiar characteristics, VANETs require the definition of specific networking techniques, whose feasibility and performance are usually tested by means of simulation. Logistic difficulties, economic issues and technology limitations make simulation the mean of choice in the validation of networking protocols for VANETs, and a widely adopted first step in development of real world technologies. A critical aspect in a simulation study of VANETs is the need for a mobility model which reflects, as close as possible, the real behavior of vehicular traffic. When dealing with vehicular mobility modelling. It is desirable for trustworthy VANETs simulation that both macro-mobility and micro-mobility descriptions be jointly considered in modelling vehicular movements. Indeed, many non-specific mobility models employed in VANETs simulations ignore these guidelines, and, thus, fail to reproduce peculiar aspects of vehicular motion, such as car acceleration and deceleration in the presence of nearby vehicles, queuing at road intersections, clustering caused by semaphores or traffic lights, vehicular congestion and traffic jams.

Manhattan Grid Mobility Model

In Bai et al. (2003), two models relevant to vehicular mobility were described, namely Freeway Mobility and Manhattan Mobility. The Freeway model limits vehicles to 1-D motion in either direction. Vehicles are tied to one of several lanes; the speed is dependent on a vehicle's previous speed; and in the 'car-following' restriction, a following vehicle cannot exceed the speed of a preceding vehicle to avoid approaching within a safety distance. The Manhattan model, an extension of the Freeway model, restricts the number of lanes in either direction to just one, but introduces a turning probability to give greater mobility. Both Freeway and Manhattan are related in that they should result in high spatial and temporal dependency. Though Bai et al. (2003) introduces maps, in our generic simulations we follow the common assumption that there is a rectangular grid of streets. This layout is characteristic of many North American city centers but differs from older cities such as London, with more irregular street layouts, or more recent North American urban areas.

Constant Speed Motion (CSM) Mobility Model

The Constant Speed Motion (CSM), the simplest of the driver models, does not produce realistic motions (as it is possible for vehicles to overlap during motion). In this model the speed of each vehicle is determined on the basis of the local state of each car and any external effect is ignored. A vehicle follows a random movement (across the road topology) in the sense that a destination is selected and vehicle moves towards it, possibly pausing at intersections. While this model may work for isolated cars, they fail to reproduce realistic movements of groups of vehicles (Fiore et al., 2007).

Fluid Traffic Model (FTM) Mobility Model

In the Fluid Traffic Model (FTM), the traffic density affects a vehicle's speed [66]. In this model the speed of car considers the presence of nearby vehicles. However the car mobility is considered for single lanes only and does not take into account the interaction between the cars on multiple lanes. This model describes the vehicle's speed as a monotonically decreasing function of the traffic density. When traffic congestion reaches a critical state, then the node speed is constrained by following equation.

$$s = max\left[s_{min}, s_{max}\left(1 - \frac{k}{k_{jam}}\right)\right] \quad (9)$$

where s is the output speed, s_{min} and s_{max} are minimum and maximum speed constraints respectively, k_{jam} is the vehicular density at which a traffic jam is declared, and k is the current vehicle density of the road the vehicle of interest is moving on. $k = n / l$, where n is the number of vehicles on that road and l is the length of the road (Fiore et al., 2007).

Intelligent Driver Model (IDM) Mobility Model

The *Intelligent Driver Model (IDM)* accords with the car-following model (Treiber et al., 2000) based on live observations. The instantaneous acceleration of a vehicle is computed by the equation below.

$$\frac{dv}{dt} = a\left[1 - \left(\frac{v}{v_0}\right)^4 - \left(\frac{s^*}{s}\right)^2\right] \quad (10)$$

where a is the maximum acceleration of the vehicle, v is the current speed of the vehicle, v_0 is the desired velocity, s is the distance from the preceding vehicle and s^* is a desired dynamical distance, as given by equation.

$$s^* = s_0 + vT + \frac{v\Delta v}{2\sqrt{ab}} \quad (11)$$

where s_0 is the minimum desirable bumper-to-bumper distance, and T is the minimum safe time between two vehicles. The speed difference with respect to front vehicle velocity is Δv, and the normal acceleration and de-acceleration rate is a and b.

Intelligent Driver Model with Intersection Management (IDM-IM)

VanetMobiSim modifies IDM by modeling intersection management (IDM-IM) as a lead vehicle approaches an intersection. Whenever a vehicle approaches an intersection with stop signs or traffic lights the parameters in following equation are used in IDM-IM to set its speed.

$$\begin{cases} S = \sigma - S \\ \Delta v = v \end{cases} \quad (12)$$

where A is now the current distance to the intersection and S is a safety distance, as the vehicle does not stop exactly at the intersection. Once the vehicle is halted at the intersection its behaviour (to pass or to wait) is handled based on other vehicles on other roads leading into the intersection and the state of the traffic lights if present. This is described in Fiore et al. (2007) in detail.

Intelligent Driver Model with Intersection Management (IDM-LC)

The IDM-IM includes lane change and overtaking behaviour in the IDM-LC model. At intersections, if a lane ceases to exist on the other side of the intersection, a vehicle will wait until a gap appears. More generally, for overtaking using lanes, a game-theoretical model (Ketsing et al., 2000) is applied. A lane change will occur if the advantage of a vehicle changing lanes to a new lane is greater than the disadvantage to following vehicles in the current lane and the new lane. This is measured in terms of acceleration, *a*, by inequalities in equations (13) and (14):

$$a^l - a \pm a_{bias} > p\left(a_{cur} - a_{new} - a^l_{cur} - a^l_{new}\right) + a_{thr} \quad (13)$$

$$a^l_{new} > -a_{safe} \quad (14)$$

Thus $\left(a^l - a\right)$ is the gain in acceleration from moving over lane l, with similar expressions for the loss in acceleration for a following vehicle in the current lane and the new lane (when the lane swapping vehicle joins it). p is a driver politeness factor, a_{bias} is a lane bias factor encouraging lane changes to a particular side, and a_{safe} restricts movements avoid a driver in the new lane having to decelerate too quickly as a result of the lane change. The term a_{thr} is a minimum acceleration below which lane changing has limited value.

FURTHER RESEARCH DEVELOPMENTS

MANETs were the subject of pioneering research but the pace of this research has slackened off in recent years. This is unfortunate, as MANETs are now being actively pursued as a way of extending the coverage and flexibility of existing cellular networks. This may be a reflection of the increased specialization that has been introduced into MANET research, which makes it difficult to keep abreast of both applications across the MANET, including multimedia streaming and ad hoc networking research in general. However, it is also the case that ad hoc network investigations are shifting to specialized variations of these networks such as VANETS.

For VANETS, the emerging IEEE 802.11p introduces rapid connection of ad hoc nodes (vehicles), higher transmitter powers (33 dBm for emergency vehicles on a highway), a channel structure that includes safety and application channels, and IEEE 802.11e traffic prioritization. However, in an urban environment node interference may restrict high transmitter powers and urban canyons confine the wireless signal. Therefore, multi-path modelling through the Rayleigh, Rician (for line-of-sight), Nakagami, and Lognormal models may well need to be introduced into detailed simulations. Alternatively, multi-ray models represent a means of estimating path loss with lower computational complexity.

There will also be a need to reconcile (asfar as the network itself is concerned) the very different conditions that exist within urban environments from those on highways. How does the VANET make the transition between open road conditions in which the network can easily become partitioned to an urban setting in which contact between vehicles is likely to be more gregarious.

CONCLUSION

In this chapter we discussed the two technologies MANET and VANET in detail. First an overview of MANET and various models at lower layers of the protocol stack was provided. Furthermore, before moving towards the application it is necessary to use an appropriate ad hoc routing protocol. As the basis of research in this area, it is necessary to have an in-depth analysis of routing protocols. This chapter classifies the routing protocols based on their different approaches. There follows an overview of well-known ad hoc routing protocols highlighting the strengths and weakness of each protocol.

Secondly an overview of VANET technology and standard follows. This leads to an overview of ad hoc and vehicular mobility models which were adapted for the simulation of ad hoc and vehicular networks. Along with this and introduction to mobility models generators was given.

REFERENCES

Abolhasan, M., Wysocki, T., & Dutkiewicz, E. (2003). A review of routing protocols for mobile ad hoc networks. *Ad Hoc Networks*, *2*(1), 1–22. doi:10.1016/S1570-8705(03)00043-X

Bai, F., Sadagopan, N., & Helmy, A. (2003). Important: A framework to systematically analyze the impact of mobility on performance of routing protocols for ad hoc networks. *Proceedings of the IEEE INFOCOM*, (pp. 825-835).

Basagni, S., Chlamtac, I., Syrotiuk, V. R., & Woodward, B. A. (1998). A distance routing effect algorithm for mobility (DREAM). *Proceedings of ACM/IEEE International Conference on Mobile Computing and Networking*, (pp. 76-84).

Biswas, S., Tatchikou, R., & Dion, F. (2006). Vehicle-to-vehicle wireless communication protocols for enhancing highway traffic safety. *IEEE Communications Magazine*, *44*(1), 74–82. doi:10.1109/MCOM.2006.1580935

Bluetooth Special Interest Group. (2001). *Specification of the Bluetooth System version 1.1 core, specification Volume 1.*

Blum, J. J., Eskandarian, A., & Hoffman, L. J. (2004). Challenges of inter-vehicle ad hoc networks. *IEEE Transactions on Intelligent Transportation Systems*, *5*(4), 347–351. doi:10.1109/TITS.2004.838218

Boukerche, A. (2004). Performance evaluation of routing protocols for ad hoc wireless networks. *Mobile Networks and Applications*, *9*, 333–342. doi:10.1023/B:MONE.0000031592.23792.1c

Broch, J., Johnson, D., & Maltz, D. (2003). *The dynamic source routing protocol for mobile ad hoc networks*. IETF Internet Draft.

Broch, J., Maltz, D. A., Johnson, D. B., Hu, Y.-C., & Jetcheva, J. (1998). A performance comparison of multi-hop wireless ad hoc network routing protocols. *Proceedings of ACM/IEEE International Conference on Mobile Computing and Networking*, (pp. 85–97).

Chiang, C. C., Wu, H. K., Liu, W., & Gerla, M. (1997). Routing in clustered multihop, mobile wireless networks with fading channel. *Proceedings of IEEE SICON*, (pp. 197-211).

Clausen, T., & Jacquet, P. (2003). *Optimized link state routing protocol (OLSR)*. IETF, RFC 3626.

Corson, S., & Macker, J. (1999). *Mobile ad hoc networking (MANET): Routing protocol performance issues and evaluation consideration*. IETF, RFC 2501.

Das, S. R., Perkins, C. E., & Royer, E. M. (2000). Performance comparison of two on-demand routing protocols for ad hoc networks. *Proceedings of the IEEE Conference on Computer Communications*, (pp. 3-12).

Eichler, S. (2007). Performance evaluation of the IEEE 802.11 p WAVE communication standard. *Proceedings of the IEEE 66th Vehicular Technology Conference*, (pp. 2199-2203).

Feeney, L. M. (1999). *A taxonomy for routing protocols in mobile ad hoc networks. Technical Report*. Swedish Institute of Computer Science.

Fiore, M., Haerri, J., Filali, F., & Bonnet, C. (2007). Vehicular mobility simulation for VANETs. *Proceedings of the 40th Annual Simulation Symposium*, (pp. 301-307).

Giordano, S. (2001). Mobile ad-hoc networks . In Stojmenovic, I. (Ed.), *Handbook of wireless network and mobile computing*. Chichester, UK: John Wiley & Sons.

Goldsmith, A. (2005). *Wireless communications*. Cambridge, UK: Cambridge University Press.

Goldsmith, J., & Wicker, S. (2002). Design challenges for energy-constrained ad hoc wireless networks. *IEEE Wireless Communications Magazine*, *9*(4), 8–27. doi:10.1109/MWC.2002.1028874

Gupta, P., & Kumar, P. R. (2000). The capacity of wireless networks. *IEEE Transactions on Information Theory*, *46*(2), 388–404. doi:10.1109/18.825799

Haas, Z. J., & Tabrizi, S. (1998). On some challenges and design choices in ad-hoc communications. *Proceedings of IEEE Military Communications Conference*, (pp. 187-192).

Hekmat, R., & Van Mieghem, P. (2004). Interference in wireless multi-hop ad-hoc networks and its effect on network capacity. *Wireless Networks*, *10*(4), 389–399. doi:10.1023/B:WINE.0000028543.41559.ed

IEEE 802.16e-2005. (2005). *IEEE standards for local and metropolitan area networks, part 16: Air interface for fixed and mobile broadband wireless access systems*.

ISO/IEC 802.11 (1999). *ANSI/IEEE Std 802.11, part 11: Wireless LAN medium access control (MAC) and physical layer (PHY) specification.*

Jiang, D., & Delgrossi, L. (2008). IEEE 802.11 p: Towards an international standard for wireless access in vehicular environments. *Proceedings of IEEE Vehicular Technology Conference*, (pp. 2036-2044).

Jiang, D., Taliwal, V., Meier, A., Holfelder, W., & Herrtwich, R. (2006). Design of 5.9 GHz DSR-based vehicular safety communication. *IEEE Wireless Communications*, *13*(5), 36–43. doi:10.1109/WC-M.2006.250356

Jiang, H., & Garcia-Luna-Aceves, J. J. (2001). Performance comparison of three routing protocols for ad hoc networks. *Proceedings of IEEE Tenth International Conference on Computer Communications and Networks*, (pp. 547 – 554).

Jiang, S., Liu, Y., Jiang, Y., & Yin, Q. (2004). Provisioning of adaptability to variable topologies for routing schemes in MANETs. *IEEE Journal on Selected Areas in Communications*, *22*, 1347–1356. doi:10.1109/JSAC.2004.829352

Joa-Ng, M., & Lu, I. (1999). A peer-to-peer zone-based two-level link state routing for mobile ad hoc networks. *IEEE Journal on Selected Areas in Communications*, *17*(8), 1415–1425. doi:10.1109/49.779923

Johansson, P., Larsson, T., & Hedman, N. (1999). Scenario-based performance analysis of routing protocols for mobile ad-hoc networks. *Proceedings of ACM/IEEE International Conference on Mobile Computing and Networking*, (pp. 195 – 206).

Johnson, D. B., & Maltz, D. A. (1996). Dynamic source routing (DSR) in ad hoc wireless networks . In Imielinski, K. (Ed.), *Mobile computing*. Kluwer Academic Publishers. doi:10.1007/978-0-585-29603-6_5

Johnson, D. B., Maltz, D. A., & Broch, J. (2001). DSR the dynamic source routing protocol for multihop wireless ad hoc networks . In Perkins, C. E. (Ed.), *Ad hoc networking*. Boston, MA: Addison-Wesley.

Jubin, J., & Tornow, J. D. (1987). The DARPA packet radio network protocols. *Proceedings of the IEEE*, *75*(1), 21–32. doi:10.1109/PROC.1987.13702

Kesting, A., Treiber, M., & Helbing, D. (1999). General lane-changing model MOBIL for car-following models. *Journal of the Transportation Research Board*, *1*, 86–94.

Ko, Y. B., & Vaidya, N. H. (1998). Location-aided routing (LAR) in mobile ad hoc networks. *Proceedings of ACM/IEEE MOBICOM.*

Ko, Y. B., & Vaidya, N. H. (2000). Location aided routing (LAR) in mobile ad hoc networks. *Wireless Networks*, *6*(4), 307–321. doi:10.1023/A:1019106118419

Lee, J. S., Hsu, J., Hayashida, R., Gerla, M., & Bagrodia, R. (2002). Selecting a routing strategy for your ad hoc network. *Proceedings of ACM Symposium on Applied Computing*, (pp. 906 – 913).

Murthy, C. S. R., & Manoj, B. S. (2004). *Ad hoc wireless networks, architecture and protocols*. New York, NY: Prentice Hall.

Murthy, S., & Garcia-Luna-Aceves, J. J. (1996). An efficient routing protocol for wireless networks. *Mobile Networks and Applications*, *1*(2), 183–197. doi:10.1007/BF01193336

O'Hara, B., & Petrick, A. (2004). *IEEE 802.11 handbook: A designer's companion*. Chichester, UK: Wiley & Sons.

Park, V. D., & Corson, M. S. (1997). A highly adaptive distributed routing algorithm for mobile wireless networks. *Proceedings - IEEE INFOCOM*, 1405–1413.

Pei, G., Gerla, M., & Chen, T.-W. (2000). Fisheye state routing: A routing scheme for ad hoc wireless networks. *Proceedings of IEEE International Conference on Communications*, (pp. 70-74).

Pei, G., Gerla, M., & Chen, T.-W. (2002). *Fisheye state routing: A routing scheme for ad hoc wireless networks.* IETF internet draft.

Perkins, C. E., & Bhagwat, P. (1994). Highly dynamic destination-sequenced distance-vector routing (DSDV) for mobile computers. *Proceedings of ACM SIGCOMM's Conference on Communications Architectures, Protocols and Applications*, (pp. 234 – 244).

Perkins, C. E., & Royer, E. M. (1999). Ad-hoc on-demand distance vector routing (AODV). *Proceedings of the 2nd IEEE Workshop on Mobile Computing Systems and Applications*, (pp. 90-100).

Perkins, C. E., Royer, E. M., & Das, S. R. (2003). *Ad hoc on demand distance vector (AODV) routing*. IETF Internet Draft.

Perkins, C. E., Royer, E. M., Das, S. R., & Marina, M. K. (2001). Performance comparison of two on-demand routing protocols for ad hoc networks. *IEEE Personal Communication, 8*(1), 16-28.

Postel, J. (1980). *User datagram protocol*. RFC 768.

Ramasubramanian, V., Haas, Z. J., & Sirer, E. G. (2002). SHARP: A hybrid adaptive routing protocol for mobile ad hoc networks. *Proceedings of the 4th ACM International Symposium on Mobile Ad Hoc Networking & Computing*, (pp. 303-314).

Rappaport, T. (2001). *Wireless communications: Principles and practice*. New York, NY: Prentice Hall.

Stallings, W. (2000). *Wireless communications and networks*. New York, NY: Prentice Hall.

Stepanov, I., Hähner, J., Becker, C., Tian, J., & Rothermel, K. (2003). A meta-model and framework for user mobility in mobile networks. *Proceedings of the 11th IEEE International Conference on Networks*, (pp. 231- 238).

Stevens, W. R. (1993). TCP/IP illustrated: *Vol. 1. The protocols*. Boston, MA: Addison-Wesley.

Stojmenovic, I. (2002). Position-based routing in ad hoc networks. *IEEE Communications Magazine*, *40*(7), 128–134. doi:10.1109/MCOM.2002.1018018

Takai, M., Martin, J., & Bagrodia, R. (2001) Effects of wireless physical layer modeling in mobile ad hoc networks. *Proceedings of the 2nd ACM International Symposium on Mobile Ad Hoc Networking & Computing*, (pp. 87-94).

Treiber, M., Hennecke, A., & Helbing, D. (2000). Congested traffic states in empirical observations and microscopic simulations. *Physical Review E: Statistical Physics, Plasmas, Fluids, and Related Interdisciplinary Topics*, *62*, 1805–1824. doi:10.1103/PhysRevE.62.1805

Trung, H. D., Benjapolakul, W., & Duc, P. M. (2007). Performance evaluation and comparison of different ad hoc routing protocols. *Computer Communications*, *30*(11-12), 2478–2496. doi:10.1016/j.comcom.2007.04.007

Weber, S., Andrews, J. G., & Jindal, N. (2007). The effect of fading, channel inversion, and threshold scheduling on ad hoc networks. *IEEE Transactions on Information Theory*, *53*(11), 4127–4149. doi:10.1109/TIT.2007.907482

Woo, S.-C. M., & Singh, S. (2001). Scalable routing protocol for ad hoc networks. *Wireless Networks*, *7*(5), 513–529. doi:10.1023/A:1016726711167

Xu, K., Gerla, M. M., & Bae, S. (2002). How effective is the IEEE 802.11 RTS/CTS handshake in ad hoc networks? *Proceedings of IEEE GLOBECOM* (pp. 72-76).

Xu, Q., Mak, T., Ko, J., & Sengupta, R. (2004). Vehicle-to-vehicle safety messaging in DSRC. *Proceedings of the 1st ACM International Workshop on Vehicular Ad Hoc Networks*, (pp. 19-28).

Yang, X., & Vaidya, N. (2005). On physical carrier sensing in wireless ad hoc networks. *Proceedings - IEEE INFOCOM*, 2525–2535.

Zheng, H., & Boyce, J. (2001). An improved UDP protocol for video transmission over Internet to wireless networks. *IEEE Transactions on Multimedia*, *3*(3), 356–365. doi:10.1109/6046.944478

Zigbee Alliance. (2003). *Zigbee specification*. ZigBee Document 053474r06, Version 1.

Zygmunt, Y. G., & Marc, R. P. (2001). ZRP: A hybrid framework for routing in ad hoc networks . In Perkins, C. M. (Ed.), *Ad hoc networking* (pp. 221–253). Boston, MA: Addison-Wesley.

ADDITIONAL READING

Aggelou, G. (2005). *Mobile ad hoc networking: Design and integration*. New York, NY: McGraw-Hill.

Conti, M., Crowcroft, J., & Passarrella, A. (2007). *Multi-hop ad hoc networks from theory to reality*. New York, NY: Nova Publishers.

Misra, S., Woungang, I., & Misra, S. C. (2009). *Guide to wireless ad hoc networks*. London, UK: Springer Verlag.

KEY TERMS AND DEFINITIONS

Ad Hoc Network: A network in which nodes communicate in a store and forward fashion between each other. Therefore, there is no centralized access point and no infrastructure linking those access points. Hence, the alternative name for such networks is infrastructureless network. Routing across these networks is usually over multiple hops and the routing algorithms work in distributed fashion between the nodes themselves. Apart from mobile networks such as MANETs and VANETs, static sensor networks can also be set up as ad hoc networks.

MANET: A generalized mobile ad hoc network.

Mobility Model: Describes the motion patterns of the nodes within a dynamic ad hoc network. The model acts as input to a mobility generator, which outputs a motion trace file that can be input to a network simulator. Mobility models can be designated by a set of rules that may include mathematical formula for the motion of nodes. Vehicular mobility models are constrained by the topology of a road system and are informed by research into driver behavior.

VANET: The translation of the MANET concept to vehicular networks.

Wireless Fading: This is produced by the combined effect of obstructions such as buildings and the motion of a wireless transceiver upon wireless propagation. The result is time varying reflection and diffraction off such obstructions. Slow fading occurs as the wireless transceiver moves into a different wireless environment, whereas fast fading occurs is superimposed upon slow fading and arises from variations in propagation over a short time span.

ENDNOTES

1 Provided there is no frequency selective fading.

2 The choice of a 10 MHz scaling of IEEE 802.11a for IEEE 802.11p ensures that the guard interval is sufficient to cover the root mean square Doppler delay spread expected in high mobility highway settings .

Section 2
Research Directions

Chapter 7
P2P Streaming over MANET and VANET

Nadia N. Qadri
COMSATS Institute of Information Technology, Pakistan

Martin Fleury
University of Essex, UK

ABSTRACT

Mobile Ad Hoc Networks (MANETs) and Vehicular Ad Hoc Networks (VANETs) as mobile wireless networks are challenging environments as there is no centralized packet routing mechanism. Packet delivery is normally multi-hop and may encounter out-of-range intermediate network nodes on the routing path. There may be problems of energy consumption in MANETs and of constrained routing paths in VANETs. Consequently, introducing real-time video streaming into these environments is problematic. Peer-to-peer (P2P) streaming from multiple sources is a way of strengthening video streaming in these circumstances. In this chapter, P2P streaming is combined with various video error resilience mechanisms that mostly take advantage of the multiple paths available in such networks. As video streams are sensitive to errors the impact of wireless channel errors should be assessed and, for VANETs, realistic mobility models should be modeled, especially in urban settings. The chapter looks in detail at how video source coding can assist in the protection of video streams, in that respect comparing various forms of multiple description coding.

INTRODUCTION

The ever increasing growth of wireless technologies together with the benefits of employing such flexible systems has made them likely to be used for real-time multimedia communication. Wireless networks allow users to communicate anywhere, anytime, and on-the-fly. *Mobile Ad hoc Networks (MANETs)* are a type of pervasive network that truly support pervasive computing. Future advances in pervasive computing rely on advancements in mobile communication, which includes both infrastructure-based wireless networks and non infrastructure-based MANETs. The traditional infrastructure-based communication model is not adequate for today's user requirements. In

DOI: 10.4018/978-1-4666-1613-4.ch007

many situations, communication between mobile hosts cannot rely on any fixed infrastructure. The cost and delay associated with installation of infrastructure-based communication model may not be acceptable in dynamic environments such as disaster scenes, battlefield, and inter-vehicular communications. MANETs are an effective solution in these scenarios.

In most emergency scenarios, whether they are man-made or natural disasters, it is essential that rescue personal can communicate with each other when they are moving around the disaster area. However, the disaster itself may well have removed the communication networks within the vicinity. Usually in the case of a natural disaster the communication infrastructure is destroyed. Therefore, in such conditions there is need for a communication network that does not depend on any infrastructure. Various technologies present themselves such as satellite, IEEE 802.16 (WiMAX), and wireless mesh networks, all of which can employ IP routing. A MANET is such a network that does not require any pre-existing fixed network, allowing rescue teams to communicate whether they are on foot equipped with handheld devices or travelling on vehicles, when the network is usually knows as *VANET (Vehicular Ad Hoc Network)* or Car-to-Car network. Real-time video transport over an ad hoc network is a challenging task, owing to the dynamic topology, the absence of an established infrastructure for centralized administration and the limited processing and power capabilities of mobile terminals.

Much research has been carried out to provide solutions to sharing information in the form of data or images. However, very little effort has been put into examining communication of video over a MANET or a VANET. Real-time video can better describe the current conditions or give a clearer picture of the disaster area than textural data or just images. The motivation of this Chapter is to assess robust *Peer-to-Peer (P2P)* video streaming over these infrastructureless networks. P2P streaming and MANETs have turned out to be two of the most active research areas for pervasive computing. P2P systems offer the means to realize decentralized networks, which can be used to share resources over the internet. On the other hand, a MANET is a spontaneous network made of mobile nodes connected wirelessly but without relying on a specific infrastructure network. These areas were developed independently of each other with the result that there is insufficient verification of whether the P2P distribution paradigm and specifically P2P real-time video streaming would work on MANETs. Along with that, there are still several issues to be addressed if MANET is to be employed for highly demanding, real-time multimedia applications. More specifically, more robust solutions are required to ensure that the delay, jitter (variation of delay) and packet loss requirement of real-time multimedia applications can be satisfied for communication over unreliable, time-varying ad hoc network. Fortunately, multimedia applications can tolerate a certain level of packet loss depending on the application, error concealment strategy at the receiver side, and compression scheme adopted.

Thus, to achieve acceptable wireless ad hoc video communication in general, and wireless video in particular, a number of key requirements need to be addressed. Firstly, there should be a solution for easy adaptability to wireless bandwidth fluctuations due to channel interference, dynamic topology and cross (competing) traffic. Secondly, it is necessary to provide robustness to partial data losses due to high packet error rates. Thirdly, energy issues are becoming important – particularly how to avoid mobile devices remaining awake for longer periods than is necessary in order to act as streaming sources. Fourthly, the problem of proving security on mobile devices with limited computational resources will require lightweight security solutions. Finally there should be support for multiple sources with multiple paths to cope with nodes leaving the network or suffering adverse channel conditions. However, this Chapter is primarily concerned with video streaming issues and cannot pretend to cover all these requirements.

The aim of this Chapter is to analyze robust P2P video streaming over MANETs and VANETs and to propose better approaches for achieving that. This can be accomplished by means of multi-path streaming combined with error-resilient video coding. Along with that video delivery using P2P streaming from multiple sources is effective in guarding against the departure of one or more of the sources from the overlay network. Careful consideration of ad hoc routing protocols was also necessary. Channel conditions across the network path needed to be planned for as well as the response to different network node densities and different node speeds. Furthermore, to model VANETs it is necessary to model mobility carefully. Developing applications, especially real-time ones, for wireless VANETs requires a reasonable assurance of the likely performance of the network, because wireless propagation strongly influences performance, especially in an urban environment.

VIDEO STREAMING TECHNIQUES

H.264/AVC Video Codec

In this Chapter we use an *H.264/AVC (Advanced Video Coding)* standard (Ozbek & Tumnali, 2005) codec to produce different video coding and error resilience techniques. In 2001, the ITU-T *Video Coding Experts Group (VCEG)* together with the ISO/IEC *Moving Picture Experts Group (MPEG)* formed the *Joint Video Team (JVT)* to develop a new video coding standard the name given ITU H.264/AVC also known as ISO MPEG-4 Part 10, with an increased compression efficiency almost twice as that of the MPEG-2 (Ozbek & Tumnali, 2005). Usually increase in compression efficiency may lead to substantial increase in complexity; this was recorded to be approximately four times and nine times for the H.264/AVC decoder and encoder as compared to MPEG-2. It is important to note (Ozbek & Tumnali, 2005) that this increase in complexity also depends on the selection of the different features and profiles described in the standard

Apart from better coding efficiency, another important feature of the Standard is enhanced error resiliency and adaptability to various networks (Kumar et al., 2006) with video representation ranging from 'conversational' (video telephony) to 'non-conversational' (storage, broadcast, or streaming) application (Narkede & Kant, 2009) In order to increase the flexibility and adaptability, H.264/AVC has adopted a two-layer structure (Sullivan & Wiegand, 2005) the *Video Coding Layer (VCL)*, which is designed to efficiently represent the video content, and a *Network Abstraction Layer (NAL)*, which formats the VCL representation of the video and provides header information for ready transmission over the network (Sullivan & Wiegand, 2005). The NAL can provide compressed video data in two formats, for the stream-based protocols like H.320, H.324 or MPEG2 and for the packet-based protocols such as the Real-Time Protocol (RTP)/Internet Protocol (IP) and the well-known TCP/IP. For the stream-based protocols the data are provided with start codes such that the transport layers and the decoder can easily identify the structure of the bitstream, while for the packet-based the data are provided without these start codes (Kumar et al., 2006). In H.264/AVC, data from the VCL are packetized into a *Network Abstraction Layer unit (NALU)*. Each NALU is encapsulated in an RTP packet, with subsequent addition of UDP/IP headers.

Video compression through video coding[1] refers to the reduction of the quantity of data used to represent video sequences. In the standard video coding method, compression is based on spatial and temporal redundancy reduction along with Entropy Coding[2] (Huffman or Arithmetic) (Ozbek & Tumnali, 2005). Spatial redundancy reduction results in Intra (I) frames, where each block of the picture is predicted from its neighboring coded block without reference to any other frame. The difference between the two blocks is transformed using a *Discrete Cosine Transform (DCT)*[3] a type of transformation matrix, followed by quantization and entropy coding. As the intra frames do not rely on other frames they are used for a number of operations, typically for random access and confining drift errors due to losses during transmission.

Inter frames exploit the temporal redundancy between frames. In this process the movements of objects in the neighboring frames are modeled using motion vectors. The resulting frame is subtracted from the original frame to be encoded, and the residue is transformed using DCT followed by entropy coding. Inter frames can be further divided in to Predictive (P) frames and Bi-predictive (B) frames. In P frames the prediction is made from earlier P or I frames, while in B frames the prediction can be made from an earlier and/or later I and/or P frame, but not a B frame. In H.264/AVC this restriction is removed and a B frame can be used as reference for predicting other frames (Narkede & Kant, 2009). Sequence of frames grouped between two I frames is referred to as Group of Pictures (GOP) as shown in Figure 1.

Scalability

Multimedia transmission over heterogeneous networks requires a high degree of flexibility from video compression systems. In this situation, scalable transmission of video data is essential to service different clients with widely varying display and processing capabilities. A scalable bitstream is the bitstream with the property that one or more bitstream subsets that are not identical to the scalable bitstream form another bitstream (Wiegand et al., 2007). In other words, scalability refers to ability of an algorithm to decode a certain part of image or video information to obtain video at the desired spatiotemporal resolution (Wien et al., 2007), in contrast to non-scalable video, in which the bitstream is required to be decoded in its entirety to obtain the video.

There are three types of scalability, SNR (Signal-to-Noise Ratio) or quality scalability, spatial scalability, and temporal scalability. The main disadvantage of scalability is the bit-rate overhead introduced to the scalable bitstreams compared to a non-scalable stream.

Several scalability tools were introduced in previous video coding standards such as H.262/MPEG-2, H.263 and MPEG-4 visual, but these tools did not become prominent due to their lower compression efficiency (Schwarz et al., 2007). The new Scalable Video Coding (SVC) was standardized as an extension of H.264/AVC, reusing most of its components as specified in Schwarz et al., (2007), and it has a better compression efficiency compared to previous standards.

In classical SVC, a base layer and one or more enhancement layers (Ghanbari, 2003) provide scalable video quality. However, unlike *Multiple Description Coding (MDC)* (as described next) for which any description can reconstruct the video, the base layer should be successfully delivered

Figure 1. A general diagram of a GOP

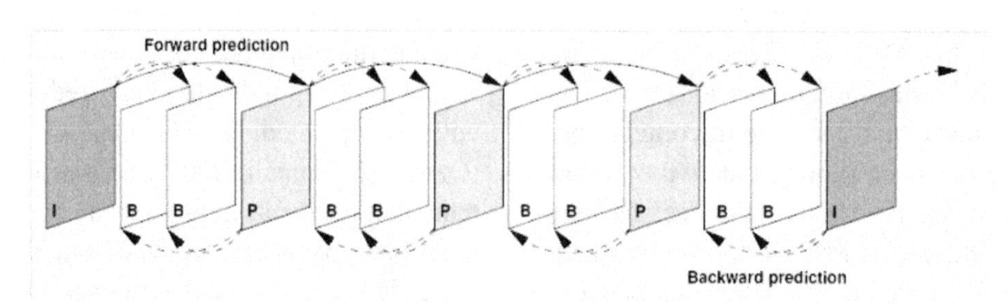

as the video cannot be reconstructed without it. In earlier experiments with layered video (Mau et al., 2001) if base layer packets were lost, then this problem had to be resolved by means of an *Automatic Repeat ReQuest (ARQ)* message.

As latency in an ad hoc network is variable and can be high, there were a limited number of settings in which layered video was competitive with MDC. However, in low bitrate SVC one option is to output one stream consisting of base layer and enhanced layer, and another stream consisting of the base layer alone. In this variety of layered video, if packets from the layered stream fail to arrive, the other stream may still serve to reconstruct the video, whereas if packets from the combined base and enhanced layer stream are lost base layer packets may still be available. This idea appears to be self-defeating but at the cost of extra bandwidth consumed by the combined stream robust video quality can be achieved.

For many video streaming applications MDC is proving to be a better solution than unicast streaming, and is also considered as a scalable video coding technique (Ketan, 2007). MDC (Wan et al., 2007) splits video streams into two or more versions or descriptions that are sent over multiple paths. Each description can serve to reconstruct the video but if more than one description is received the quality can be enhanced by combining the descriptions. MDC differs from layered video coding because in the latter the base layer must be received in order to reconstruct the video for display. Therefore, MDC has become a popular technique for real-time applications as it provides graceful video quality degradation without the need for retransmission and is focus of this research. If packet loss or delay occurs on one of the paths then this can be compensated for by the encoded bit stream from other paths. MDC may also reduce the bandwidth requirement for any one route through an ad hoc network (Chen et al., 2004) (though obviously not the overall bandwidth requirement across multiple paths), at a cost in increased coding redundancy. Various forms of splitting between the descriptions can occur including in the spatial (Franchi et al., 2005) and the frequency domain (Reibman et al., 2001) but this Chapter will examine practical temporal and spatial MDC schemes that can be implemented on mobile devices without customized codecs or excessive complexity. Specifically, temporal MDC is tested using either redundant frames or *Video Redundancy Coding (VRC)*, and spatial MDC with and without *FMO (Flexible Macro-block Ordering)* slicing (to be explained later) and error concealment. Though practical versions of MDC are used, MDC still adds complexity to the sending and receiving devices. For some applications, especially over MANETs the added complexity may still present a challenge. Jitter levels may also be increased if path diversity is employed.

Error Resilience

H.264/AVC along with its efficient compression has strong error resilient features. Error resilience is a form of error protection through source coding. Due to the growing importance of multimedia communication over wireless the range of these techniques has been expanded in the H.264/AVC codec (Wenger, 2003). Error resilience introduces limited delay and as such is suitable for real-time, interactive video streaming, especially video-telephony, and video conferencing. It is also suitable for one-way streaming over cellular wireless networks and broadband wireless access networks to the home. As physical-layer *Forward Error Correction (FEC)* is normally already present at the wireless physical layer, application-layer FEC may duplicate its role. The exception is if application-layer FEC can be designed to act as an outer code after inner coding at the physical layer, in the manner of concatenated channel coding. Various forms of ARQ are possible but in general their use has been limited in ad hoc networks, because the mobility of the nodes and multi-hop nature of routing tends to introduce more delay. For example, in Mau et al. (2001)

if packets from the base layer of a layered video stream (sent over multiple paths) failed to arrive then only a single ARQ is permitted.

H.264/AVC has combined the error-resiliency schemes of the previous coders along with some techniques introduced newly or implemented differently such as IntraMB (Intra MacroBlock), DP (Data partitioning), and slicing borrowed from H.261, H.262 and MPEG 1 and 2. Of the newly introduced techniques of H.264/AVC, this Chapter considers Redundant Slices or Frames and FMO.

H.264/AVC can divide a picture into slices, whose size can be as small as a MB and as large as one complete picture. MBs are assigned to slices in raster scan order, unless FMO (explained later) is used (Wenger, 2003) as shown in Figure 2, which shows a picture representation with three slices. Intra prediction across the slice boundaries is not allowed making (Osterman et al., 2004) slices as self contained units that can be decoded without referring to other slices of the frame and thus can prevent error propagation (Son & Jeong, 2008). However, the non- availability of intra prediction across the slice boundaries reduces the compression efficiency which can be further decreased with the increase in number of slices per frame. Slices can be decoded independently. However, some information from other slices may be needed for applying de-blocking filters (Sullivan & Wiegand, 2005).

Figure 2. Picture partitioned into slices (no FMO used).

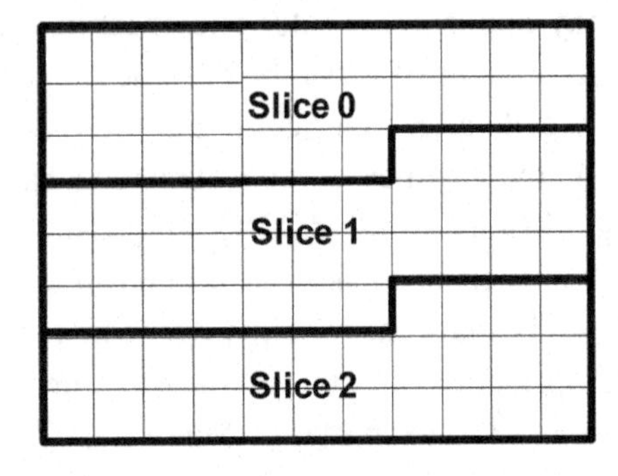

To enhance error robustness in H.264/AVC, it is possible for the encoder to send duplicate copies of some or all parts of a picture (Sullivan & Wiegand, 2005). Redundant frames (Baccichet et al., 2006) or strictly redundant slices making up a frame) are coarsely quantized frames that can avoid sudden drops in quality marked by freeze frame effects if a complete frame (or slice) is lost. To improve error resilience redundant pictures intended for error resilience in H.264/AVC, can also serve to better reconstruct frames received in error. The main weakness of the redundant frame solution is that these frames are discarded if not required but they are still an efficient solution compared to including extra I-frame synchronization, as redundant frames are predictively coded and require less bits compared to I-frames. A subsidiary weakness of this scheme is the delay in encoding and transmitting redundant frames, making it more suitable for one-way communication. If the redundant frame/slice replaces the loss of the original frame/slice there will still be some mismatch between encoder and decoder. This is because the encoder will assume the original frame/slice was used. However, the effect will be much less than if no substitution took place.

FMO allows different arrangements of MBs in a slice by utilizing the concept of slice groups. The MBs are arranged in a slice in a different order to the scan order (Wenger, 2003) enhancing the error resilience (Sullivan & Wiegand, 2003). In each slice group the MBs are arranged according to macroblock to slice group map. In H.264/AVC, by varying the way in which the macroblocks are assigned to a slice (or rather group of slices), FMO gives a way of reconstructing a frame even if one or more slices are lost. Within a frame up to eight slice groups are possible. H.264/AVC provides different macroblock classification patterns. Assignment of macroblocks to a slice group can be general (type 6) but the other six types pre-define an assignment formula, thus reducing the coding overhead from providing a full assignment map. Pre-defined types are interleaved, checker board, foreground, Box out, Raster scan and wipe (Thomos et al., 2005). In the explicit case the parameter "slice_group_id" is transmitted for each *Macroblock (MB)* in the picture specifying the slice group to which it belongs.

The checkerboard type stands apart from other types, as it does not employ adjacent macroblocks as coding references, which decreases its compression efficiency and the relative video quality after decode. However, if there are safely decoded macroblocks in the vicinity of a lost packet error concealment can be applied. Consequently, the rate of decrease in video quality with an increase in loss rate is lower than for the other pre-set types.

Arranging MBs in multiple slice groups' increases error resilience, for example if one of the slice groups in the dispersed map is 'lost' due, the missing slice can be concealed by interpolation from the available slices. Experiments show (Son & Jeong, 2008) that at a loss rate of 10% in case of video conferencing, the impairments due to losses can be kept so small that it is very difficult to be observed.

In FMO the possible errors are scattered to the whole frame to avoid its accumulation in a limited area. In this way the distance between the correctly recovered block and the erroneous block is reduced. As the distance between correct and recovered block is reduced, the distortion is reduced in recovered blocks and vice versa (Kumar et al., 2006). Therefore, it is easier to conceal scattered errors as compared to the errors concentrated in a region.

P2P SYSTEMS

P2P can be broadly defined as a self organizing, decentralized distributed system that consists of potentially un-trusted, unreliable nodes with symmetric roles (Wu, 2006). P2P systems are designed for the sharing of resources such as CPU cycles, storage, files and data directly without any centralized control or hierarchical organization. P2P architectures are characterized by their ability to adapt to failures, handling and accommodating large number of nodes while maintaining acceptable connectivity and performance (Androutsellis-Theotokis & Spinellis, 2004). P2P networks are mainly based on application-level overlays built over the physical network, enabling a range of P2P applications such as P2P file sharing and P2P streaming. In P2P networks, each peer can act as server or as a client. Each peer can connect to one or more peer at any time to provide resource sharing.

The P2P paradigm has largely adopted a layered approach. A P2P overlay network built on top of the Internet provides a general purpose substrate that provides many common properties desired by distributed applications, such as self-organization, decentralization, diversity and redundancy. Such an overlay shields distributed application designers from the complexities of organizing and maintaining secure overlays, tolerating node failures, load balancing, and locating application objects.

P2P network architectures can be classified in three main categories: centralized, decentralized and hybrid (Francisciani et al., 2003) (Minar, 2001). Table 1 shows some categories, derived

Table 1. Classification of P2P systems

Centralized	Decentralized		Hybrid
	Unstructured	Structured	
Napster	Freenet	Chord	KazaA
SETI@Home	Gnutella	CAN	Morpheus
Bit Torrent		Pastry	Overnet/eDonkey2000
		Tapastry	
		Kademlia	

Table 2. Advantages and disadvantages of P2P systems based on different properties

	Centralized	**Decentralized**		**Hybrid**
		Unstructured	**Structured**	
Manageable	Yes	No	Yes	No
Coherent	Yes	No	No	partially
Extensible	No	Yes	Yes	Yes
Fault-Tolerant	No	Yes	Yes	Yes
Secure	Yes	No	No	No
Lawsuit-Proof	No	Yes	Yes	Yes
Scalable	Depends on server	May be	Yes	Depends on super peers

from (Jurca et al., 2007), whereas Table 2 shows the advantages and disadvantages of the P2P categories considering different properties.

In a centralized P2P network, the topology is based on a central indexing server (or servers) that maintains a directory of the resources available on the peers and coordinates the interaction between peers. However, after receiving the information from the central server, peers communicate directly. The advantage of server based P2P systems is that they can be more easily managed and therefore fewer security issues are likely to be involved. However they are less fault-tolerant as some of the information is only held by the central server. If anything goes wrong with the server, the whole system goes down. The extensibility and scalability of these systems depends on the capacity of the server which is usually limited by its processing power capabilities.

In decentralized P2P networks, all peers act as both server and client equally. The communication is made through multiple unicasts, where peers forward messages on behalf of others peer. The decentralized topology is further divided into two categories i.e. Unstructured P2P overlays, is formed when the overlay links are established arbitrarily by peers. Such networks can be easily constructed as a new peer that wants to join the network can copy existing links of another node and then form its own links over time and structured P2P overlays, which is precisely controlled and determined by some algorithm and were developed to improve the search mechanism to discover data. The advantage of decentralized systems is that they can be more easily extended and are more fault-tolerant than centralized P2P. However, security is a big issue in these systems.

Decentralized structured P2P system can be managed more easily and are more scalable than their unstructured counterpart. For instance, DHT may be used to manage the P2P overlay (Lua et al., 2005). In unstructured P2P systems scalability is difficult to measure due to the lack of deterministic overlay management mechanism. Increasing the number of nodes supports the resource sharing system but also increases signalling overheads as most unstructured protocols periodically broadcast messages.

In hybrid (centralized + decentralized) P2P networks, there are some super-nodes or super-peers that play a more important role than others. Peers forward their queries to super-peers which communicate with each other in a decentralized manner. Hybrid P2P systems have the same advantages as decentralized P2P systems but have difficulty in managing the peers and in this case the scalability depends on the actual number of super peers and their capabilities.

P2P over MANET

Although conceptually and practically appealing, deploying P2P over MANETs is not straightforward since many of the classic assumptions that can be made for conventional P2P over wired networks do not hold any longer. In fact, the ad-hoc network's flexibility and convenience comes with new issues (Goldsmith & Wicker, 2002). MANETs not only inherit the traditional problems of wireless networking such as bandwidth optimization, power control, and transmission quality enhancement (Giordano, 2001) but, due to the lack of a fixed infrastructure suffer from a number of complexities and design constraints that are specific to ad-hoc networks (Corson et al., 1999; Chiasserini et al., 1999). Examples are: dynamically changing network topology (Chlamtac & Redi, 1998), energy constraints, signalling overheads, variation in link and node capabilities (Vaidya, 2004), misbehaving nodes, and unreliable links.

There is a magnitude of potentially useful applications for mobile ad-hoc networks such as video streaming, but application development for mobile ad-hoc networks is not easy. This issue is exacerbated in the case of P2P applications, which are highly distributed and generate traffic surges from and to unpredictable locations.

Recently, the synergy between MANETs and P2P networks was recognized. P2P is broadly classified into two main applications: 1) P2P file sharing such as Gnutella, Bit Torrent, Chord (Stoica et al., 2001), Pastry (Rowstron & Druschel, 2001), Free-Pastry, KaZaA, and Tapestry (Zhao et al., 2001); 2) P2P streaming such as PPlive, Peercast (Zhang et al., 2008), Sopcast, Joost (Fu et al., 2007), Zattoo and Coolstreaming (Zhang et al., 2005). A number of studies have been devoted to bringing P2P file sharing protocols into ad-hoc networks. Many of these studies (Conti et al., 2004) (Datta, 2003) (Eberspacher eta l., 2004) (Yan et al., 2004) (Hu et al., 2004) are mainly conceptual, presenting architectural proposals but not evaluating them. Others (Oliveira et al., 2005) (Rajogopalan & Chien-Chung, 2006) (Cramer & Fuhrmann, 2006) (Pucha et al., 2004) try to evaluate the performance of P2P file sharing over MANETs using Gnutella, Chord, Pastry, Free pastry and Bit torrent, or propose their own (P2P file sharing algorithms e.g. (Papadopouli, 2002) (Kortuem & Schneider, 2001) Little effort has been made to experiment with P2P streaming on top of MANETs (Le et al., 2006).

Comparison Between P2P and MANET

By analyzing the synergy of MANET and P2P networks, we summarize the similarities between them as follows.

- Dynamic network topology. Nodes in P2P networks may randomly join and leave the network without any signs, which causes the network topology to change frequently with time. It is the same with MANET. Furthermore, the mobility character of mobile nodes in MANET can make network topology changes to be even more frequent.
- Multi-hop connection. Nodes in both types of networks connect with each other via multi-hop routing. Hop-to-hop connections in P2P network are typically via TCP links with physically unlimited range, while hop-to-hop connections in MANETs are via wireless links, which are limited by the radio transmission range.
- Decentralized control (infrastructure less network). Both P2P networks and MANETs have a decentralized structure. And there is no central administration point in the network.
- Node's multi-identity. Nodes in P2P networks and MANETs act both as client and a server, as well as implementing routing functionality. Each node can route and

transfer messages independently and their nodes have equivalent functionalities and capabilities. They can also provide resource downloading and communicate with each other directly.

Besides, unstructured P2P networks such as Gnutella share additional similarities with MANET, including flooding-based routing protocols and limited scalability due to bandwidth consuming traffic from flooding.

Despite these similarities there are some major differences between both the networks that are described in Table 3. A detailed discussion on similarities and differences can be found in Schollmeier et al. (2002).

Challenges in Converging P2P and MANET

The similarities and differences between P2P and MANET discussed in previous sub section lead to interesting although challenging research issues on P2P systems over MANET, such as:

- Lack of Infrastructure.
- Frequent topology changes.
- Node Churn.
- Bandwidth limitation
- Limited Battery Power.
- Scalability.

Table 3. Differences between p2p and manets

	P2P	MANET
Layer level	Application layer overlay	Network layer
Reason of creating a network	To provides service over logical infrastructure	To provides Connectivity over physical infrastructure
Purpose of communication b/w nodes	Search data do not necessarily communicate	Communicate with other users
Communication pattern	Multiple unicast with virtual broadcasting	Physical broadcasting
Nodes behavior	Static	Mobile
Communication link	Direct	Indirect
Relay nodes	Not required	Required intermediate nodes
Connection medium	Fixed wired	Wireless
Connection establishment	Hop by Hop via TCP links, Whereas the single hop path length is not physically limited	Hop by Hop via radio links, which are thus limited by the radio transmission
Connection maintenance	Comparatively Easy	Difficult
Node location	Anywhere on internet	Limited area
Network Structure	Logical structure apart from physical structure	Logical structure corresponds to physical Structure
Physical position of the node	Difficult to find	Can be roughly estimated
Routing	Reactive only (reliable algorithms not implemented yet) stops when TTL field is 0	Proactive, Reactive and Hybrid (reliable algorithms exist), Stops when destination is found
Network Topology	Changes less frequently	Changes more frequently due to mobility
Network Size	Large networks (millions of nodes)	Usually small networks (few nodes)
Network connectivity	Not effected by nodes joining or leaving	Effected by nodes joining and leaving
Bandwidth	Large bandwidth	Small bandwidth
Available Resources	Practical unlimited	limited

Both P2P and MANET have recently become popular research areas due to the wide deployment of P2P applications over the Internet and rapid progress of wireless communication. Many decentralized applications have been built upon P2P overlays and these same ideas may be useful in developing higher level services in an ad hoc setting.

P2P Video Streaming

The P2P streaming concept has nowadays been developed into several trial P2P streaming systems, such as Joost, Sopcast, Zattoo, PPlive, and Coolstream. The online broadcasting arena is evolving in response to the clear commercial interest for these new technologies in support of IPTV (TV over IP-framed networks). P2P streaming architectures can be categorized according to their distribution mechanisms. The various approaches to P2P streaming have been surveyed by (Liu et al. 2008). Two main topologies have emerged, Figure 3, i.e. tree-based (Padmanabhan et al., 2003) and mesh-based P2P (for a comparison see Sentinelli et al. (2007)) although this work is particularly concerned with the second one. In a MANET, the network topology changes randomly and unpredictably over time. Hence, an application that can easily adapt to the dynamic behavior of the ad hoc network will be an effective solution. Mesh-P2P streaming is flexible and can be managed easily in comparison to a tree-based topology. Moreover, it is not affected by the churn of peers or the effects of handoff. A mesh-based topology can also overcome the bandwidth heterogeneity present in a MANET. Mesh-based distribution is becoming more widespread than tree-based distribution and has been adopted by most successful P2P streaming systems (Fu et al., 2008) (Zhang et al., 2005) (Reza, 2006).

In the tree-based architecture, the peers are ordered hierarchically by the source, known as the parent. The parent node, in turn, sends data packets to intermediate nodes, and these nodes relay them iteratively until leaf nodes are reached. An example of tree-based streaming application is Peercast, open-source software for streaming both audio and video. A peculiarity of Peercast is that any node can specify the maximum number of incoming connections allowed. Despite introducing a good level of parallelism and distribution, this approach suffers from a number of limitations. The root, or data source, is a single point of failure, which limits the robustness of the system. The other problem is that if peers join and leave frequently the tree has to be rebuilt too often, which has a negative impact on signaling overheads, latency, and stability. Hence, implementing this kind of architecture over a MANET will be inefficient.

Figure 3. P2P overlay over MANET with (a) Tree-based multicast topology (b) Mesh-based multicast topology. Arrowed lines represent logical connections, which may be over multiple wireless hops.

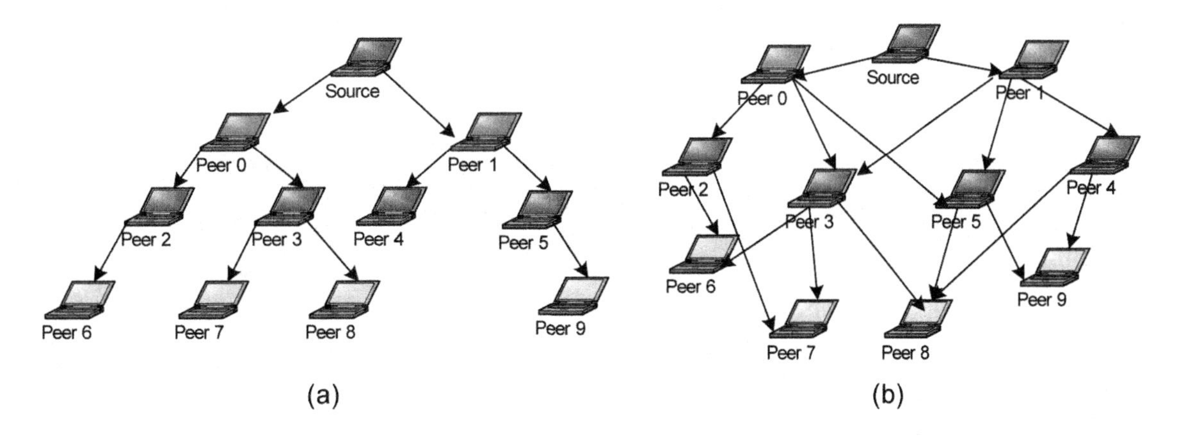

The very successful BitTorrent P2P file distribution protocol (Shah & Pâris, 2007) has been the inspiration behind mesh-P2P streaming. In the all-to-all connectivity of a mesh, the overlay network supporting the stream distribution incorporates the swarm-like content delivery introduced by BitTorrent. To deliver a video stream, the video is divided into chunks or blocks in such a way that allows a peer to receive portions of the stream from different peers and assemble them locally, leading to the delivery of good quality streams to a large number of users. The original video stream from a source is distributed among different peers (Jurca et al., 2007). A peer joining the mesh retrieves video chunks from one or more source peers. The peer also receives information about the other receiver peers from the same sources that serve the video chunks. Each peer periodically reports its newly available video chunks to all its neighbors. The chunks requested by each peer from a neighbor are determined by a packet scheduling algorithm based on available content and bandwidth from its neighbors (Jurca et al., 2007). Figure 4 shows an example of chunk distribution in mesh P2P architecture from source peer to neighboring peers. This approach is more robust than the tree-based architecture, since when a stream comes from various sources communication does not break when only a subset of peers disconnect.

P2P VIDEO STREAMING OVER MANET

Introduction

As discussed earlier, P2P streaming and MANETs have turned out to be two of the most active research areas for pervasive computing. These areas were developed independently of each other with the result that there is insufficient verification of whether the P2P distribution paradigm and specifically P2P real-time video streaming would work on MANETs. This Section considers multicast video clip streaming in an ad hoc network. Efficient delivery of video is accomplished by means of a P2P overlay network. To exploit multi-path transport the Section introduces MDC and layered video streaming without the need for base layer protection. First it is demonstrated that Mesh-based P2P streaming together with MDC over MANETs effectively provides real-time video streaming. This Section then shows that mesh P2P when combined with MDC results in improvement in delivered video quality making it acceptable for ad hoc networks. In an urban environment with higher node densities and lower node speeds, the network performance is shown to be stable and superior compared to a simple mesh with H.264/AVC distribution.

Figure 4. An example of P2P stream chunk distribution

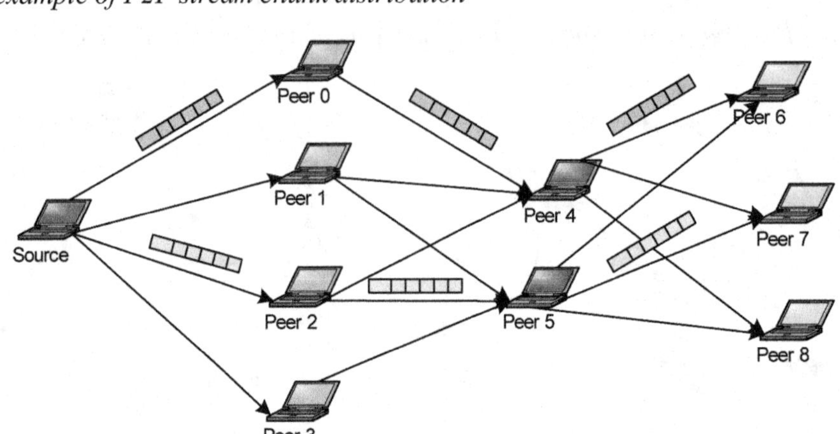

Real-time video transport over an ad hoc network is a challenging task, owing to their dynamic topology, the absence of an established infrastructure for centralized administration and the limited processing and power capabilities of mobile terminals. Computational intelligence may even be required (Natsheh & Wan, 2008) to estimate link lifetime within an ad hoc network based on signal strength and node affinity. Several decentralized P2P streaming systems such as mesh-P2P streaming (Yong et al., 2008) have been deployed to provide live and on-demand video streaming services on the Internet and the same ideas may be useful in providing real-time video streaming in ad hoc networks. Both MANETs and P2P networks are decentralized, autonomous and highly dynamic in a fairly similar way. In both cases, network nodes contribute to the overall system performance in an intermittent and unpredictably manner but nonetheless exhibit a high level of resilience and availability. Figure 5 illustrates a P2P application overlay (Oliveira et al., 2005) over a MANET, in which an overlay network is placed over the network layer. As discussed earlier, the overlay node placement is logically different to that of the physical placement of the nodes.

Effective P2P Streaming Using Multiple Description Coding

For many P2P streaming applications particularly within a mesh-P2P-type architecture, MDC is proving to be a better solution than single streams, and is also a scalable video coding technique (Ketan, 2007) (Zhengye eta l., 2007) (Leung et al., 2006). The Scalable Video Coding (SVC) extension (Schwartz et al., 2007) of H.264/AVC also allows lightweight creation of quality layers suitable for transport over multiple paths. A key problem with SVC is that there are complex interdependencies between the layers. In particular, the base layer component at the lowest temporal rate must arrive intact, before reconstruction of enhancements layers can take place. At a cost in overhead, application layer FEC can be applied. However, without knowledge of the FEC present at the PHY layer application layer FEC may simply overlap what is already present. Alternatively, negative acknowledgements can be sent (Mau

Figure 5. An example of a P2P application overlay over MANET

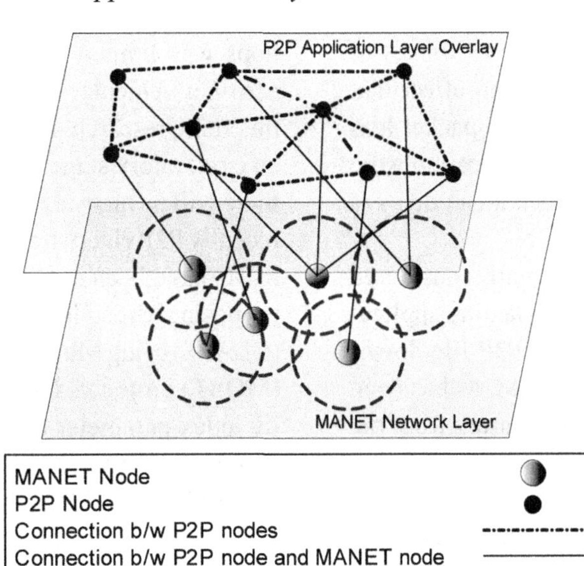

et al., 2001) but this only adds to the generally high end-to-end delays in this type of network. Therefore, in this work, though SVC remains a promising alternative, spatial Multi-Description Coding (MDC) to counter the effect of packet loss on video streams is introduced. MDC has emerged as an alternative way to improve the performance of streaming video in both P2P streaming (Zhengye et al., 2007) and over ad hoc networks (Chow & Ishii, 2007).

Providing real-time video streaming for search and rescue applications will help rescue teams to get a clearer picture of a disaster area than through text messages or still images. In these situations, a video clip will probably originate from a single source but then be distributed to other peers, whereupon the video can be streamed through the P2P overlay network. Because of the display resolution and processing power of hand-held or wearable devices used in these applications the Quarter Common Intermediate Format (QCIF) pixel resolution at a maximum of 30 frame/s (fps) and possibly as low as 10 fps is likely (Karlsson et al., 2005). This is convenient as supportable data rates across multi-hop paths could be low. However, encoded video streams are fragile as temporal redundancy is removed through the processes of motion estimation and compensation (Ghanbari, 2003). Because loss of packets from a reference frame within the 12 or 15 frames of a *Group of Pictures (GOP)* [4] has an effect that endures to the end of the GOP, the packet loss ratio is important. If video communication were to be two-way or interactive then mean delay is also important.

Little prior research has apparently considered the possibility of running P2P streaming applications across a MANET, though P2P file download has been actively investigated, as has been mentioned earlier and will be apparent from the survey in Yan (2005), which mentions eight contributions to file sharing across ad hoc networks. However, streaming applications have additional latency constraints. P2P overlay networks allow physical networks to better cope with the higher bitrates required for video streaming by increasing the number of peers and Gurses & Kim (2008) investigated ways of facilitating peer selection to optimize throughput across a MANET. To do so, congestion information across the underlay network (the physical network underlying the P2P overlay network) was utilised to assign rates between peers. Network-wide optimisation of throughput was achieved through the method of Lagrange multipliers. However, there was no investigation of node speed and limited investigation of the effect of network size and distribution protocol. In contrast, a naturally-inspired peer search algorithm, based on the behaviour of ant colonies, was investigated in Zuo et al. (2007). A simulation with 1600 nodes supported the idea, though clearly this work was at a preliminary stage in the investigation. It should also be mentioned that unless in a stadium or other crowded venue it is unlikely that there will be so many nodes available. Similarly, in Liu et al. (2008) a hierarchical arrangement of MANETs was introduced to reduce distribution latency. P2P streaming was given a practical investigation in Giordano et al. (2008), when vehicles were driven around a building in a university campus to check the feasibility of video streaming. The main finding was that provided the number of hops was limited then streaming was possible across a vehicular network. Like other work in this field, research in Giordano et al. (2008) is at an exploratory stage but it is certainly likely that there will be many cars able to act as peers. More recently P2P video streaming over MANET using MDC have been studied in Niraula et al. (2009) proposing Cross-layer and P2P based Solution (CLAPS) using Multicast Overlay Spanning Tree (MOST) protocol. The paper computes continuity index parameter to show the performance of their proposed scheme, though latency and video quality is not investigated.

Figure 6. Example of frame compensation scheme for MDC

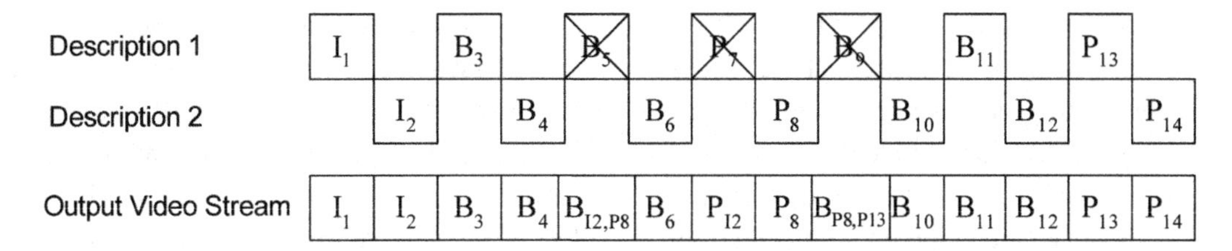

Simulation Case Study

A simplified version of MDC (Wenger et al., 1998) was used, in which two independent streams are formed from encoding odd and even frame sequences and sent over different paths. The same streams which were adopted in this Section. The well-known QCIF video clip 'Foreman' was encoded for the tests. The H.264/AVC Constant Bit Rate (CBR)-encoded data rates for description 1 and description 2 were 51.95 kbps and 51.93 kbps respectively. The frame rate of the video stream was set to be 15 fps.

Figure 6 illustrates an example of the frame compensation scheme (Chow & Isshii, 2007) tested for MDC. The frame numbers associated with each frame do not refer to a decoding sequence but to the original frame order as produced by the video camera. Suppose B5, P7 and B9 are lost then B5 can be reconstructed from I2 and P8 of description 2. As it is a bi-predictive frame it requires at least two reference frames. P7 can be decoded from I2 as it is a predictive frame so it needs only one reference and similarly B9 can be decoded from P8 and P13. In this case, closest frames will be selected for decoding to reduce the error. However if only one stream is used and suppose P7 is lost then it would not be possible to decode all the frames following P7 and this error (drifting error) will propagate until the next synchronization point i.e. intra-coded (I) frame is received, because an I-frame can be decoded independently. (An *Instantaneous Decoding Refresh (IDR)* frame is equivalent in H.264/AVC to what was previously known as an I-frame.)

Mesh-Based P2P Streaming Setup

In this Section, a mesh-P2P-type architecture is simulated, as shown in Figure 7. Here seven nodes

Figure 7. Mesh-based P2P topology sending MDC video streams from sources to receiver

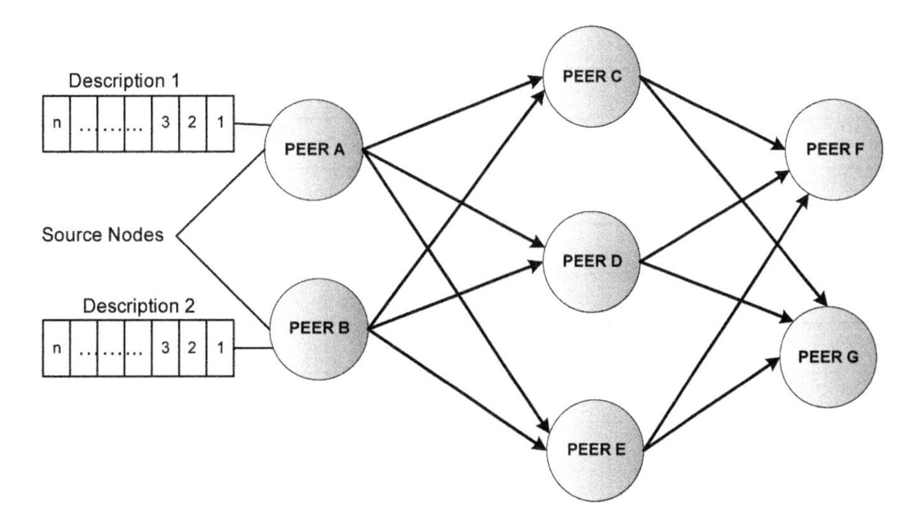

are used to form a mesh out of which two nodes are source nodes i.e. peer A and peer B each have an independent video description. Three nodes (peer C, peer D and peer E) join with these source nodes to retrieve the video content. These three nodes then connect to two further nodes (peer F and peer G) to serve the video contents that they are to receive from peers A and B. Here, nodes C, D and E download and upload at the same time.

In order to identify the issues and limitation of P2P over MANETs, GloMoSim simulator (Zheng et al., 1998) have been extended to support P2P traffic sources mimicking a mesh architecture. The simulation time was set to be 500 s. IP framing was employed with UDP transport. At the data-link layer, CSMA/CA Medium Access Control (MAC) was set up to emulate an IEEE 802.11 wireless system. The parameters for the simulations are summarized in Table 4. The random waypoint mobility model (Broch et al., 1998) was employed with 50 nodes in a roaming area of 1000×1000 m^2. The minimum speed was 0 m/s, while the maximum node speed ranged from 1 to 35 m/s, i.e. from a slow walk to fast motorbike speeds. All results were repeated 50 times and the average taken.

Figure 8 illustrates the steps followed (Chow & Ishii, 2007) to simulate MDC video streaming. In the first step, the raw video was split and encoded into even and odd frame sequences and then divided into packets from which a trace file was generated. This trace file formed the input to the GloMoSim simulation of the ad hoc network. Packet sizes and their sending times were recorded in a trace file. The video streams were sent as CBR streams with two sources sent over different paths to different receivers. After determining which frames were lost, the received frames from the two descriptions were merged and decoded separately with the JM H.264 software to generate the output video for display.

Before assessing the performance issues of mesh P2P streaming with MDC over MANETs, it is necessary to identify the network parameters that could affect the Quality-of-Experience (QoE) of a streaming service (Agboma et al. 2008). Herein, the focus of the study was on three key parameters which can better reveal the impact of the video streaming techniques:

- Packet loss ratio: The ratio between dropped and transmitted data packets.

Table 4. Simulation parameters for experiments

Parameter	Value
Wireless technology	IEEE 802.11
Channel model	Two-ray
Max. range	250 m
Data Rate	2 Mbps
Roaming area	1000×1000 m^2
Pause time	5 s
No. of nodes	50 (10-100)
Min. speed	0 m/s
Max. speed	10m/s (1 – 35 m/s)
Mobility model	Random waypoint
Routing protocol	AODV

Figure 8. H.624 and GloMoSim simulation environment for MDC

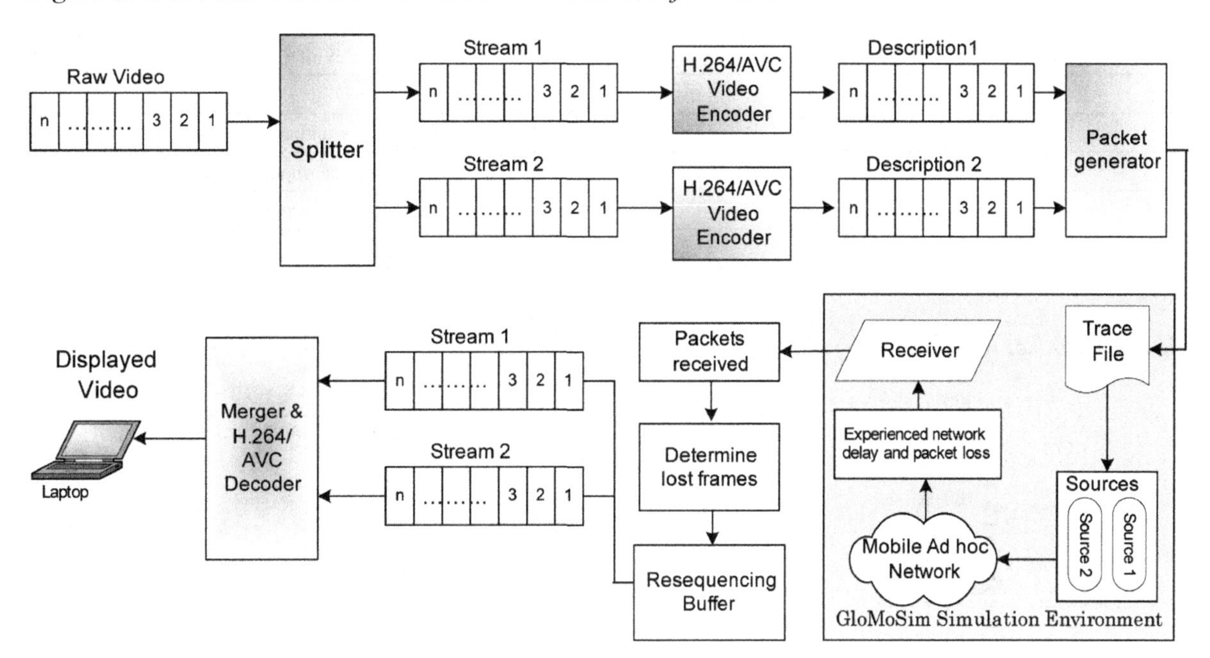

This gives an account of efficiency and the ability of the network to discover routes. Table 2 summarizes findings published in Agboma et al. (2008) correlating packet loss ratio with quality of experience. Note QoE is extremely sensitive to packet loss, and, in the next Section, it will be observed that this is one of the major impediments to P2P streaming over MANETs.

- Average end-to-end delay: The average time span between transmission and arrival of data packets. This metric includes all possible delays introduced by intermediate nodes for processing and querying of data. End-to-end delay has a detrimental effect on real-time video streaming. Its effect can be countered only to some extent by increasing buffering.
- Control overhead: The routing overhead measures the protocol's internal efficiency. It is calculated as the total number of routing (control) packets transmitted divided by the number of data packets delivered successfully at destination. Herein, routing overhead is calculated in terms of packets.
- For the video source described previously, each frame was placed in a single packet, unless the packet was from an IDR-frame, in which case two packets were employed. An IDR-frame may occupy as much as 1 KB, whereas a B-frame will commonly be encoded in less than 100 B. This implies that though encoder CBR mode is selected, an encoder output is never completely CBR. For simple mesh-P2P each packet of size approximately 500 B was sent every 60 ms to form a CBR stream.

Experimental Results

Scenario: Varying Network Size

In this scenario, the numbers of nodes was varied from 10 to 100 nodes. Otherwise, the simulation parameters were as in Table 5. Figure 9 shows the average packet loss ratio against number of nodes for mesh-P2P with and without MDC. Noticeably,

Table 5. Quality-of-Experience acceptability thresholds

Packet loss ratio [%]	QoE acceptability [%]	Video quality playback
0	84	Smooth
14	61	Brief interruptions
18	41	Frequent interruptions
25	31	Intolerable interruptions
31	0	Stream breaks

Figure 9. Packet loss ratio, varying node numbers

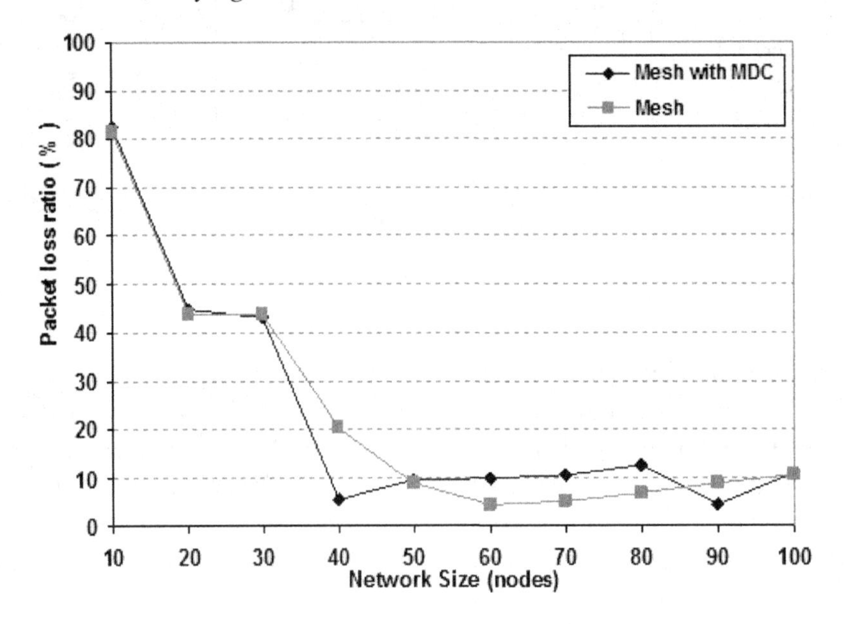

packet loss decreases considerably in the high node density configuration because there will be greater opportunities for packet transfer. In both cases, for a network size of more than 50 nodes the packet losses were within the acceptable range for P2P streaming, i.e. below 14% as per Table 5.

The results of packet losses for separate receivers are shown in Table 6, where result shown for mesh-P2P with MDC is total loss in percentage terms after compensation between the two streams. These results show that in the case of mesh-P2P with MDC most of the receivers in many configurations were able to achieve the required QoE acceptability range as compared to simple mesh-P2P. In the case of MDC, one stream can compensate the losses from the other stream, which is not possible with simple mesh-P2P.

Figure 10 shows the average end-to-end delay incurred when altering the network size (the number of nodes). For both schemes, no clear trends are apparent in terms of number of nodes but at least 0.25 s of start-up delay occurs when streaming video one-way. Interactive video (two-way) streaming would incur unacceptable time lags in the response, especially if the video was accompanied by speech. If the playback buffer can absorb at least 0.2 s of video stream, then it would be enough to avoid frame loss. Turning to Figure 11, again there is no clear trend. However, the control overhead incurred by simple mesh-P2P

Table 6. Packet loss rate (%) of receiver peers with simple mesh-P2P and mesh-P2P with MDC

Packet loss rate (%) at receiver nodes										
	Simple MESH P2P					MESH P2P with MDC				
Node	**Peer C**	**Peer D**	**Peer E**	**Peer F**	**Peer G**	**Peer C**	**Peer D**	**Peer E**	**Peer F**	**Peer G**
10	50	100	58	100	100	0	100	24	100	100
20	45	53	58	14	50	3	4	22	4	11
30	50	53	58	14	45	3	2	3	2	13
40	38	14	1	0	50	4	3	2	0	11
50	10	15	1	13	5	6	5	13	3	0
60	3	8	10	3	0	0	11	3	9	4
70	3	15	3	6	0	4	16	0	22	2
80	4	25	4	1	0	4	2	11	7	0
90	13	28	4	0	0	3	2	4	2	0
100	14	16	16	1	5	4	20	7	9	0
Speed										
1	10	5	4	10	0	4	0	0	4	4
5	6	3	0	0	0	0	0	4	18	0
10	10	15	1	13	5	6	5	13	3	0
15	0	25	0	4	0	12	0	4	0	2
20	3	21	16	4	23	8	19	5	2	0
25	14	14	25	4	23	4	11	6	3	18
30	20	6	11	10	0	0	18	27	5	0
35	24	20	26	6	5	3	26	0	0	16

Figure 10. Average end-to-end delay, varying node numbers

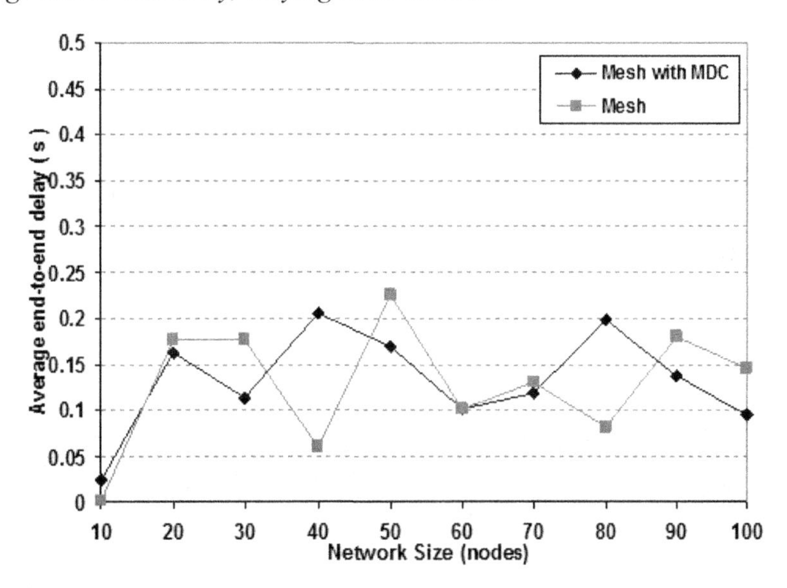

Figure 11. Control overhead, varying node numbers

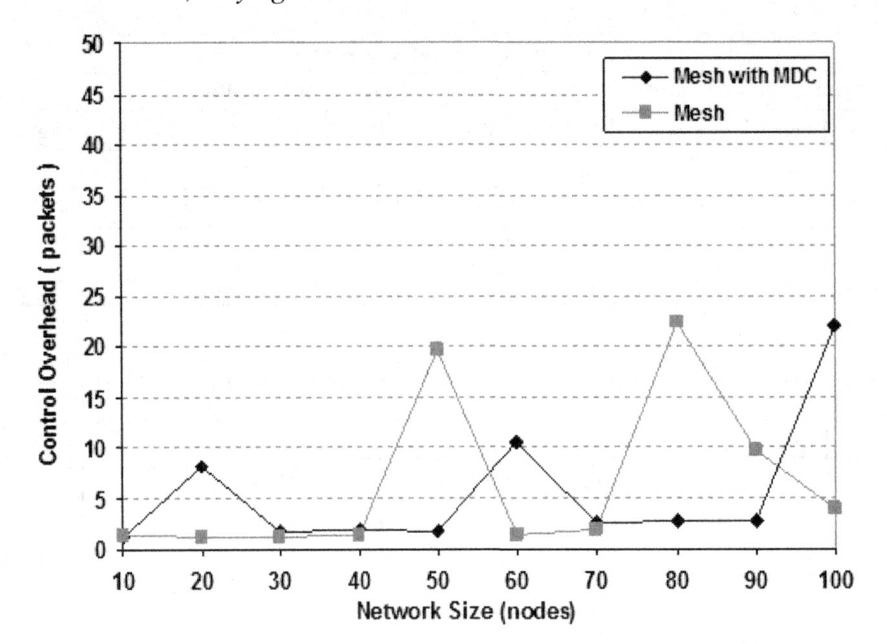

with H.264/AVC (Single) is higher than mesh-P2P with MDC, except for a network size of 100 nodes.

In Figure 12, the resulting video quality is calculated for a sample receiver, i.e. Peer E with the different packet loss ratios recorded in Table 6. An objective measure of the delivered video quality, Peak Signal to Noise Ratio (PSNR)5, is reported, with a logarithmic scale as this coincides with the human visual system. PSNR is used because network statistics can be misleading. For example, if the packets that are lost from reference frame then the impact is greater than indicated by loss statistics alone. Conversely, if packets are lost from bi-predicted B-pictures then the loss

Figure 12. Video quality for H.264(AVC) and MDC by packet loss ratio

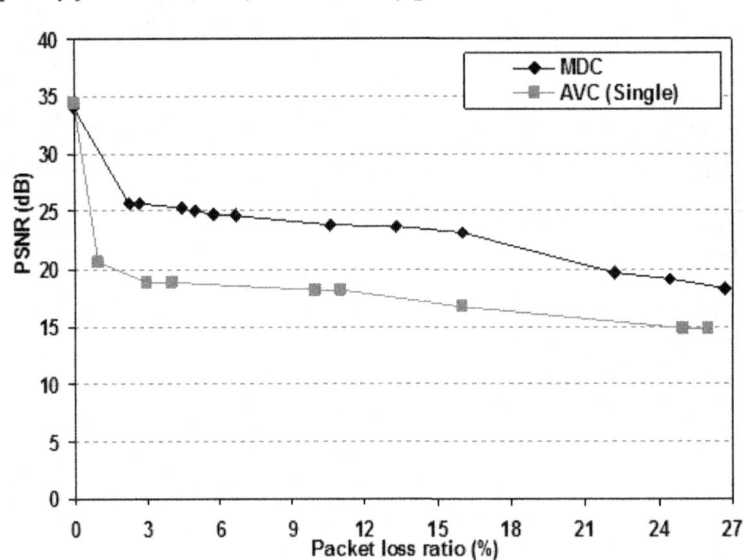

will not impact upon later frames and will be barely noticed by the viewer, especially if error concealment allows the missing data to be reconstructed. In Figure 11, though a quality greater than 25 dB is advisable, users will generally accept levels between 20-25 dB in mobile wireless scenarios (Sadka, 2002). From the Figure such levels are achieved with packet loss rates as high as 15%. Were it not for the presence of MDC, then packet losses above 5% are likely to lead to poor quality video.

P2P VIDEO STREAMING OVER VANET

This Section considers the concept of resilience multi-source streaming using P2P overlay network over a VANET. P2P overlay networks with multiple sources have proven to be robust, distributed solutions to multimedia transport, including streaming. To achieve video streaming over a VANET overlay, this Section introduces a spatial partition of a video stream based on Flexible Macroblock Ordering. Furthermore the Section examines the impact of differing traffic densities and road layouts upon a P2P overlay network's performance. The work modeled the emerging IEEE 802.11p for wireless VANETs. The research demonstrates that the vehicles' mobility pattern and their drivers' behavior need to be carefully modeled to determine signal reception. The Section also considers the impact of the wireless channel, which also should be more realistically modeled.

A multi-source P2P streaming system is an overlay network that has been organized in such a way that a number of destination vehicles receive the same streaming data from multiple peers within the overlay. The presence of multiple sources means that if one source fails during the streaming process another can step in. In a VANET a source could easily fail if the vehicle was parked or moved out of the area. Here multi-sources are also required, as the video stream is split into several streams in order to exploit path diversity using MDC. This Section seeks to establish under what circumstances (vehicle mobility, network size, and wireless channel conditions) streaming is feasible within a P2P overlay network. Ad hoc networks and peer-to-peer (P2P) overlays are decentralized, autonomous and highly dynamic in a fairly similar way. Therefore, it is natural to suppose (Yan, 2005) that the P2P paradigm of data distribution conveniently maps onto a VANET. In the wired Internet, overlay networks have proved to be a robust way of coping with server bottlenecks (Wu, 2006). Though file download is a common P2P application, streaming from multiple sources is effective (Yong et al., 2008) in guarding against the departure of one or more of the sources from the overlay network. Similarly, in a VANET vehicles leave an area or are simply parked. There is also a high risk of broken links, as a vehicle goes out of range. Multiple delivery paths within the P2P network reduce the risk of adverse channel conditions and allow bandwidth sharing across the VANET.

VANETs bring several advantages to video streaming within an ad hoc network. Battery power is no longer such a problem, implying that larger buffers (with passive and active energy consumption) can serve to absorb any latency arising from multi-hop routing. Satellite navigation systems in cars already use GPS devices and consequently these devices can assist in location-aware routing. In this work urban VANETs is considered. Within a city, because of traffic congestion, high speeds do not generally arise. Therefore, connections are on average longer and Doppler effects are limited. Vehicle motion is indeed restricted by the road geometry but, compared to a highway VANET; vehicle motion is no longer linear.

This Chapter introduces spatial MDC to counter the effect of packet loss on video streams. MDC can be achieved through temporal decomposition of the frame sequence (Apostopoulos, 2001) but this may lead to additional anchor frames or encoder-decoder drift, unless specialist codecs are employed as explained earlier. Instead, spatial decomposition is considered in this Section. By employing checkerboard FMO (Lambert, 2006), an error-resilience feature of the H.264/AVC, each frame can be separated into two or more slices. These slices can aid the reconstruction of adjacent macroblocks in another slice in the event of the packet bearing that slice being lost. Checkerboard FMO is the only one of the H.264/AVC FMO types that has this property. To achieve reconstruction error, error concealment (Varsa et al., 2001) at the decoder is necessary, which might tax the processor available to a battery-powered device but is not a barrier to its use within a vehicle. Therefore, the application of spatial FMO to MDC within a VANET is the final contribution of this Section.

In the context of P2P streaming, this work shows that greater realism is needed in modeling road layouts and driver behavior. Some mobility modeling software such as BonnMotion[6] incorporate an ideal model of road layouts and as such do not account for traffic signals at intersections or other obstacles that give rise to queuing. While these models may be perfectly adequate for generic comparisons, clustering of vehicles within wireless range results in increased access contention. Interestingly, behavior at bottlenecks (Treiber, 2009) in the presence of obstacles of whatever form[7] is similar in road systems across the globe. Such behaviour results in synchronized traffic, evidenced by clusters of vehicles passing along a road. Five different clustering patterns (Trieber, 2009) have been identified from mathematical modeling. Vehicular mobility can be split into macro-mobility effects such as road layout, number of lanes, and speed limits, and micro-mobility effects, especially driver behavior, which should account for the presence of other vehicles, both nearby and due to traffic congestion. This implies that the fixed speed simulations that are common in ad hoc network modeling are no longer applicable to VANETs, whereas vehicle density has a more important role to play. It is difficult to conduct repeated live experiments and because the complex vehicle mobility models that arise are unlikely to be represented analytically, simulation is the main tool for research on VANETs (Tonguz, 2009). The recent trend amongst research supported by the manufacturers (Tonguz, 2009) is for detailed simulations (Liu, 2009) to represent the large number of variables (vehicle density, car speeds, driver behavior, road obstacles, road topologies ...) that can occur.

Similarly to mobility, this work carefully models wireless channel conditions (Matolak, 2008). General-purpose VANET simulations tend to restrict modeling of the wireless channel to either free space or two-ray, line-of-sight (LOS) modeling and do not account for wireless fading through multi-path, as can obviously occur, especially in a built environment. Moreover, it is possible that the simulator may not properly account for inter-node interference. Modeling of ad hoc routing protocols at higher layers of the protocol stack but neglecting physical layer modeling (Takai et al. 2001) may result in misleading rankings of the protocols. In fact, the well-known ns2 simulator may well have deficiencies (Wellens et al., 2005) in that respect, which is why GloMoSim (Zheng et al., 1998) is selected for network simulations. Both Rayleigh and Rician fading models are used for wireless channel conditions.

The Rayleigh distribution represents a situation in which nodes are highly mobile. The restrictions of an urban topology make highly mobile nodes unlikely. In communication, between vehicles in a city there will often be a LOS component, as vehicles will be aligned along road segments and mobility is often restricted. Here Rician fading channel is considered, as the Rician distribution includes a LOS component. Setting K = 3 (or 4.7 dB) by default (Doukas et al., 2006) results in a moderate departure from the Rayleigh distribution and, in fact, in many cases K does not exceed 7 dB.

Simulations may also mislead if they take no account of packet length by thresholding the Signal-to-Interference-Noise Ratio (SINR) level rather than determining the Bit Error Rate (BER). Notice that signal reception thresholding is acceptable if all packets are long (Takai et al., 2001). However, the latter is not normally the case for video communication. In summary closer modeling of wireless conditions in this Chapter includes in the results: 1) an additive node interference model; 2) fading and a simple path loss model, as remains common; and 3) appropriate treatment of signal reception.

Related Research

Prior work by others such as Oliveira et al. (2005) and Niraula et al. (2009) considered the possibility of combining P2P overlay networks with mobile ad hoc networks (MANETs) but did not considered an extension to VANETs, in which quite different mobility patterns occur, different assumptions are made, such as the possibility of buffering, and more rigorous channel modeling is necessary. The work in Oliveira et al. (2005) applied the random waypoint model (Broch et al., 1998), typical of MANET modeling, to find routing protocol performance under file download. The research employed a shadowing wireless model but, as the simulator was ns2, a simple capture threshold was in use. The paper does, however, summarize prior work on overlays combined with MANETs. Similarly the work in [Niraula et al. (2009) simulated P2P video streaming over MANET using MDC using Multicast Overlay Spanning Tree (MOST) protocol. However this work fails to identify the issues related to P2P video streaming, on latency and video quality.

Previously, MDC over the P2P overlay is considered but again this was in the context of a MANET. However, the work for a MANET applied temporal MDC and not spatial MDC. In this work spatial MDC is used, which more efficient for the reasons outlined earlier. In Chow & Ishii (2007), simple temporal splitting occurred with the simulation environment. In fact, in what may be termed the pioneering stage of ad hoc networks, there was considerable investigation of video streaming (Zhang, 2005) but the simulation environments were simplified. For example, in Djenouri et al. (2008) the inadequacy of general-purpose MANET mobility modeling for VANETs is highlighted. The FMO facility of H.264/AVC is giving rise to a number of ways of exploiting spatial MDC. The closest research to this work, use of spatial FMO was independently outlined in Lamy-Bergot (2009). However, the analysis in Lamy-Bergot (2009) was for a very general mobile environment and only examined the resulting video quality. As a further example, in Yong & Gong (2009), FMO is used to isolate *regions of interest (ROIs)*, while redundant slices are sent to protect the ROI. A problem with this scheme may indeed be the extra data sent in the redundant slices, while checkerboard FMO does not entail such overheads or the need to identify ROIs.

Therefore, this work represents a realistically modeled scheme for streaming video over a VANET, by means of a P2P overlay network with multi-sources. In combining spatial FMO, MDC, VANET, and P2P overlay, it most likely arrives at a unique combination. As such it demonstrates a way forward for 'infotainment' within automotive networks, which are a growing sector of the wireless communications industry.

P2P Overlay Streaming

A P2P application overlay functioning with in VANET can be mapped onto the physical network in the same way as it does in a MANET. In this work, a mesh architecture with all-to-all connectivity is simulated, Figure 13. This approach is more robust than a tree-based architecture, because, when a stream comes from various sources, communication does not break-down when a subset of peers disconnects. In the example scenario illustrated, seven nodes form a mesh within which two nodes are the original source nodes for the same video. A likely way that two nodes in a VANET could acquire the same video clip is that both could have passed a roadside unit offering informational/advertising video clips. However, it is also possible that these two nodes previously could have acquired the same video from a single node, prior to the distribution process illustrated in Figure 13. Three nodes (node C, node D and node E) receive video from these two source nodes. The three nodes in turn also act as sources to two further nodes (node F and node G). Hence, nodes C, D and E download and upload at the same time i.e. act as receivers and sources at the same time. All receiving nodes, C to G, are served by multiple sources. Moreover in Figure 13, nodes F and G can receive data from an alternative source if a source ceases to be available.

Peer selection for streaming purposes can either be achieved in one of three ways. Firstly, it is achieved in a hybrid fashion, by including some server nodes in the manner of commercial P2P streaming. Secondly, a structured overlay can be organized that allows quicker discovery of source peers than in an unstructured overlay. However, structured overlays require the source data to be placed in particular nodes, which may not be practical in a VANET. Therefore, this work assume an unstructured and decentralized overlay such as Gnutella, KaZaa, or GIA (Fiore et al., 2007).

Modeling a VANET

GloMoSim was employed to generate the simulation results. For video transport IP framing was employed with UDP transport. The default network configuration consisted of seven nodes forming an overlay network, as arranged in Figure 13, with these nodes selected from a total of 100 nodes (vehicles) forming the complete VANET. GloMoSim was altered so that nodes start at random locations rather than at the origin, to avoid biased results. As a default, a two-ray path loss model is used. IEEE 802.11p was used as wireless technology. In IEEE 802.11p transmission is at 5.9 GHz, when the two-ray crossover distance is about 556.5 m, which is calculated from (1).

$$d_{cross} = \frac{4\pi h_t h_r}{\lambda} \quad (1)$$

Figure 13. Mesh-based overlay topology sending video streams from sources to receivers

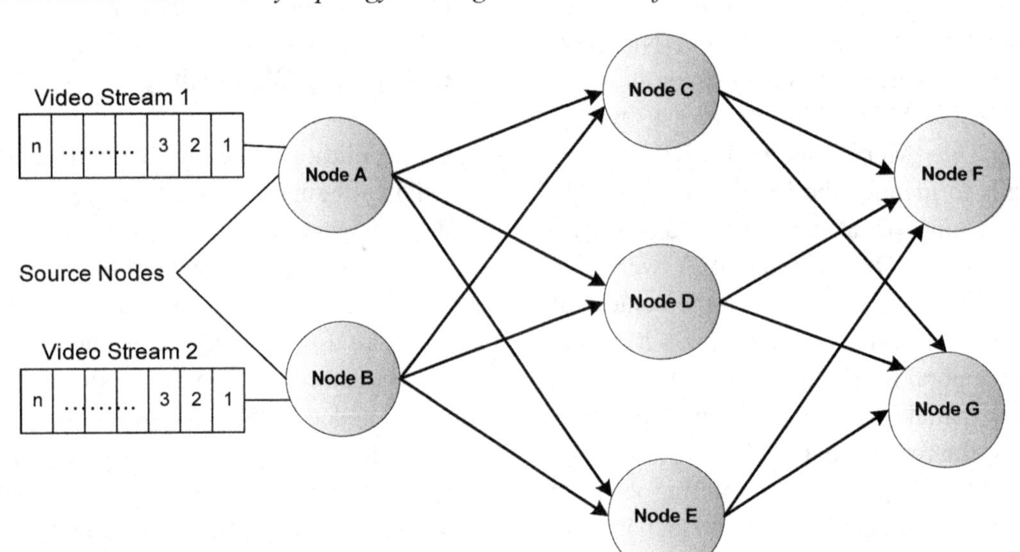

where h_t and h_r are the transmitter and receiver antenna heights respectively, and λ is the transmission wavelength. Receiver sensitivity in the simulation was set to -92 dBm. IEEE 802.11p's Binary Phase Shift Keying (BPSK) modulation mode with 1/2 coding rate was simulated. Accordingly, the data-rate was set to 3 Mbps. (Request to Send / Clear to Send) (RTS/CTS) signaling was turned on.

The *Location Aware Routing (LAR)* (Ko & Vaidya, 2000) protocol is used at the routing layer. The LAR protocol is able to restrict the area for route propagation by virtue of *Global Positioning System (GPS)* information gathered from the vehicles in the VANET. The advent of satellite navigation systems has shown the benefits of GPS provision in vehicles and if a WLAN transceiver is available within the vehicle, GPS will also most likely be present. The result is that LAR will incur less control packet overhead compared to other well-known protocols such as *Ad Hoc On-Demand Distance Vector (AODV)* (Perkins et al., 20003). The main weakness of LAR appears to be (Qadri & Liotta, 2008) increased end-to-end latency but, provided streaming is one-way, then this will only be reflected in longer start-up times at the receiver. In multi-source streaming, interactive video is not envisaged and, hence, the extra latency is not an issue.

Mobility Modeling

In VanetMobiSim (Fiore et al., 2007), as used by us, a variety of urban road layouts can be generated of increasing density by means of the random backbone mode using Voronoi tessellations. In the simulations, a square 1000 m^2 area was defined and nodes (vehicles) were initially randomly placed within the area. Other spatial settings to do with road clusters, intersection density, lanes (2) and speeds are given in Table 7. The number of clusters refers to the number of rectangular areas with a road density of 2 obstacles/ 100 m2 (2e-4) for the Downtown model (to create randomly a non-homogeneous simulation terrain), while the remainder of the terrain is at the minimum road density, which was the same as that for a residential area, i.e. 2e-5. The number of traffic lights (at intersections) and time intervals between changes are also defined. In the following paragraph, the equation variable annotations in Table 7 are identified, with some uncommented terms explained at the end of this Section. These values in Table 7 were the defaults for VanetMobiSim.

The real advantage of VanetMobiSim is the variety of driver behavior models. A summary of mobility configurable settings applied in the simulations is also given in Table 7. In Table 7, "Recalculation of movement parameters interval" is the interval after which the simulator recalculates parameters and makes a decision on those parameters. This time interval does not have a direct impact on the simulation time but indicates the update frequency. In the CSM model, pause time is set to zero because, unlike in a general ad hoc network, it is not expected that a vehicle will malinger for any reason. The minimum and maximum stay times are times spent parking or stopping for any reason. These are set to low values by default.

Therefore, the micro-mobility models presented by VanetMobiSim are of increased sophistication in driver behavior. Notice though that when driver behavior is introduced into simulations it is no longer possible to easily examine node speed dependencies, as the vehicles will have a range of speeds depending on local conditions, even though the minimum and maximum speeds are not exceeded.

Video Configuration

The results were generated using two test video sequences to allow for content variation. Both sequences were encoded at *Quarter Common Intermediate Format (QCIF)* video resolution (176 × 144 pixel/frame) @ 15 Hz (frame/s). Encoding was with the reference JM 15 software for H.264/

Table 7. VanetMobiSim configurable settings with road layout and mobility models

Global Parameters	
Terrain dimension	1000 m^2
Graph type	Space graph (Downtown model)
No. of road clusters	4
Min. intersection density	$2e^{-5}$
Max. traffic lights	6
Time interval between traffic lights change	10 s
Number of lanes	2
Min. stay	10 s
Max. stay	100 s
Nodes (vehicles)	50, 100
Min. speed (s_{min})	3.2 m/s (7 mph)
Max. speed (s_{max})	13.5 m/s (30 mph)
CSM model	
Min. and max. pause time	0 s
FTM Model	
Density of traffic jam	0.2 cars/m
Recalculation of traffic parameters interval	0.1 s
IDM Model, IDM-IM Model, IDM-LC Model	
Length of vehicle	5 m
Max. acceleration (a)	0.6 m/s^2
Normal deceleration (b)	0.5 m/s^2
Traffic jam distance	2 m
Node's safe time headway (T)	1.5 s
Recalculation of movement parameters interval	0.1 s
Other Parameters of IDM-LC Model	
Safe deceleration (a_{safe})	4 m/s^2
Politeness factor of drivers when changing lane (p)	0.5
Threshold acceleration for lane change (a_{thr})	0.2 m/s^2

AVC. The *Foreman* sequence, with 300 frames, is a short well-known sequence typical of one taken by a handheld camera with jerky motion and a rapid pan towards the end. The *Paris* sequence consists of 1065 frames taken from a TV studio set with significant spatial coding complexity. A trace file for the longer *Paris* clip served to simulate the impact of VANET transport. Both sequences were used to ascertain video quality experienced by the user.

Evaluation

Network Performance

In this Section, averaged (arithmetic mean) results are considered from the destination nodes in the overlay network presented in Section 7.5. All results were repeated 50 times and the average taken. In repeating each run, it is important to realize that the road layout topology changes each time. Therefore, an average is across a set of road topologies, as well as node placements. Of course, for packets to reach a destination vehicle from the source vehicles, those packets will be routed across multiple hops to reach their destination. The effective wireless range was set to 625 m with a transmission power of 19.3 dBm[8]. In most urban settings, the presence of buildings will impede transmission (Oishi et al., 2006) through reflections, diffraction and absorption of signals. The reduced power level is a concession to that fact.

In Figure 14a, the packet loss ratios (ratio of packets sent to packets lost) would make unprotected streaming of video feasible in the denser network if lane changing was permitted, as losses below 10% occur in that scenario. It is possible that for packet loss rates up to 20% but no more than 30% (see Section 7.6.2), inclusion of error resilience or application-layer forward error correction could compensate for high packet loss rates. Figure 14a shows that more careful modeling of driver behavior can actually decrease the predicted level of packet loss (going from CSM to IDM-LC). However, lane-changing (IDM-LC) which allows increased mobility is not always possible within a city.

A feature of Figure 14b and 14c is that more careful wireless channel modeling results in a drop in packet loss. However, the introduction of fading models causes the simulator to report greater packet losses in the denser network (100 nodes). This is because, when path loss only is modeled, then the distance between nodes is the principal effect and the sparser network (50 nodes) results in greater losses. However, if the nodes are closer together then interference at road obstacles due to traffic queuing increases the number of packet losses. In addition, both sparse and dense networks suffer from additional packet loss from fading. Figure 15 varies the fading parameter or factor K arising from the Rician distribution for reasonable values of K (Doukas & Kalivas, 2006) with the IDM-IM mobility model. When K is lower than 3, then an increase in packet loss occurs and the packet loss behavior begins to resemble that for the Rayleigh distribution, Figure 14c. When K is higher than the default value of 3, the increasing dominance of the LOS component means that the packet loss regime depends more on the LOS component. As K reaches 6 (7.8 dB) then the sparser network begins to suffer more packet losses, perhaps because of a relative increasing impact of path loss rather than fading.

The impact of cross-traffic on packet loss was assessed. Four nodes that were not part of the overlay were introduced, each sending 100 packets of packet length 400 bytes as a CBR stream, again with the IDM-IM mobility model. Such messages could represent text safety alerts. From Figure 14 in which the basic two-ray ground propagation model is used, it is apparent that moderate cross traffic does not greatly affect the packet loss ratios compared to the default values (extracted from Figure 14a), though it might prove important to video quality as the losses reach above 20% (as it is difficult to reconstruct even with error resilience once losses reach above 20%). However, varying the overlay size certainly does have a significant effect on packet loss ratios. In the 'More Peers' scenario of Table 8, a total of 14 peers were part of the overlay, compared to the default scenario previously used and presented in Section 7.5.1. Two peers were source peers initially and six peers acted as receivers. These same six peers served four more peers. These four peers sent streams to two more peers. This means that 12 peers acted as source and 12 peers were receivers. Amongst these, ten peers simultaneously acted as receivers and senders. The mobility model was again IDM-IM and the two-ray channel model was used. As can be seen from Table 8, the packet loss ratio increases significantly with this amount of traffic passing across the overlay. Therefore, in a VANET the size of an overlay network must be tightly controlled. This contrasts with an Internet overlay, when there are few or no restrictions on joining the overlay network.

As might be expected, when the node density increases, then end-to-end-delay increases. The

Figure 14. Packet loss ratio compared by mobility model for different network sizes for (a) no fading, (b) Rician fading, and (c) Rayleigh fading, with routing by the LAR protocol

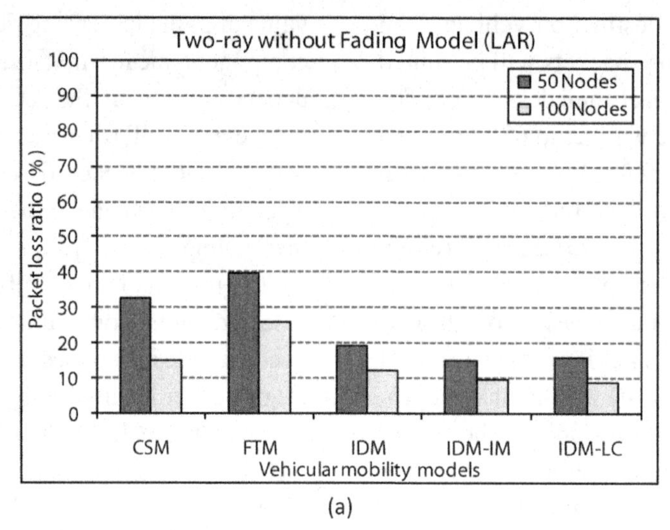

(a)

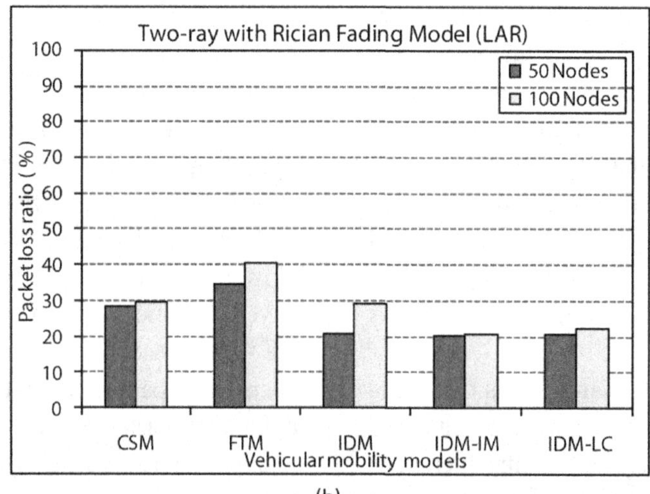

(b)

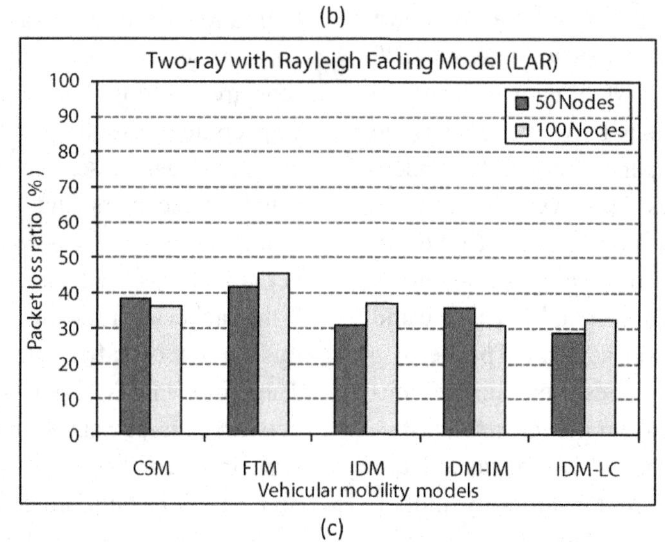

(c)

Figure 15. Packet loss ratio with Rician fading and varying K factor for an IDM-IM mobility model and LAR routing

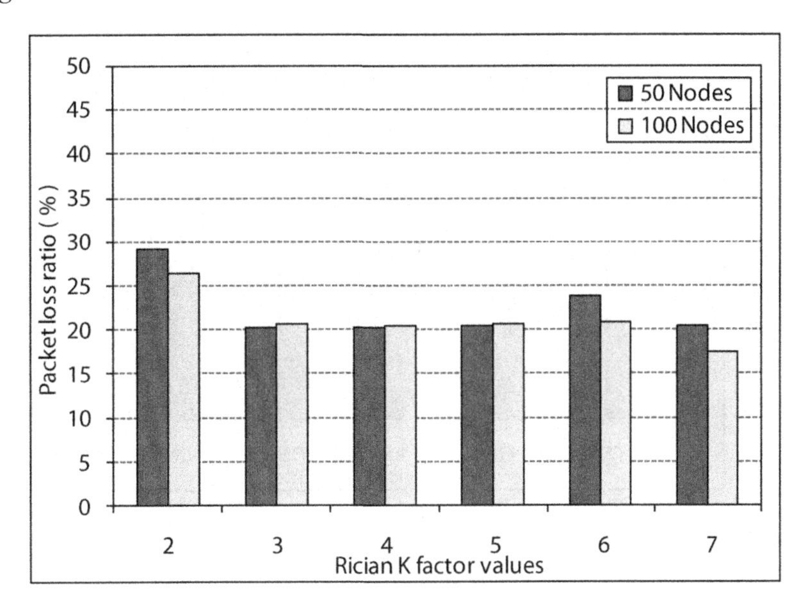

Table 8. Packet loss ratios when either introducing cross-traffic or more peers into the overlay for an IDM-IM mobility model, two-ray channel model, and LAR routing

Scenarios	Default	Cross Traffic	More Peers
50 Nodes	15.26%	22.22%	45.75%
100 Nodes	9.42%	21.45%	46.43%

number of hops traversed by a packet increases on average, resulting in longer delay. This is illustrated by the differences between the results for different network sizes, Figure 16, for the original overlay scenario with seven nodes. However, an interesting feature of these results is that more realistic modeling actually indicates that the delay for the sparser networks is less than that predicted by the coarser mobility models. That is the effect of network sparseness is equalized to some extent when the effects of driver behavior are taken into account. For the denser network of 100 nodes, delay is forecast to be less under a two-ray model. When fading is taken into account, Figure 16b, c, then predicted delay increases. At the levels of delay reported, only one-way video streaming is possible9 but this is not a problem as overlay network applications are generally not interactive. Therefore, the main purpose of estimating the delay is to assist in buffer dimensioning, which for the *Paris* video stream at 15 Hz obviously requires at least a 45 frame buffer. The size of the buffer will affect the start-up delay experienced by the user. If the user consciously selects a clip then delay will be perceptible, as it would need to be less than 20 ms for that to occur, which it clearly is not. If delivery is automatic, e.g. in a congestion monitoring application, then the user will not be aware of the start time of streaming.

Figure 17 considers the efficiency of the routing process by charting the per-packet control packet overhead. The LAR protocol has reduced the levels of overhead below those normally

Figure 16. End-to-end delay compared by mobility model for different network sizes for (a) no fading, (b) Rician fading, and (c) Rayleigh fading, with routing by the LAR protocol

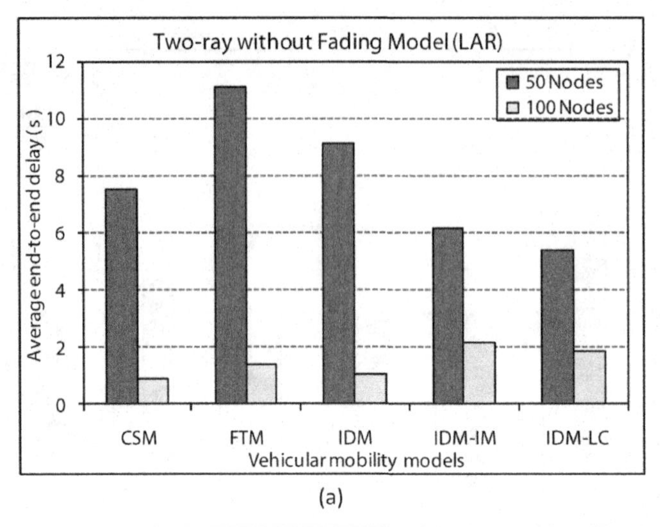

(a)

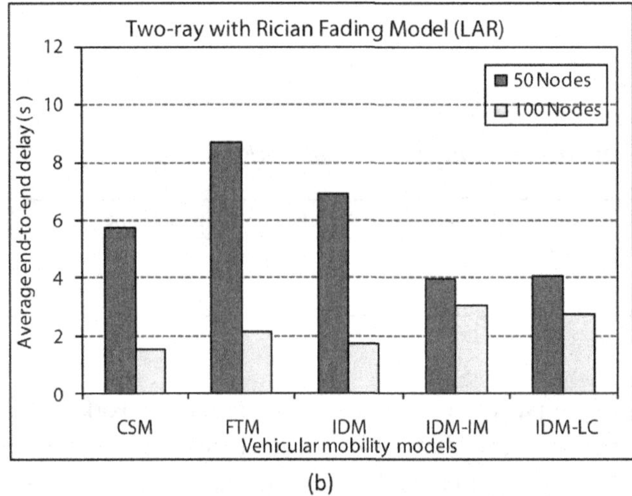

(b)

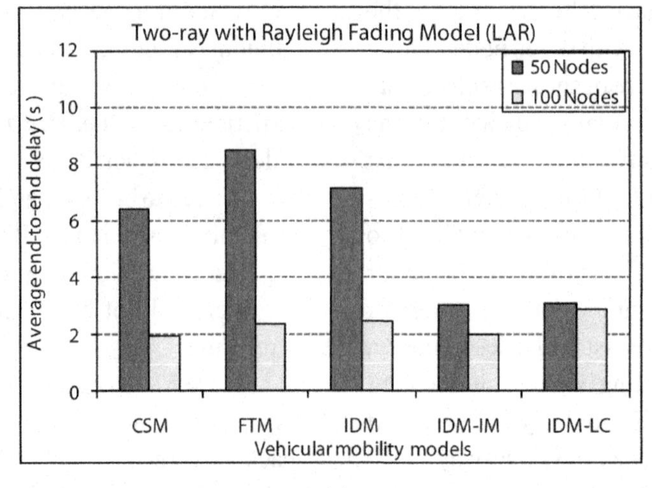

(c)

Figure 17. Control overhead in packets for different mobility models with 50 and 100 nodes with two-ray propagation path-loss model with (a) no fading, (b) Rician fading, and (c) Rayleigh fading, with routing by the LAR protocol

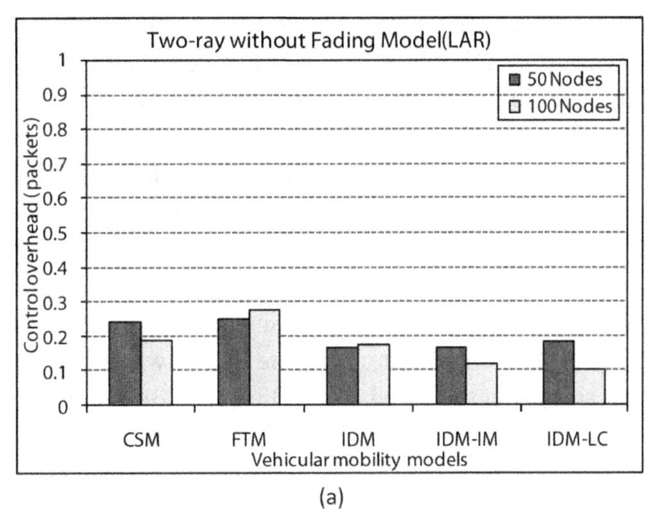

(a)

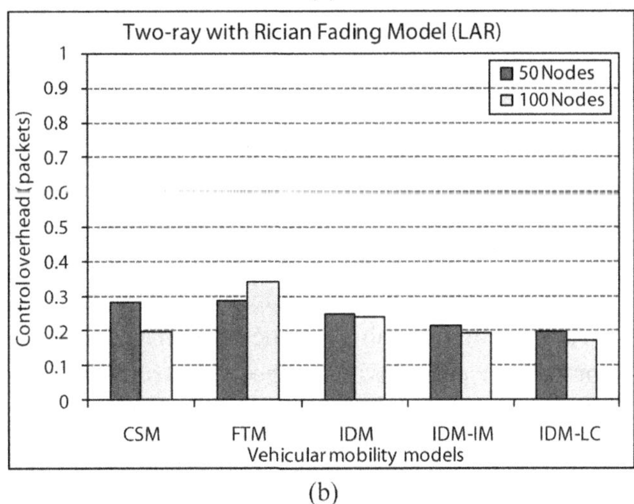

(b)

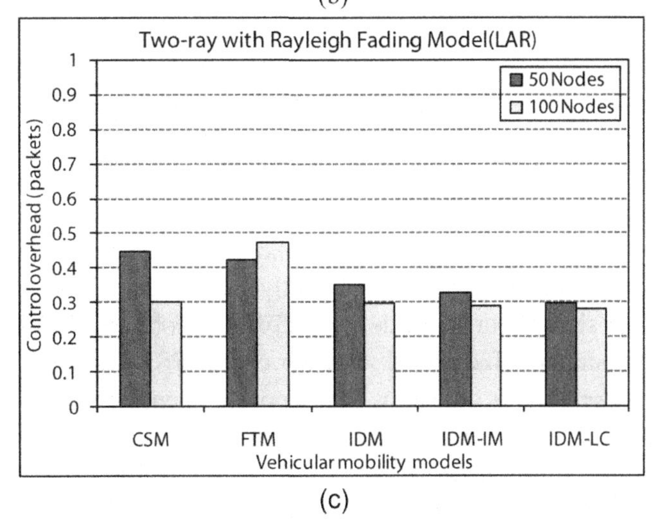

(c)

experienced. However, when fading is present overhead does increase. There is also an energy consumption implication, as transmission for a battery-driven device can consume as much as 80% of the energy. However, for VANETs, assuming the transceivers are re-chargeable from the engine's alternator, there is not such an implication.

Video Performance

In this work, the consequences of applying spatial multi-description coding to the dual streams are considered. For FMO error resilience, compressed frame data is normally split into a number of slices each consisting of a set of macroblocks (Ghanbari, 2003). Slice resynchronization markers ensure that if a slice is lost then the decoder is still able to continue with entropic decoding. Therefore, a slice is a unit of error resilience and it is normally assumed that one slice forms a packet, after packing into a *Network Abstraction Layer unit (NALU)* in H.264/AVC (Wenger, 2003). Each NALU is further encapsulated by a Real-Time Protocol (RTP) header (Wenger et al., 1998) within an IP/UDP packet, resulting in an additional 40 bytes. For comparison purposes, one simple form of spatial 'MDC' is to employ slicing (without FMO) in which the top part of the encoded frame forms one slice and the bottom half forms the other slice. Then each set of slices (top and bottom) form a description but these cannot be used to reconstruct the other if the packet bearing that slice is lost. Instead, previous frame replacement must be employed by the decoder.

In checkerboard FMO, the macroblocks equivalent to the white squares of a checker or chess board form one slice while the remaining macroblocks form the other slice. Error concealment is a non-normative feature of H.264/AVC. In the experiments, the motion vectors of macroblocks from correctly received slices are utilized to reconstruct macroblocks from missing FMO slices. This takes place if the average motion activity is sufficient (more than a quarter pixel). Research in Varsa et al. (2001) gives details of which motion vector to select to give the smoothest block transition. It is also possible to select the intra-coded frame method of spatial interpolation, which provides smooth and consistent edges at an increased computational cost. However, temporal error concealment is preferable, unless 'high motion activity' or scene cuts are present. The method of concealment can be selected depending on the smoothness of reconstructed macroblock border transitions. As the lower complexity H.264 Baseline Profile was employed the Group of Pictures (Ghanbari, 2003) frame structure was IPPPPP… To reduce error propagation both MDC schemes employed Gradual Decoder Refresh (Wenger, 2003) by insertion of an intra-coded row of macroblocks into each encoded frame, cycling the replaced row through the video sequence.

In Figure 18, packet losses at rates up to 30% were generated from a Uniform distribution and the Peak Signal-to-Noise Ratio (PSNR) was calculated. To ensure convergence each data point was the mean of fifty tests. From the Figure it is apparent that apart from zero packet loss, simple slicing (MDC) fares badly in comparison to checkerboard FMO with MDC (MDC with Error Res.). At zero packet loss the extra overhead from including the FMO macroblock mapping in a packet results in a drop in quality (if the datarate is the same as for simple slicing). However, from 8% packet loss onwards, the video quality for both sequences using simple slicing would be unacceptable. For *Paris*, the video quality using checkerboard FMO in conjunction with error concealment results in reasonable video quality. Though the PSNR drops below 30 dB, it is generally the case that users will tolerate PSNRs above 25 dB for mobile applications. In fact, it is possible to equate PSNR to the ITU-R's Mean Opinion Scores, when the range 25 to 31 dB inclusive is approximately equivalent to a score of 3 or "fair" (from a range 1 to 5 with 5 being "excellent". For *Foreman*, the drop in quality is steeper reflecting

Figure 18. Applying MDC with and without error resilience to dual path video streaming of (a) the Foreman sequence, (b) the Paris sequence

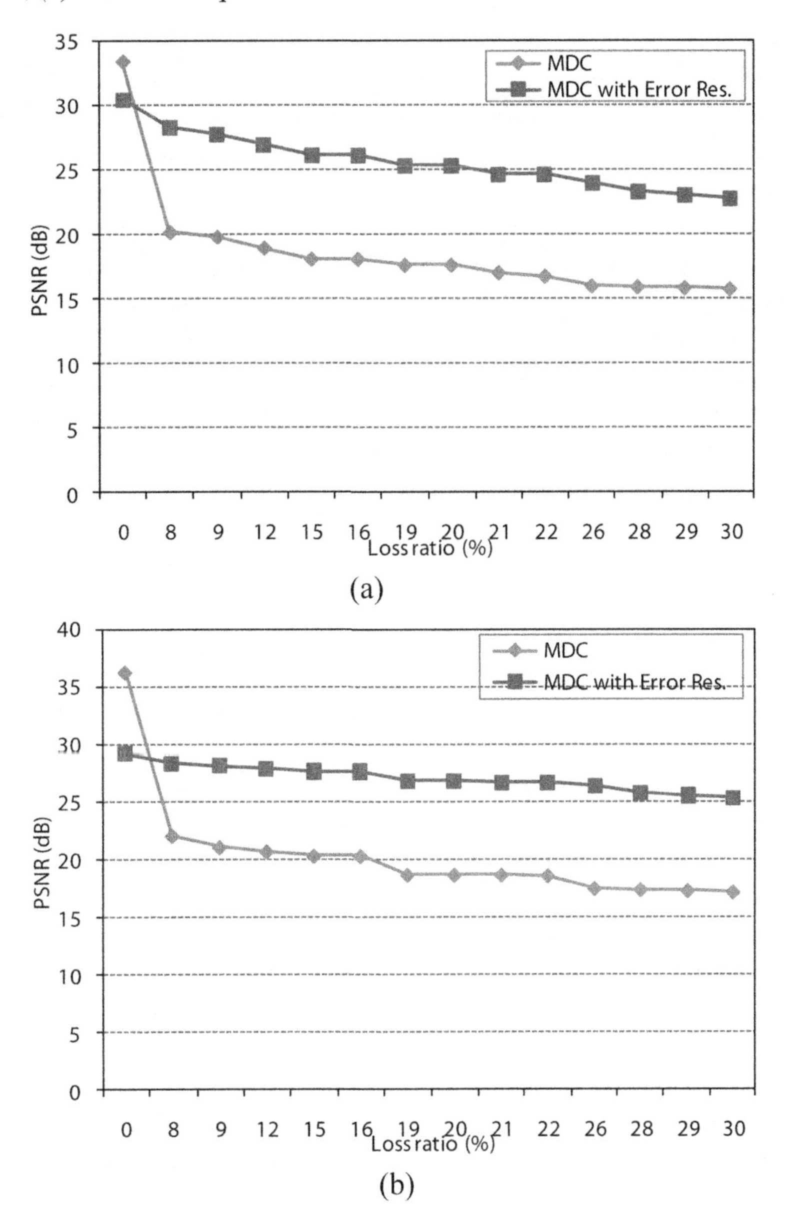

the more complex source coding task. However, below 20% packet loss ratio, the quality would be tolerable. From Figure 15b, this occurs under Rician channel conditions and IDM models when there are fifty vehicles in the network.

FUTURE RESEARCH DEVELOPMENTS

This Section presents a summary of possible avenues of future research as follows.

- There is a need for peer selection algorithms for ad hoc networks to reduce the impact of packet loss during streaming. A network-aware P2P selection algorithm for ad hoc network based on fewer hop counts and reliable source peers can be developed for effective delivery of video chunks over ad hoc networks. An algorithm can select a peer (to download video) based on its physical location and shortest path. This information can be determined using network hop information from routing tables. This approach will significantly reduce the packet loss experienced by nodes when the overlay were created without prior knowledge of peer locations. Along with that an efficient querying mechanism can also be developed for ad hoc applications.

- Reliability in video streaming over an ad hoc network is one of the important challenges facing the technical community. Multipath routing algorithms reduce the impact of packet loss, delay and overhead. Perhaps, further improvements can be made by considering a reliability factor for route selection along with reduced hop count. This can be done by applying a feedback mechanism which can take round-trip time or a similar factor to calculate the reliability of the routes.
- A goal in VANETs is to enable the dissemination of traffic and road conditions such as local congestion and surface ice as detected by independently moving vehicles. This is also useful for vehicles on the highway and enables early reaction. Video communication can permit warning incoming vehicles about approaching a possibly dangerous area. Future work can consider the experimental analysis of video communication over VANET highway scenarios, where such hazardous conditions exist.
- In addition, improvements can be made to video coding techniques such as applying scalable video coding or rateless channel error codes to protect the streamed video. These techniques are suitable for the delivery of multimedia traffic over ad hoc wireless networks and they should be compared with using MDC.
- End-to-end delay remains a concern for interactive applications. Delay may be addressed by the reduction of hop counts or can be absorbed by suitable buffering.
- Finally, an important research issue is the establishment of realistic propagation and channel modeling. Improvements made in propagation modeling to portray real urban network conditions results in higher packet loss. One way forward is to propose a new routing protocol which exploits the accurate predictions of channel modeling in evaluating the next hop in a multi-hop path.

CONCLUSION

Multi-source P2P streaming across an overlay networks is potentially a robust form of video delivery. However, the size of such overlays will need to be strictly controlled in an ad hoc network to prevent an increase in competing traffic amongst the overlay nodes, leading to unacceptable increase in packet loss rates. P2P streaming is adapted to support multisource to overcome the node churn problem. If one of the sources effectively left the overlay because it is out-of-range or has poor connectivity then another source can provide the source data.

To exploit multipath transfer, MDC was introduced. This work has established that if video chunks or blocks are distributed to two or more peers within a MANET or VANET, than MDC is a way of streaming the same segments of video from any two of the peers in such a way that the quality is at an acceptable level despite significant packet loss rates.

Robust video communication was achieved using two error resilience techniques, namely redundant frame coding and FMO. Surprisingly, given the amount of packet loss was higher, inserting redundant frames or introducing different types of FMO, allows lost or dropped predictive frames to be reconstructed, resulting in a considerable improvement in delivered video quality. For a variety of reasons, intra-coded macroblocks placed within predictively-coded P-frames should replace periodic IDR frames.

REFERENCES

Agboma, F., Smy, M., & Liotta, A. (2008). QoE analysis of a peer-to-peer television system. *Proceedings of IADIS International Conference on Telecommunications, Networks and Systems*, (pp. 114-119).

Androutsellis-Theotokis, S., & Spinellis, D. (2004). A survey of peer-to-peer content distribution technologies. *ACM Computing Surveys*, *36*(4), 335–371. doi:10.1145/1041680.1041681

Apostolopoulos, J. (2001). Reliable video communication over lossy packet networks using multiple state encoding and path diversity. *Proceedings of SPIE, Visual Communications and Image Processing*, (pp. 24-26).

Baccichet, P., Shantanu, R., & Bernd, G. (2006). Systematic lossy error protection based on H. 264/AVC redundant slices and flexible macroblock ordering. *Springer Journal of Zhejiang University-Science A*, *7*(5), 900–909. doi:10.1631/jzus.2006.A0900

Berspacher, J., Schollmeier, R., Zols, S., & Kunzmann, G. (2004). Structured P2P networks in mobile and fixed environments. *International Working Conference on Performance Modeling and Evaluation of Heterogeneous Networks*, (pp. 1-25).

Broch, J., Maltz, D. A., Johnson, D. B., Hu, Y.-C., & Jetcheva, J. (1998). A performance comparison of multi-hop wireless ad hoc network routing protocols. *Proceedings of the 4th Annual ACM/IEEE International Conference on Mobile Computing and Networking*, (pp. 85 – 97).

Chen, J., Chan, S., & Li, V. (2004). Multipath routing for video delivery over bandwidth-limited networks. *IEEE Journal on Selected Areas in Communications*, *22*(10), 1920–1932. doi:10.1109/JSAC.2004.836000

Chiasserini, C. F., & Rao, R. R. (1999). Pulsed battery discharge in communication devices. *Proceedings of the 5th Annual ACM/IEEE International Conference on Mobile Computing and Networking*, (pp. 88-95).

Chlamtac, I., & Redi, J. (1998). Mobile computing: Challenges and opportunities . In Ralston, A., Hemmendinger, E. R. D., & Reilly, E. (Eds.), *Encyclopedia of computer science* (4th ed.). Chichester, UK: Wiley & Sons.

Chow, C.-O., & Ishii, H. (2007). Enhancing realtime video streaming over mobile ad hoc networks using multipoint-to-point communication. *Computer Communications*, *30*(8), 1754–1764. doi:10.1016/j.comcom.2007.02.004

Conti, M., Gregori, E., & Turi, G. (2004). Towards scalable P2P computing for mobile ad hoc networks. *Proceedings of the Second IEEE Annual Conference on Pervasive Computing and Communications Workshops*, (pp. 109–113).

Corson, M. S., Macker, J. P., & Cirnicione, G. H. (1999). Internet-based mobile ad hoc networking. *IEEE Internet Computing*, *3*(4), 63–70. doi:10.1109/4236.780962

Cramer, C., & Fuhrmann, T. (2006). Performance evaluation of chord in mobile ad hoc networks. *Proceedings of the 1st International Workshop on Decentralized Resource Sharing in Mobile Computing and Networking*, (pp. 48 – 53).

Datta, A. (2003). MobiGrid: Peer-to-Peer overlay and mobile ad-hoc network rendezvous - A data management perspective. *Proceedings of the 15th Conference on Advanced Information Systems Engineering.*

Ding, E. G., & Bhargava, B. (2004). Peer-to-peer file-sharing over mobile ad hoc networks. *Proceedings of the Second IEEE Annual Conference on Pervasive Computing and Communications Workshops (PERCOMW '04)*, (pp. 104–108).

Djenouri, D., Nekka, E., & Soualhi, W. (2008). Simulation of mobility models in vehicular ad hoc networks. *Proceedings of the 2008 Ambi-Sys Workshop on Software Organisation and MonI-Toring of Ambient Systems*, (pp. 1-7).

Doukas, A., & Kalivas, G. (2006). Rician K factor estimation for wireless communication systems. *Proceedings of the IEEE International Conference on Wireless and Mobile Communications*, (pp. 69-73).

Fiore, M., Haerri, J., Filali, F., & Bonnet, C. (2007). Vehicular mobility simulation for VANETs. *Proceedings of 40th Annual Simulation Symposium*, (pp. 301-307).

Franchi, N., Fumagalli, M., Lancini, R., & Tubaro, S. (2005). Multiple description video coding for scalable and robust transmission over IP. *IEEE Transactions on Circuits and Systems for Video Technology*, *15*(3), 321–334. doi:10.1109/TCSVT.2004.842606

Franciscani, F. P., Vasconcelos, M. A., Couto, R. P., & Loureiro, A. A. F. (2003). *Peer-to-peer over ad-hoc networks: (Re)Configuration algorithms.* International Parallel and Distributed Processing Symposium.

Fu, X., Lei, J., & Shi, L. (2007). *An experimental analysis of Joost peer-to-peer VoD service. Technical Report*. Institute of Computer Science, University of Goettingen.

Ghanbari, M. (2003). *Standards codecs: Image compression to advanced video coding*. Stevenage, UK: IET Press. doi:10.1049/PBTE049E

Giordano, E., Ghosh, A., Pau, G., & Gerla, M. (2008). *Experimental evaluation of peer-to-peer applications in vehicular ad-hoc networks.* First Annual International Symposium on Vehicular Computing Systems.

Giordano, S. (2001). Mobile ad-hoc networks . In Stojmenovic, I. (Ed.), *Handbook of wireless network and mobile computing*. Chichester, UK: John Wiley & Sons.

Goldsmith, J., & Wicker, S. (2002). Design challenges for energy-constrained ad hoc wireless networks. *IEEE Wireless Communications Magazine*, *9*(4), 8–27. doi:10.1109/MWC.2002.1028874

Gurses, E., & Kim, A. (2008). Maximum utility peer selection for P2P streaming in wireless ad hoc networks. *Proceedings of the IEEE GLOBECOM*, (pp. 1-5).

Hu, Y. C., Das, S. M., & Pucha, H. (2003). Exploiting the synergy between peer-to-peer and mobile ad hoc networks. *Proceedings of HotOS IX Workshop*, (pp. 37–42).

Jurca, D., Chakareski, J., Wagner, J. P., & Frossard, P. (2007). Enabling adaptive video streaming in P2P systems. *IEEE Communications Magazine*, *45*(6), 108–114. doi:10.1109/MCOM.2007.374427

Karlsson, J., Li, H., & Eriksson, J. (2005). Realtime video over wireless ad-hoc networks. *14th International Conference on Computer Communications and Networks*, (pp. 596-607).

Ko, Y. B., & Vaidya, N. H. (2000). Location aided routing (LAR) in mobile ad hoc networks. *Wireless Networks*, *6*(4), 307–321. doi:10.1023/A:1019106118419

Kortuem, G., & Schneider, J. (2001). *An application platform for mobile ad-hoc networks*. Workshop on Application Models and Programming Tools for Ubiquitous Computing, International Conference on Ubiquitous Computing.

Kumar, S., Xu, L., Mandal, M., & Panchanathan, S. (2006). Error resiliency schemes in H.264/AVC standard. *Elsevier Journal of Visual Communication and Image Representation, 17*, 425–450. doi:10.1016/j.jvcir.2005.04.006

Lambert, P., De Neve, W., Dhondt, Y., & Van de Walle, R. (2006). Flexible macroblock ordering in H.264/AVC. *Journal of Visual Communication and Image Representation, 17*(2), 358–375. doi:10.1016/j.jvcir.2005.05.008

Lamy-Bergot, C., & Candillon, B. Pesquet-Popescu, B., & Gadat, B. (2009). A simple, multiple description coding scheme for improved peer-to-peer video distribution over mobile links. *Proceedings of IEEE Packet Coding Symposium.*

Leung, M.-F., Chan, S.-H., & Au, O. (2006). COSMOS: Peer-to-peer collaborative streaming among mobiles. *IEEE International Conference on Multimedia and Expo*, (pp. 865-868).

Li, X. Y. Z., Yao, P., & Huang, J. (2006). Implementation of P2P computing in design of MANET routing protocol. *Proceedings of the First International Multi-Symposiums on Computer and Computational Sciences,* (pp. 594 – 602).

Liu, B., Khorashadi, B., Du, H., Ghosal, D., Chuah, C., & Zhang, M. (2009). VGSim: An integrated networking and microscopic vehicular mobility simulation platform. *IEEE Communications Magazine, 47*(5), 134–141. doi:10.1109/MCOM.2009.5277467

Liu, Y., Guo, Y., & Liang, C. (2008). A survey on peer-to-peer video streaming systems. *Journal of P2P Networking and Applications, 1*(1), 18-28.

Liu, Y.-S., Nuang, Y.-M., & Hsieh, M.-Y. (2006). Adaptive P2P caching for video broadcasting over wireless ad hoc networks. *Proceedings of Joint Conference on Information Sciences.*

Lua, E. K., Crowcroft, J., Pias, M., Sharma, R., & Lim, S. (2005). A survey and comparison of peer-to-peer overlay network schemes. *IEEE Communications Surveys & Tutorials*, *7*(2), 72–93. doi:10.1109/COMST.2005.1610546

Mao, S., Lin, S., Panwar, S., & Wang, Y. (2001). Reliable transmission of video over ad-hoc networks using automatic repeat request and multipath transport. *Proceedings of IEEE Vehicular Technology Conference*, (pp. 615-619).

Matolak, D. W. (2008). Channel modeling for vehicle-to-vehicle communications. *IEEE Communications Magazine*, *46*(5), 76–83. doi:10.1109/MCOM.2008.4511653

Mayer-Patel, K. (2007). Systems challenges of media collectives supporting media collectives with adaptive MDC. *Proceedings of the 15th International Conference on Multimedia*, (pp. 625-630).

Minar, N. (2001). Distributed systems topologies. *Proceedings of the O'Reilly P2P and Web Services Conference.*

Narkhede, N. S., & Kant, N. (2009). The emerging H.264/AVC advanced video coding standard and its applications. *Proceedings of International Conference on Advances in Computing, Communication and Control*, (pp. 300-309).

Natsheh, E., & Wan, T. C. (2008). Adaptive and fuzzy approaches for nodes affinity management in wireless ad-hoc networks. *Mobile Information Systems*, *4*(4), 273–298.

Niraula, N. B., Kanchanasut, K., & Laouiti, A. (2009). Peer-to-peer live video streaming over mobile ad hoc network. *Proceedings of the 2009 International Conference on Wireless Communications and Mobile Computing: Connecting the World Wirelessly*, (pp. 1045-1050).

Oishi, J., Asukura, K., & Watanabe, T. (2006). A communication model for inter-vehicle communication simulation systems based on properties of urban area. *International Journal of Computer Science and Network Security*, *6*, 213–219.

Oliveira, L. B., Siqueira, I. Q., & Loureiro, A. A. (2003). Evaluation of ad-hoc routing protocols under a peer-to- peer application. *Proceedings of IEEE Wireless Communications and Networking,* (pp. 16-20).

Ostermann, J., Bormans, J., List, P., Marpe, D., Narroschke, M., & Pereira, F. (2004). Video coding with H. 264/AVC: Tools, performance and complexity. *IEEE Circuits and Systems Magazine*, *4*(1), 7–28. doi:10.1109/MCAS.2004.1286980

Ozbek, N., & Tumnali, T. (2005). A survey on the H.264/AVC standard. *Turk Journal of Electrical Engineering*, *13*, 287–302.

Padmanabhan, V., Wang, H., & Chou, P. (2003). Resilient peer-to-peer streaming. *Proceedings of 11th IEEE International Conference on Network Protocols*, (pp. 16- 27).

Papadopouli, H. S. M. (2002). Effects of power conservation, wireless coverage and cooperation on data dissemination among mobile devices. *Proceedings of ACM Symposium on Mobile Ad Hoc Networking and Computing*, (pp. 117-127).

Perkins, C. E., Royer, E. M., & Das, S. R. (2003). *Ad hoc on demand distance vector (AODV) routing*. IETF Internet Draft, 2003.

Pucha, H., Das, S. M., & Hu, Y. C. (2004). Ekta: An efficient DHT substrate for distributed applications in mobile ad hoc networks. *Proceedings of the 6th IEEE Workshop on Mobile Computing Systems and Applications*, (pp. 163- 173).

Qadri, N. N., & Liotta, A. (2008). A comparative analysis of routing protocols for MANETs. *Proceedings of IADIS International Conference on Wireless Applications and Computing* (pp. 149-154).

Rajagopalan, S., & Chien-Chung, S. (2006). A cross-layer decentralized BitTorrent for mobile ad hoc networks. *Proceedings of the 3rd Annual International Conference on Mobile and Ubiquitous Systems - Workshops*, (pp. 1-10).

Reibman, A., Jafarkhani, H., Wang, Y., & Orchard, M. (2001). Multiple description video using rate-distortion splitting. *Proceedings of the IEEE International Conference on Image Processing*, (pp. 978-981).

Reza, R. (2006). Anyone can broadcast video over the internet. *Communications of the ACM*, *49*(11), 55–57. doi:10.1145/1167838.1167863

Rowstron, A., & Druschel, P. (2001). Pastry: scalable, distributed object location and routing for large-scale peer-to-peer systems. *IFIP/ACM International Conference on Distributed Systems (Middleware),* (pp. 329-350).

Sadka, A. H. (2002). *Compressed video communications*. Chichester, UK: Wiley & Sons. doi:10.1002/0470846712

Schollmeier, R., Gruber, I., & Finkenzeller, M. (2002). Routing in mobile ad hoc and peer-to-peer networks, a comparison. *Revised Papers from the NETWORKING 2002 Workshops on Web Engineering and Peer-to-Peer Computing*, (pp. 172 – 186).

Schwarz, H., Marpe, D., & Wiegand, T. (2007). Overview of the scalable video coding extension of the H. 264/AVC standard. *IEEE Transactions on Circuits and Systems for Video Technology*, *17*(1), 1103–1120. doi:10.1109/TCSVT.2007.905532

Sentinelli, A., Marfia, G., Gerla, M., Kleinrock, L., & Tewari, S. (2007). Will IPTV ride the peer-to-peer stream? *IEEE Communications Magazine*, *45*(6), 86–92. doi:10.1109/MCOM.2007.374424

Shah, P., & Pâris, J. F. (2007). Peer-to-peer multimedia streaming using BitTorrent. *IEEE International Performance, Computing, and Communications Conference,* (pp. 340-347).

Son, N., & Jeong, S. (2008). An effective error concealment for H.264/AVC. *IEEE 8th International Conference on Computer and Information Technology Workshops*, (pp. 385-390).

Stoica, I., Morris, R., Karger, D., Kaashoek, M. F., & Balakrishnan, H. (2001). Chord: A scalable peer-to-peer lookup service for internet applications. *Proceedings of ACM SIGCOM.*

Sullivan, G., & Wiegand, T. (2005). Video compression—From concepts to the H. 264/AVC standard. *Proceedings of the IEEE*, *93*(1), 18–31. doi:10.1109/JPROC.2004.839617

Takai, M., Martin, J., & Bagrodia, R. (2001). Effects of wireless physical layer modeling in mobile ad hoc networks. *Proceedings of the 2nd ACM International Symposium on Mobile Ad Hoc Networking & Computing*, (pp. 87 – 94).

Thomos, N., Argyropoulos, S., Boulgouris, N., & Strintzis, M. (2005). Error-resilient transmission of H. 264/AVC streams using flexible macroblock ordering. *Second European Workshop on the Integration of Knowledge, Semantic, and Digital Media Techniques*, (pp. 183-189).

Tonguz, O., Viriyasitavat, W., & Bai, F. (2009). Modeling urban traffic: A cellular automata approach. *IEEE Communications Magazine*, *47*(5), 142–150. doi:10.1109/MCOM.2009.4939290

Treiber, M., Hennecke, A., & Helbing, D. (2000). Congested traffic states in empirical observations and microscopic simulations. *Physical Review E: Statistical Physics, Plasmas, Fluids, and Related Interdisciplinary Topics*, *62*, 1805–1824. doi:10.1103/PhysRevE.62.1805

Vaidya, N. H. (2004). Mobile ad hoc networks: Routing, MAC and transport issues. *Proceedings of the IEEE International Conference on Computer Communication*, tutorial.

Varsa, V., Hannuksela, M., & Wang, Y. (2001). Non-normative error concealment algorithms. *ITU-T VCEG Doc: VCEG-N62,* Vol. 62.

Wang, Y., Reibman, A., & Lin, S. (2005). Multiple description coding for video delivery. *Proceedings of the IEEE*, *93*(1), 57–70. doi:10.1109/JPROC.2004.839618

Wellens, M., Petrova, M., Riihijarvi, J., & Mahonen, P. (2005). Building a better wireless mousetrap: Need for more realism in simulations. *Proceedings of the Second Annual Conference on Wireless On-demand Network Systems and Services*, (pp. 150-157).

Wenger, S. (2003). H.264/AVC over IP. *IEEE Transactions on Circuits and Systems for Video Technology*, *13*(7), 645–656. doi:10.1109/TCSVT.2003.814966

Wenger, S., Knorr, G., Ott, J., & Kossentini, F. (1998). Error resilience support in H.263. *IEEE Transactions on Circuits and Systems for Video Technology*, *8*(7), 867–877. doi:10.1109/76.735382

Wiegand, T., Sullivan, G., Reichel, J., Schwarz, H., & Wien, M. (2007). *Joint draft ITUT rec. H. 264–ISO/IEC 14496-10/Amd. 3 scalable video coding.* ISO/IEC JTCI/SC29/WG11 and ITU-T SG16 Q.

Wien, M., Schwarz, H., & Oelbaum, T. (2007). Performance analysis of SVC. *IEEE Transactions on Circuits and Systems for Video Technology*, *17*(1), 1194–1203. doi:10.1109/TCSVT.2007.905530

Wu, J. (Ed.). (2006). *Handbook on theoretical and algorithmic aspects of sensor, ad hoc wireless, and peer-to-peer networks*. Boca Raton, FL: Auerbach Publications.

Yan, L. (2005). Can P2P benefit from MANET? Performance evaluation from users' perspective. *Proceedings of the International Conference on Mobile Sensor Networks*, (pp. 1026-1035).

Yan, L., Sere, K., & Zhou, X. (2004). Towards an integrated architecture for peer-to- peer and ad hoc overlay network applications. *Proceedings of the 10th IEEE International Workshop on Future Trends of Distributed Computing Systems*, (pp. 312–318).

Yang, J., & Gong, S. (2009). A content-based layered multiple description coding scheme for robust video transmission over ad hoc networks. *Proceedings of the Second International Symposium on Electronic Commerce and Security*, (pp. 21-24).

Yong, L., Guo, Y., & Liang, C. (2008). A survey on peer-to-peer video streaming systems. *Peer-to-Peer Networking and Applications*, *1*, 18–28. doi:10.1007/s12083-007-0006-y

Zeng, X., Bagrodia, R., & Gerla, M. (1998). GloMoSim: A library for parallel simulation of large-scale wireless networks. *Workshop on Parallel and Distributed Simulation,* (pp. 154-161).

Zhang, J., Liu, L., Ramaswamy, L., & Pu, C. (2008). PeerCast: Churn-resilient end system multicast on heterogeneous overlay networks. *Journal of Network and Computer Applications*, *31*(4), 821–850. doi:10.1016/j.jnca.2007.05.001

Zhang, Q. (2005). Video delivery over wireless multi-hop networks. *Proceedings of the International Symposium on Intelligent Signal Processing and Communication Systems*, (pp. 793-796).

Zhang, X., Liu, J., Li, B., & Yum, Y.-S. P. (2005). CoolStreaming/DONet: A data-driven overlay network for peer-to-peer live media streaming. *Proceedings of the 24th Annual Joint Conference of the IEEE Computer and Communications Societies (INFOCOM),* (pp. 2102-2111).

Zhao, B. Y., Kubiatowicz, J. D., & Joseph, A. D. (2001). *Tapestry: An infrastructure for fault-resilient wide-area location and routing.* Technical report UCB//CSD-01-1141, U.C. Berkeley.

Zhengye, L., Shen, Y., Panwar, S. S., Ross, K. W., & Wang, Y. (2007). P2P video live streaming with MDC: Providing incentives for redistribution. *Proceedings of the IEEE International Conference on Multimedia and Expo*, (pp. 48-51).

Zuo, D.-H., Du, X., & Yang, Z.-K. (2007). Hybrid search algorithms for P2P media streaming distribution in ad hoc networks. *Proceedings of the 7th International Conference on Computational Science*, (pp. 873-876).

ADDITIONAL READING

Antonopoulos, N., Exarchakos, G., & Liotta, A. (Eds.). (2009). *Handbook of research on p2p and grid systems for service-oriented computing: Models, methodologies and applications*. Hershey, PA: Information Science Reference.

Biswas, J., Tatchikou, R., & Dion, F. (2006). Vehicle-to-vehicle wireless communication protocols for enhancing highway traffic safety. *IEEE Communications Magazine*, *44*(1), 74–82. doi:10.1109/MCOM.2006.1580935

Blum, J. J., Eskandarian, A., & Hoffman, L. J. (2004). Challenges of inter-vehicle ad hoc networks. *IEEE Transactions on Intelligent Transportation Systems*, *5*(4), 347–351. doi:10.1109/TITS.2004.838218

Ding, W. (2010). *Synergy of peer-to-peer networks and mobile ad hoc networks: Bootstrapping and routing*. New York, NY: Nova Science.

Goldsmith, A. (2005). *Wireless communications*. Cambridge, UK: Cambridge University Press.

Hartensein, H., & Laberteaux, K. (Eds.). (2009). *VANET: Vehicular technologies and internetworking technologies*. New York, NY: John Wiley & Sons Inc.

Loo, J., Mauri, J. L., & Ortiz, J. H. (2011). *Mobile ad hoc networks: Current status and future trends*. Boca Raton, FL: CRC Press. doi:10.1201/b11447

Papadopouli, M., & Schulzrinne, H. (2010). *Peer-to-peer computing for mobile networks: Information discovery and dissemination*. New York, NY: Springer Verlag.

Rappaport, T. (2001). *Wireless communications: Principles and practice*. Upper Saddle River, NJ: Prentice Hall.

Richardson, I. (2003). *H.264 and MPEG-4 video compression: Video coding for next-generation multimedia*. Chichester, UK: John Wiley & Sons Inc. doi:10.1002/0470869615

Shen, X., Yu, H., Buford, J., & Akon, M. (Eds.). (2010). *Handbook of peer-to-peer networking*. New York, NY: Springer Verlag. doi:10.1007/978-0-387-09751-0

Simpson, W. (2008). *Video over IP: IPTV, internet video, H.264, P2P, Web TV, and streaming: A complete guide to understanding the technology*. Burlington, MA: Focal Press.

Toh, C.-K. (2001). *Mobile ad hoc networks: Protocols and systems*. Upper Saddle River, NJ: Prentice Hall.

van der Schaar, M., & Chou, P. A. (Eds.). (2007). *Multimedia over IP and wireless networks*. Amsterdam, The Netherlands: Elsevier.

Watfa, M. (2010). *Advances in vehicular ad hoc networks: Developments and challenges*. Hershey, PA: Information Science Reference. doi:10.4018/978-1-61520-913-2

Wu, J. (Ed.). (2006). *Handbook on theoretical and algorithmic aspects of sensor, ad hoc wireless, and peer-to-peer networks*. Boca Raton, FL: Auerbach Publications.

KEY TERMS AND DEFINITIONS

De-blocking Filter: In the H.264/AVC codec standard a de-blocking filter was reintroduced into the coding cycle. This filter smooths the boundaries between coding blocks as discontinuities between blocks is otherwise one of the main sources of video quality degradation.

Discrete Cosine Transform: Is an orthogonal transform that in the context of a video codec transforms a pixel from the spatial to the frequency domain. It succeeds in compacting the signal energy in an efficient way and unlike a Fourier Transform works upon real-valued pixels.

Entropy Coding: Is a lossless coding method that in a hybrid video codec occurs immediately prior to encoder output of the compressed bitstream. Normally a unique prefix code is assigned to each output symbol. This mapping is arranged to assign the shortest codes to the most common symbols. Adaptive entropy coding varies the mapping according to the frequency of symbols occurring over time.

Error Resilience: Generally, a source coding set of techniques that allow error concealment at the video decoder better construct a transmitted video sequence in the event of transmission errors.

Error Concealment: Reconstruction by a video codec decoder to try and remove remaining errors after forward error correction and/or error resilience have been applied. Simple replacement from a previous frame is normal but this does not account for motion between the frames. If motion vectors survive transmission then a copy can be made from the matching macroblock in a previous frame. Many other forms of intelligent error concealment exist but they can introduce latency from the computational overhead.

Fading: A wireless signal may be affected by reflections resulting in multi-path propagation. As a result, multiple versions of a signal may arrive at the receiver offset in time. The receiver will thus receive a distorted version of the signal. This distortion can vary over time as a result of constructive and destructive interference. Fading is usually divided between fast fading in which short term variations occur, which are commonly modeled by a random process, and long term variations, slow fading, when a mobile user enters a different wireless environment, for example going from an open to built-up environment. The result of fading is variations in the amplitude and phase of the signal.

Forward Error Correction (FEC): (also known as channel coding) is a form or error protection in which the sender adds redundant data which better enables the information to be reconstructed if any part of the combined information and redundant data are lost. The number of errors that can be corrected is given by a finite limit. FEC is usually used when requesting repair data introduces excessive latency (as for interactive video streaming) or when too many repair requests might occur (as in multicast). Ideally to avoid unnecessary overhead the FEC level should be adaptive.

H.264/AVC (Advanced Video Coding): is the most recent specification for video codecs in the MPEG and ITU H.26x series. It is known as MPEG-4 in the broadcasting industry,

MANET: A Mobile Ad Hoc Network is an infrastructureless network of mobile devices: meaning that there is no central access point and no core network linking those access points. Normally, the mobile nodes are wireless devices. Routing across a network is often multi-hop and uses intermediate nodes as routers. The nodes are self-configured for that purpose.

Mobility Model: In a MANET, random motion is often assumed. However, in a VANET, mobility models specify the motion characteristics of the nodes according to highway topology and driver behavior.

Multiple Description Coding: A form of error resilient transport of video streams. Two or more versions of a video stream are transmitted, hopefully across different routes through a network. Either one of the streams may serve to display the video. However, a higher quality version results from combining multiple description streams.

Peer-to-peer Streaming: Designed to distribution video or TV utilizing peer-to-peer networks. Nodes not only download video chunks from other nodes in the network but also act as sources for video chunks. These networks are either organized in a tree structure or in a mesh topology.

Propagation Model: A wireless signal's attenuation is modeled by a deterministic path loss model, which varies according to distance, and a propagation model, which also varies according to a number of factors such as the frequency response to precipitation, reflection and diffraction from obstructions, and scattering. In addition, wireless signals are affected by Doppler frequency shifts due to node speed and co-channel interference.

VANET: Vehicular Ad Hoc Networks are an extension of MANETs for vehicles. The main differences are that high speeds are possible and battery usage is no longer a key determinant of performance. Vehicles normally move across road networks and along highways, whereas it is commonly assumed that in a MANET, node motion is random.

WiMAX: Worldwide Interoperability for Microwave Access is a broadband wireless technology that provides fixed and mobile Internet access. It has been standardized as IEEE 802.16 with a series of parts such as d, e, and now m.

ENDNOTES

1. Video compression and video coding will be used interchangeably.
2. Entropy coding is a way to compress (and losslessly recover) digital data.
3. DCT is now replaced by a reversible integer transform in H.264 to avoid drift during the inverse transform.
4. The distinction between picture and frame is only relevant for interlaced video and the terms are inter-changeable when progressive video is considered.
5. PSNR = 10 log (MAX^2/MSE), where MAX is the maximum intensity value possible for a pixel, and ME is the pixel-wise mean square error between a reference frame and the frame under test.
6. BonnMotion is available from http://iv.cs.uni-bonn.de/wg/cs/applications/bonnmotion/ (accessed 21/10/09)
7. Obstacles include lane closures, uphill gradients, and potholes.
8. On a highway using IEEE 802.11p, 33 dBm with a potential range of 1300 m is more appropriate.
9. For interactive video one-way delays up to 250 ms are possible, though delays of around 100 ms are ideal.

Chapter 8
Recent Advances in Peer-to-Peer Video Streaming by Using Scalable Video Coding

Dan Grois
Ben-Gurion University of the Negev, Israel

Ofer Hadar
Ben-Gurion University of the Negev, Israel

ABSTRACT

Scalable Video Coding provides important functionalities, such as the spatial, temporal, and SNR (quality) scalability, thereby significantly improving coding efficiency over prior standards such as the H.264/AVC and enabling the power adaptation. In turn, these functionalities lead to the enhancement of the video streaming over Peer-to-Peer networks, thereby providing a powerful platform for a variety of multimedia streaming applications, such as video-on-demand, video conferencing, live broadcasting, and many others. P2P systems are considered to be extremely cost-effective, since they utilize resources of the peer machines (e.g., CPU resources, memory resources, and bandwidth). However, since bandwidth is usually not constant and also since Peer-to-Peer networks suffer from the packet loss, there is no guarantee for the end-user video presentation quality. In addition, due to different server and end-user hardware configurations, it will be useful to specify the quality of the media (e.g., the bit-rate, spatial/temporal resolution, and the like). As a result, the Scalable Video Coding approach is an excellent choice, since the media streaming can be adjusted to a suitable stream to fit a particular Peer-to-Peer network and particular end-user requirements.

INTRODUCTION

This chapter comprehensively covers the topic of the Peer-to-Peer (P2P) media streaming by using Scalable Video Coding (SVC), which is the extension of the H.264/AVC, while making a special emphasis on the Peer-To-Peer video streaming in a wireless environment, which is currently a very important issue due to the recent technological achievements in the mobile device field and due to the recent worldwide trends.

The advent of cheaper and more powerful mobile devices having the ability to play, create,

DOI: 10.4018/978-1-4666-1613-4.ch008

Figure 1. Example of the region-of-interest dynamic adjustment and scalability (e.g., for mobile devices with different spatial resolution)

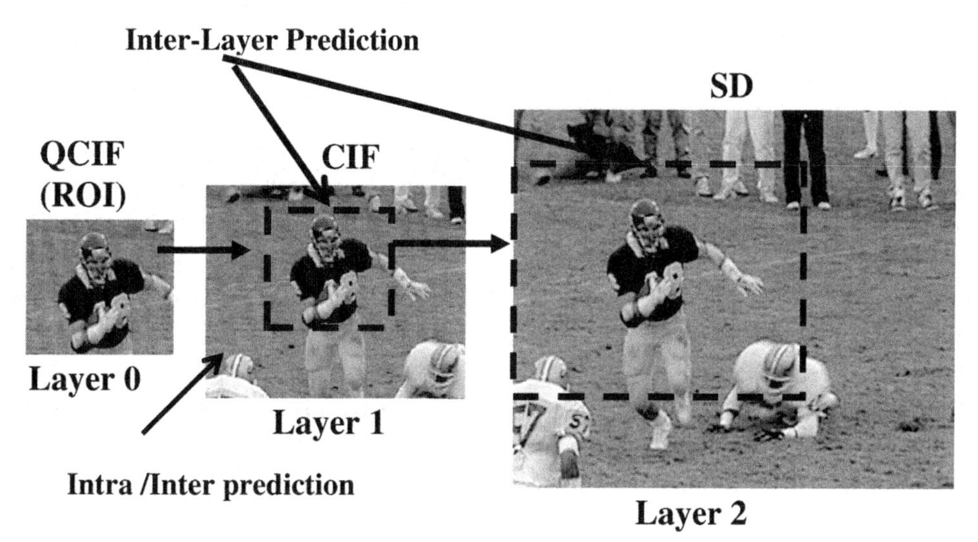

and transmit video content, thereby maximizing a number of multimedia content distributions on various mobile networks, such as peer-to-peer wireless networks, has placed unprecedented demands for the high capacity, low-latency, and low-loss communications paths. The reduction of cost of digital video cameras along with the development of user-generated video sites (e.g., iTunes™, Google™ Video and YouTube™) have stimulated the new user-generated content sector. The growing premium content coupled with advanced video technologies, such as the Mobile TV, will mainly replace in the near future the conventional technologies (e.g., the cable or satellite TV). In this context, high-definition, highly interactive networked media applications pose challenges to network operators.

The variety of end-user devices with different capabilities, ranging from cell phones with small screens and restricted processing power to high-end PCs with high-definition displays, have stimulated a significant interest in effective technologies for video adaptation for spatial formats, consuming the power and bit rate. As a result, much of the attention in the field of video adaptation is currently directed to the Scalable Video Coding (SVC), which was standardized in 2007 as an extension of H.264/AVC (Schwarz et al., 2007; Wiegand et al., 2006), since the bitstream scalability for video is currently a very desirable feature for many multimedia applications to be used on heterogeneous devices (Grois & Hadar, 2011a; Grois & Hadar, 2011b). Figure 1 presents, for example, the usage of the SVC Region-of-Interest (ROI) scalability for end-user devices having different spatial resolutions (e.g., QCIF, CIF, and SD).

The need for the scalability arises from the need for spatial formats, bit rates and power (Wiegand & Sullivan, 2003; Grois et al., 2010a; Grois et al., 2010b). To fulfill these requirements, it would be beneficial to simultaneously transmit or store video in variety of spatial/temporal resolutions and qualities, leading to the video bitstream scalability. The major requirement for the Scalable Video Coding is to enable encoding of a high-quality video bitstream that contains one or more subset bitstreams, each of which can be transmitted and decoded to provide video services with lower temporal or spatial resolutions, or to provide reduced reliability, while retaining reconstruction quality that is highly relative to the rate of the subset bitstreams.

As a result, the SVC provides important functionalities, such as the spatial, temporal and SNR (quality) scalability, thereby enabling the power adaptation (Schwarz et al., 2007; Wiegand et al., 2006). Also, the SVC has achieved significant improvements in coding efficiency compared to the scalable profiles of prior video coding standards. In turn, these functionalities and improvements lead to enhancements of video streaming, and particularly Peer-to-Peer (P2P) media streaming.

In the following sections, we review in detail the recent advances in P2P media streaming by using SVC. Particularly, in the following sections, we discuss in detail:

- SVC streaming over P2P networks, including: a) rate-distortion adaptation in SVC P2P systems, b) adaptive SVC P2P video streaming, c) combined network coding and Scalable Video Coding for the P2P streaming, and d) layer extraction in the SVC P2P streaming;
- Scheduling and priorities in SVC P2P streaming;
- Hybrid mesh and tree overlays in P2P SVC systems;
- Wireless perspectives of streaming over P2P networks by using Scalable Video Coding;
- Combined Scalable Video Coding and Multiple Description Coding schemes;
- Quality control and adaptation in SVC-based Peer-to-Peer systems; and
- Scalable Video-on-Demand P2P systems.

A special attention in this chapter is paid to the Peer-to-Peer video streaming in a wireless environment, which is currently an extremely relevant issue due to the recent technological achievements in the mobile device field and due to the recent worldwide trends. Finally, this chapter is concluded by providing future research directions.

SVC STREAMING OVER PEER-TO-PEER NETWORKS

Internet-based video streaming is becoming more and more popular and attract millions of online viewers every day (Shen et al., 2010; Hei et al., 2007). The number of unique viewers of online video increased 5.2% year-over-year, from 137.4 million unique viewers in January, 2009, to 142.7 million in January, 2010. For example, 120.5 million viewers watched videos on YouTube™ in the month of August of 2009, and the number is expected to rise to at least one billion viewers worldwide in 2013. Live streaming applications provide live broadcasting streams from live channels such as TV and live events. For instance, YouTube™ live streaming of the Ireland U2™ concert was watched by 10 million people (Shen et al., 2010; Hei et al., 2007).

P2P is a powerful platform to enable a variety of multimedia streaming applications over the Internet, such as video-on-demand, video conferencing, and live broadcasting, etc. P2P system is extremely cost-effective since it utilizes the resources (e.g., CPU resources, storage space, and bandwidth) of the peer machines (Hossain et al., 2009). Another reason for P2P's success is its instant deployment: it allows almost ubiquitous network coverage in the absence of Content Dependent Network (CDN) services and Internet Protocol (IP) multicast. As a result, peer-to-peer techniques have attracted significant interests for live video broadcasting over the Internet, especially due to its high scalability (Shen et al., 2010; Hei et al., 2007). In a P2P live video streaming system, a streaming media server generates a series of chunks, each of which is a small video stream fragment containing the media contents of a certain length. The peers watching the same video program form an overlay for video stream sharing between each other in the form of chunks. The P2P paradigm dramatically reduces the bandwidth burden on the centralized content provider and generates more available bandwidth as the number of viewers increase. Typical P2P video streaming applications include PPLive™, Joost™, SopCast™, UUSee™, ESM™ (Chu et al., 2000) and CoolStreaming™ (Zhang et al., 2005). As an example, UUSee™ simultaneously sustains 500 live stream channels and routinely serves millions of users each day (Wu & Li, 2008).

However, as users spend more and more time watching videos online, they are becoming increasingly unsatisfied with the Quality-of-Service (QoS), i.e., image freezes and poor resolution (Picconi & Massoulie, 2008; Wang et al., 2008). As a result, the important question now is how the SVC streaming can be used in P2P networks, in which each host has an equal position, especially, in overlay tree-based P2P streaming, in which the host forwards all the data it receives to the next-level hosts. Since the bandwidth is usually not constant and also since the P2P networks suffer from the packet loss, there is no guarantee for the end-user service quality. For example, if the actual bandwidth is smaller than that of the required media, then although some data segments are received, they are useless. Besides, due to the different host and end-user hardware configurations, a given host may wish to specify the quality of the media (e.g., the bit-rate, spatial/temporal resolution, and the like). The novel SVC standard enables efficient usage of the network capacity by allowing intermediate high capacity nodes in the overlay network to dynamically extract layers from the scalable bit stream to serve less capable peers (Baccichet et al., 2007). As a result, the SVC approach is an excellent choice, since the media streaming can be adjusted to a suitable stream to fit a particular P2P network and a particular end-user requirement.

The novel H.264/SVC standard supports the encoding of a video signal at different qualities within the same layered bit stream (Schwarz et al., 2007; Baccichet et al., 2007). This allows performing efficient on-the-fly rate adaptation, while achieving compression efficiency comparable to H.264/AVC single-layer coding. Furthermore, SVC allows a more efficient usage of the network bandwidth on a P2P network by enabling intermediate high capacity nodes in the overlay to dynamically extract layers from the scalable bit stream to serve less capable peers (Baccichet et al., 2007).

The SVC stream usually has one Base-Layer (Layer 0) and at least one Enhancement Layer, while each Enhancement Layer cannot be decoded independently. Instead, for decoding the received video stream, the Enhancement Layer must be combined with other low-level Enhancement Layers and also with the Base-Layer, since not all portions of the SVC stream are equally important to the video presentation quality. Thus, the most essential portion of the SVC stream (i.e., the SVC Base-Layer) should be transmitted in a highest priority, based on which the user can further select one or more enhancement layers to obtain the better visual presentation quality. As a result, when using the Scalable Video Coding applications, users that have the higher bandwidth, benefit from better visual presentation quality by receiving more SVC layers, while other users having the lower bandwidth, obtain the quality proportional to a number of SVC layers they are able to receive (Schwarz et al., 2007; Baccichet et al., 2007).

In the last several years, many researchers around the world have combined their efforts to develop efficient P2P streaming solutions by using Scalable Video Coding. Thus, for example, (Itaya et al., 2005) developed a heterogeneous asynchronous multi-source streaming (HAMS) model, according to which each communication channel may support different Quality-of-Service (QoS) and each peer may support different transmission rate. Packets of a multimedia content are transmitted in parallel to a leaf peer from multiple content peers, as schematically shown in Fig. 2. Every active content peer asynchronously starts transmitting a subsequence of the packets of content to each leaf peer independently of the other content peers. Each content peer autonomously selects some packets of the multimedia content

Figure 2. Multi-source streaming model of Itaya et al., (2005)

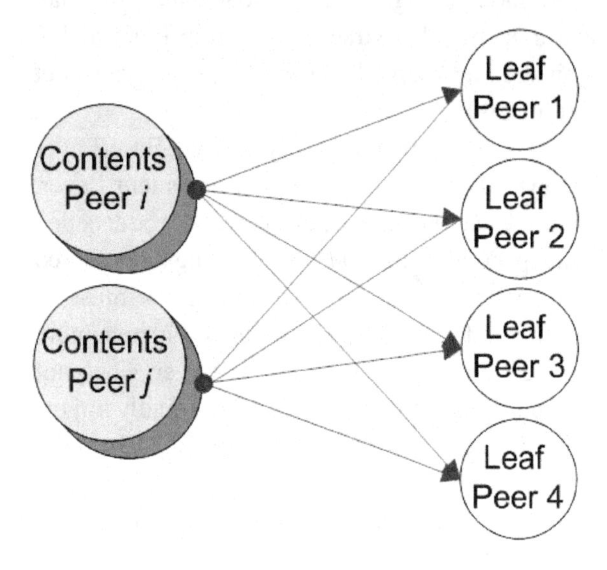

by exchanging information on what packets they have sent with other content peers.

In addition, (Baccichet et al., 2007) proposes a model that allows evaluating the trade-off of using a SVC with respect to single-layer coding, given the distribution of the receivers capacities in an error-free network. Also, (Baccichet et al., 2007) proposes a mathematical framework that allows quantifying the advantage of using the SVC for P2P distribution from a resource-constrained server in an error-free network, with respect to single-layer simulcasting.

Also, one of the solutions of the P2P media streaming is *Scalable Island Multicast (SIM)* for Peer-to-Peer Media Streaming, proposed by (Jin et al., 2006; Lin et al., 2008). SIM is a fully distributed protocol, which can effectively integrate IP multicast and *Application-Layer Multicast (ALM)* for media streaming. The result of using SIM shows that compared with traditional protocols, SIM can achieve low end-to-end delay and link stress. In P2P streaming model, each host has an equal position. Especially, in overlay tree-based P2P streaming, host forwards all the data it receives to next level hosts (Jin et al., 2006; Lin et al., 2008). However, because of the bandwidth variability and network packet loss, there is no guarantee on service quality to the uses. For example, in some condition, the actual bandwidth is smaller than the media required. So although some data segments are received, they are useless. Besides, due to the difference of host hardware ability and user payment ability, a given host may want to specify the quality of media (e.g., bit- rate, resolution). All of the issues above require that the media streaming should become scalable, and should be adjusted to a suitable stream to fit the network and user requirement. Since the SVC uses a layered method, thereby providing media by using a base-layer and one or more enhancement layers, it is an excellent choice for it. At this way, the scalability of media can be implemented. Thus, Lin et al., (2008) focuses on how to integrate SVC and P2P model, and proposes SPM (i.e., SVC technology for media streaming) to integrate SVC and SIM to achieve the advantages of these two technologies. In SPM, the media for distributing is encoded in a SVC form, being divided into several layers. The data packets of each layer are transferred in a corresponding overlay tree.

Also, Harmanci et al. (2009) proposes a method for peer assisted on demand delivery of scalable video using an optimal, distributed caching algorithm for cache constrained peers. The authors of Harmanci et al. (2009) assume that there are one or more content delivery servers that are provisioned to support delivery of the base layer for all of the peers, and the enhancement layer is only supported for a fraction of peers. It has been shown that in such an application, peer assisted content distribution improves the delivery of the content. Harmanci et al. (2009) uses peer-assisted delivery of the enhancement layer and shows that in a cache constrained video-on-demand environment, availability of the cached content becomes a bottleneck. By using peer arrival rates and supply-demand analysis of the peers, the authors propose better caching policies at each peer.

Further, Fesci-Sayit et al. (2009) proposes a tree-formation method, which takes the available resources (e.g., bandwidth) of the peers into the account, such that peers with higher resources are placed near the source and those with lower resources are placed near the leaves of the trees. This feature nicely fits with scalable video (SVC) streaming since once an enhancement layer is dropped by a peer it is not available to any peer underneath it in the tree.

The following summarizes algorithm steps (Fesci-Sayit et al., 2009) that are run at all hierarchical layers simultaneously:

a. *The control leader periodically solicits video requests.* Any node willing to download a video sends a request to its control leader containing the name and quality of the video together with its available bandwidth. The control leader broadcasts this message to all other nodes in the cluster from which it collects replies containing whether a node possesses/requests the video or not and the quality at which the video supplied/requested together with the available bandwidth. When replies are completed, the control leader checks to see if the video is available within the cluster.
b. *If the video is not available within the cluster, the control leader declares the highest bandwidth video requesting node as the download streaming leader of the cluster, which carries out two tasks.* First, it sends the video request to upper layer control leader. In this request, the quality is chosen as the maximum of the qualities requested and available bandwidth is the available bandwidth of the download streaming leader. Second, it calls step (e) with top control leader and source being itself.
c. *If the video is available in the cluster but not at requested quality, the procedure is the same as in step (b)* with the only difference that the source information is also conveyed to the upper layer.
d. *If the video is available in the cluster at the requested quality, the control leader declares itself as top control leader and with individual sources being separate roots.* No request is forwarded to upper layer and the algorithm proceeds to step (e).
e. *Top control leader broadcasts a message containing identities of each individual source node together with its available video quality and available bandwidth.* In addition, this broadcast message contains the identities of download streaming leaders together with their available bandwidths and requested video rates.
f. *Upon receipt of this broadcast message, nodes exchange delay information among them after which they form a unique bandwidth-delay requested/offered quality table.* Each individual source becomes the root of a different multicast tree.

Therefore, according to (Fesci-Sayit et al., 2009), the dynamical multiple source multicast tree formation algorithm allows exclusion of the nodes that do not wish to download the video and/or take part in the video forwarding process.

Moreover, Abbasi & Ahmed (2010) propose the *Small Word Overlay (SWOR)*, architecture for delivering SVC contents over P2P network. The SWOR architecture is based on two-fold mechanisms: (1) organization of peers in *Small-World (SW)* overlay networks; and (2) delivering SVC content using push-pull mechanism. Also, Capovilla et al. (2010) aims to distribute an open-source system with full SVC support by presenting a producer- and consumer-site architecture. The producer-site architecture includes the encoding of the SVC bitstream and splitting of the bitstream into layers, the packetizing of the base-layer and the audio stream into an MPEG-TS, the creation of the scalability metadata based on the SEI information at the beginning of the bitstream, and the ingest of the content into the core of the P2P system. On the consumer-site, the modules for retrieving and consuming the content were presented. The retrieved content is provided to the demuxer, utilizing an HTTP socket, and the demuxed audio and video streams are forwarded to the media consumption solution using RTP. In turn, the de-packetizing and decoding of the SVC content is subsequently performed by the customized SVC tools utilizing the corresponding scalability information.

It should be also noted that the P2P content delivery is also considered to be a promising technique for video group communication, for which the main requirement is a low-delay (Sanchez et al., 2010). Combining the low-delay encoding and low-delay P2P application layer multicast makes it possible to fulfill the delay constraints for interactive group communication applications. For such an application, congestion is a considerable problem, since it causes packet loss or late arrival of the packets, degrading the quality of the service. With this regard, Sanchez et al. (2010) shows how the rate adaptation in combination with the SVC helps to overcome problems in the network, providing a better solution than when the non-adaptive single layer coding is transmitted. A congestion state comprises two different phases: the detection phase and the rate-adaptation phase. For the first phase, random losses and late arrivals are considered depending on the degree of congestion. In other words, depending on this degree of congestion, a rate reduction is selected and a given amount of packets, which corresponds to this reduction in the throughput, are discarded due to packet loss or because their play-out time has already passed. After the detection phase, i.e. time in which packet loss or late packet arrival is detected by the sender, the rate adaptation phase takes place until the congestion is over. The congestion detection is performed based on *acknowledgement (ACK)* packets and *Round Trip Time (RTT)* calculation at the side of the sender, as shown in Figure 3. Failing ACKs and significant increases in the RTT will indicate the congestion in the network, and senders should start the rate-adaptation phase (Sanchez et al., 2010). During this phase the sender adapts the sent media content based on the packet importance and the estimated available throughput, and the mentioned rate-shaping is performed.

In a more recent work, (Medjiah et al., 2011) proposes a new pull approach for efficient layered video content (e.g., the SVC bitstream) distribution in a mesh-based peer-to-peer architecture. The scheme proposed by Medjiah et al. (2011) optimizes different distribution criteria such as request diversity, request reliability, the number of requests and the request overhead. In addition, Abboud et al. (2011b) presents a media-aware network solution based on the router virtualization that aims at striking a balance between intelligence and adaptation at the edge and in the core of the network. Using an extensive simulative study, Abboud et al. (2011b) demonstrates that the media-aware network not only helps in enhancing streaming performance during bottlenecks, but also minimizes the side effects of congestions on

Figure 3. Rate adaptation and congestion detection of Sanchez et al. (2010)

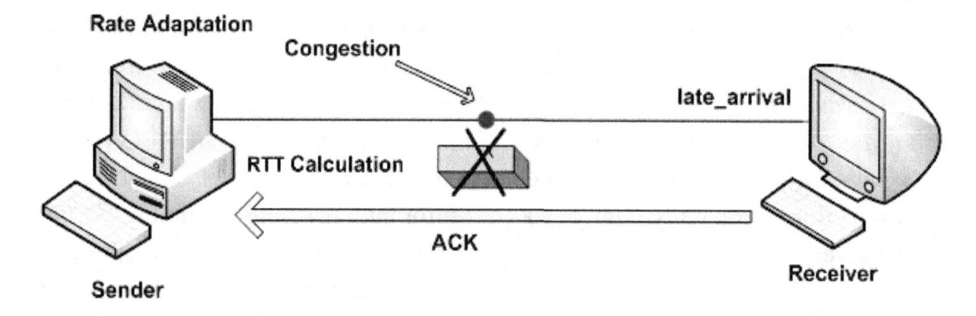

user perceived quality, making it a need for future Internet multimedia applications. Also, having a media-aware network helped in achieving 52% less stalling delay and 34% less SVC quality switches during bottlenecks. Also, Asioli et al. (2010) exploits the characteristics of Scalable Video Coding and P2P network in order to propose an efficient streaming mechanism. The scalable video is divided into chunks and prioritized with respect to its significance in the sliding window by an efficient piece picking policy. Furthermore, a neighbor selective policy is also proposed to receive the most important chunks from the good peers in the neighborhood to maintain smooth content delivery of a certain *Quality of Service (QoS)* for the received video. Further, Nunes et al. (2010) describes the architecture and the evaluation of a SVC and Peer-to-Peer Player prototype, compatible with various terminal hardware setups, with adaptive video streaming capabilities for heterogeneous access networks, using a novel algorithm, the Prioritized Sliding Window, to maintain a stable QoS. The solution showed to be stable, achieving a robust real-time streaming service in P2P environments and discouraging free-riders.

In the next section, we overview recent advances in the developing the SVC P2P networks with regard to the rate-distortion adaptation.

Rate-Distortion Adaptation in SVC Peer-To-Peer Systems

The *rate-distortion (R-D)* analysis is useful not only for efficient source coding, but also for the SVC video streaming, such as the SVC Peer-to-Peer video streaming. While it is well-known that R-D based compression approaches can adaptively select quantization steps and maximize video quality under given buffer constraints, the R-D analysis can also be used during the SVC P2P streaming rate control to optimally allocate bits in joint source-channel coding, and allocate constant visual presentation quality at the decoder end (Mehrotra & Weidong, 2009; Dai et al., 2006; Chenghao et al., 2010; Baccichet et al., 2007; Abanoz & Tekalp, 2009; Hossain et al., 2009). The accurate R-D modeling of video encoders and channel characteristics of communication systems is challenging due to a variety of images and videos, and due to the inherent complexity of network channels (such as the wire/wireless Internet).

Generally, there are two basic approaches for the R-D modeling, i.e. the empirical and analytical approach. Each of these approaches has its benefits and drawbacks. Such, according to the empirical approach, the R-D modeling is performed by interpolating samples of a given encoder. However, the empirical approach lacks theoretical insight into the structure of the coding system. On the other hand, according to the analytical approach, the R-D modeling is performed based on the information and quantization theory, while considering some statistical and correlation properties of the source. Thus, for accurately considering the statistical structure and source correlation of video encoders, a third type of the R-D modeling, namely the operational approach is used, according to which the operational R-D model obtains the basic structure from an analytical expression, but then parameterizes the equation with regard to several parameters sampled from the real communication system (Mehrotra & Weidong, 2009; Dai et al., 2006; Chenghao et al., 2010; Baccichet et al., 2007; Abanoz & Tekalp, 2009; Hossain et al., 2009).

Several researches have been recently carried out with regard to the rate-distortion adaptation. For example, Hossain et al. (2009) presents an optimal rate allocation solution for P2P video streaming applications that minimizes the aggregate rate distortion for all peers. The authors propose a distributed algorithm, in which each peer adjusts its own streaming rate to reach the global optimum. The optimization problem formulation takes into account peer relaying, a constraint unique in P2P distribution scenario in which a peer is both receiver and sender. Peer relaying constraint ensures that the receiving rate of a peer does not exceed the receiving rate of its parent peer. This is because during the rate adaptation, the video quality, once lost, cannot be recovered. As such, the rate change occurred on one peer not only changes the video quality for itself, but also for all of its descendant peers. Therefore, price based resource allocation that considers peer relaying, ensures that peers with more children receives higher bandwidth compared to peers with fewer children. Hossain et al. (2009) combines the solution with two video adaptation techniques: video transcoding and scalable coding.

In addition, Abanoz & Tekalp (2009) propose a multiple objective optimization (MOO) framework for selection of the best encoding configuration for SMDC from a set of candidates, which will strike the best balance between minimizing average end-to-end rate-distortion performance of each description given a set of packet loss probabilities, while points of each scalable description. The optimization variables are some SVC encoding parameters and MD generation alternatives that result in different levels of redundancy at a fixed total rate for all descriptions.

In the next section, we overview recent advances in the adaptive SVC P2P video streaming.

Adaptive SVC Peer-To-Peer Video Streaming

Recently, several adaptive P2P streaming systems that use Scalable Video Coding have been also proposed in the literature. For example, Zhang & Yuan, (2010) leverage the characteristics of Scalable Video Coding and Peer-to-Peer networks in order to propose an unstructured self-adaptive P2P streaming for heterogeneous networks and bandwidth fluctuation. The system of Zhang & Yuan, 2010) integrates P2P features and SVC together for video streaming and is constructed based on BitTorrent-like mesh-pull mechanism. In the system proposed by Zhang & Yuan, (2010), every peer relates to a specified layer, which can be self-adaptive according to the available bandwidth; peers have to request packet data from peers related to the same or higher layer. Deployment and evaluation shows that the system of Zhang & Yuan, (2010) can achieve good self-adaption and remarkable throughput. The system contains a tracker, upload server, cache server and normal peer, as schematically illustrated in Figure 4:

- Tracker, which records the peer information and video streaming information; compared to a traditional P2P video streaming system, an additional attribute for the SVC format video streaming is required (e.g., a "maximal-Spatial-Layer", "maximal-Temporal-Layer", "maximal-Quality-Layer");
- Upload server that reads a video file, packs each Network Abstraction Layer (NAL) unit into an RTP packet, and sends the packets to the cache server; and
- Cache server that stores the RTP packets and acts as a source of the video streaming delivering.

Also, (Abboud et al., 2009) presents a P2P video streaming system based on the Scalable Video Coding with an inherent support for adaptation. The authors make use of a mesh-based streaming architecture that is applicable to both live streaming and video-on-demand. The key feature of the system design is that it allows three degrees of freedom for adaptation - receivers can have different:

- Screen sizes and resolutions;
- Connections with variable downlink bandwidth and delay; and
- Processing capabilities.

The architecture for quality adaptive video streaming, which is proposed by Abboud et al. (2009), is presented in Figure 5.

Figure 4. Architecture of the system proposed by Zhang & Yuan (2010)

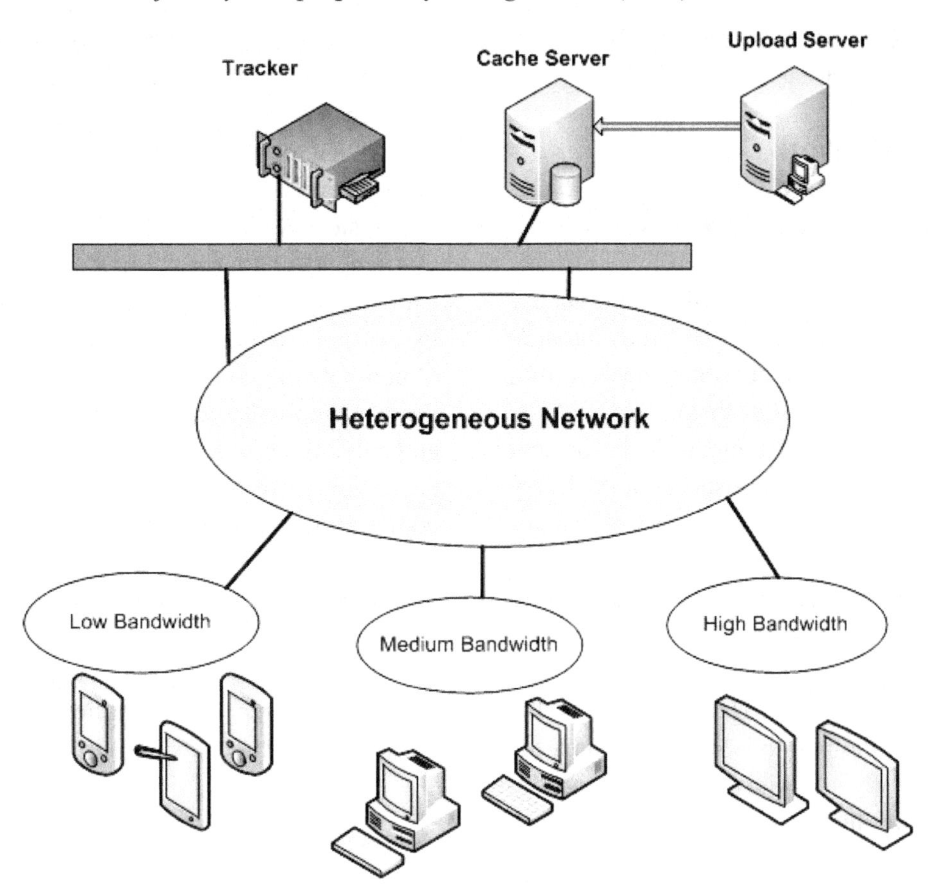

Figure 5. Quality adaptive P2P streaming architecture proposed by Abboud et al. (2009)

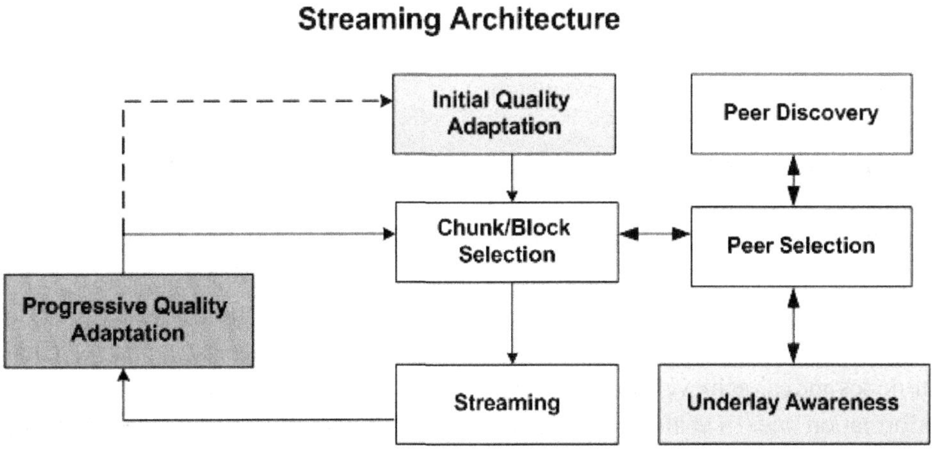

Streaming Architecture

Initial Quality Adaptation

Peer Discovery

Chunk/Block Selection

Peer Selection

Progressive Quality Adaptation

Streaming

Underlay Awareness

In the next section, recent advances with regard to the combined network coding and SVC for the P2P streaming are discussed in detail.

Combined Network Coding and Scalable Video Coding for Peer-To-Peer Streaming

Due to the mass increase of using high quality multimedia streaming on the Internet, broader bandwidth is required to distribute the data from the service provider (Choi et al., 2009). Also, most service requires the media to be present while downloading, making successful delivery of stream time critical. In file distribution, a guaranteed receiving rate is not given and a receiving failure can be overcome with retransmission, allowing reliable distribution with a low upload capable connection, such as the *Asymmetric Digital Subscriber Line (ADSL)*. However, the distribution of the real-time media requires the guaranteed bandwidth with a delay-sensitive delivery to ensure the required Quality of Service (Choi et al., 2009). For developing broadband connections at home, peer-to-peer infrastructure for media distribution is much more adequate for peer-to-peer media streaming. Convergence of diverse networks with many types of devices makes many combinations possible, increasing the need for more flexible methods that are compatible in all cases (Choi et al., 2009).

Network coding has emerged as a promising approach to improve the performance of network transmissions. It is effective due to intermediate nodes' additional abilities of encoding and decoding data at the packet level (Si et al., 2009). Network coding allows packets to be coded in intermediate nodes and can achieve the maximum multicast information rate (Li et al., 2009b). The application of network coding to video delivery has two clear advantages. First, it can be used to improve the throughput of data transmission. It has been proved that the optimal multicast capacity is the minimum of all minimum cuts for all multicast audience. Second, it is more resilient to the packet loss by generating redundant packets in a rateless way (Li et al., 2009b).

In the recent years, several researches have proved that the network coding can also improve the performance of P2P networks. For example, (Gkantsidis & Rodriguez, 2005) demonstrated that network coding can be used in the large-scale content distribution, and the throughput provided by network coding is 2-3 times larger than that of non-network coding. Moreover, network coding can improve the robustness of the system and solve the problem caused by "rare blocks". Generally, there are two main types of network coding strategies (Zeng & Jiang, 2010):

- **Multi-layer Network Coding:** each layer is encoded separately, which means that the receiver can decode each layer separately when obtain sufficient number of independent encoded packets of that layer; and
- **Intra-layer Network Coding:** as is known, the main difference between the traditional video coding (e.g., H.264/AVC) and SVC relates to the prioritization of different part of the video. In SVC, the layers form a clear hierarchy from base-layer to last enhancement layer. An enhancement layer can only add details to the video by combining lower layers, instead of forming video independently. Thus, the most essential part (i.e., the base-layer) should be distributed and decoded first.

In this connection, Si et al.(2009) investigates the *Layered Network Coding (LNC)* and *Hierarchical Network Coding (HNC)* for the P2P live multimedia streaming. Si et al., (2009) implements network coding according to the importance order of encoded source layers. Then, the important layers are recovered early. At the playback deadline, if currently received blocks are not enough to recover the original data completely, the most

Figure 6. System architecture proposed by Li et al. (2009b)

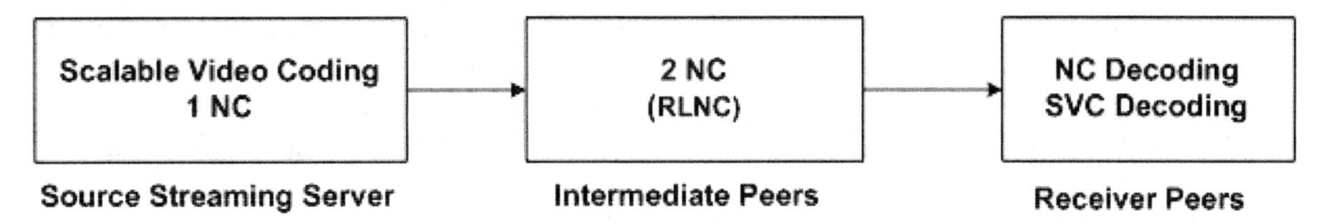

important layers can be recovered to maintain the smoothness of playback. Also, Li et al. (2009b) shows how to apply network coding technology to multicast scalable video content over P2P network. The block diagram of the LNC-based SVC video multicast system, which is proposed by Li et al., (2009b), is shown in Figure 6. According to Li et al. (2009b), firstly, the SVC video coding and Source-Network Coding (1-NC) are performed at the source streaming server. Then, the *random linear network coding (RLNC)* is conducted at intermediate peers within a multicast tree. Finally, the Gaussian elimination method is used for network decoding, followed by SVC video decoding at receiver peers.

Further, Nguyen et al., (2010a) explores the feasibility of using the network coding to make layered P2P streaming much more realistic, by combining network coding and SVC in a fine-granularity manner. To fully evaluate the approach, a complete adaptive P2P streaming protocol "Chameleon" is designed. Nguyen et al., (2010a) uses benefits of the network coding in mitigating peer coordination problems, also taking into the account a scalability structure specified by the SVC standard. "Chameleon" can adapt to bandwidth variations to provide the best possible quality, while maintaining efficiency and scalability of a P2P system. In summary, the main contributions of Nguyen et al., (2010a) include:

- An effective and complete P2P streaming protocol, also considering the problems of neighbor selection, quality adaptation, receiver-driven peer coordination, and sender selection with different design option; and
- A segmentation method to use SVC in P2P streaming in combination with the network coding.

To counteract with unreliable network conditions, also the *Forward Error Correction (FEC)* can be used to guarantee the *Quality-of-Service (QoS)*. The Reed-Solomon, which is a reliable FEC code, is very efficient in the recovering of loss. Given k source symbols to encode n encoded symbols, the Reed-Solomon code can recover lost symbols with k, or more than k received symbols. However, the Reed-Solomon code is limited in the number of encoding symbols due to the required computation. On the other hand, the Raptor code is an FEC with the very low complexity (even with a infinite number of n). With this regard, Choi et al., (2009) uses Raptor-coded symbols to distribute SVC in a peer-to-peer network to reduce the complex and time-consuming peer-to-peer coordination among peers, while still maintaining the advantages of P2P delivery. Choi et al., (2009) also differentiates the amount of each SVC layer sent by the peers for asymmetrical protection of video layers in order to achieve the maximum QoS. Further, Zeng & Jiang, (2010) propose a P2P streaming system that employs both scalable video coding and network coding. Simulation results show that the system proposed by Zeng & Jiang, (2010) can achieve lower average latency and more robustness against poor network environment, thereby providing an improvement in the throughput.

In summary, by using the network coding, P2P multimedia system can obtain at least the following advantages (Zeng & Jiang, 2010):

- Increase the throughput of the P2P system;
- Improve the robustness of the system and solve the problem caused by "rare blocks"; and
- Reduce the complexity of scheduling.

An additional work of Mirshokraie & Hefeeda (2010) also proves that the integration of the network coding and SVC in P2P live streaming systems yields better performance than traditional systems that use the single-layer streams, or traditional systems that use only either network coding or SVC. The design proposed by Mirshokraie & Hefeeda, (2010) enables flexible customization of video streams to support heterogeneous receivers, highly utilizes upload bandwidth of peers, and quickly adapts to network and peer dynamics. The authors also conduct an extensive quantitative analysis to demonstrate the expected performance gain from the proposed design. The analysis uses actual scalable video traces and realistic P2P streaming environments with high churn rates, heterogeneous peers, and flash crowd scenarios. The results show that the system of Mirshokraie & Hefeeda, (2010) can achieve:

- Significant improvement in the visual quality perceived by peers;
- Smoother and more sustained streaming rates;
- Higher streaming capacity by serving more requests from peers; and
- Better robustness against high churn rates and flash crowd arrivals of peers.

The recent layer extraction techniques in the SVC P2P networks are discussed in the next section.

Layer Extraction in SVC Peer-To-Peer Streaming

By taking an advantage of the Scalable Video Coding scalability, several end-to-end SVC streaming schemes enable to perform a multilayered streaming, in which the end-user controls a number of multimedia layers that are provided from the streaming server. In such a way, the end-user visual presentation quality is optimized by allowing each user to retrieve a customized multimedia stream, which is based on both varying network and user conditions. The SVC allows a more efficient usage of the network bandwidth on a P2P network by enabling intermediate high capacity nodes to dynamically extract layers form scalable bit stream to serve less capable peers (Li et al., 2009a). As a result, it can solve the problem of the limited uplink bandwidth at peers. After a peer makes a request, the system attempts to set up a session, which consists of multiple substreams sent from different nodes to the peer. The sample workflow is presented in Figure 7.

In this connection, Li et al., (2009a) presents a R-D optimization substream extraction process for scalable video streaming transmission over the P2P network:

- Calculate the rate and PSNR values for each valid extraction point;
- Establish the PSNR-Rate curve and compute average PSNR for all possible paths;
- Choose the curve with the maximum area as the optimal path - the points lying on the curve are the corresponding extraction points; and
- Assign a larger priority ID for Network Abstraction Layer (NAL) units (extraction points) determined in the previous step, which will have a priority transmission. Then these points will be delivered to the peer, which has sent requests.

As a result, the extraction process with better performance is achieved, while not dramatically increasing the computational cost. The issues with regard to scheduling in the SVC P2P networks are discussed in the following section.

Figure 7. Sample workflow of a scalable video coding P2P system (Li et al., 2009a)

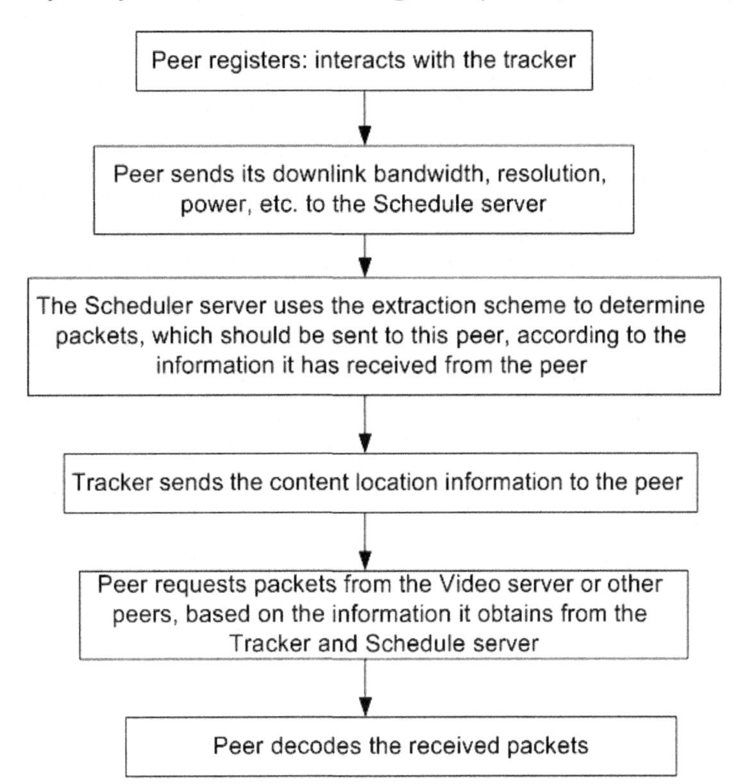

SCHEDULING AND PRIORITIES IN SVC PEER-TO-PEER STREAMING

When the transmission rate is constrained, it is not obvious which SVC stream packets should be transmitted first, and according to what priority: for example, whether all base-layer packets should be transmitted first, or at least a part of the enhancement layer packets. Therefore, there is a need to decide which packets to transmit and when, in order to optimize the visual presentation quality, while considering the bit-rate constraint and the ongoing delivery performance of the P2P channels (Yoon et al., 2006). With this regard, Liu et al., (2009) presents and validates the LayerP2P video streaming, which is a P2P live streaming system that combines layered video, mesh P2P distribution, and a tit-for-tat-like algorithm, in a manner such that a peer contributing more upload bandwidth receives more layers, and consequently, better video quality. According to Liu et al., (2009), the LayerP2P has the following key characteristics:

- *Built-in incentives:* With the layered video, more received video chunks in the order of their importance lead to the higher video quality. LayerP2P exploits this property, together with a tit-for-tat-like strategy, to provide incentives for uploading. Specifically, each peer measures its download rates from its neighbors, and reciprocates by providing a larger fraction of its upload rate to the neighbors, from which it is downloading at higher rates.
- *Adaptation to available upload bandwidth:* LayerP2P dynamically adapts the system demand to the system supply. As the system bandwidth supply evolves due to peer churn, LayerP2P automatically adjusts video quality for the individual peers.

- *Graceful video quality degradation:* With the LayerP2P, lost packets in an enhancement layer do not affect the decoding of lower layers.

As a result, the LayerP2P video streaming system of Liu, Z. et al., (2009) has high efficiency, provides differentiated service, adapts to bandwidth deficient scenarios, and provides protection against free-riders.

Also, Xiao et al, (2009) presents a new scheduling approach for the LayerP2P video streaming, in which the three-stage scheduling mechanism is designed to request absent blocks, where the min-cost flow model, probability decision mechanism and multi-window remedy mechanism are employed in "Free Stage", "Decision Stage" and "Remedy Stage", respectively. Each stage has different scheduling objective to achieve the high throughput, high layer delivery ratio, low useless packet ratio and low subscription jitter. According to Xiao et al, (2009), there are four basic requirements to be considered when performing the data scheduling:

- **Throughput and Delay:** Maximizing the overlay throughput, as well as keeping low packet delay.
- **Layer Delivery Ratio:** In the non-layered streaming, maximizing the node delivery ratio is almost equal to maximizing the throughput, which is not the case in the layered streaming. In the layered streaming, subscribing many layers, but with low delivery ratio for each layer, can also result in high throughput. However, the video quality cannot be high due to the layer dependency.
- **Useless Packets Ratio:** The decoding of upper layers depends on the availability of lower layers. If some lower layer packets are missed, the packets with the same sequence identifications in the upper layers cannot be correctly decoded, and thus become useless. The useless packet ratio should be kept low.
- **Jitter Prevention:** The bandwidth variation is common in today Internet communication. Therefore, if a node subscribes more layers immediately after the bandwidth increase, it may have to drop the highest layers. This short term drop is called the jitter, which not only brings fluctuations in node's Quality-of-Service (QoS), but also causes buffer overflow or underflow.

In addition, Hu et al., (2011) develops a utility maximization model to understand the interplay between efficiency, fairness and incentive in-layered P2P streaming. According to Hu et al., (2011), peers periodically exchange chunk availability with their neighbors using buffer-maps. Neighbors help each other to retrieve missing chunks. Chunk scheduling decides how to issue chunk requests to neighbor peers, and how to serve the chunk requests from neighbor peers. The goal is to properly utilize peers' uplink bandwidths, so that peers always receive the entitled layers and receive the subscribed excess layers with high probability. Since in the SVC, the lower layer bit-stream is more important than higher layer bit-stream, lower layer chunks should be requested before higher layer chunks. In order to increase the data chunk diversity and improve the chance that two peers always have chunks to exchange, Hu et al., (2011) assumes that data chunks belonging to the entitled layers are equally important. The chunks are requested in the order of their importance: from entitled layer chunks to excess layer chunks. A peer selects one neighbor peer that owns the missing chunk to request for the chunk. The probability of choosing a specific peer is proportional to its serving rate to that peer (Hu et al., 2011). On the other hand, in the chunk serving process, individual peers maintain two FIFO queues for each neighbor. One queue is called an entitled queue and the other is called an excess queue. Entitled queue holds chunk requests for entitled layers, while excess queue holds chunk requests for excess layers. The chunk requests in excess queues are sorted in ascending order of video layers, with lowest layer chunk requests at the head. The excess queues are not served unless all entitled queues become empty.

Further, Luo et al., (2009) formulates the quality driven P2P scheduling algorithm into a distributed distortion-delay optimization problem, where the expected video distortion is minimized under the constraint of a given packet playback deadline to select the optimal combination of system parameters residing in different network layers. Then, Luo et al., (2009) provides the algorithmic solution to the formulated problem based on dynamic programming. As a result, the distributed optimization running on each partner node adopted in the scheduling algorithm greatly reduces the computational intensity. According to Luo et al., (2009), the experimental results demonstrate 5-15dB quality enhancement in terms of the PSNR.

In the following section, we present recent developments in using the combined mesh and tree overlays in P2P SVC systems.

HYBRID MESH AND TREE OVERLAYS IN P2P SCALABLE VIDEO CODING SYSTEMS

The fundamental differences between structured architectures and P2P systems must be addressed to provide efficient P2P streaming solutions to existing media applications. P2P systems present the advantage of low-cost service deployment and the flexibility of resource aggregation through multiple path transmission, as schematically shown in Figure 8 (Jurca et al., 2007).

Since a P2P system does not provide any guaranteed support to streaming services, such services must rely on self-organized and adaptive network architectures to meet their stringent quality requirements (Jurca et al., 2007). Two following types of architectures are mainly considered for providing the organization for streaming applications:

Figure 8. Unstructured P2P network: The receiving peers connect and retrieve data from other peers

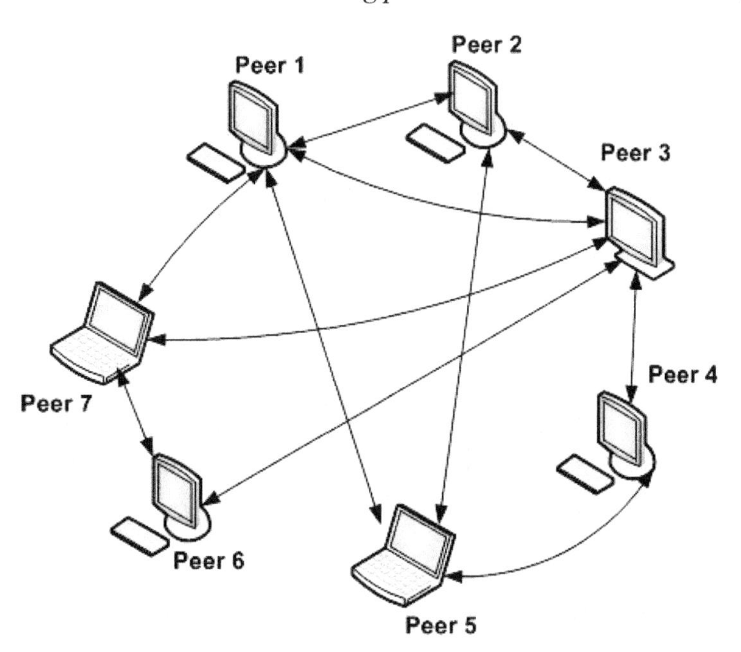

- ***Tree-based overlay:*** *For streaming sessions from media sources to client peers; and*
- ***Mesh-based overlay:*** *For parallel content distribution among peers.*

The tree-based overlays organize the peers as a single or multiple tree overlay that connects the source of the media content to the clients (Jurca et al., 2007), as schematically shown in Figure 9. A peer can simultaneously be a leaf in some distribution trees and an intermediate node in others. However, they are limited by the following two factors:

- Due to the high rate of peers joining/leaving the system, the architecture suffers from high instability; and
- The received media quality is limited by the minimum upload bandwidth of the intermediate peers in the branch.

On the other hand, mesh-based overlay architecture is based on self-organization of nodes in a directed mesh that is used for media delivery to clients (Jurca et al., 2007), as schematically shown in Figure 10. The advantages of such architecture reside in the low cost, simplicity of structural maintenance, and in the resilience of the topology to node failure or departure due to the increased probability of available distinct network paths. However, due to the inherent sequential media encoding and play-out, packet dissemination and data requests must follow closely the temporal ordering of the content at the source (Jurca et al., 2007).

Therefore, a fundamental problem in the Internet-based media distribution can be defined as a routing problem: how to establish an optimal routing structure (e.g., a tree, mesh), which delivers the content from the sender to all receivers, simultaneously achieving certain optimization objective, such as throughput, or delay (Dai et al., 2007). As noted above, although many tree-based and mesh-based streaming protocols have been

Figure 9. Tree architecture for media delivery in P2P systems

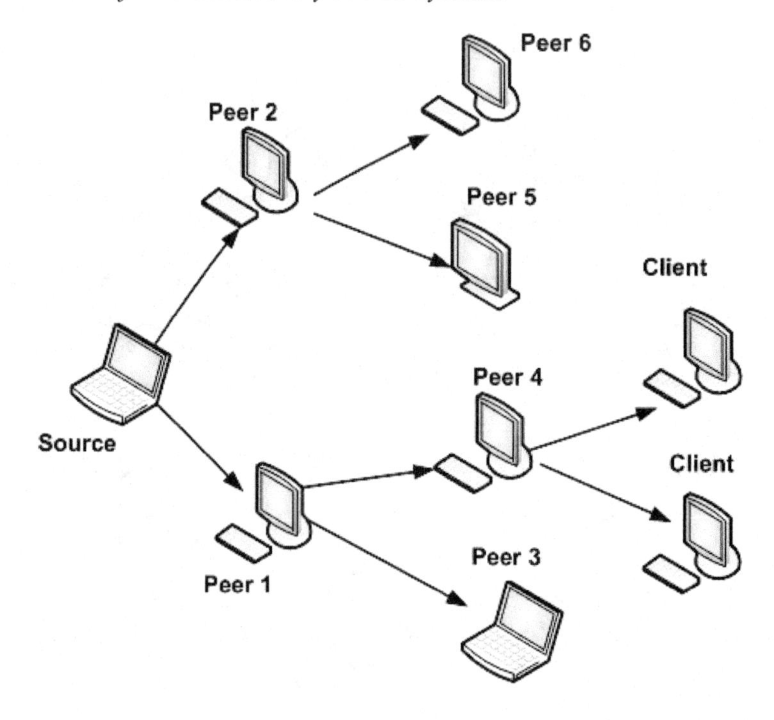

Figure 10. Mesh architecture for media delivery in P2P systems

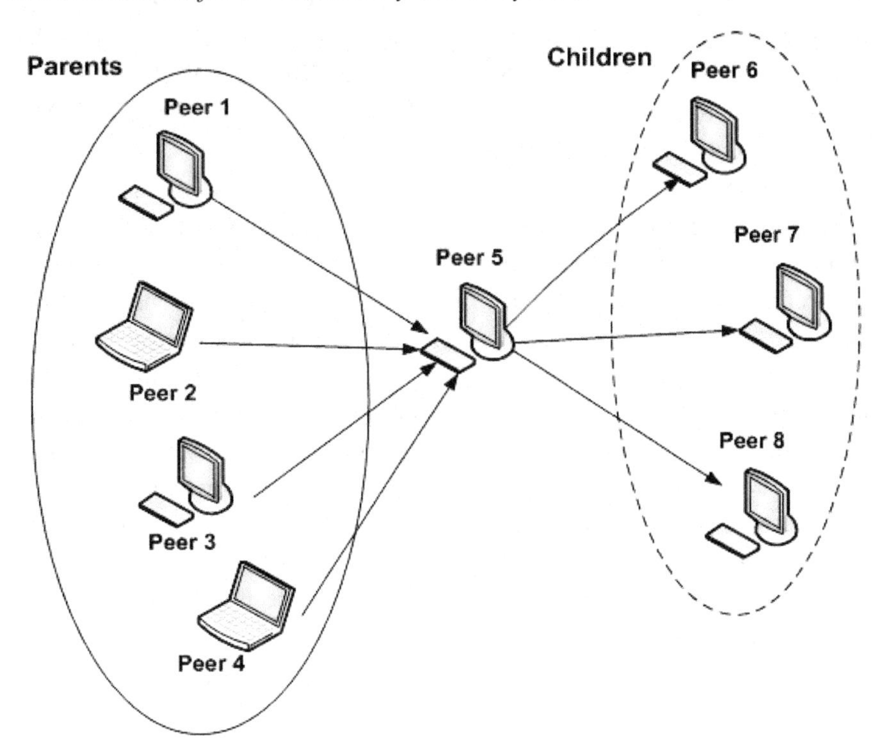

proposed for the P2P networks, each has its own drawbacks such as unreliability and unfairness in tree based and long startup delay and complex scheduling in mesh based protocols (Moshref et al., 2010). According to Moshref et al., (2010), a new video streaming protocol is proposed, called "LayeredCast". The LayeredCast's tree network pushes the base-layer to all peers, while the enhancement layers and missing base-layer segments are pulled over a mesh network by peers with extra bandwidth using a new data-driven scheduling scheme (Moshref et al., 2010). The main features of the LayeredCast protocol are:

- **Hybrid:** Drawbacks of the simple approaches are compensated using a hybrid of mesh and tree overlays;
- **Layered Video:** Provides an adaptive scheme to enhance the video quality using a layered video codec for heterogeneous clients; and
- **Quality-of-Service (QoS):** LayeredCast scheduling aims at moving complexity of multi-service network core to the network clients' application layer, thus providing better QoS over simple regular networks.

As known, the tree-based systems have low-delay but are vulnerable to churn, while the mesh-based systems are churn-resilient, but suffer from high-delay and overhead; both systems cannot make full utilization of the bandwidth in the system. To overcome these problems, Shen et al., (2010) proposes a DHT-aided *Chunk-driven Overlay (DCO)* by introducing a scalable DHT-ring structure into a mesh-based overlay to efficiently manage the video stream sharing. The DCO includes a hierarchical DHT-based infrastructure and a chunk sharing algorithm. Aided by DHT, DCO guarantees stream chunk availability. In this way, the DCO flexibly takes full advantage of available bandwidth in the system and at the same time provides high scalability and low-latency. According to Shen et al., (2010), simulation results show the superiority of DCO compared with the mesh-based and tree-based systems.

Also, Jahromi et al., (2010) proposes a hybrid mesh and tree overlay by selecting stable peers in order to form a tree and "inject" the base-layer of a layered coded media to the mesh overlay. According to (Jahromi et al., 2010), in the first phase of data delivery, prior to the tree construction, both base and enhancement layers of video are transmitted over the mesh overlay and each peer has to request chunks of video from its neighbors. The second phase of data delivery begins after the tree construction: the base-layer is transmitted over the tree overlay for reducing the delay. Peers, who wish to play the video with better quality, should request the enhancement layers through the mesh overlay, so the enhancement layers and also missed frames of video are pulled through the mesh overlay.

In addition, Dai et al., (2007) considers different layers of the video content as the "commodities" (layers) to be distributed. Each layer is routed via a *Minimum Spanning Tree (MST)*, spanning all its receivers in the P2P network. The metric for the MST routing is an artificial length, assigned to each overlay edge, as an asymptotic function of its traffic load. Evidently, the attempt to find the MST equals to finding the most lightly-loaded paths. As a result, Dai et al., (2007) presents an optimal algorithm, which proves to return the maximum achievable throughputs for all layers in proportion to their demands. The key component of the algorithm is the definition of the overlay edge length function.

As seen from the above, providing the combined tree and mesh-based P2P network, by using Scalable Video Coding, may introduce clear advantages over traditional solutions. In the next chapter, we overview various wireless perspective of the streaming over P2P networks by using the Scalable Video Coding.

WIRELESS PERSPECTIVES OF STREAMING OVER P2P NETWORKS BY USING SCALABLE VIDEO CODING

Streaming the SVC video to a variety of decoding devices (e.g., having different spatial/temporal resolution, such as cellular phones) over heterogeneous P2P networks considered to be a relatively new challenge, especially, when the streaming is performed over P2P wireless networks. Considering the flexibility given by scalable bitstreams within P2P overlays, it is clear that P2P streaming systems supporting SVC technology will play an important role in the Internet of the future (Ramzan et al., 2011).

Recent results on the practical throughput and packet loss analysis of multi-hop wireless networks have shown that the incorporation of appropriate utility functions that take into account specific parameters of the protocol layers, such as expected retransmissions, loss rate and bandwidth of each link (de Couto et al., 2003), as well as expected transmission time (Draves et al., 2004) can significantly impact the actual end-to-end network throughput (Mastronarde et al., 2007). While significant contributions have been made to enhance the separate performance of the various OSI layers, no integrated and realistic cross-layer optimization framework exists for efficient multi-user multimedia transmission over multi-hop mesh networks. Different centralized and distributed approaches have been adopted to solve the resource management problem for wireless networks. Centralized approaches solve the end-to-end routing and path selection problem as a combined optimization using multi-commodity flow algorithms, as this ensures that the end-to-end throughput is maximized while constraints on individual link capacities are satisfied. In contrast, distributed approaches use fairness or incentive policies to resolve resource allocation issues in a scalable manner (Liao et al., 2003; La & Anantharam, 2002). However, the traditional research has not considered the benefits of dynamic resource and information exchanges among wireless peers. In order to solve this issue, Mastronarde et al., (2007) focuses on delay-sensitive multimedia transmission among multiple peers over wireless multi-hop enterprise mesh networks. Mastronarde et al., (2007) proposes a distributed and efficient framework for resource exchanges that enables peers to collaboratively distribute available wireless resources among themselves based on their quality of service requirements, the underlying channel conditions, and network topology. The resource exchanges are enabled by the scalable coding of the video content and the design of cross-layer optimization strategies, which allow efficient adaptation to varying channel conditions and available resources. According to Mastronarde et al., (2007), 2-5 dB improvement is achieved in the decoded PSNR for each peer due to the deployed cross-layer strategy.

Peer-to-Peer collaboration paradigms fundamentally change the passive way wireless stations currently adapt their transmission strategies to match available resources, by enabling them to influence system dynamics through exchange of information and resources (Mastronarde et al., 2007).

The problems related to the video transmission in wireless peer-to-peer networks arise from the decentralized and dynamic behavior of these networks, drawbacks of wireless environment and diversity of user bandwidth and signal-to-noise ratio (Ozbilgin & Sunay, 2008). Scalable video coding offers simple and highly flexible solutions for video transmission over heterogeneous networks. Recently, several solutions for using scalable video coding in wireless P2P networks have been proposed. Thus, for example, Ozbilgin & Sunay, (2008) investigate the effects of Scalable Video Coding and user cooperation (which is a technique in which users coordinate to mitigate the system insufficiencies) on the wireless transmission in wireless peer-to-peer networks and determine that both of them improve the quality of the received video.

Also, Peltotalo et al., (2009) presents an effective real-time peer-to-peer streaming system for the mobile environment. The basis for the system is a scalable overlay network, which groups peers into clusters according to their proximity using RTT values between peers as a criterion for the cluster selection. The actual media delivery in the system is implemented using the partial *Real-Time Protocol (RTP)* stream concept: the original RTP sessions related to a media delivery are split into a number of so-called partial streams according to a pre-defined set of parameters in such a way that it allows low-complexity reassembly of the original media session in real-time at the receiving end. Also, according to Peltotalo et al., (2009), the partial stream can be replaced by one description, when using *Multiple Description Coding (MDC)*, or by one layer when using SVC, without affecting the clustered overlay network architecture. In such a case, the above-mentioned partial stream would have much lower complexity than MDC or SVC, thereby enabling a fast proof-of-concept implementation.

In addition, Li & Chan, (2010) present a wireless P2P streaming for scalable interactive streaming, as schematically illustrated in Fig. 11. In the network, videos are divided into segments and collaboratively cached and accessed among mobile devices. The major challenge is to determine which segment to cache at each mobile device to achieve efficient access, in terms of low segment access cost. Li & Chan, (2010) first formulate the problem of segment caching to minimize segment access cost. Then, a distributed algorithm is presented, which achieves collaborative and efficient segment caching, given heterogeneous caching capacities of the participating users.

It should be noted that the resource management problem for wireless networks involves distributing available wireless resources among users given their QoS requirements, underlying channel conditions and the network topology (Mastronarde et al., 2006). Different centralized and distributed approaches have been proposed to solve the resource management problems. Thus,

Figure 11. P2P network for mobile interactive streaming (Li & Chan, 2010)

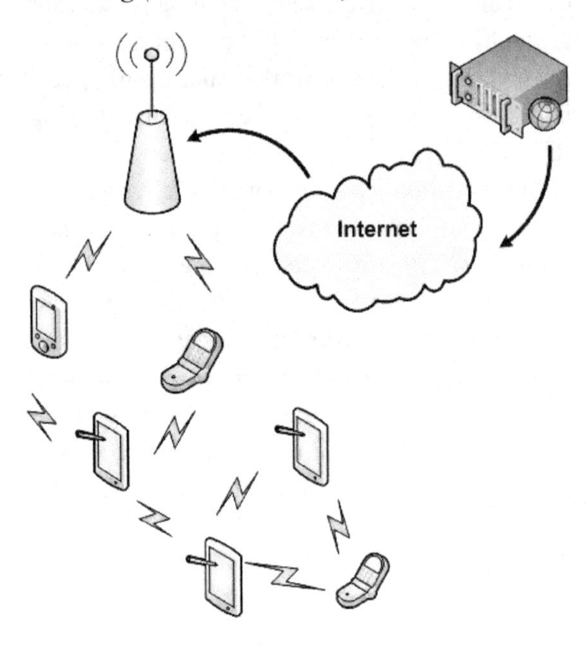

for example, Mastronarde et al., (2006) considers the problem of real-time multimedia transmission among several wireless peers. The peers use a heterogeneous wireless multi-hop mesh network for the delivery of these high-bandwidth streams. One of the main challenges of the considered problem is the division of the scarce wireless resources among the various peers. To address this problem, Mastronarde et al., (2006) proposes an efficient, distributed and collaborative framework for wireless resource exchanges that enables peers to divide available wireless resources among themselves based on their QoS requirements, the underlying channel conditions and network topology. The scalable coding of the video content and decomposition of video flows into various sub-flows (priorities) allow peers to transfer the video at different quality levels, depending on the network load.

Also, Garcia et al., (2009) describes novel forms of delivering seamless content services over P2P networks by using SVC or Multiview Video Coding (MVC). According to Garcia et al., (2009), the ciphering technology is applied to each SVC layer separately. Then, each layer is transported in its own RTP flow, which supports both point-to-multipoint and P2P topologies. By this way, it is possible to apply layer-specific policies to the content: for example, a content creator can decide to distribute a base-layer to anyone for free and to restrict an access to enhancement layers for licensed users. Further, Schierl et al., (2006) presents a multisource streaming approach to increase the robustness of real-time video transmission in *Mobile Ad Hoc Networks (MANETS)* by introducing video coding as well channel coding techniques on the application layer and by exploiting the multisource representation of the transferred media. The source coding is based on the SVC and the channel coding is based on a novel unequal packet loss protection scheme, which is based on Raptor forward FEC codes. While in the presented approach, the reception of a single stream guarantees base quality only, the combined reception enables playback of video at full quality and/or lower error rates. Also, an application layer protocol is introduced by Schierl et al., (2006) for supporting the P2P based multisource streaming in MANETs.

The basic idea of the multisource media coding approach and protocol is to reduce failures in video transmission caused by route losses on the transmission path (Schierl et al., 2006). Route losses are the main problem when comparing ad hoc networks with networks using fixed infrastructures. With this regard, Schierl et al., (2006) proposes to increase a number of used sources for enhancing reliability in server availability, while keeping the overall used network transmission rate/bandwidth as small as possible. The protocol for source monitoring and selection probes available sources cyclically. The assumption is that the addresses of source nodes available in the ad hoc network area are introduced by an external instance. The monitoring of sources is achieved by sending probing packets (inquiry packets) to all known sources for collecting path quality information per source. The link/route quality

Figure 12. Simplified client scheme for frequent server evaluation (Schierl et al., 2006)

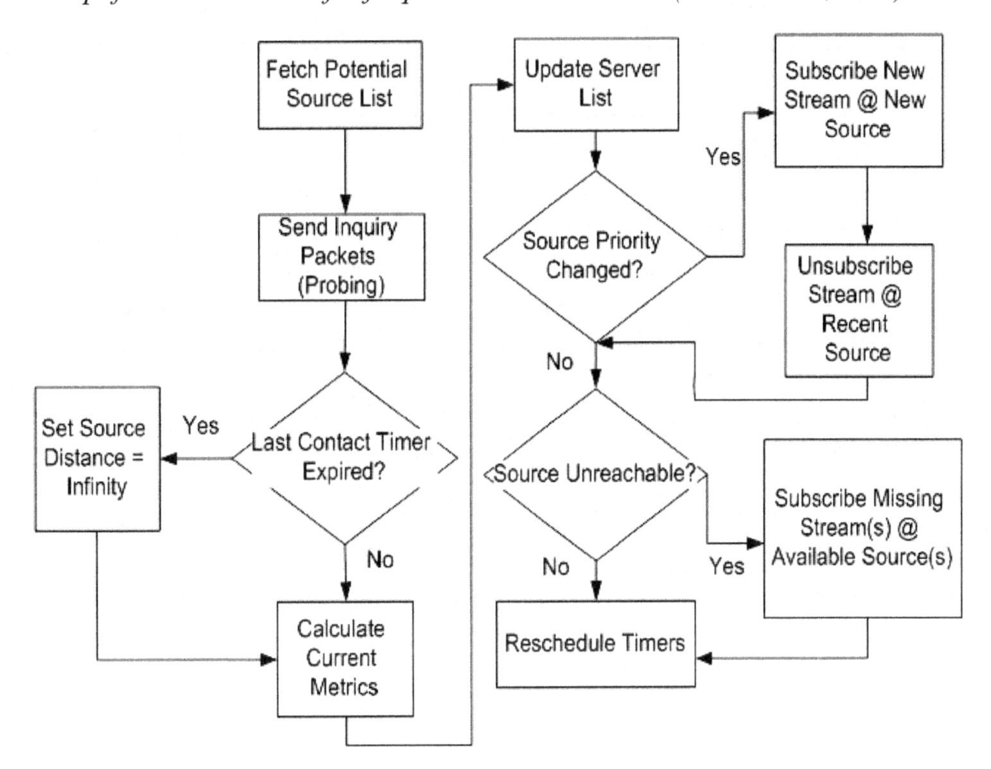

information collected by the inquiry process is called the metric information. In Figure 12, the basic control scheme of the protocol is shown.

The advantages of the combined SVC and MDC schemes are further presented in the following section.

COMBINED SVC AND MDC CODING SCHEMES

MDC addresses the problem of encoding source information using more than one independently decodable and complementary bitstreams, which when combined, can provide the highest level of quality and when used independently, can provide still an acceptable level of quality. This is made possible by introducing some redundancy in each description, which can be discarded (Akyol et al., 2007).

The main difference between MDC and SVC is that, in MDC, video quality is improved with the number of descriptions received in parallel, while in SVC the enhancement layers are applied to improve stream quality (Lopez-Fuentes, 2011). Scalable video has been used to adapt the same video quality for different videos distributed from different sources to multiple requesting peers (López-Fuentes & Steinbach, 2008). The traditional studies conclude that MDC has advantages over SVC for applications with very stringent delay constraints or for networks with a long RTT.

The error-resilience capabilities of MDC and SVC have been also studied by Lee et al., (2003) through extensive experiments. The results provide a comprehensive performance comparison between these two techniques. The performance of specific implementations of MDC and SVC over error-prone packet switched networks was also examined by Chakareski et al., (2003), who compares the performance of MDC and SVC techniques by using different transmission schemes. Chakareski et al., (2003) concludes that the performance between MDC and SVC depends on the employed transmission scheme. On the other hand, in other works, SVC and MDC are combined in order to exploit the individual benefits of these schemes. For example, a combination of scalable video and erasure coding to generate multiple descriptions is introduced by Taal et al., (2004). In this work, the system estimates the bandwidth of the nodes to calculate an optimal allocation of rates for all layers.

It is also well-known that in P2P streaming from a single source, single point of failure can be avoided by MDC over multiple multicast trees, since it provides path diversity. In this scenario, Abanoz & Tekalp, (2009) propose using *Scalable Multiple Description Coding (SMDC)*, where each description is scalable, so that all descriptions can be efficiently adapted to the available rate of each link for effective congestion control.

In addition, Akyol et al., (2007) proposes a novel *Flexible MDC (F-MDC)* framework, within which the highly scalable video encoder of Secker & Taubman, (2003) and Taubman et al., (2004) is extended to MDC in order to provide the efficient adaptation of the P2P streaming. The P2P video streaming system is designed using the proposed F-MDC method by varying:

- Number of base and enhancement descriptions;
- Redundancy of each individual description;
- Rate of each description/layer; and
- Rate allocation among descriptions (i.e., balanced/unbalanced MDC or base/enhancement descriptions).

One advantage of the system of (Akyol et al., 2007) is that it enables a low pre-roll delay, since multiple descriptions, generated with the initial F-MDC parameter set, are received and displayed while the quality of each path is measured.

Also, Lopez-Fuentes, (2010) and Lopez-Fuentes, (2011) proposes and evaluates a combined SVC-MDC video coding scheme for P2P video multicast. The scheme is based on a full cooperation established between the peer sites, which contribute their upload capacity during video distribution. The source site splits the video content into many small blocks and assigns each block to a single peer for redistribution. The video content is encoded by using the SVC, which is combined with MDC to alleviate the packet loss problem. Lopez-Fuentes, (2010) presents a flow control mechanism that allows to dynamically optimize the overall throughput and to automatically adjust the video quality for each peer. Thus, peers with different upload capacity receive different video quality.

The P2P video streaming applications are widely popular and emphasized, e.g., PPstream™ and PPlive™ (Zhang et al., 2005), because of higher transmission speed and data availability; however, in the heterogeneous P2P network environment, users are able to utilize PDA, notebook or desktop computer through distinct network interfaces to get on-demand videos ubiquitously (Huang et al., 2009). To provide distinct spatial-resolution/fidelity videos and flexible video transmission (playback) over P2P networks, a new coding architecture is presented by Huang et al., (2009), who proposes a combined SVC-MDC video coding scheme using the multi-core parallel programming paradigm for P2P video streaming, which is denoted Co-SVCMDC. In the Co-SVC-MDC coding scheme, distinct MDC descriptions contain distinct portions of raw video frames, and each raw frame can be compressed as base-layer and SVC enhancement layers.

Further, Guo et al., (2008) proposes a peer-based cooperative streaming framework with distributed caching for delivering media content effectively over the Internet. The architecture proposed by (Guo et al., 2008) combines the strengths of many techniques such as SVC, MDC, peer-to-peer networking, multisource/multipath streaming, and distributed proxy caching. Specifically, Guo et al., (2008) tries to improve the streaming quality of a later joined client through the help of one or more earlier joined clients in the same neighborhood, given that there is no dedicated powerful proxy server.

The quality control and adaptation techniques by using the SVC over the P2P networks are shown in the next section.

QUALITY CONTROL AND ADAPTATION IN SVC-BASED PEER-TO-PEER SYSTEMS

Video encoding schemes play vital role for real-time streaming applications. A number of video encoding schemes for real-time applications like layered encoding and multiple description coding are suitable for many applications over heterogeneous networks. At the present, the SVC is deemed most promising video format for streaming applications over heterogeneous networks (Mushtaq & Ahmed, 2008).

As already mentioned, the SVC standard is capable to produce highly compressed bit-streams with a wide variety of bit-rates, thereby adapting to the end-user available download capacity. Quality control is an important concern for end-users, since the human eyes are very sensitive to quality fluctuations, which are typical for the video streaming over *Constant Bit Rate (CBR)* (and, in some cases, in *Variable Bit Rate (VBR)*) channels. Mushtaq & Ahmed, 2008) leverage the characteristics of SVC and P2P networks in order to propose an efficient and adaptive video streaming mechanism to maintain smooth content delivery of certain acceptable QoS for the received video. The system proposed by Mushtaq & Ahmed, (2008) is composed of an adaptive streaming mechanism, which is based on the efficient SVC content scheduling to coordinate among receiver and sender peers for smooth media streaming.

The traditional studies with regard to the neighbor selection in the single-layer P2P streaming are insufficient when applied to the layered P2P streaming because in the layered P2P streaming, the video stream is encoded into quality layers, and peers aim to receive the maximum number of layers according to their available bandwidth capacity (Nguyen et al., 2010b). Therefore, while the average playback skip rate is the main performance metric in single-layer P2P streaming, the ratio of the experienced quality level and the expected quality level determined by the bandwidth capacity, called quality satisfaction, is also an important metric in layered P2P streaming. In this scenario, Nguyen et al., (2010b) believes that quality- and context-aware neighbor selection will boost the layered streaming protocol, and finds out factors that impact the quality satisfaction of each peer with experiments on the new adaptive streaming protocol, named "Chameleon" (Nguyen et al., 2010a).

Also, the system of Abboud et al., (2010) is based on the idea of dividing quality adaptation into two stages, as schematically shown in Figure 13. The first stage, called Initial Quality

Figure 13. Quality adaptive P2P streaming system architecture (Abboud et al., 2010)

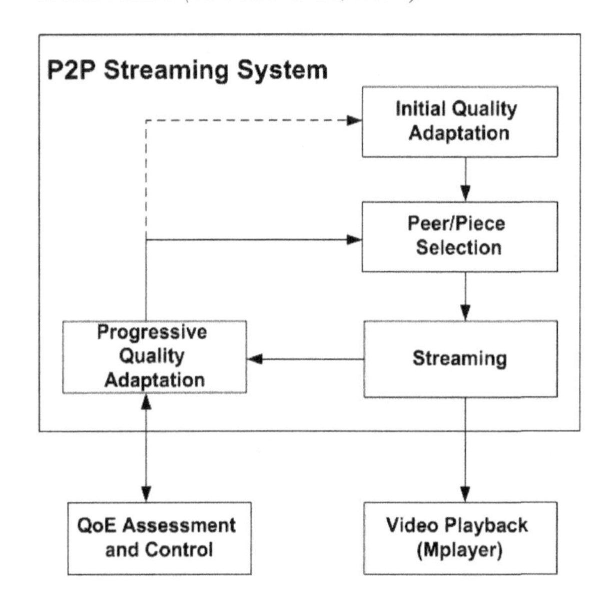

Adaptation, allows adapting to static resources at the peers (e.g. screen resolution, bandwidth, and processing power).

After the initial layer has been selected, peers that can provide the selected layer are contacted and required pieces are requested. The system further employs a closed loop adaptation algorithm, called Progressive Quality Adaptation, that constantly monitors playback performance and throughput and will change the selected layer accordingly. This relies on the *Quality-of-Experience (QoE)* Assessment and Control module that takes the QoS into the consideration. For example, this module provides indications with regard to best layer combinations and layer switching frequency that do not deteriorate the required QoE.

In addition, Abboud et al., (2011a) presents the SVC-based quality adaptive *Video-on-Demand (VoD)* system. The authors assume a mesh-based pull approach for VoD, as presented by Abboud et al., (2009). Also, it is assumed that there is a tracker that keeps track of all peers in the network. To ensure a certain quality of service, servers with modest resources are deployed, which additionally inject the initial content. Figure 14 below depicts the basic architecture of the quality adaptation workflow of Abboud et al., (2011a).

The quality adaptation is achieved by adjusting quality according to the different peer resources and network dynamics. It is performed by two modules: the *Initial Quality Adaptation (IQA)* and the *Progressive Quality Adaptation (PQA)* (Abboud et al., 2011a). Both modules form algorithms that match layers with the resources, which are available at the peer. On the one hand, the IQA is used for determining the highest possible layer that a peer can retrieve and play, and is performed at session start. Also, the PQA is performed periodically to adjust the layer according to the changes of the network environment. After the playback is initiated, the IQA is first called to make a decision on the feasible quality level based on local resources. Based on this decision, peer selection and block selection are performed. Peers are selected in such a way that they are able to provide the selected layer. To ensure continuous playback, the PQA is performed regularly, and if required, it may increase or decrease the selected layer accordingly.

In the next section, the recent advances in the development of efficient scalable VoD systems are discussed in detail.

SCALABLE VIDEO-ON-DEMAND P2P SYSTEMS

Compared with file sharing and live streaming, VoD generally supports richer user interactions,

Figure 14. Quality adaptation workflow proposed by Abboud et al. (2011a)

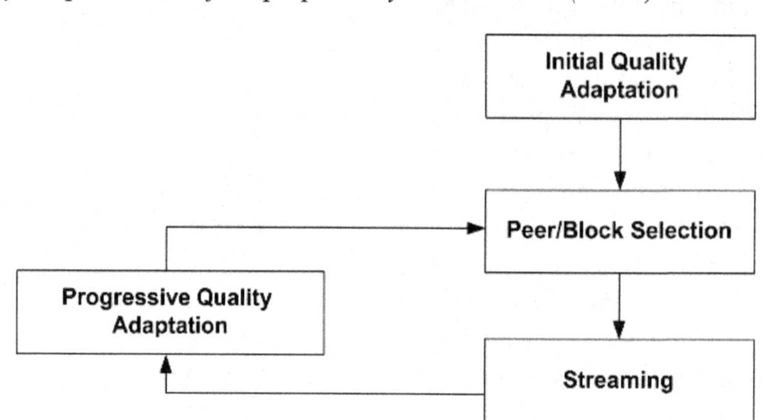

and users switch among videos more frequently (Ding et al., 2010). As a result, the P2P VoD systems face much more challenges, and existing solutions are yet to be perfect to compete with the conventional client/server models. In particular, it is well-known that the startup delay of the traditional P2P VoD systems remains much longer than powerful client/server-based systems. For example, for the on-demand mode in PPLive™ (Zhang et al., 2005), the startup delays are often longer than 10 seconds, and sometimes go beyond half a minute, while the average delay of YouTube™ is about 6.5 seconds. The situation is only getting worse when operations, such as the fast forward, rewind, and random seek are introduced.

Several researches have been recently performed on constructing efficient overlays for P2P VoD (Ding et al., 2010; Cheng et al., 2007; Wang & Liu, 2008; Qiu et al., 2009). They optimize the indexing structures for organizing the peers of similar playback progresses, so as to speed up the video segment discovery. Yet the video buffering time also constitutes an important part in the startup delay, which can hardly be minimized even with an advanced indexing scheme. This is mainly because, in the existing systems, the video stream structures are not flexible and generally with a fixed rate while users have heterogeneous capacities (i.e., the upload and download bandwidths).

To this end, Ding et al., (2010) presents a P2P VoD system that efficiently utilizes SVC:

- Starting from the base-layer only, the start-up delay for a peer to join the system can be reduced; and
- By dynamically adding or dropping layers, the occurrences of frame freezing due to temporal network congestion or the insufficient peer bandwidths can be minimized.

It should be noted that the VoD service in a wireless network is considered to be very important to achieve the goal of providing video services anywhere anytime. Typically, carrier mobile networks are used to deliver videos wirelessly. Since every video stream comes from the base station, regardless of what bandwidth sharing techniques are being utilized, the media stream system is still limited by the network capacity of the base-station. The key to overcome the scalability issue is to exploit resources available at mobile clients in a peer-to-peer setting. In this connection, Do et al., (2009a) and Do et al., (2009b) provide the on-demand service to archived videos in wireless networks, proposing the "PatchPeer", which leverages the network characteristic to allow the video-on-demand system scale beyond the bandwidth capacity of the server. Mobile clients in "PatchPeer" are no longer passive receivers, but also active senders of video streams to other mobile clients. Patching is actually a technique developed for the Internet that enables video-on-demand service to utilize the multicast service at the network layer. The basic idea of patching (Cai & Hua, 2003) is to allow a client to join an existing regular multicast for the remainder of the video, and download the missed portion over a dedicated patching stream (Do et al., 2009a; Do et al., 2009b). A straightforward application of the patching technique in a wireless environment is to implement the patching technique at the media server.

Also, additional researches with regard to VoD P2P systems by using the Scalable Video Coding have been recently conducted. For example, Xie et al., (2009) presents an efficient weight-based caching mechanism, including a pre-fetching algorithm and cache replacement algorithms to satisfy these needs for P2P based VoD systems. In addition, based on the requirements of the high-quality video application, Peng et al., (2009) presents an unstructured Peer-to-Peer Video-on-Demand system with asynchronous data transfer among peers and centralized directory service. Further, Abbasi et al., (2009) presents an adaptive video streaming mechanism that constructs overlay networks based on SW of peers, as schematically illustrated in Figure 15, combining the key characteristics

Figure 15. Small world overlay organization of peers (Abbasi et al., 2009)

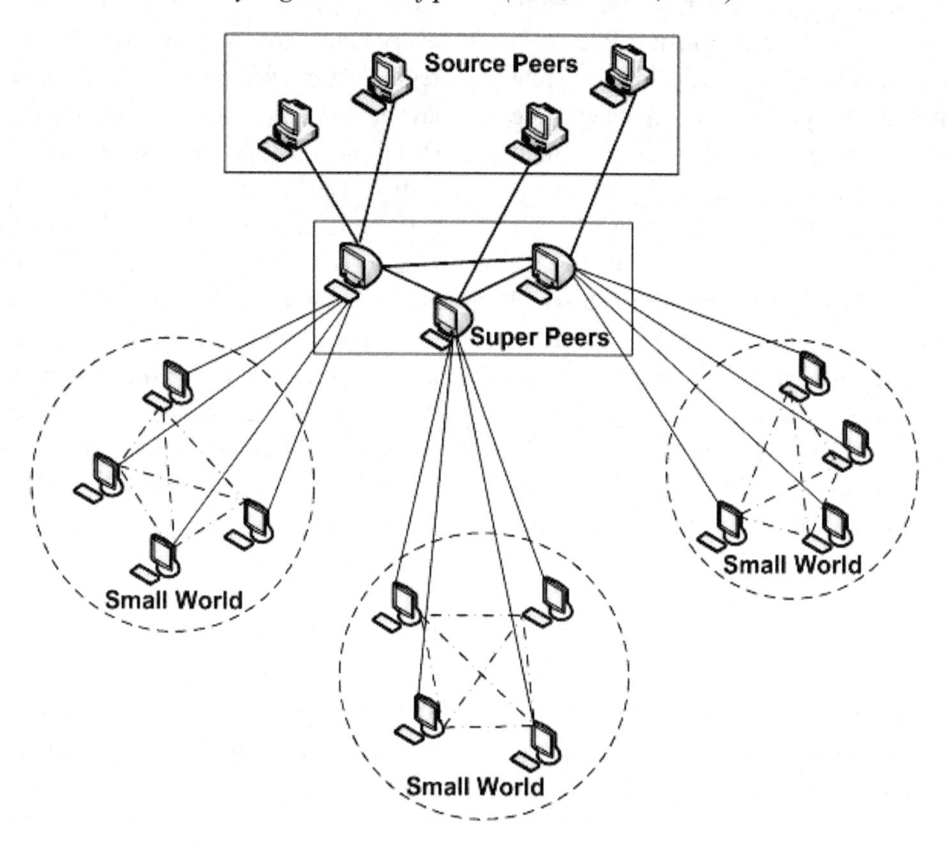

of the P2P push and pull mechanisms to improve the packet delivery ratio and the overall quality of service, and using the SVC that is considered to be very promising for real-time applications over heterogeneous networks.

The following section briefly summarizes future research directions.

FUTURE RESEARCH DIRECTIONS

Although, many efficient techniques have been recently proposed for using the Scalable Video Coding over P2P networks, there are still several issues to be solved.

For example, the current approaches do not consider *Digital Rights Management (DRM)*. The performance of the P2P systems are expected to be decreased when working with the encrypted content, because of the encryption and decryption processes on the source and receiver ends (Muller et al., 2009).

For this, different existing DRM systems can be applied to measure and evaluate the P2P stream performance. In this connection, the challenge will be to efficiently apply the Scalable Video Coding for the DRM systems over the P2P networks.

CONCLUSION

This chapter comprehensively covered the recent advances in the Peer-to-Peer media streaming by using Scalable Video Coding, especially emphasising the Peer-To-Peer video streaming in the wireless environment. As shown in detail, the Scalable Video Coding approach is an excellent choice for the media streaming over P2P networks, since the data can be dynamically adjusted to a suitable stream to fit: a particular Peer-to-Peer network, varying bandwidth conditions, and particular end-user requirements.

As a result, it is very clear that the usage of the Scalable Video Coding over P2P networks will play a very important role in future Internet applications.

REFERENCES

Abanoz, T. B., & Tekalp, A. M. (2009). Optimization of encoding configuration in scalable multiple description coding for rate-adaptive P2P video multicasting. *16th IEEE International Conference on Image Processing*, (pp. 3741-3744).

Abbasi, U., & Ahmed, T. (2010). SWOR: An architecture for P2P scalable video streaming using small world overlay. *IEEE Consumer Communications and Networking Conference*, (pp. 1-5).

Abbasi, U., Mushtaq, M., & Ahmed, T. (2009). Delivering scalable video coding using P2P small-world based push-pull mechanism. *Global Information Infrastructure Symposium*, (pp.1-7).

Abboud, O., Pussep, K., Kovacevic, A., & Steinmetz, R. (2009). Quality adaptive peer-to-peer streaming using scalable video coding. *Proceedings of the 12th IFIP/IEEE International Conference on Management of Multimedia and Mobile Networks and Services*, (pp. 41-54).

Abboud, O., Pussep, K., Stingl, D., & Steinmetz, R. (2011). Media-aware networking for SVC-based P2P streaming. *Proceedings of Network and Operating System Support for Digital Audio and Video (NOSSDAV)*, (pp. 15-20).

Abboud, O., Zinner, T., Pussep, K., Oechsner, S., Steinmetz, R., & Tran-Gia, P. (2010). A QoE-aware P2P streaming system using scalable video coding. *IEEE Tenth International Conference on Peer-to-Peer Computing*, (pp. 1-2).

Abboud, O., Zinner, T., Pussep, K., & Steinmetz, R. (2011). On the impact of quality adaptation in SVC-based P2P video-on-demand systems. *ACM Conference on Multimedia Systems*, (pp. 223-232).

Akyol, E., Tekalp, A. M., & Civanlar, M. R. (2007). A flexible multiple description coding framework for adaptive peer-to-peer video streaming. *IEEE Journal of Selected Topics in Signal Processing*, *1*(2), 231–245. doi:10.1109/JSTSP.2007.901527

Asioli, S., Ramzan, N., & Izquierdo, E. (2010). *A novel technique for efficient peer-to-peer scalable video transmission*. 18th EURASIP European Signal Processing Conference.

Baccichet, P., Schierl, T., Wiegand, T., & Girod, B. (2007). Low-delay peer-to-peer streaming using scalable video coding, *International Workshop on Packet Video*, (pp. 173-181).

Cai, Y., & Hua, K. A. (2003). Sharing multicast videos using patching streams. *Multimedia Tools and Applications*, *22*(2), 125–146. doi:10.1023/A:1025516608573

Capovilla, N., Eberhard, M., Mignanti, S., Petrocco, R., & Vehkapera, J. (2010). An architecture for distributing scalable content over peer-to-peer networks. *Second International Conference on Advances in Multimedia (MMEDIA)*, (pp. 1-6).

Chakareski, J., Han, S., & Girod, B. (2003). Layered coding vs. multiple descriptions for video streaming over multiple paths. *Proceedings of the 11th ACM International Conference on Multimedia*, (pp. 422–431).

Cheng, B., Jin, H., & Liao, X. (2007). Supporting VCR functions in P2P VoD services using ring-assisted overlays. *Proceedings of the IEEE International Conference on Communications*, (pp. 1698–1703).

Chenghao, L., Bouazizi, I., & Gabbouj, M. (2010). Advanced rate adaption for unicast streaming of scalable video. *IEEE International Conference on Communications*, (pp.1-10).

Choi, B. S., Suh, D. Y., Park, G. H., Kim, K., & Park, J. A. (2009). Peer-to-peer scalable video streaming using Raptor code. First International Conference on Ubiquitous and Future Networks, (pp.137-141).

Chu, Y.-H., Rao, S. G., Seshan, S., & Zhang, H. (2002). A case for end system multicast. *IEEE Journal on Selected Areas in Communications*, *20*(8), 1456–1471. doi:10.1109/JSAC.2002.803066

Dai, L., Cui, Y., & Xue, Y. (2007). Maximizing throughput in layered peer-to-peer streaming. *IEEE International Conference on Communications,* (pp.1734-1739).

Dai, M., Loguinov, D., & Radha, H. M. (2006). Rate-distortion analysis and quality control in scalable internet streaming. *IEEE Transactions on Multimedia*, *8*(6), 1135–1146. doi:10.1109/TMM.2006.884626

de Couto, D. S. J., Aguayo, D., Bicket, J., & Morris, R. (2003). A high throughput path metric for multi-hop wireless routing. *Proceedings of the ACM Conference om Mobile Computing and Networking,* (pp. 134-146).

Ding, Y., Liu, J., Wang, D., & Jiang, H. (2010). Peer-to-peer video-on-demand with scalable video coding. *Computer Communications*, *33*(14), 2010. doi:10.1016/j.comcom.2010.04.025

Do, T., Hua, K., Aved, A., Liu, F., & Jiang, N. (2009). Scalable video-on-demand streaming in mobile wireless hybrid networks. *IEEE International Conference on Communications,* (pp.1-6).

Do, T., Hua, K., Jiang, N., & Liu, F. (2009). PatchPeer: A scalable video-on-demand streaming system in hybrid wireless mobile peer-to-peer networks. *Peer-to-Peer Networking and Applications*, *2*(3), 182–201. doi:10.1007/s12083-008-0027-1

Draves, R., Padhye, J., & Zill, B. (2004). Routing in multi-radio, multi-hop wireless mesh networks. *Proceedings of the ACM Conference on Mobile Computing and Networking,* (pp. 114-128).

Fesci-Sayit, M., Tunali, E. T., & Tekalp, A. M. (2009). Bandwidth-aware multiple multicast tree formation for P2P scalable video streaming using hierarchical clusters. *16th IEEE International Conference on Image Processing,* (pp.945-948).

Garcia, L., Arnaiz, L., Alvarez, F., Menendez, J. M., & Gruneberg, K. (2009). Protected seamless content delivery in P2P wireless and wired networks. *IEEE Wireless Communications*, *16*(5), 50–57. doi:10.1109/MWC.2009.5300302

Gkantsidis, C., & Rodriguez, P. (2005). Network coding for large scale content distribution. *Proceedings of the IEEE INFOCOM,* (pp.2235-2245).

Grois, D., & Hadar, O. (2011a). Efficient adaptive bit-rate control for scalable video coding by using computational complexity-rate-distortion analysis. *IEEE International Symposium on Broadband Multimedia Systems and Broadcasting,* (pp.1-6).

Grois, D., & Hadar, O. (2011b). Recent advances in region-of-interest coding . In Del Ser Lorente, J. (Ed.), *Recent advances on video coding* (pp. 49–76). Vukavar, Croatia: Intech Open Access Publisher.

Grois, D., Kaminsky, E., & Hadar, O. (2010a). *ROI adaptive scalable video coding for limited bandwidth wireless networks* (pp. 1–5). IFIP Wireless Days.

Grois, D., Kaminsky, E., & Hadar, O. (2010b). Adaptive bit-rate control for region-of-interest scalable video coding. *IEEE 26th Convention of Electrical and Electronics Engineers in Israel,* (pp.761-765).

Guo, H., & Lo, K.-T. (2008). Cooperative media data streaming with scalable video coding. *IEEE Transactions on Knowledge and Data Engineering, 20*(9), 1273–1281. doi:10.1109/TKDE.2008.18

Harmanci, O., Kanumuri, S., Kozat, U. C., Demircin, U., & Civanlar, R. (2009). Peer assisted streaming of scalable video via optimized distributed caching. *6th IEEE Consumer Communications and Networking Conference,* (pp. 1-5).

Hei, X., Liang, C., Liang, J., Liu, Y., & Ross, K. W. (2007). A measurement study of a large-scale P2P IPTV system. *IEEE Transactions on Multimedia, 9*(8), 1672–1687. doi:10.1109/TMM.2007.907451

Hossain, T., Cui, Y., & Xue, Y. (2009). Minimizing rate distortion in peer-to-peer video streaming. *6th IEEE Consumer Communications and Networking Conference,* (pp. 1-5).

Hu, H., Guo, Y., & Liu, Y. (2011). Peer-to-peer streaming of layered video: Efficiency, fairness and incentive. *IEEE Transactions on Circuits and Systems for Video Technology, 21*(8), 1013–1026. doi:10.1109/TCSVT.2011.2129290

Huang, C.-M., Lin, C.-W., Yang, C.-C., Chang, C.-H., & Ku, H.-H. (2009). An SVC-MDC video coding scheme using the multi-core parallel programming paradigm for P2P video streaming. *IEEE/ACS International Conference on Computer Systems and Applications,* (pp. 919-926).

Itaya, S., Hayashibara, N., Enokido, T., & Takizawa, M. (2005). *Scalable peer-to-peer multimedia streaming model in heterogeneous networks*. Seventh IEEE International Symposium on Multimedia.

Jahromi, N. T., Akbari, B., & Movaghar, A. (2010). A hybrid mesh-tree peer-to-peer overlay structure for layered video streaming. *5th International Symposium on Telecommunications,* (pp. 706-709).

Jin, X., Cheng, K.-L., & Chan, S.-H. G. (2006). SIM: Scalable island multicast for peer-to-peer media streaming. *IEEE International Conference on Multimedia and Expo,* (pp.913-916).

Jurca, D., Chakareski, J., Wagner, J.-P., & Frossard, P. (2007). Enabling adaptive video streaming in P2P systems. *IEEE Communications Magazine, 45*(6), 108–114. doi:10.1109/MCOM.2007.374427

Kao, J.-Y. (2010). The method of SVC over P2P in video streaming system. *International Conference on E-Business and E-Government,* (pp.3479-3482).

La, R., & Anantharam, V. (2002). Optimal routing control: Repeated game approach. *IEEE Transactions on Automatic Control, 47*(3), 437–450. doi:10.1109/9.989076

Lee, Y.-C., Kim, J., Altunbasak, Y., & Mersereau, R. M. (2003). Performance comparisons of layered and multiple description coded video streaming over error-prone networks. *Proceedings of the IEEE International Conference on Communications,* (pp. 35–39).

Li, C., Yuan, C., & Zhong, Y. (2009a). A novel substream extraction for scalable video coding over P2P networks. *11th International Conference on Advanced Communication Technology,* (pp.1611-1615).

Li, C., Yuan, C., & Zhong, Y. (2009b). Robust and flexible scalable video multicast with network coding over P2P network. *2nd International Congress on Image and Signal Processing,* (pp.1-5).

Li, J., & Chan, S.-H. G. (2010). Optimizing Segment caching for mobile peer-to-peer interactive streaming. *IEEE International Conference on Communications,* (pp.1-5).

Liao, R.-F., Wouhaybi, R., & Campbell, A. (2003). Wireless incentive engineering. *IEEE Journal on Selected Areas in Communications*, *21*(10), 1764–1779. doi:10.1109/JSAC.2003.815014

Lin, K., Liu, N., & Luo, X. (2008). An optimized P2P based algorithm using SVC for media streaming. *Third International Conference on Communications and Networking in China,* (pp.569-573).

Liu, Z., Shen, Y., Ross, K. W., Panwar, S. S., & Wang, Y. (2009). LayerP2P: Using layered video chunks in P2P live streaming. *IEEE Transactions on Multimedia, 11*(7), 1340–1352. doi:10.1109/TMM.2009.2030656

Lopez-Fuentes, F. A. (2010). Adaptive mechanism for P2P video streaming using SVC and MDC. *International Conference on Complex, Intelligent and Software Intensive Systems*, (pp.457-462).

Lopez-Fuentes, F. A. (2011). P2P video streaming combining SVC and MDC. *International Journal of Appied Mathematical Computer Science*, *21*(2), 295–306. doi:10.2478/v10006-011-0022-1

López-Fuentes, F. A., & Steinbach, E. (2008). Adaptive multisource video multicast. *Proceedings of the IEEE International Conference on Multimedia and Expo,* (pp. 457–460).

Luo, H., Ci, S., & Wu, D. (2009). A cross-layer optimized distributed scheduling algorithm for peer-to-peer video streaming over multi-hop wireless mesh networks. *6th Annual IEEE Communications Society Conference on Sensor, Mesh and Ad Hoc Communications and Networks,* (pp.1-9).

Mastronarde, N., Turaga, D. S., & van der Schaar, M. (2006). Collaborative resource management for video over wireless multi-hop mesh networks. *IEEE International Conference on Image Processing,* (pp. 1297-1300).

Mastronarde, N., Turaga, D. S., & van der Schaar, M. (2007). Collaborative resource exchanges for peer-to-peer video streaming over wireless mesh networks. *IEEE Journal on Selected Areas in Communications*, *25*(1), 108–118. doi:10.1109/JSAC.2007.070111

Medjiah, S., Ahmed, T., Mykoniati, E., & Griffin, D. (2011). Scalable video streaming over P2P networks: A matter of harmony? *IEEE 16th International Workshop on Computer Aided Modeling and Design of Communication Links and Networks (CAMAD)*, (pp.127-132).

Mehrotra, S., & Weidong, Z. (2009). Rate-distortion optimized client side rate control for adaptive media streaming. *IEEE International Workshop on Multimedia Signal Processing,* (pp. 1-6).

Mirshokraie, S., & Hefeeda, M. (2010). Live peer-to-peer streaming with scalable video coding and networking coding. *Proceedings of the First Annual ACM SIGMM Conference on Multimedia Systems,* (pp. 123-132).

Moshref, M., Motamedi, R., Rabiee, H. R., & Khansari, M. (2010). LayeredCast - A hybrid peer-to-peer live layered video streaming protocol. *5th International Symposium on Telecommunications,* (pp. 663-668).

Muller, J., Magedanz, T., & Fiedler, J. (2009). NNodeTree: A scalable peer-to-peer live streaming overlay architecture for next-generation-networks. *Network Protocols and Algorithms, 1*(2).

Mushtaq, M., & Ahmed, T. (2008). Smooth video delivery for SVC based media streaming over P2P networks. *5th IEEE Consumer Communications and Networking Conference,* (pp.447-451).

Mykoniati, E., Landa, R., Spirou, S., Clegg, R., Latif, L., Griffin, D., & Rio, M. (2008). Scalable peer-to-peer streaming for live entertainment content. *IEEE Communications Magazine*, *46*(12), 40–46. doi:10.1109/MCOM.2008.4689206

Nguyen, A. T., Eliassen, F., & Welzl, M. (2011). Quality-aware membership management for layered peer-to-peer streaming. *IEEE Consumer Communications and Networking Conference,* (pp. 720-724).

Nguyen, A. T., Li, B., & Eliassen, F. (2010a). Chameleon: Adaptive Peer-to-peer streaming with network coding. *Proceedings of the IEEE INFOCOM,* (pp.1-9).

Nguyen, A. T., Li, B., & Eliassen, F. (2010b). Quality- and context-aware neighbor selection for layered peer-to-peer streaming. *IEEE International Conference on, Communications*, (pp.1-6).

Nunes, R. P., & Cruz, R. S. (2010). Scalable video coding distribution in peer—to-peer architecture. *10a Conferencia sobre Redes de Computadores Conference*, (pp. 95-100).

Ozbilgin, T., & Sunay, M. O. (2008). Scalable video streaming in wireless peer-to-peer networks. *IEEE 16th Signal Processing, Communication and Applications Conference*, (pp. 1-4).

Peltotalo, J., Harju, J., Saukko, M., Vaatamoinen, L., Bouazizi, I., Curcio, I. D. D., & van Gassel, J. (2009). A real-time peer-to-peer streaming system for mobile networking environment. *Proceedings of the IEEE INFOCOM Workshops,* (pp.1-7).

Peng, Z., Cheng, G., Yang, Z., & Chen, J. (2009). A scalable peer-to-peer video-on-demand system with asynchronous transfer. *Fourth International Conference on Communications and Networking in China,* (pp. 1-6).

Picconi, F., & Massoulie, L. (2008). Is there a future for mesh-based live video streaming? *Eighth International Conference on Peer-to-Peer Computing,* (pp. 289-298).

Qiu, X., Wu, C., Lin, X., & Lau, F. (2009). InstantLeap: Fast neighbor discovery in P2P VoD streaming. *Proceedings of the 18th International Workshop on Network and Operating Systems Support for Digital Audio and Video,* (pp. 19–24).

Ramzan, N., Quacchio, E., Zgaljic, T., Asioli, S., Celetto, L., Izquierdo, E., & Rovati, F. (2011). Peer-to-peer streaming of scalable video in future internet applications. *IEEE Communications Magazine*, *49*(3), 128–135. doi:10.1109/MCOM.2011.5723810

Sánchez, Y., Schierl, T., Hellge, C., & Wiegand, T. (2010). P2P group communication using scalable video coding. *17th IEEE International Conference on Image Processing,* (pp. 4445-4448).

Schierl, T., Ganger, K., Hellge, C., Wiegand, T., & Stockhammer, T. (2006). SVC-based multisource streaming for robust video transmission in mobile ad hoc networks. *IEEE Wireless Communications*, *13*(5), 96–103. doi:10.1109/WC-M.2006.250365

Schwarz, H., Marpe, D., & Wiegand, T. (2007). Overview of the scalable video coding extension of the H.264/AVC standard. *IEEE Transactions on Circuits and Systems for Video Technology*, *17*(9), 1103–1120. doi:10.1109/TCSVT.2007.905532

Secker, A., & Taubman, D. (2003). Lifting-based invertible motion adaptive transform framework for highly scalable video compression. *IEEE Transactions on Image Processing*, *12*(12), 1530–1542. doi:10.1109/TIP.2003.819433

Shen, H., Zhao, L., Li, Z., & Li, J. (2010). A DHT-aided chunk-driven overlay for scalable and efficient peer-to-peer live streaming. *39th International Conference on Parallel Processing,* (pp. 248-257).

Si, J., Zhuang, B., Cai, A., & Cheng, Y. (2009). Layered network coding and hierarchical network coding for peer-to-peer streaming. *Pacific-Asia Conference on Circuits, Communications and Systems,* (pp. 139-142).

Taal, J. R., Pouwelse, J. A., & Lagendijk, R. L. (2000). Scalable multiple description coding for video distribution in P2P networks. *Proceedings of the Picture Coding Symposium.*

Taubman, D., Reji, M., Maestroni, D., & Tubaro, S. (2004). *SVC core experiment 1- Description of UNSW contribution, MPEG doc. m11441.*

Wang, D., & Liu, J. (2008). A dynamic skip list-based overlay for on-demand media streaming with VCR interactions. *IEEE Transactions on Parallel and Distributed Systems, 19*(4), 503–514. doi:10.1109/TPDS.2007.70748

Wang, J., Huang, C., & Li, J. (2008). On ISP-friendly rate allocation for peer-assisted VoD. *Proceeding of the 16th ACM International Conference on Multimedia,* (pp. 279-288).

Wiegand, T., & Sullivan, G. (2003). *Final draft ITU-T recommendation and final draft international standard of joint video specification* (ITU-T Rec. H.264 ISO/IEC 14 496-10 AVC).

Wiegand, T., Sullivan, G., Reichel, J., Schwarz, H., & Wien, M. (2006). *Joint draft 8 of SVC amendment.* ISO/IEC JTC1/SC29/WG11 and ITU-T SG16 Q.6 9 (JVT-U201), 21st Meeting, Hangzhou, China, Oct. 2006.

Wu, C., Li, B., & Zhao, S. (2008). Exploring large-scale peer-to-peer live streaming topologies. *ACM Transactions on Multimedia Computing . Communications and Applications, 4*(3), 1–23.

Xiao, X., Shi, Y., Gao, Y., & Zhang, Q. (2009). LayerP2P: A new data scheduling approach for layered streaming in heterogeneous networks. *Proceedings of the IEEE INFOCOM,* (pp.603-611).

Xie, H., Gao, L., Zhang, L., Zhang, Z., & Yang, M. (2009). An efficient caching mechanism for video-on-demand service over peer-to-peer network. *Eighth International Conference on Embedded Computing Scalable Computing and Communications,* (pp.251-256).

Yoon, S., Mao, M., & Kalman, M. (2006). Rate-distortion optimized video streaming for scalable H.264. *IEEE International Conference on Multimedia and Expo,* (pp. 2157-2160).

Zeng, P., & Jiang, Y. (2010). Robust scalable video with network coding for peer-to-peer streaming. *International Conference on Multimedia Technology,* (pp.1-5).

Zhang, G., & Yuan, C. (2010). Self-adaptive peer-to-peer streaming for heterogeneous networks using scalable video coding. *12th IEEE International Conference on Communication Technology,* (pp.1390-1393).

Zhang, X., Liu, J., Li, B., & Yum, Y.-S. P. (2005). CoolStreaming/DONet: A data-driven overlay network for peer-to-peer live media streaming. *Proceedings of the IEEE INFOCOM,* (pp. 2102-2111).

ADDITIONAL READING

Abboud, O., Pussep, K., Mohr, K., Kovacevic, A., Kaune, S., & Steinmetz, R. (2011). Enabling resilient P2P video streaming: Survey and analysis. *Multimedia Systems, 17*(3), 177–197. doi:10.1007/s00530-011-0229-x

Dai, M., Loguinov, D., & Radha, H. M. (2006). Rate-distortion analysis and quality control in scalable internet streaming. *IEEE Transactions on Multimedia, 8*(6), 1135–1146. doi:10.1109/TMM.2006.884626

Gomaa, H., Messier, G. G., Davies, R., & Williamson, C. (2010). Peer-assisted caching for scalable media streaming in wireless backhaul networks. *IEEE Global Telecommunications Conference,* (pp. 1-5).

Itaya, S., Hayashibara, N., Enokido, T., & Takizawa, M. (2006). Distributed coordination protocols to realize scalable multimedia streaming in peer-to-peer overlay networks. *International Conference on Parallel Processing,* (pp.569-576).

Liu, C., Bouazizi, I., & Gabbouj, M. (2010). Advanced rate adaption for unicast streaming of scalable video. *IEEE International Conference on Communications,* (pp. 1-5).

Mehrotra, S., & Zhao, W. (2009). Rate-distortion optimized client side rate control for adaptive media streaming. *IEEE International Workshop on Multimedia Signal Processing,* (pp.1-6).

Srinivasan, S. K., Vahabzadeh-Hagh, J., & Reisslein, M. (2010). The effects of priority levels and buffering on the statistical multiplexing of single-layer H.264/AVC and SVC encoded video streams. *IEEE Transactions on Broadcasting, 56*(3), 281–287. doi:10.1109/TBC.2010.2049610

Yoon, S., Mao, M., & Kalman, M. (2006). Rate-distortion optimized video streaming for scalable H.264. *IEEE International Conference on Multimedia and Expo,* (pp. 2157-2160).

Zhu, X., Schierl, T., Wiegand, T., & Girod, B. (2011). Distributed media aware rate allocation for video multicast over wireless networks. *IEEE Transactions on Circuits and Systems for Video Technology, 21*(9), 1181–1191. doi:10.1109/TCSVT.2011.2129690

KEY TERMS AND DEFINITIONS

Adaptation: Adaptation refers to a transformation of the media content to a desired quality and/or bit-rate.

H.264/AVC (Advanced Video Coding): H.264/AVC is a video coding standard of the ITU-T Video Coding Experts Group and the ISO/IEC Moving Picture Experts Group, which was officially issued in 2003. H.264/AVC has achieved a significant improvement in rate-distortion efficiency relative to prior standards.

H.264/SVC (Scalable Video Coding/SVC): SVC is an extension of H.264/AVC video coding standard; SVC enables the transmission and decoding of partial bit streams to provide video services with lower temporal or spatial resolutions or reduced fidelity, while retaining a reconstruction quality that is high relative to the rate of the partial bit streams.

Scalability: In video coding, the term scalability relates to providing diffierent qualities in various domains, such as in the spatial domain (by varying the video resolution), temporal domain (by varying the frame rate), and fidelity domain (by varying the SNR/quality), which are embedded into a single SVC bit stream.

Chapter 9
Quality of Experience in Mobile Peer-to-Peer Streaming Systems

Florence Agboma
University of Essex, UK

ABSTRACT

This chapter considers the various parameters that affect the user's Quality-of-Experience (QoE) in mobile peer-to-peer streaming systems, which are a form of content delivery network. Network and content providers do not necessarily focus on users' QoE when designing the content delivery strategies and business models. The outcome of this is quite often the over-provisioning of network resources and also a lack of knowledge in respect to the user's satisfaction. The focus is the methodology for quantifying the user's perception of service quality for mobile video services and user contexts. The statistical technique of discriminant analysis is employed in defining prediction models to map Quality-of-Service (QoS) parameters onto estimates of the user's QoE ratings. The chapter considers the relative contribution of the QoS parameters to predicting user responses. The chapter also demonstrates the value of the prediction models in developing QoE management strategies in order to optimize network resource utilization. To investigate the versatility of the framework, a feasibility study was applied to a P2P TV system. P2P systems continue to develop and as such, not a lot is known about their QoE characteristics, which situation this chapter seeks to remedy.

INTRODUCTION

With the growth in the availability of multimedia services, coupled with the technological advances in the mobile access devices, *Quality of Service (QoS)* has become the adopted set of technologies used in managing network traffic in the delivery systems that provide these services. The aim of QoS has been to manage the performance of networks and to provide availability guarantees to network traffic. QoS enables the measurement of network parameters, and the detection of changing network conditions (such as congestion or availability of bandwidth). This information is

DOI: 10.4018/978-1-4666-1613-4.ch009

utilized in resource management by prioritizing traffic. However, QoS processes by themselves are not adequate enough in that they do not take into account the user's perception of network performance and service quality. There is now a realization that measuring and studying users' *Quality of Experience (QoE)* should be an important metric to be employed during the design and management of content delivery systems and other engineering processes. This is because QoE is a metric that refers to a measure of the end-to-end performance at the service level from the user's perspective. This metric is measured at the end devices and can conceptually be seen as the remaining quality after the distortions introduced during the preparation of the content and the delivery through the network until it reaches the decoder at the end device.

This Chapter presents a QoE-based framework as an invaluable tool in resource management. There are many definitions of QoE (Siller, 2006), (Soldan, et al., 2006), (Patrick, et al., 2004), (Nokia, 2004), (ITU-T Study Group 12, 2007) all of which share a similar concept *i.e.,* QoE relates to the user satisfaction of the offered service. The definition provided by (Soldani *et al.*, 2006) is the favored one in the context of this Chapter: "QoE is the perception of the user about the quality of a particular service or network". QoE strongly depends on the expectations of the users on the offered service. The provision of all the appropriate QoS conditions does not by itself guarantee a satisfied user. Thus, there is the need to understand users' QoE in order to use QoS effectively. Put another way, understanding of the user QoE will lead to better resource management strategies, and the discovery of ways of increasing and maintaining user satisfaction

The ability to manage service quality has become an essential part of the service delivery chain, and it is an important differentiator between the service qualities being offered by service and content providers. The types of online services that are likely to emerge include Video-on-Demand (VoD) and live streaming. As a result of this, service and content providers would be very attracted to the prospects of being able to offer user specific charging schemes and service quality in order to satisfy a large and varied user or customer base. The providers would be able to implement such user specific services only if there is a thorough understanding of users' QoE.

The proliferation of different types of access devices further highlights the importance of QoE research. As an illustration; the QoE for a user watching a news clip on a PDA will most likely differ from another user watching that same news clip on a 3G mobile phone. This is because the two terminals come with different display screens, bandwidth capabilities, frame rates, codecs and processing power. Therefore, delivering multimedia contents or services to these two terminal types without carefully thinking about the users' quality expectations or requirements for these terminal types, might lead to service over provisioning and network resource wastage. Thus, it is essential to determine the thresholds of quality acceptability across contents and terminal types. With this information, content and network providers will have the capability to minimize storage and network resources by allocating only the resources that are sufficient to maintain a specific level of user satisfaction.

The focus of this Chapter is on streaming applications and the results are applicable to video streaming service and *peer-to-peer television (P2P TV)*. Streaming paradigms in content delivery systems fall under two main categories; on-demand and live streaming. In live streaming, the media streams are usually broadcast in real time to the users. In this paradigm, it is essential to maintain constant service availability, since a disruption in a broadcast session often means that a subscriber (user) cannot go back and replay the missing portion of the session due to the strict broadcast viewing timelines. On the other hand, on-demand streaming offers a more personalized service where individuals can control the flow of the media contents they want to watch. The service availability requirements here are not as stringent as those of live streaming. A subscriber can interact with the streams using the 'start', 'stop', 'pause' and 'replay' function on the device thereby making the user's viewing behaviour more flexible. On-demand streaming allows the playback of archived media contents and also facilitates the new growing demand for user-generated contents.

The *peer-to-peer (P2P)* streaming architecture is a type of a content delivery system that uses a non-centralised approach. The users (peers) act as both clients and servers, whereby a peer not only downloads media streams from the network, but also uploads the downloaded media streams to other users in the network. In P2P streaming, the users' terminals collaborate with each other, with the purpose of making the stream distribution process more efficient by reducing the number of client-server interactions to the minimum. Depending on the type of the distribution method used to build and maintain the application layer overlay, P2P streaming can be classified into two general categories. They are the; tree-based architectures and mesh-based architectures. This Chapter now describes these two architectures.

CONTEXT

Tree-Based Streaming Architecture

The tree-based architecture is modelled on a tree distribution graph (Marfia *et al.*, 2007) whereby the peers are arranged in a hierarchical order. This is illustrated in Figure 1. Peers participating in a streaming session form a tree-like overlay distribution that is rooted at the multimedia source server. The source server forwards media packets to its leaf nodes (peers), which in turn relay the

Figure 1. P2P tree-based streaming architecture

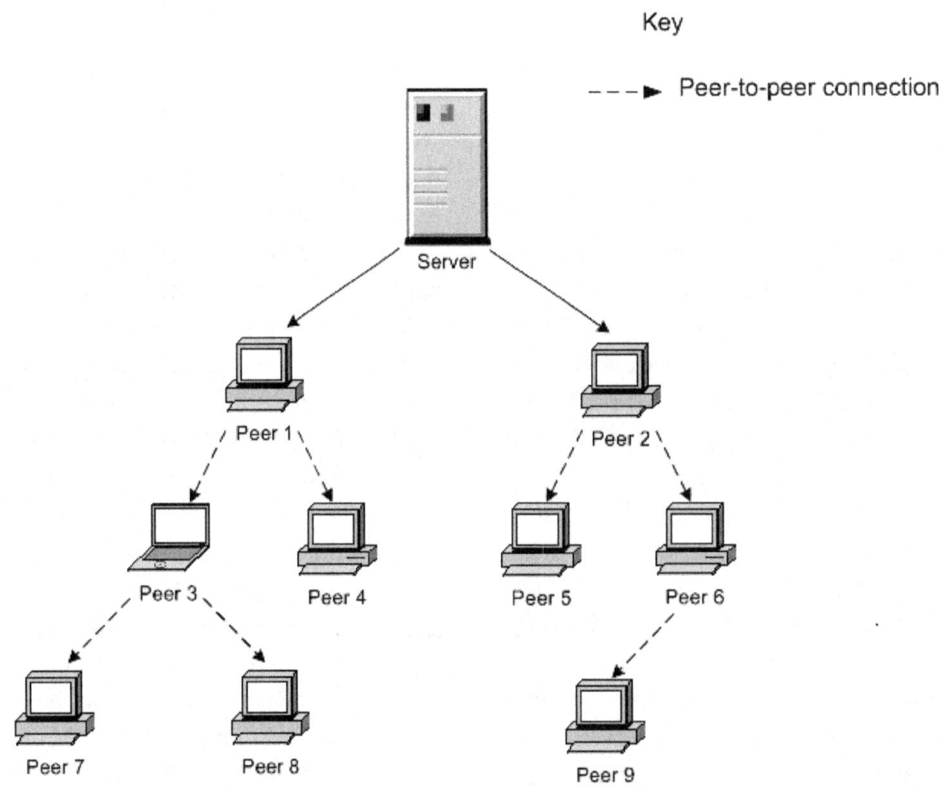

media packets to intermediate peers. An example of a P2P application built on this architecture is PeerCast (Diot, et al., 2000), which is an open-source software.

Although the tree-based architecture scales effectively, peers can only receive media packets from a single source at a given time, thus subjecting the system to the single-point of failure limitation. Another limitation of the tree-based architecture is that, it usually takes a long time for a receiving peer to rebuild a tree when other peers destroy it by rapidly joining and leaving the networks. This is because the receiving peer needs to first find a new source when rebuilding the tree, before continuing with the streaming session from where it was interrupted. This lack of continuity in the streaming, often leads to a repeat or a skip during playback of the multimedia content. Rebuilding of a tree generates a lot of control traffic, which can overload the network (Diot, et al., 2000).

The multi-tree architecture is similar to the single tree-based one. This architecture is used in overcoming the potential single point of failure, by using multiple trees to provide redundancy on network paths. The multi-tree architecture leads to a balanced tree that makes the system self-organizing and more resilient as compared to the single-tree architecture. The main limitation of the multi-tree architecture is that peers can only receive data packets from a single source even though the data packets go along different paths.

Mesh-Based Architecture

The mesh-based architecture is built over a mesh distribution graph (Marfia, et al., 2007). Here, a peer receives the media streams from multiple sources in chunks as shown in Figure 2. The mesh-based approach is based on the popular BitTorrent P2P approach (Qiu & Srikant, 2004) used in file-sharing applications.

The selection of trusted sources that a peer will receive media streams from forms the main challenge in this approach. There is the additional stringent requirement of the re-ordering of data packets at the receiving peers, as media packets are transmitted and received from multiple sources. Examples of applications using the mesh-based architecture are CollectCast (Hefeeda, et al., 2003), CoolStreaming (Zhang, et al., 2005), PPLive (Hei, et al., 2007), and Joost (Moreira, et al., 2008).

P2P systems look promising as a way of solving the bandwidth scalability issues, and to provide more robustness to central failures as well as being more cost effective in deployment. The universal characteristic of P2P systems is that, the more the users, the better the quality of service (Tang, et al., 2007). One of the key features in a P2P system is that each node contributes resources including bandwidth, storage space, and computing power, and thus, the total system capacity can actually increase as more nodes join a system (Li & Yin, 2007). This is in contrast to the client-server architecture in which the addition of new clients always degrades its overall performance.

Introducing QoE

The ultimate measure of a network and the services it offers is how subscribers perceive the performance (Nokia, 2004). QoE is the term used to describe this perception. Its concept comprises the QoS and psychological factors (such as the user expectations and requirements), as well as the overall performance of the service as perceived by the end user. The concept of QoE in addition to QoS mechanisms has been proposed. This approach can be found in (Zapater & Bressan, 2007), where a layered model is proposed, which combines QoS and QoE in one integrated framework. (López, et al., 2006) describes QoE as an extension of the traditional QoS in the sense that QoE provides information regarding the delivered services from an end-user point of view. In (Nokia, 2004), QoS is a subset of the overall QoE scope. The goal of QoS is to deliver a desired QoE.

Figure 2. P2P mesh-based architecture

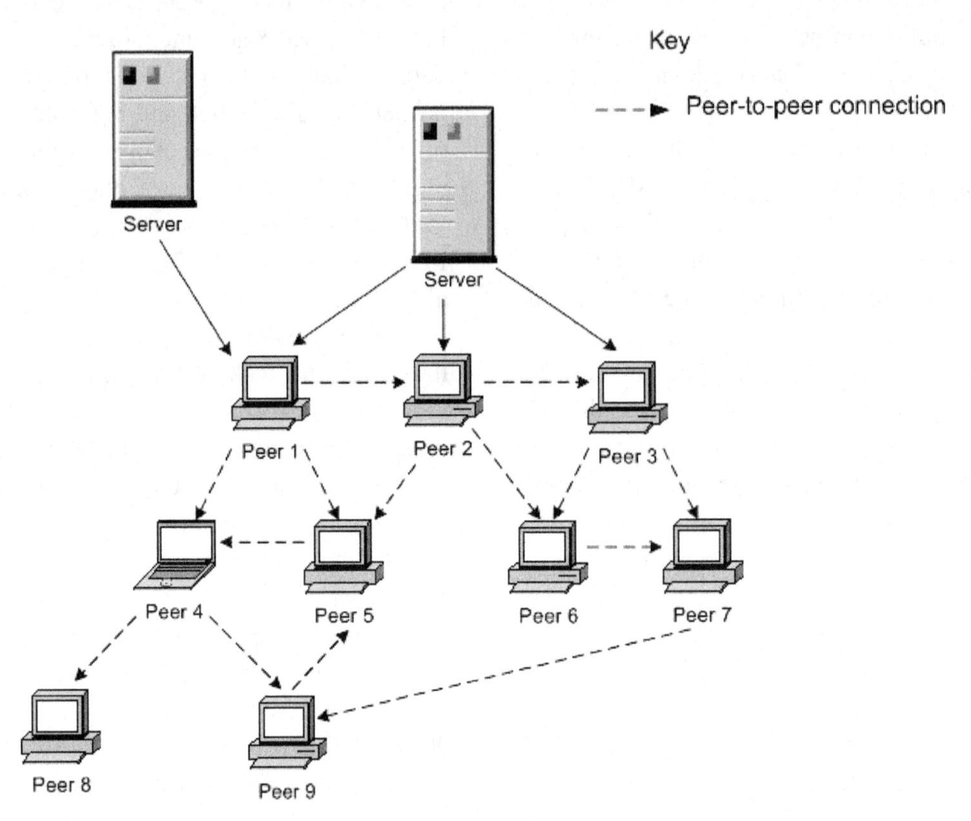

Delivering a desirable QoE depends on gaining an understanding of the factors that contribute to the overall user experience. These different factors can be combined into one integrated QoS/QoE layered model (Siller, 2006), (Goodchild, 2005), (Zapater & Bressan, 2007) as illustrated in Figure 3. For any given service, multiple QoS parameters contribute to the overall user's percep-

Figure 3. QoS/QoE layered model

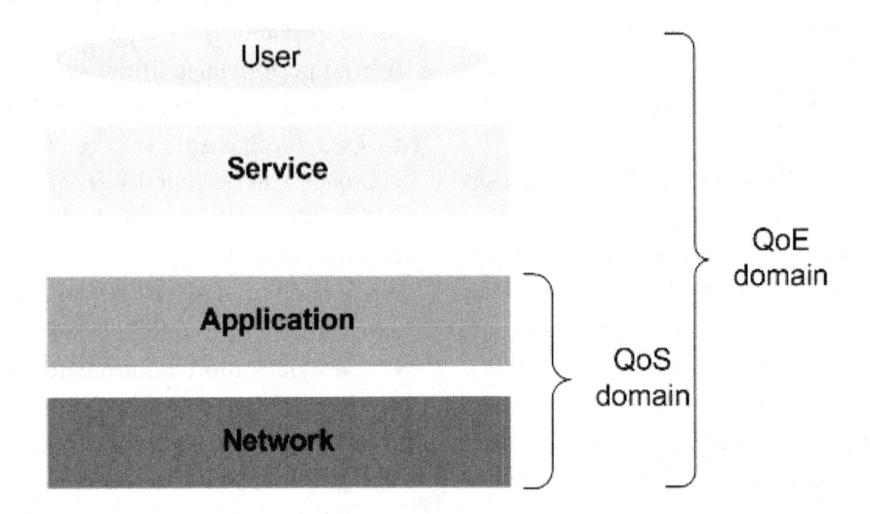

tion of the service quality. A classification of these parameters can be made by grouping them into two levels: *application-level QoS (AQoS)* and the *network-level QoS (NQoS)* (Hong & Hong, 2003), (Siller, M., 2006), (Goodchild, 2005), (Zapater & Bressan, 2007).

Service

This level is like a pseudo-layer where the user's experience of the overall performance of the service can be measured (Siller, 2006). It is the level that is exposed to the user, and where QoE assessments are carried out. The user's experience for video services for example, will include the video and audio quality, the interactive responsiveness (start-up delay, channel change delay), the cost of using the service, the type of multimedia content, and the type of terminal used to access the service.

Application

This is the *Application QoS (AQoS)* level which deals with application specific parameters such as the video resolution, frame rate, codec type, and encoding bitrate. At this level, QoS is driven by the human perception of audio and video. For example, the maximum bitrate may be fixed at a certain value by the access network. However, this does not necessarily fix the quality of the media content at say, 'high'. This is because there are numerous ways the multimedia content could have been encoded, giving rise to differing perceived content qualities. Understanding the user requirements at the application level is essential so that only the needed amounts of network resources are allocated.

Network

This is the *Network QoS (NQoS)* level which is concerned with the low level network parameters such as bandwidth, delay, jitter, and packet loss. Real-time applications are very intolerant of transmission errors. QoS mechanisms such as Integrated and Differentiated Services could be employed to mitigate the effects that the transmission errors have on service quality. The QoS/QoE layered model is designed to aid in ensuring end-to-end user satisfaction. Taking the layers as a whole unit rather than individual entities, might aid in providing better QoEs.

Factors Affecting QoE

Measuring and ensuring good QoE in a mobile environment, especially a P2P streaming one, is non-trivial as it is very subjective in nature. QoE includes all the factors of a service that can be experienced by the user. The main factors include terminal types, usability, image resolution, viewing distance, contents types, delay, and media quality.

Terminal Types and Display Characteristics

There are multitudes of different terminal types such as mobile phones, pagers, PDAs, and laptops, each possessing varying display characteristics (screen sizes, device capabilities and resolutions). An application may be intended to run on more than one terminal type e.g. a web site requiring to be displayed on a desktop computer and a mobile phone. Such an application was seen with the introduction of *Wireless Application Protocol (WAP)* services. However, users experienced difficulties in navigating to specific parts of the service, due to its poor design. It was time consuming even in accomplishing the simplest tasks on these WAP services. The consequence was a commercial failure. Unlike conventional television viewing, mobile content viewing is constrained by the requirement of small display screens. Thus, the QoE of a user watching a football game on a laptop is likely to differ from another user watching that same football game on a mobile phone.

It is generally acknowledged that larger image sizes can positively influence the viewer's evaluation of contents. Larger images positively affect arousal, and are better remembered (Reeves & Nass, 1998), (Detenber & Reeves, 1996). They also generate responses of more intense viewing experience (Lombard, et al., 1997), and produce positive evaluations (Lombard, 1995) in addition to increasing a sense of presence (Lund, 1993). Users want as large screens as possible for viewing but they do not want their devices (e.g. mobile phones) to be impracticably large. Content providers generally refrain from using shot types that depict subjects from a great distance because of the intrinsic small screen sizes found in mobile devices. Shot types where the object of interest fills the screen are deemed to be more appropriate for mobile devices. The author in (Knoche, et al., 2006) studied how shot types used in standard broadcast television are affected when shown on mobile devices at different levels of resolution. Their results show that the user's acceptability of the different shot types depends on the type of content and at the resolution the content is displayed. The extreme long shot type was generally the least favorable shot type.

Usability

The user friendliness of both the device and the services are of paramount importance to the user. This user friendliness will have a major impact on QoE. If the design of the device interface lacks the ability to properly render the service, it will result in users abandoning the service. The device used to access a service plays a major role on the user experience as this dictates the interaction between the user and the service. In (ISO 9241-11, 1998), usability is defined as "the extent to which a product can be used by specified users to achieve specified goals with effectiveness, efficiency and satisfaction in a specified context of use". A survey carried out by (Olista, 2007) found that 55% of first-time users abandoned the value-added service of mobile data offered by mobile operators because of usability problems. These problems include difficulty in navigating through the menus, inability to find downloaded content and confusing terminology like 'streaming' and 'download'. This reflects the need to have service applications that are easy to use and do not require thorough understanding of the service logic (Raatikainen, 2005). The interface of the mobile devices should enable the user to easily interact with the service such as using the start, stop, pause and replay functions on the device. Also, the provision of *Electronic Programme Guide (EPG)* functionality should inform the user of the contents that are available, their time duration, as well as allowing content preview. If the user's interaction with this service is poor, the consequence will be a decline in the use of the service. Ease and intuitive service usability provides a foundation for the adoption of mobile content services.

Image Resolution and Viewing Distance

From mobile applications' point of view, reducing the image resolution provides bitrate savings. The delivery of high resolution content demands more resources. A study on resolution requirements carried out by Knoche, et al. (2006) found that contents shown on mobile devices at higher resolutions are generally more acceptable than lower resolutions at identical encoding bit rates. Specifically, the mobile contents received poor ratings when presented at resolutions smaller than 168×126. The viewing distance always affects the device display performance, and is a key design factor for determining the optimum perceptual quality. The optimum viewing distance is taken as the shortest distance at which a person with normal vision of 1.0 is unable to recognize the pixel structure on a screen (Sugawara, et al., 2006). When the viewing distance is reduced relative to the image resolution of the device, subjects begin to notice the picture quality degrade, and may also experience visual fatigue (Satamoto, et al., 2008). In an experiment carried out by Kato, et al. (2005) to determine the viewing distance on mobile phones, the authors found typical viewing distance to be approximately 30 cm. Compared to traditional TV viewing, mobile contents are viewed at shorter distances usually not longer than the length of the arm. Therefore, accommodating the preferred viewing distance during the initial design of the application in the intended environment may have a direct impact on the user satisfaction and viewing experience.

Content Types

The choice of available mobile video contents is a key factor in the success of the evolution of mobile content services. The content distributed to mobile devices ranges from interactive content, specifically tailored for the mobile devices, to material that is produced for standard TV consumption. A survey carried out by Gilliam (2006) to determine the types of content consumers desire to watch on their mobile terminals revealed that consumers do have preferences as to the type of content they are interested in and would like to watch on their mobile terminals. Examples of these contents in order of preference are news, movies, comedy, sports and travel. Hands (1997) and Moore, et al. (2002) studied the effects that the subjects' level of interest in the content types would have on their perception of picture quality evaluation. Their results support a conclusion that the relationship between the users' interests and picture quality evaluation is insignificant. There is the unanswered question about the user-preferred length of contents. For example, it is not clear whether user would like to watch a full movie of about 90 minutes on a small screen. The issue of battery performance and user's usage behaviour on a small screen is still not well defined.

Cost

Pricing is crucial to the success of mobile content services. For example, the initially slow uptake of *Multimedia Messaging Service (MMS)* was due to the high cost of using the service, coupled with interoperability issues amongst the handsets. The long-established practice of judging quality by price implies that expectations are price dependent (Robert, 1986). If the cost is high and user's expectations are not met, it might cause users to stop using the service. The uptake of existing mobile TV services falls behind expectations, possibly because customers are not willing to pay high premiums for content (KPMG, 2006). Another barrier to the uptake is the high cost of the mobile terminals. An additional problem is how to appropriately price the services, as users always assume that services on the Internet ought to be free. A study by comScore (2007) reveals that 71% of their respondents said 'cost' is a top consideration.

Service providers are always looking for ways to differentiate their products and offerings from their competitors. The battle for market share will be won on 'value' by creating a superior entertainment experience and not on 'pricing' (Rainey, et al., 2005). Four possible pricing strategies that are available are: flat-rate subscription, pay-per-view, one time fee[1] and ad-based[2]. As pricing strategies are developed, it is important to translate them into a marketable message that potential consumers will find easy to understand. For example, in South Korea early payment models greatly influenced the use of mobile TV. Mobile TV usage was billed in the amount of kilobytes received, and each one-minute part of a programme made, especially for mobile TV, had to be confirmed for delivery. This method of billing resulted in a discontinuous viewing experience (Knoche, & Sasse, 2008). Results from initial field trials (Nokia, 2005) (Lloyd, et al., 2006) suggested that users preferred flat-rate subscriptions, i.e. a single payment for unlimited access to contents during a billing period.

Delay

Three types of delay can be distinguished in video streaming applications: end-to-end delay, start-up delay and channel-switching delay. In real-time conversational video communication (e.g., video conferencing), users start becoming intolerant at delays longer than 150 ms. Delays higher than 300 ms are completely unacceptable. This requirement for low end-to-end delay prohibits the use of playout buffers and retransmission of lost packets (Balaouras, & Stavrakakis, 2005).

The start-up delay is the total time taken in connecting to the media server (system) until the content playout by mobile terminals. A start-up delay between 5 and 15 seconds is usually acceptable for most video streaming applications (Balaouras, & Stavrakakis, 2005). In fact, this start-up delay is longer than that experienced in traditional TV viewing. During the start-up delay period, the media packets are received from the network stored in a playout buffer before their playback deadline, allowing for a smooth viewing experience when the playback starts. Also, the implementation of playout buffer mitigates the effects that jitter and end-to-end delay has on media quality. The effects of delay and jitter results in the user experiencing jerky video and lack of lip synchronization between what is heard (audio) and what is seen (video) (Ito, et al., 2004), (Steinmetz, 1996), (Kouvelas, et al., 1996).

The channel-switching delay is the time it takes for the system to respond to a channel change request from a user. The channel-switching time for digital broadcast services (traditional TV) is about 1 to 1.5 seconds (Benham, D., 2005). The users' experiences from traditional TV has set a high benchmark for streaming services since users have become accustomed to the quick response performance on traditional TV. The ultimate challenge will be to provide a similar match to that presently experienced in traditional TV viewing

Another consideration is the handover delay over heterogeneous networks in the case of user mobility: A user on the move should be able to have a continuous service experience without having to worry about service failure or interruptions due to the different network characteristics. Ensuring a continuous service quality at the boundaries of heterogeneous networks becomes quite challenging, as the characteristics suddenly change in terms of service capabilities such as data rate, transmission range, and access cost.

Media (Audio-Visual) Quality

Media (audio-visual) quality is perceived as a significant factor affecting QoE, as this is the part of a service that is most noticeable by the user. There have been studies on the influence of audio on video and vice versa on the overall audiovisual quality. Aldridge (1996) studied the influence of audio on subjective video quality. The results from experiments showed that unimpaired audio has a positive effect on the video quality ratings. However, the author hypothesises that impaired audio could cause a severe negative effect. Good audio quality tends to produce a better video quality experience (Joly, et al., 2001), (Reeves, & Nass, 1998). Other authors (Hands, 2004), (Winkler, & Faller, 2005) investigated the interactions between audio and video in terms of perceived audiovisual quality. The integration of audio and video quality tends to be content dependent (Hands, 2004) so that for less complex scenes (e.g., head and shoulder content), the importance of audio quality is slightly more important than video quality. By contrast, for high motion content, video quality is significantly more important than audio quality. The results from Winkler, & Faller (2005) suggests that the optimum audio/video bitrate allocation depends on scene complexity. For instance, visually complex scenes would benefit from the allocation of a higher bitrate with relatively more bits allocated towards audio, since a high audio bitrate seems to produce the best overall quality.

Frame rates also play an important role in the perceived video quality. If the video has been encoded with a lower frame rate with respect to the terminal capability, the user perceives jerkiness. Apteker, et al., (1995) studied the effect of frame rate for different classifications of video within a multitasking environment. It was observed that subjects perceived video of a high temporal nature more acceptable at a lower frame rate than video of a low temporal nature. This was consistent with another experiment carried out by Ghinea, et al. (1999) to investigate the interaction between *Quality-of-Perception (QoP)* and QoS. In Ghinea, et al. (1999), a reduction in frame rate did not proportionally reduce the subject's perception and their assimilation of the multimedia materials. However, in some cases, users seem to assimilate more information. This is also similar to the results found in McCarthy, et al. (2004), when the authors found subjects to be more sensitive to reductions in frame quality than to changes in frame rates for small screen devices.

Network parameters such as bandwidth, jitter, and packet loss also affect the quality of media services and applications. The effect of packet loss tends to be the dominant factor that affects media services by visibly affecting the user's viewing experience. To reduce the effects that packet loss and bit errors may have on perceptual video quality, error control mechanisms are employed to increase robustness and resilience against transmission errors. Three types of error control mechanisms can be identified: error concealment techniques, error resilience techniques, and transport-based error control techniques (Sadka, 2002). The interested reader is referred to Sadka (2002) and Ghanbari (2003) for detailed descriptions of these error control mechanisms. An empirical study carried out by Murphy, et al. (2004) evaluated how packet loss and jitter impact on media quality for different types of video content. Their results illustrated that the different content types (news, music video and movie trailer) performed differently under the same network conditions, with the movie trailer affected the most. Another study carried out by Claypool & Tanner (1999) concluded that the effect of jitter can be nearly as important as packet loss in influencing perceptual video quality.

EVALUATION OF QOE FOR P2P STREAMING

The experiments in this Section are intended to identify the potential for QoE-based management of mobile videos and P2P TV. The results and findings from these experiments are used to show that the method of management proposed in this Chapter is sufficiently general in terms of its applicability to QoE-based management.

Users' expectations of service quality have increased over the years. The network operator's aim is to provide an acceptable user experience at minimal network resource usage. It is therefore important from the network operator's perspective to be aware of the thresholds at which the user's perception of service quality becomes unacceptable. The user's QoE is often measured via carefully controlled subjective tests (ITU-R BT500; ITU-T P.910). However, some experimental and field studies (Watson & Sasse, 1998), (Teunissen, 1996), (Narita, 1993), (Jones, & McManus, 1986) have raised concerns about the appropriateness of the ITU quality scales for the subjective assessment of quality. The main concern is that the quality scales are not interpreted by subjects as having equal step intervals (Jones & McManus, 1986), (Narita, 1993), (Teunissen, 1996), which poses problems with the computation of means. International use of the quality scales to facilitate the comparison of information across different research laboratories has raised questions about their suitability for the purpose. This is because the labelling of the 5-level quality scale (*i.e.,* Excellent, Good, Fair, Poor and Bad) is interpreted differently in different languages (Narita, 1993). Also, subjects tend to avoid extreme end-points on the quality scale when giving their quality responses.

It should be noted that, despite their shortcomings, these scales do have their usefulness and have proved their worth in many communication arenas (Bouch, et al., 2000). For example, the quality-rating scale provides a standardized scale on which quantitative data relating to one level of quality can be directly compared to another. Also, by allowing the same subjects to evaluate different levels of quality, one can ensure that the same conceptual value is associated with internals of the scale (Bouch, et al., 2000). Weighing up their pros and cons, these scales were judged not to be well suited when evaluations are clearly targeted at customer services (Jumisko-Pyykkö, et al., 2008).

Since there are also no one-to-one mappings between the Mean Opinion Score (MOS) values onto users' acceptance of service quality for a given QoS condition, the MOS method does not provide a clear indication of service quality acceptability. For instance, is a MOS rating of 3 acceptable or unacceptable to the user? (McCarthy et al., 2004). This Chapter focuses on the 'threshold of acceptance' methodology for quantifying the user's QoE. The threshold of acceptance refers to the minimum acceptable quality that fulfils user expectations and needs for a certain application or system (Jumisko-Pyykkö, et al., 2008). This methodology has been implemented in Jumisko-Pyykkö, et al. (2006), McCarthy, et al. (2004), Knoche & Sasse, (2008), and Knoche, et al. (2005).

QoE Evaluation on the User Requirements of Mobile Video Services

The psychophysics technique 'Method of Limits' was employed in this Chapter to evaluate and determine the subject's thresholds of acceptability. QoE assessment techniques evolved through the adaptation and application of psychophysics methods during the early stages of television systems (Gescheider, 1997). Psychophysics is concerned with the measurement and quantification of sensations due to physical stimuli. Sensations can differ on at least four basic dimensions– intensity, quality, extension, and duration. The dimension of quality refers to the fact that sensations may be different in kind. The different sensory modalities have unique kinds of sensations; for example, seeing is an entirely different experience to hearing. Within the sensory modalities, sensations also vary in quality. A sound becomes higher or lower in pitch as the vibration frequency of the stimulus changes (Gescheider,1997). Central to psychophysics, is the theory of sensory thresholds – the minimum value of a stimulus that is required to elicit a perceptual response. Gustav Theodor Fechner (Fechner, 1966), the founder of psychophysics, provided methods for measuring the subjects' perceptual response (thresholds) to stimuli. Two types of sensory thresholds can be identified: absolute and difference thresholds. The *absolute* threshold is the smallest amount of stimulus energy necessary to produce a sensation. The *difference* threshold is the amount of change in a stimulus required to produce a *just noticeable difference* (jnd) in the sensation (Gescheider, 1997).

Psychophysical experiments have traditionally used three methods to measure the *absolute* and *difference* thresholds. These methods are the Method of Limits, the Method of Constant Stimuli, and the Method of Adjustment (Fechner, 1966). In the Method of Constant Stimuli, the examiner randomly presents a set of stimuli with fixed, predetermined values above and below threshold. In the Method of Limits, the examiner sequentially presents a set of stimuli with fixed values. The Method of Adjustment is a variant to the Method of Limits, where the subject controls and adjusts the stimulus values. There are no fixed thresholds, as these differ among individuals. However, the threshold must be specified as a statistical value and it is typically found between two sets of parameters in the form of a psychometric function (Levine & Shefner, 2000).

Psychophysics response measurement techniques embrace the four basic types of measurement scales (nominal, ordinal, interval and ratio) to measure and quantify the subject's sensitivity to stimuli. The scaling methods employed to measure and quantify subjects' sensitivity to stimuli can take a variety of forms. For example, subjects can be asked to discriminate between sensations' magnitude of stimuli (*paired comparison*), or to directly judge their sensations to stimulus (*detection of perceptual threshold*). Another psychophysics scaling method is the category scales. In category scaling, the subject is presented with a large number of stimuli and told to assign all of them to a specified number of categories. The number of categories is usually somewhere between 3 and 20. Categories are usually specified either as numbers (such as 1, 2, and 3) or as adjective (such as low, medium or high).

In the early years of television assessment, experimenters (e.g., in (Fredendall & Behrend, 1960) tended to use their own preferred set of response scales for assessing picture quality characteristics. As well as different response scales, there were inconsistencies and differences in the viewing conditions and in the type of subjects used. These made comparison of results difficult. The need for a common standard for the subjective assessment of television was necessary to enable the exchange and comparison of information between various laboratories. At this time, the *paired comparison,* the *detection of perceptual threshold* and the *category* psychophysics scales were in common use for subjective assessment of television pictures (Aldridge,1996). The International Telecommunications Union (ITU), an active standardization organisation, published its first recommendations and standards for carrying out subjective assessment of television pictures in 1972, known as the: "*ITU-R Recommendation BT.500-6: Methodology for the subjective assessment of the quality of television pictures*". The major points of the current version (ITU-R Recommendation BT500-11, 2002), whilst being refined, have not changed significantly since their introduction. Since then, the ITU have set recommendations guidelines on how to perform subjective assessments intended for audio systems (ITU-R Recommendation BS.1116, 1997; ITU-R Recommendation BS.1679, 2004), voice (ITU-T Recommendation P.800., 1996) and multimedia applications (ITU-T Recommendation P.910., 1999; ITU-T Recommendation P.911., 1998). These recommendations were formalized in order to maintain consistency in the data being analyzed from different subjective quality assessments.

The method of applying psychophysics measurements in this Chapter is now discussed. During the measurement of absolute thresholds, the stimuli were manipulated in either an ascending series or descending series. In the descending series, the experiments started by presenting to the subject, a stimulus value that was well above the threshold. The threshold was approached by decreasing the value of the stimulus in successive steps until the subject reported the presence of the sensation. For the ascending series, the experiment started by presenting a very weak sub-threshold stimulus to the subject. The intensity of the stimulus was increased in successive steps until the subject reported the presence of the sensation. A series was terminated when the intensity of the stimulus became detectable. The subjects gave a binary response of 'yes' or 'no' when the stimulus was perceived. The mean value of the transition points (where the response changed from 'yes' to 'no' or 'no' to 'yes') in both series was then considered as an estimation of the threshold. The method of limits can be adapted, *e.g.,* by using only one series depending on the research goal. The Snellen eye chart is an example of the use of the descending methods of limits coupled with correct identification of the target letters (Gale, 1997, p. 128).

In adopting the method of limits and guided by Knoche & McCarthy, (2005), Kuang, & Williamson (2001) and UMTS Forum report 11 (2000), the quality of the video parameters was gradually decreased to determine the minimum level (threshold) of quality at which the user's QoE became unacceptable. The audio quality was kept constant, because audio consumes less bandwidth relative to video. Also, previous research on audiovisual quality (Aldridge, 1996; Joly, *et al.*, 2001; Reeves & Nass, 1998) suggests that good audio quality tends to produce a better video quality experience. Hands, (2004) and Jumisko-Pyykkö, et al. (2006) have used the subjective rating of acceptance to determine acceptance thresholds of perceptual quality. The common goal is to determine the minimum level of quality that subjects find acceptable. This methodology generates data that exhibits a logistic relationship to the perceived quality metric (McCarthy, et al., 2004).

To test this experimental strategy, preliminary experiments were conducted. First, experiments were carried out using the mobile devices emulators such as the Microsoft Windows Mobile 5.0 Software Development Kit (for PocketPC and Smart phone) and S60 Software Development Kit for Nokia N70. The observations from this preliminary experiment were: the emulated mobile devices on the computer monitor appeared larger than the actual physical handheld devices. In addition, the emulated devices required high computational resources from the desktop machines to work properly. Learning from this, it was concluded that using the physical devices instead of the emulated devices will provide the participants the freedom to move the screen closer to suit their needs, thus yielding results that would correlate well to real-world scenarios rather than feeling constrained to watching the contents on a fixed monitor.

Secondly, the test materials were initially limited to two types of video contents (News and Football clips). To achieve a more detailed analysis of the user's QoE on various types of video contents, more video clips were added. Lastly, when the quality of the audio parameter was varied, the comments gathered from the participants indicated that bad audio quality tends to produce a bad video quality experience.

Three terminals which are functionally similar but having different characteristics in terms of display screen sizes and device capabilities were selected to represent the common devices used for mobile streaming services. The specifications of the terminals are given in Table 1.

Selection of Test Materials

The test materials used in experiments were chosen such that they were representative of the types of video contents consumers desire to watch on their mobile terminals (Gilliam, 2006). The Moving Picture Expert Groups (MPEG) uses similar categories of contents in their subjective testing (MPEG-4 AVC Verification Tests). However, the content types in MPEG-4 AVC Verification Tests were not suitable for these experiments because they contain no audio counterpart and their short duration is not sufficient to capture user's QoE. Table 2 shows a summary of the test materials used.

In accordance with the ITU-T Recommendation P.910 (1999) on scene characteristics, the test sequences used in these experiments contained different amounts of spatial and temporal informa-

Table 1. Specifications of the mobile terminals

Terminal type	*Specifications*
Mobile phone	Nokia N70 3G mobile phone with display of 28x35mm, 18-bit colour depth, resolution of 176x220 pixels and a Nokia HS-3 headphone for audio playback
PDA	HP Ipaq rx1950 PDA with display type of 3.5 in TFT active matrix, 16-bit colour support, maximum resolution of 320x240 pixels and Goodmans PRO CD 3100 headphone for audio playback
Laptop	Sony FR315B Laptop with a 15-inch TFT display, resolution of 640x480 pixels and Goodmans PRO CD 3100 headphone for audio playback.

Table 2. Descriptions and screen shots of the test sequences

Multimedia Contents	*Genre*	*Description*
News	BBC news	Man sitting reading news story with headlines scrolling text underneath.
Sports	Football	Cut scene of a football match without the replay effect
	Cricket	Cut scene of a cricket match without the replay effect
Animation	Video game	A walkthrough video game
	Cartoon	Mixture of full and half shots of animation
Music	Music concert	Mixture of panoramic view and full shots with high motion
	Music video	Music video clip with high motion
Entertainment	Top gear	Racing cars with fast changing scenes
	Comedy	Mixture of full, half and head and shoulders shots with moderate motion
Movie	Action	Action movie trailer with fast changing scenes
	Romance	Romance movie trailer with low changing scenes

a). News Clip b). Football Game Clip c). Cricket Game Clip d). Video Game Clip

e). Cartoon Clip f). Music Concert Clip g). Music Video Clip h). Top Gear Clip

i). Comedy Clip j). Action Movie Clip k). Romance Movie Clip

tion, which spanned a wide range of coding complexity. Details on the calculation and measurement of the spatial and temporal information can be found in ITU-T Recommendation P.910 (1999).

The calculation and measurement of the spatial information (SI) is based on the Sobel filter, and was implemented by convolving two 3 × 3 kernels over the video frame (luminance plane). The convolution process is described as follows (ITU-T Recommendation P.910, 1999):

Let $G_v(i, j)$ be the result of the first convolution that denotes the pixel of the input image at the ith row and jth column and let $G_h(i, j)$ be the result of the second convolution for the same pixel.

Then, the output of the Sobel filter at the ith row and jth column in the nth frame $y_n(i, j)$, is the square root of the sum of the squares of the results of both convolutions i.e.

$$y(i,j) = \sqrt{[Gv(i,j)]^2 + [Gh(i,j)]^2} \quad (1)$$

The spatial information (SI) value is the standard deviation (std_{space}) over all pixels in the nth frame, and is computed as follows (ITU-T Recommendation P.910, 1999):

$$SI = std_{space}[y_n] \quad (2)$$

This process was repeated for each video frame in the video sequence, resulting in a time series of spatial information of the scene. The maximum value in the time series $(\max_{time})$ was chosen to represent the spatial information of the scene. This process can be represented in equation form as (ITU-T Recommendation P.910, 1999):

$$SI = \max_{time}\left\{std_{space}[Sobel(F_n)]\right\} \quad (3)$$

The temporal information (TI) is based on the motion difference feature in successive frames. The motion difference feature $M_n(i,j)$, is the difference between the pixel values in the ith row and jth column of nth frame in time $F_{n-1}(i,j)$.. The measure of TI was computed as the maximum over time $(\max_{time})$, of the standard deviation over all pixels in space (std_{space}). It was computed as follows (ITU-T Recommendation P.910, 1999):

$$M_n(i,j) = F_n(i,j) - F_{n-1}(i,j) \quad (4)$$

$$TI = \max_{time}\left\{std_{space}[M_n]\right\} \quad (5)$$

Together, both the SI and TI values produce a time varying measure of the spatial and temporal complexity of a piece of content. The values of SI and TI were computed using MATLAB software. The spatial and temporal complexities of the test materials are shown in Figure 4.

Along the Temporal Information axis, scenes having substantial motion activity are found towards the high end of the axis, while those with minimal motion activity are found towards the low end. Along the Spatial Information axis, scenes with the most spatial details are found towards

Figure 4. Spatial-temporal plots of the test sequences

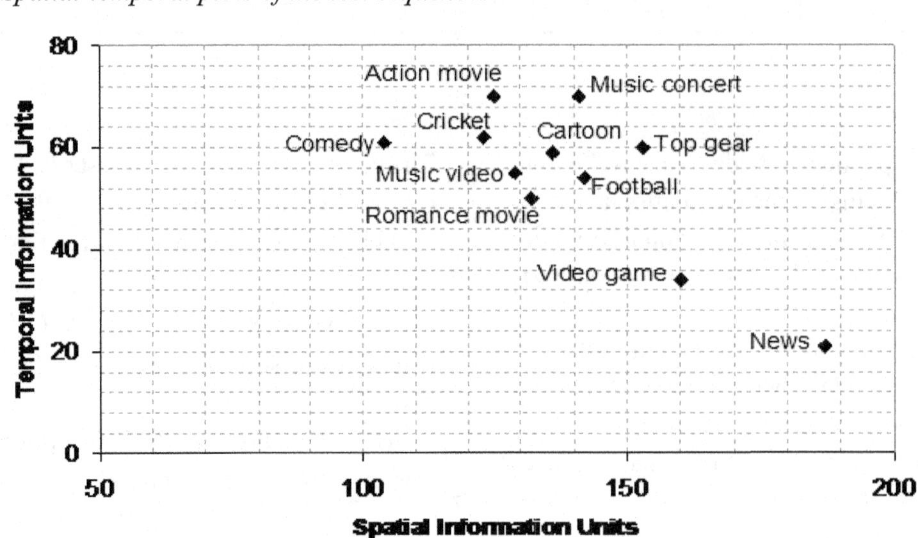

the high end, whilst scenes with minimal spatial details found towards the lower end. From Figure 4, it can be noticed that the choice of the test materials covers a broad range of video content types.

Preparation of Test Sequences for Mobile Terminals

The duration of the test sequences used for quality evaluation in previous studies, varied between 8 to 30 seconds (Webster, et al., 1993), (Hands, 2004), (MPEG-4 AVC Verification Tests), (VQEG, 2003), (Winkler & Dufaux, 2003). However, based on results from a preliminary test, it was found in this experiment that this range of duration was insufficient to capture the user's QoE. The test sequences used in these experiments were prepared as follows: Sections of TV programs were recorded for approximately 2 minutes and 40 seconds. This seemed appropriate since, according to studies by Knoche, et al. (2005) and Sodergard (2003) the watching time of mobile television is very short, usually within 2 to 5 minutes. The Virtualdub software was used to divide each of the recorded contents into 8 segments (labelled 1 to 8 in Tables 3 to 5). Each segment was encoded using the constant bitrate coding technique (CBR) at different bit rates. At a given target bit rate in the CBR coding technique, the encoder dynamically adjusts the quantization parameter QP, to maintain that target bit rate value. QP is a measurement of how compressed a frame is; low values mean low compression ratios, and high values mean high compression ratios (Brun, et al., 2004).

The video parameters manipulated were the video encoding bit rates and frame rates for the:

Mobile phone, using the trial version of Helix Mobile Producer (video codec: MPEG 4 and audio codec: AMR-NB) PDA and Laptop, using Windows Media Encoder series 9 (video codec: windows media video 9 and audio codec: windows media audio 9.1). The segments labelled 1 to 8 were encoded as illustrated in Tables 3 to 5. After encoding, each of these segments was concatenated using TMPGEnc 3.0 Xpress software to produce a continuous stream. The idea was to gradually decrease the values of the video parameters, to determine the thresholds of quality acceptability.

The Subjects

Sixty non-expert university students participated in this experiment. Their ages ranged between 22 and 36 years. Prior to the assessment, each subject completed and passed a two-eyed Snellen test for 20/20 vision and an Ishihara test for color blindness.

Table 3. Test-bed combinations for the mobile phone terminal (Descending series)

Segment number	*Time (seconds)*	*Video encoding bit rates (kbps)*	*Audio encoding bit rates (kbps)*	*Frame rate (FPS)*
Descending Series				
1	1- 20	384	12.2	25
2	21- 40	303	12.2	25
3	41- 60	243	12.2	20
4	61- 80	194	12.2	15
5	81- 100	128	12.2	12.5
6	101- 120	96	12.2	10
7	121- 140	64	12.2	6
8	141- 160	32	12.2	6

Figure 5. Contents acceptability for the mobile phone terminal with error bars indicating the 95% confidence intervals

Test Procedure

The duration of the experiment, which included a briefing about the experiment's requirements and a training session, lasted less than 30 minutes. The subjects were asked to indicate when they found the video quality to be unacceptable. The subjects had no prior knowledge of the test-bed combinations. They were required to indicate when they felt that video quality had just become unacceptable. Customized media players (software) were implemented for each mobile terminal to collect the subjects' details (name, age and gender) along with their respective ratings.

Results and Discussion

The bar chart in Figure 5 shows the acceptability thresholds for the content types on the mobile phone terminal. The numbers used in representing the mean QoE correspond to the encoding parameters of a segment at which the contents were encoded (first column of Table 3). The Mean QoE is obtained by taking the average of the transition points upon which the quality became unacceptable.

From Figure 5, it is evident that users were more tolerant to content such as news, romance movies and cartoon. This was because of the low temporal motion in these contents. Conversely, contents types such as football, cricket, video game, and music concert were less tolerated due to their intrinsically higher level of temporal motion. (Cricket is a bat and ball team game widely played in the UK and Commonwealth countries. It is played with a small hard ball.) Example comments from subjects regarding reasons for poor quality were: "loss of visual details, especially text details" and "it was also impossible identifying players for contents like football and cricket except when the players were zoomed in".

Figure 6 shows that the news and the cartoon contents led to better QoE on the PDA, just as for the mobile phone terminal. The numbers used in representing the mean QoE correspond to the first column of Table 4.

It was noticed that high temporal motion content such as action movies, music concerts and sports were rated with higher QoE for the mobile phone terminal as compared to the PDA terminal. Judging from the feedback from the subjects, for the news clips, they preferred better audio quality. For the action movie and sport contents, they preferred high frame rates because jerkiness in the video resulted in it being out-of-synch with the audio.

The acceptability thresholds of contents on the laptop are illustrated in Figure 7. The subjects' perception of news, romance movie and cartoon contents were similar to those obtained for the

Table 4. Test-bed combinations for the PDA terminal (Descending series)

Segment number	Time (seconds)	Video encoding bit rates (kbps)	Audio encoding bit rates (kbps)	Frame rate (FPS)
Descending Series				
1	1- 20	448	32	25
2	21- 40	349	32	25
3	41- 60	285	32	20
4	61- 80	224	32	15
5	81- 100	128	32	10
6	101- 120	96	32	10
7	121- 140	64	32	6
8	141- 160	32	32	6

Table 5. Test-bed combinations for the laptop terminal (Descending series)

Segment number	*Time (seconds)*	*Video encoding bit rates (kbps)*	*Audio encoding bit rates (kbps)*	*Frame rate (FPS)*
Descending Series				
1	1- 20	512	32	29
2	21- 40	420	32	29
3	41- 60	363	32	25
4	61- 80	224	32	25
5	81- 100	180	32	20
6	101- 120	128	32	15
7	121- 140	64	32	12.5
8	141- 160	32	32	10

Figure 6. Contents acceptability for the PDA terminal with error bars indicating the 95% confidence intervals

Figure 7. Contents acceptability for the laptop terminal with error bars indicating the 95% confidence intervals

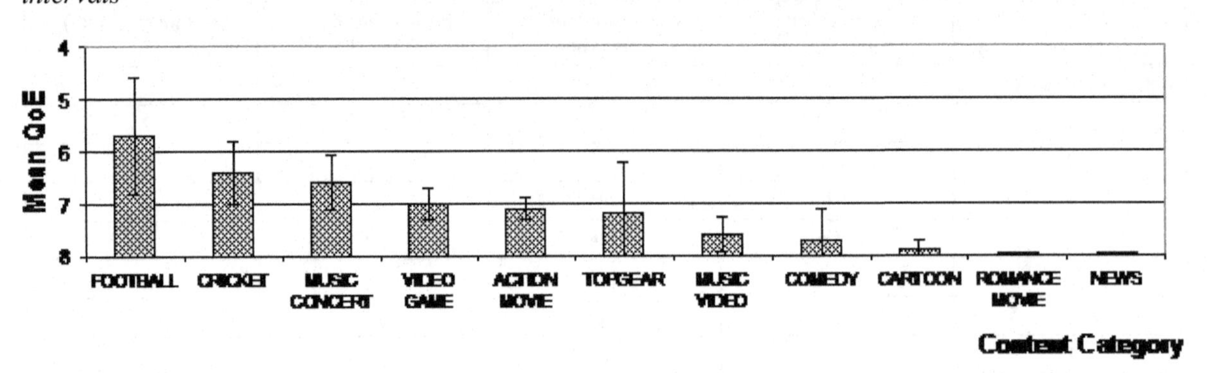

mobile phone terminal. It was noticed that the QoE for football was rated higher when compared to the other mobile terminals used. Perhaps, users preferred a larger display screen for contents like sports. Jerkiness of video was more pronounced for all content types on the laptop as compared to the other terminals.

Summarizing Experiment 1: Acceptability Thresholds via Descending Series

Initial results indicated that QoE does change considerably not only with the encoding parameters but also with the type of video contents and terminals. Although these results were somewhat expected, it is only via a thorough and quantitative assessment of perception thresholds that one can optimize the service provisioning process. It was also noticed that the users' expectations were considerably lower when using a small terminal such that they were more tolerant to visual impairments when using, for example, a mobile phone terminal. These 'psychological' factors cannot be taken into account via conventional QoS management studies.

In the next experiment, the values of the video parameters were gradually increased in order to determine if the results will differ from those obtained in this Section.

EXPERIMENT 2: ACCEPTABILITY THRESHOLDS VIA ASCENDING SERIES

Aim of Experiment

The purpose of this experiment was to determine the user acceptability thresholds in the ascending series (i.e., step-wise increment in the values of the video parameters), and to compare them with the thresholds for the descending series, to determine if there were any correlations. The same procedure described previously to prepare the test sequences, is applied to this experiment. Tables 6 to 8 show the test-bed combinations.

The Subjects

As before, thirty-six non-expert university students participated in this experiment. Their age range was between 22 and 36 years. Prior to the assessment, each subject completed and passed a two-eyed Snellen test for 20/20 vision and an Ishihara test for color blindness.

Test Procedure

A training session was given to make sure the subjects understood what was required. The subjects were asked to indicate when they found the video quality to be acceptable. The subjects had

Table 6. Test-bed combinations for the mobile phone terminal (ascending series)

Segment number	Time (seconds)	Video encoding bit rates (kbps)	Audio encoding bit rates (kbps)	Frame rate (FPS)
Ascending Series				
8	1-20	32	12.2	6
7	21- 40	64	12.2	6
6	41- 60	96	12.2	10
5	61- 80	128	12.2	12.5
4	81- 100	194	12.2	15
3	101- 120	243	12.2	29
2	121- 140	303	12.2	25
1	141- 160	384	12.2	25

Table 7. Test-bed combinations for the PDA terminal (ascending series)

Segment number	Time (seconds)	Video encoding bit rates (kbps)	Audio encoding bit rates (kbps)	Frame rate (FPS)
Ascending Series				
8	1-20	32	32	6
7	21- 40	64	32	6
6	41- 60	96	32	10
5	61- 80	128	32	10
4	81- 100	224	32	15
3	101- 120	285	32	20
2	121- 140	349	32	25
1	141- 160	448	32	25

Table 8. Test-bed combinations for the laptop terminal (ascending series)

Segment number	Time (seconds)	Video encoding bit rates (kbps)	Audio encoding bit rates (kbps)	Frame rate (FPS)
Ascending Series				
8	1-20	32	32	10
7	21- 40	64	32	12.5
6	41- 60	128	32	15
5	61- 80	180	32	20
4	81- 100	224	32	25
3	101- 120	363	32	25
2	121- 140	420	32	29
1	141- 160	512	32	29

no prior knowledge of the test-bed combinations and were asked to indicate when they felt that video quality had become acceptable.

Results and Discussion

The analysis presented below addresses the aim of this experiment i.e., to find the correlation, if any, between the ascending and descending series acceptability thresholds.

The acceptability thresholds reported in Experiment 1 "Acceptability thresholds via descending series" are compared against the Experiment 2 "Acceptability thresholds via ascending series". The bar chart in Figure 8 shows the comparison between acceptability thresholds on the mobile phone terminal. It is clearly seen that in the descending series, the subjects' satisfaction was generally lengthened before quality became unacceptable to them. In the ascending series, the subjects kept on wanting further improvements to the video quality. However, using the two methods together will give the actual bounds on the QoE threshold[3]. The observed asymmetry in users' responses when comparing acceptability thresholds between the descending and ascending series (Figures 8 to 9) was similar to that reported in Aldridge (1996) and Hands (1997).

In the case for the PDA (Figure 9), over provisioning for the descending series was not so pronounced when compared visually to the mobile phone terminal. For all content types, the audio quality was acceptable.

The comparison between the acceptability thresholds for the laptop is depicted in Figure 10.

It was noticed for all terminals that were used in this experiment that the video game content had the highest requirements in the ascending series. This may be because users are generally

Figure 8. Comparison of acceptability thresholds for the mobile phone terminal, with error bars indicating the 95% confidence intervals

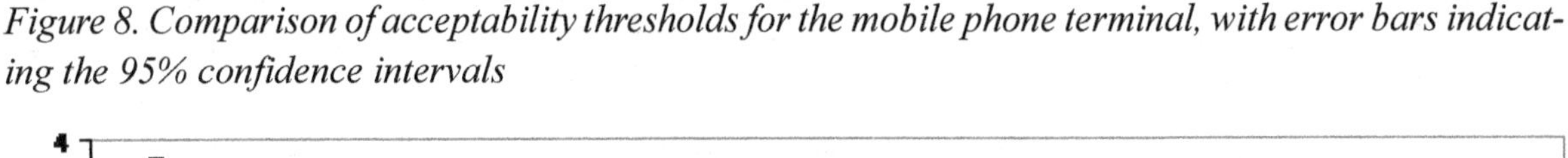

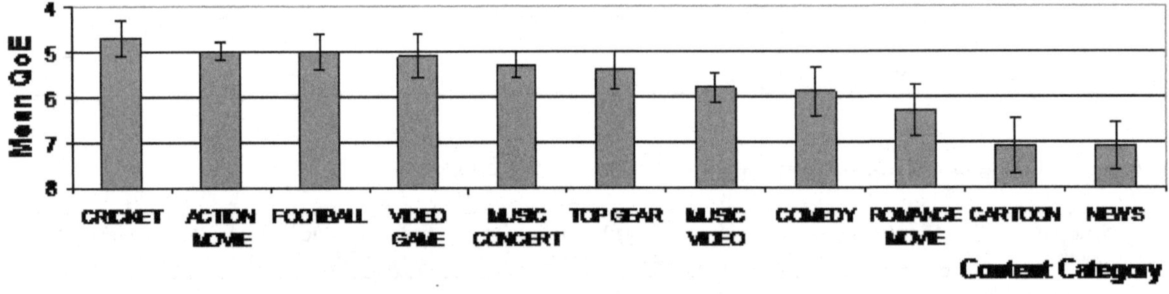

Figure 9. Comparison of acceptability thresholds for the PDA terminal, with error bars indicating the 95% confidence intervals

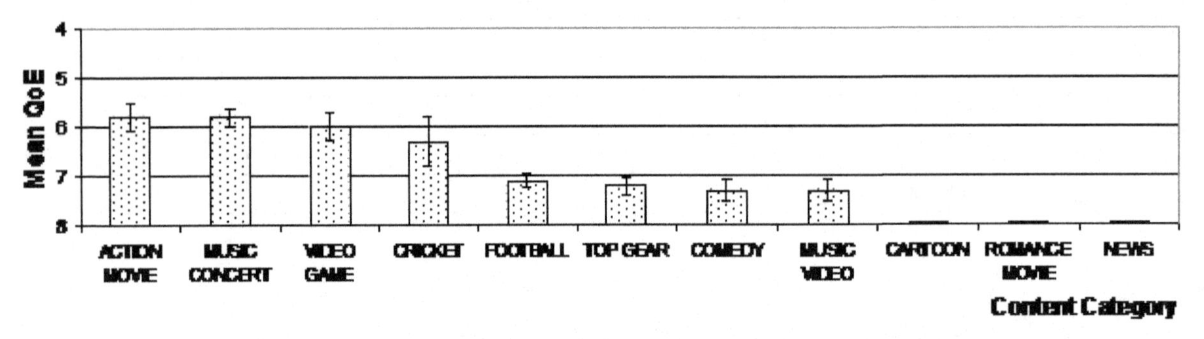

Figure 10. Comparison of acceptability thresholds for the laptop terminal, with error bars indicating the 95% confidence intervals

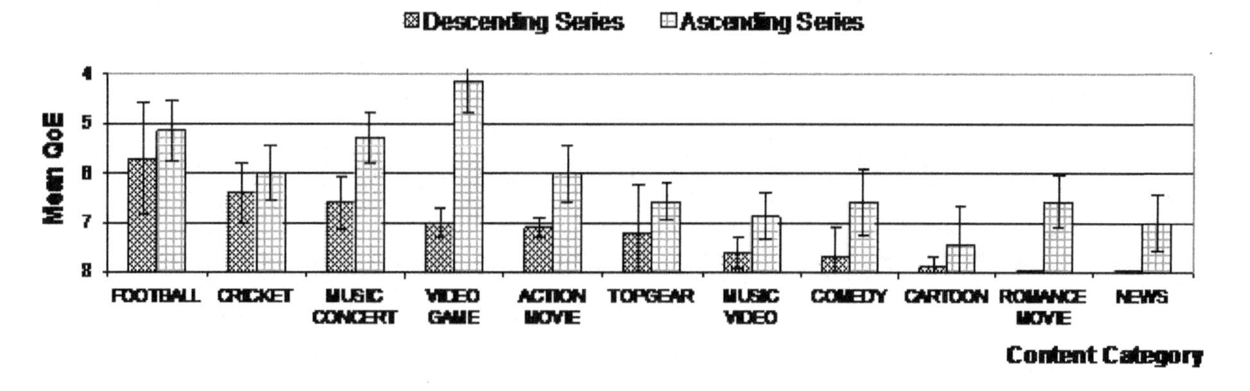

accustomed to watching and playing high quality video games.

Summarizing Experiment 2: Comparisons between Acceptability Thresholds

The results clearly indicate that QoE-based QoS management will lead to better network resource utilization if implemented for the descending series (i.e., step-wise decrement in the values of the video parameters) as opposed to the ascending series. The next experiment was conducted to determine independently the video parameters, which led to the subjects' acceptability thresholds.

EXPERIMENT 3: PER PARAMETER ACCEPTABILITY THRESHOLDS

Aim of Experiment

To determine which parameters contributed the most to the degradation of video quality, a full matrix of tests was carried out by varying the video parameters independently. The experiment was divided into two sessions. In the first session, the frame rate parameter was gradually decreased, whilst keeping the video encoding bit rates constant. In the second session, the video encoding bit rates was gradually decreased, keeping the frame rate constant. The test-bed combination is illustrated in Table 9. For instance, for the mobile phone terminal, in presenting the test sequences in the first session, the frame rate parameter was

Table 9. Test-bed combination for the per parameter degradation.

Target Device	Encoding bit rates		Frame rate (FPS)
	Video (kbps)	Audio (kbps)	
Mobile phone	384	12.2	25 → 3
	384→ 32	12.2	25
PDA	448	32	25 → 6
	448→ 32	32	25
Laptop	512	32	29 →10
	512→ 32	32	29

Figure 11. Acceptability thresholds of frame rates, with error bars indicating the 95% confidence intervals

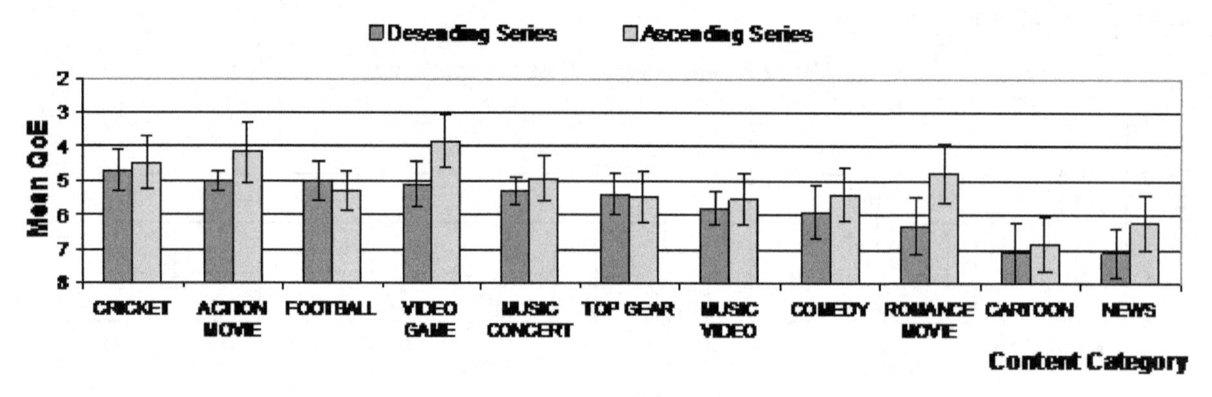

gradually decreased in discrete steps from 25 fps to 3 fps, while the video encoding bitrate was not changed. In the second session, the video encoding bitrate was gradually reduced in discrete steps from 384 kbps to 32 kbps, while the frame rate was not changed. The audio quality was kept constant throughout the presentation of test sequences.

The Subjects

Seventy-two non-expert university subjects participated in this experiment. Their ages ranged from 19 to 36 years. Prior to the assessment, each subject completed and passed a two-eyed Snellen test for 20/20 vision and an Ishihara test for color blindness.

Test Procedure

A training session was given to make sure subjects understood what was required. The subjects had no idea about the combination parameters. Subjects rated the video quality in the first session (only frame rate parameter changing), had a short break and then rated the video quality in the second session (only video encoding parameter changing).

Subjects were told to indicate by clicking on the "Not Accept" button when the quality became unacceptable. The subjects could also click on the "Accept" button if they got to the end of the video, and they still found the video quality to be acceptable.

Results and Discussion

The subjects' acceptability thresholds for the frame rate degradation are illustrated in Figure 11. At these thresholds, video quality became unacceptable. As expected for the mobile phone, the frame rates requirements were significantly reduced. This is because of the small screen inherently found in these devices.

The chart (Figure 11) clearly shows that subject's sensitivity to frame rates is based on the type of video content and the terminal type used to display the content. The qualitative results obtained from subjects concerning the quality were lack of synchronization between audio and video, jerkiness and loss of visual details, especially for the mobile phone terminal. In the case of cricket content, locating the ball was difficult for the subjects and most of the subjects said they could only guess where the ball was. The subjects also found it difficult identifying the teams that were playing. For the cricket and action movies, jerkiness during panning of the camera led to video quality being generally unacceptable. But this was different for the football content, as subjects preferred low frame rates (to some extent), because this enabled them to watch the game easily.

Acceptability thresholds for the overall video bit rate degradation are illustrated in Figure 12. The variations in acceptability thresholds for the laptop were at first puzzling. However, when

Figure 12. Acceptability thresholds of video encoding bitrate, with error bars indicating the 95% confidence intervals

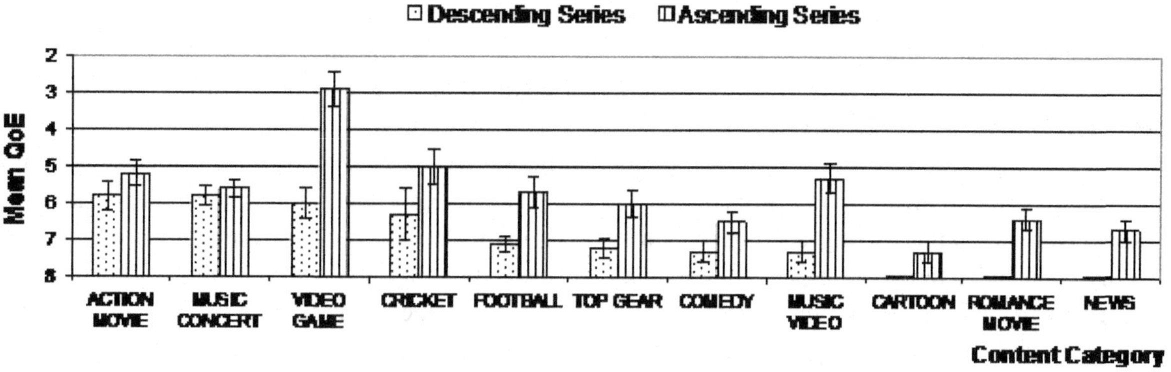

subjects were questioned about their ratings for the laptop, the general response was that for the news content, their primary interest was on the audio quality, so they were more tolerant. The responses from subjects regarding the contents with intrinsically high thresholds of acceptability (such as comedy, music video, music concert, and action movie) were the inability to identify facial expressions and region of interests due to the smearing effect and pixelation.

Summary of Mobile Video Services

The experiments carried out, helped in identifying the high potential of using QoE management as opposed to QoS management, by capturing the users' expectations and subsequently identifying their requirements. Service and content providers may work together with the network operators to maximize user satisfaction in relation to the type of terminal, access network, and type of video content being offered to the user.

Another lesson learned from these experiments was that media adaptation cannot be achieved merely by acting on encoding video bit rates, quantization parameters, frame rate and so forth. Content must be edited specifically (e.g. larger text size for smaller screen size) for the type of terminal that will be used to access the video content. This is a further level of optimization which can make a significant difference to the end user. During the experiments, the test sequences were created from recordings from TV programs, and prepared. This approach would hardly satisfy the user watching, for instance, a football match from a mobile terminal. This is because the loss of visual details and the difficulty in identifying players, or in detecting ball movement is bound to substantially decrease viewing satisfaction. Matters could be improved if contents are specifically edited for mobile terminals.

Zooming out when viewing a video content did not lead to a measurable improvement in QoE, since the small screen sizes did not allow for the identification of details. Increasing the usage of close-up shots and optimized textual rendering in mobile video will allow for a TV-like experience on small screens. These findings will find immediate applications in the improvement of QoE and also in the minimization of the consumed network resources.

Service providers can use the lessons learnt to determine the best resource trade-offs. They can achieve this by utilizing information about the thresholds (based on the type of contents and terminal being used) to gradually reduce the values of certain video quality parameters without considerably affecting the user's QoE.

The next section describes the methods used for collating the user's QoE for P2P TV systems.

Although the QoE approach has been recognized as a major tool in understanding user perceptions of content delivery systems, not a lot is known about the users' QoE on P2P streaming systems.

QoE Analysis of a P2P TV System

Carrying out QoE analysis on P2P streaming applications is challenging because such systems are usually large and heterogeneous, i.e., they span a large geographical area, involving a large number of peers (users). Thus, their performances are considerably affected by the user's behavior and the statuses of the sub-networks making up the system. The user behavior and network statuses are unpredictable, thus making them difficult to parameterize and model in simulation software. Because of the above features of P2P systems, the experiments carried out on them on a small scale (e.g., in a laboratory) are often meaningless. As such, these small-scale P2P systems do not adequately reflect the real-life P2P systems. Experiments carried out on a large scale require a large-scale deployment and user base. The latter approach is hard to follow since many P2P systems are proprietary and also, it is often too late to identify their major architectural limitations since the system would already have been deployed.

The key starting point to resolving these problems is an assessment based on existing QoE subjective methods. However, carrying out QoE analysis requires some reverse engineering in order to understand and to gather sufficient knowledge about the system itself. It also requires an investigation into which QoS parameters are most sensitive to the system, and how these parameters affect the user's QoE. During the QoE evaluation presented here, a method was introduced for assessing P2P TV systems, by starting with an analysis of the most sensitive QoS parameters that affect QoE, and leading to the QoE subjective study itself.

The P2P TV system chosen for evaluation was Joost, which at the time of writing, represented the most up to date commercial P2P TV system. The Joost P2P system distributes licensed television channels and is the originators of its contents. Thus, all new content is introduced through its own servers. Carrying out QoE experiments on Joost required an in-depth knowledge of its architecture, protocols and mechanisms. But these were not readily available since it is a proprietary system.

Two experimental methods were used in studying Joost. The first method (experimental measurement) enabled the unravelling of the Joost architecture and how it worked. The second method (a subjective study) enabled the identification and isolation of the NQoS4 parameters that the system was most sensitive to.

EXPERIMENT 4: QOE EVALUATION OF JOOST P2P TV SYSTEM

Experimental Measurement Methodology

Joost (Beta version 1.04) was inspected at its packet level in order to resolve its network architecture and communication protocols. By running its client software, it was possible to measure the start-up delay and channel switching delay, to get an insight into its internal working mechanisms.

In this experiment, traces of Joost packets were collected within two different settings – a university campus environment and a residential environment. The PCs used in both environments were Windows XP based and had processor speeds of at least 2.4 GHz. Each PC (peer) ran Wireshark Version 0.99.8 (a network protocol analyzer) and Netpeeker Version 3.10 (a network traffic monitor and control tool) to capture all inbound and outbound traffic of Joost.

Subjective Assessment Methodology

In order to determine the sensitivity of the Joost system to changes in the network parameters, an initial experimental assessment (see below)

Table 10. Initial experimental QoS matrix

Delay (ms)	Bandwidth (kbps)	Packet loss (%)
200	600	0
700	700	14
1300	750	18
2200	800	25
3000	900	31

based on traffic analysis was followed, in order to obtain the experimental QoS matrix depicted in Table 10. This provided a basis for identifying and isolating the network parameters which were sensitive to the Joost P2P TV system.

Figure 13 shows a basic diagram of the experimental test-bed. A network emulator was used to reproduce each network parameter.

From the experimental test-bed (Figure 13), the network emulator acted as a proxy server and was used in introducing the delay, bandwidth and packet loss parameters into the network.

The sensitivity of the delay parameter was first determined: Between the values of 200 ms to 700 ms, no significant observations were noticed. At a delay of 1300 ms and above, it took longer to buffer video packets before playback commenced. But once video playback started, the playback continuity was smooth.

Next, the sensitivity of the bandwidth parameter was determined: The system could not support video playback at bandwidth lower than 600 kbps. Between the values of 600 kbps to 700 kbps, the system would connect and begin rendering the video, but there were recurrent freezes in the video playback due to buffer starvation (lack of video packets in the buffer). At 800 kbps and above, the video play-back was smooth.

Packet loss was the final network parameter to be studied: At an ideal rate of 0%, the video playback was smooth. At packet loss rates between 14% and 25%, video playback was constantly interrupted with the system needing to buffer video packets. The system could not tolerate packet loss rates above 31%.

From this initial experimental assessment based on traffic analysis, it was seen that the negative effects of the delay parameters were mitigated by the buffer implementation. Also, the variations in the effects caused by the bandwidth values in the range 600 to 700 kbps were negligible.

Figure 13. Experimental test-bed for Joost P2P television

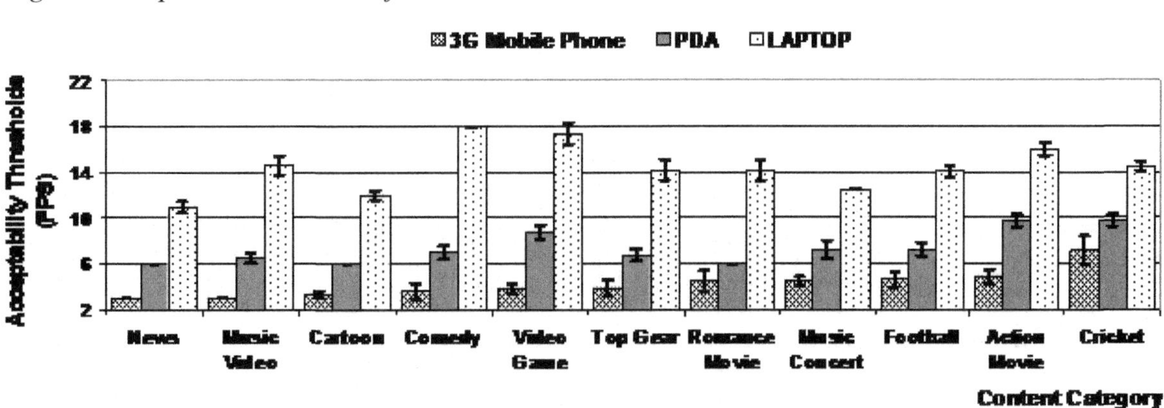

Table 11. Final experimental QoS matrix

Test Conditions	Packet loss (%)
1	0
2	14
3	18
4	25

The Joost P2P TV system was still sensitive to packet losses, in spite of the buffer implementation and the use of packet loss recovery technique (FEC) to mitigate this effect. It was concluded that packet loss rate was the only parameter that the system was most sensitive to. Hence, packet loss was the only parameter that was focused on. Table 11 shows that final QoS matrix.

Packet losses over the Internet are unpredictable and dynamic, and can follow the degradation trend as shown in Table 11. This experiment aimed at determining the packet loss rates the viewing experience became unacceptable for the subjects. This packet loss rate is realistic for a wireless networks, which are more susceptible to losses as compared to wired line networks. A packet loss rate of 31% was excluded from the subjective assessment since from the initial experimental assessment based on traffic analysis, it was noticed that the Joost system could not tolerate packet loss rates at 31% and above. The QoS matrix defined above (Table 11) was put forward to subjects to generate QoE responses, which can be used for evaluating its performance, and identifying architectural limitations of the system.

On the PC running the Joost client software, PHP Application Server software ran as well, to enable the Joost client to run within the web application used in conducting this experiment. The web application was implemented as a collection of PHP scripts which enabled the Joost client to be embedded and started within a web page. The other pages in the application, had forms which enabled the collection of the user's information (i.e., name, age, gender, email address, details about previous experience with Internet and P2P streaming etc.) and the collection of the user's responses regarding their streaming experience. The PHP script structure is shown in Figure 14.

- The scripts '1_intro.php, 2_confirm.php and 3_welcome.php' implemented the web page forms where the users' information

Figure 14. Diagram of the script structure

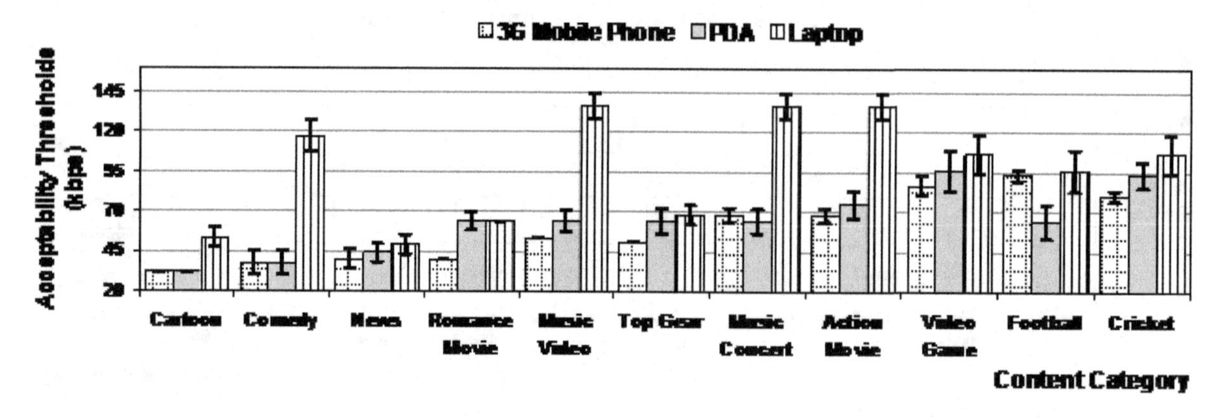

were collected and also provided instructions for the user.

- The scripts '4_conditions.php and 5_timer.php' implemented the functionality where a packet loss rate was introduced into the network.
- The script '5_timer.php' was used in limiting each test condition to 4 minutes.
- The scripts '6_results.php and 6_results_b.php' were used in collecting the user ratings (*i.e.,* acceptable or unacceptable) at the end of each test condition.
- The pages looped a total of three times from '4_conditions.php' to '7_continue.php' in order to set the four different test conditions and to collect the results. Every time a new test condition begins, the Joost cache was cleared in order to avoid biased results.
- After the four test conditions are executed, the script '8_cancel.php' automatically shut down the Joost client to mark the end of the experiment.

Figures 15 to 18 show the screenshots of the test environment.

The Subjects

Twenty-four non-expert university students participated in this experiment. The age range was between 20 and 22 years. The sample size included those who had had previous experience in using Internet streaming and/or P2P streaming applications.

Test Procedure

The subjects were briefed about the experiment's requirements and a training session was given to make sure subjects understood what was required. Each test condition lasted for 4 minutes and at the end of each test condition, the subject rated the test condition as acceptable or unacceptable. The total duration of the experiment lasted approximately 20 minutes. The experiment was performed without the users' having any knowledge of the values of the packet loss parameter.

Results from the Experimental Measurements

- **Basic operation of Joost**

At initialization, a peer initiated encrypted handshakes with three servers at the following IP addresses: (89.251.4.178), (89.251.4.175) and (89.251.2.85). Next, the peer sent an HTTP GET request to the site (89.251.2.87) to check for the cur-rent version of the Joost software. Finally, the peer contacted some super nodes at the sites (89.251.4.71), (89.251.0.17) and (89.251.0.16) in order to obtain lists of the available peers from which to download video chunks.

Joost uses the UDP protocol for the distribution of video chucks. To bypass Network Address Translators (NATs) and firewall restrictions on peers, Joost uses the Simple Transversal of UDP through NAT (STUN) and Interactive Connectivity Establishment (ICE) protocol (Maccarthaigh, 2007) to allow peers behind fire-walls to participate in a P2P streaming systems, thus providing network transparency. Joost also uses the Forward Error Correction (FEC) packet loss recovery technique to handle live peer packet losses (Maccarthaigh, 2007).

After analyzing the incoming and outgoing traffic, it was inferred that Joost belonged to a mesh-based architecture. It pulls video chunks from multiple video sources thus providing more resilience to churn rate as compared to the tree-based system, where peers can only receive video packets from a single source at a given time. However, a disadvantage of a mesh-based architecture is the extra processing time in re-ordering the video chunks at the receiving peer, since these video chunks are received from multiple sources.

Figure 15. User details entry form

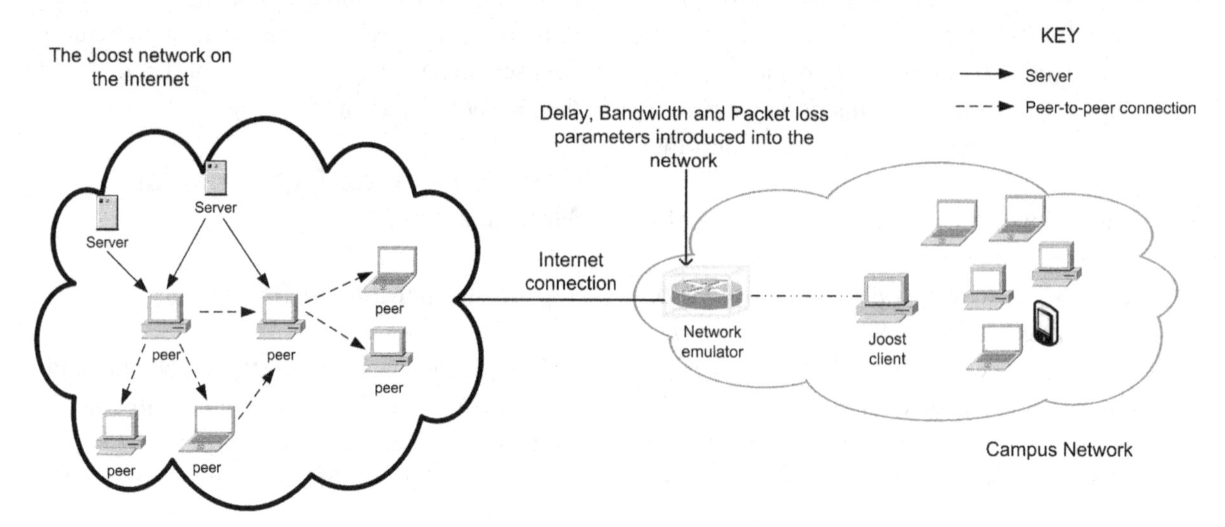

Figure 16. PHP script initializing Joost start-up

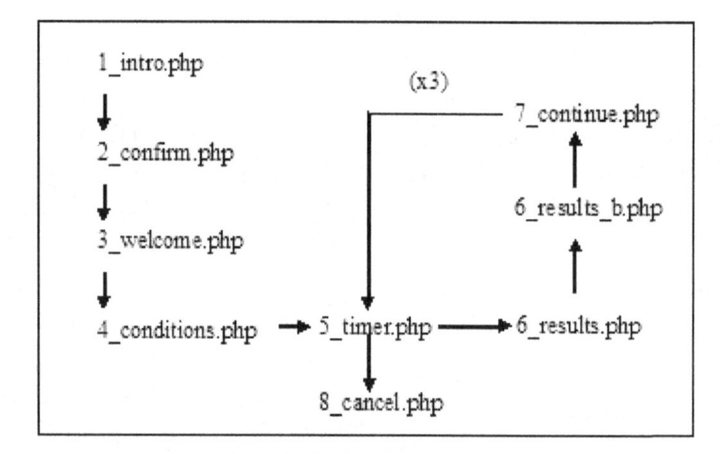

Figure 17. An example of Joost playing a popular channel

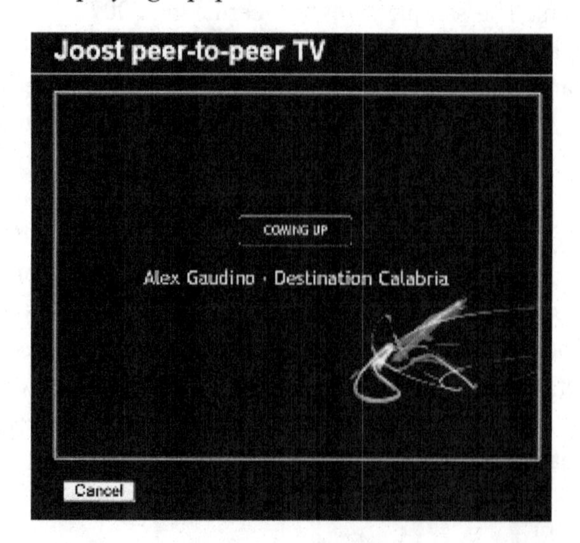

Figure 18. Qualitative feedback form

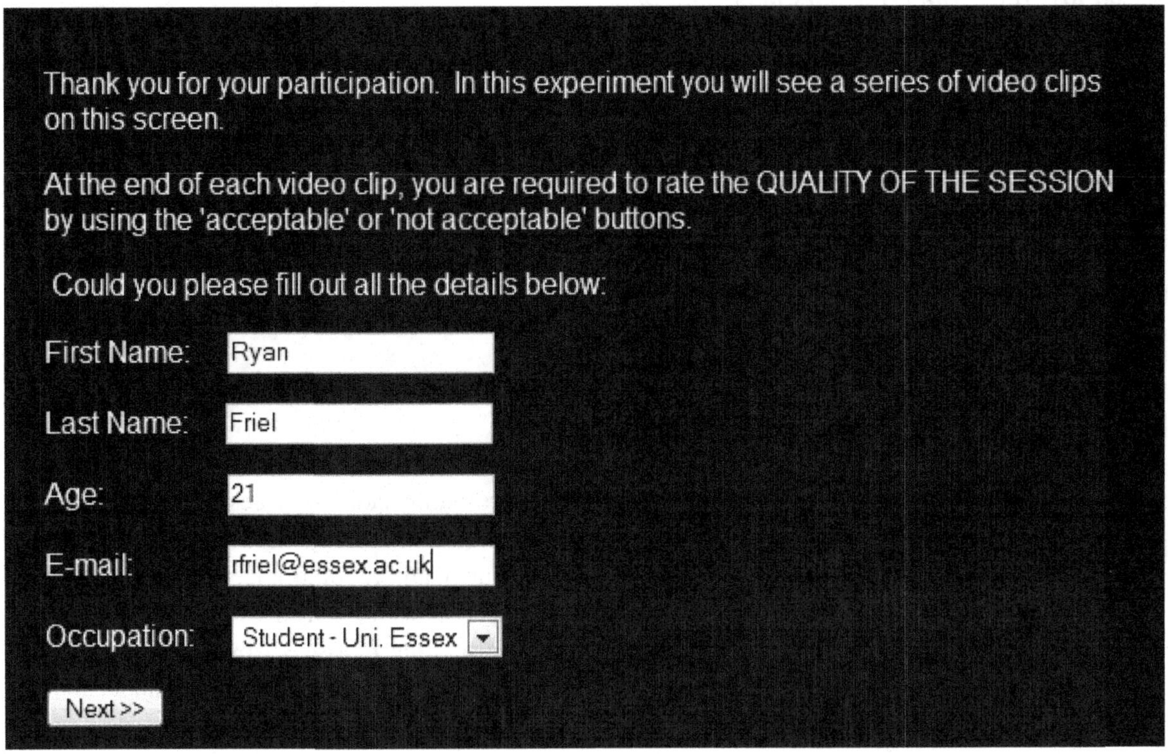

Thank you for your participation. In this experiment you will see a series of video clips on this screen.

At the end of each video clip, you are required to rate the QUALITY OF THE SESSION by using the 'acceptable' or 'not acceptable' buttons.

Could you please fill out all the details below:

First Name: Ryan

Last Name: Friel

Age: 21

E-mail: rfriel@essex.ac.uk

Occupation: Student - Uni. Essex

Next >>

- **Start-up delay**

The start-up delay was measured. i.e., the delay from when a peer gets connected to Joost until video playback started. By taking the average over several trials (each time, clearing the local video cache in order to eliminate biases), the start-up delay of a popular Joost channel on the campus settings was estimated to be 25 seconds, with minimum and maximum values of 16 and 25 seconds respectively.

For the residential settings, the average start-up delay was 28 seconds, with minimum and maximum values of 15 and 38 seconds respectively. These values are still higher than those experienced in client-server architectures, which is typically between 5 and 15 seconds.

- **Channel switching delay**

To measure the channel switching delay, a random channel was left running until the video stream was considered stable. Then navigating through the 'Joost Channel Explorer', another channel was selected. The time it took the system to respond to this request was recorded. Taking the average over several trials, for different selection of channels, the average channel switching delay was estimated to be 5 seconds, with minimum and maximum values of 4 and 6 seconds respectively. The channel switching delays were significantly longer than those experienced in traditional TV which are between 1 to 1.5 seconds.

Figure 19. Subjects acceptance thresholds of packet loss rate, with error bars indicating the 95% confidence intervals

Results from the Subjective Assessment

The user's acceptability of the Joost service at a packet loss rate is depicted in Figure 19 Acceptability in this context refers to the proportion of the subjects who found viewing quality at a given packet loss rate acceptable all of the time. As can be seen from the graph, the users were relatively accurate in predicting degradation in service quality i.e., their acceptability decreased as the packet loss rate increased. From observations, it was noticed that the subjects who had previous experience on Internet streaming and/or P2P streaming had a higher acceptability threshold because they had a general idea of what to expect. They perceived the P2P services as free (i.e., they were not paying for it) on the Internet, so they were more tolerant to poor service qualities. However, at a packet loss rate of 25%, the majority of subjects said viewing quality was unacceptable.

The qualitative responses from the subjects suggested that at higher packet loss rates, the resulting longer start-up delays and the lack of smooth continuity during video playback, significantly affected their viewing experience. The error bars were quite large due to the small sample size. It is clear that a larger number of subjects with more diverse ages and a lower granularity of packet loss between 0 to 14% will be needed in future studies. The aim was to illustrate that the "threshold of acceptance" methodology, which has been the focus of this chapter can be applied to other systems (such as the one presented here) in quantifying the user's QoE.

Summary of P2P TV System

Although P2P television offers a cheap platform for providing video services, understanding the user's QoE in such systems could increase their popularity and revolutionize this evolving technology. As seen from this experiment, some of parameters may be assessed by direct measurement on the platform (e.g., start-up or channel switching delays). On the other hand, the assessment of the quality perceived by the user requires more sophisticated subjective studies. In either case, the assessment of QoE requires the determination of the sensitivity of the various factors in relation to the architectural and functional properties of the platform. This was evident during the experiment, in how buffering mechanisms reduced the sensitivity to delay. On the other hand, packet loss was more critical, although thanks to the mesh-based approach there is a good level of resilience.

The significance of the results obtained in this experiment can be further increased by extending the study to a larger user population, which will allow a more fine-grained assessment of the relationship between network quality and user's experience. This in turn will lead to QoS-to-QoE mapping algorithms that can considerably improve overlay management. P2P streaming is still at its infancy and the subject of intensive studies aimed at improving overlay management algorithms, resilience, and the ability to work in a dynamic environment.

Extending the assessment of P2P streaming to a mobile environment may unveil interesting unexpected findings. QoE subjective studies will also help taking into account psychological factors. For instance, users may have lower expectations if they connect with an inexpensive (or low-power) terminal.

PREDICTION MODELS FOR QOE-BASED MANAGEMENT

In this Section, the statistical technique of Discriminant Analysis is used in developing models to make predictions of what users' estimates of QoE are likely to be, for given QoS conditions.

Introduction

The main statistical methods used in the prediction of group membership are: Discriminant function analysis, logit analysis and logistic regression. Logit analysis tends to be used when all independent variables are all discrete. Logistic regression is used when the independent variables are a mix of continuous and discrete and/or poorly distributed. The discriminant analysis tends to be used when all the independent variables are continuous and nicely distributed. It was thus chosen to be used in building the QoE models since the values of the variables (video encoding bitrates, frame rate etc) were continuous. Discriminant Analysis is a technique for classifying a known set of observations into predefined groups, in order to produce discriminant functions that will be used to predict the group of a new observation with unknown group.

In this chapter, the technique was employed in providing answers to the following questions:

- Will a chosen QoS parameter lead to a reliable user perception of service quality?
- What is the relative importance of the QoS parameters in predicting the user's QoE?

To use discriminant analysis, a set of observations (or test cases which comprise values of quantitative variables) whose group memberships are already known is needed. Then, a set of linear functions of the quantitative variables, known as discriminant functions, is constructed. These functions best separate the group memberships. The maximum number of discriminant functions for an analysis with N group memberships and q quantitative variables, is either (N-1) or q whichever is the smallest. The first discriminant function maximises the differences between the group means. Subsequent functions achieve the same goal but with the added constraint that they are not correlated with the values from the previous functions.

The classification procedure provides a means used to predict the group to which a set of observation most likely belongs. This process is based on either the use of the discriminant functions or the linear combination of the quantitative variables. The latter uses the theory of maximum group differences to derive the classification functions. The formula is as follows (Klecka, 1980):

$$h_k = b_{k0} + b_{k1}X_1 + b_{k2}X_2 + ... + b_{kn}X_n \qquad (6)$$

where h_k = the classification score for group *k*

b_{ki} = the coefficient for variable *i* in the equation corresponding to group *k*

X_i = the value of the quantitative variable for variable *i, i=1...,n*

b_{k0} = is a constant

In simpler terms, the classification functions can be used to determine to which group each response most likely belongs. There are as many classification functions as there are vote categories. For example, a two-group membership with

Table 12. Discriminant function 1 analysis showing Wilks's lambda value and its significance for P2P TV

Content type	Wilks's lambda (Λ)	Significance (p)
Music video	0.764	< 0.001

two quantitative variables, Equation (6) can be expanded to

$$h_1 = b_{10} + b_{11}X_1 + b_{12}X_2$$
$$h_2 = b_{20} + b_{21}X_1 + b_{22}X_2 \quad (7)$$

A case is classified as belonging to the group for which it has the highest classification score. It is possible to classify new cases into group membership based on their classification scores.

The Wilks's lambda and the standardized discriminant coefficients are other important statistics produced when performing discriminant analysis. These statistics help to determine how accurate a prediction is, and how much influence each quantitative variable has on predicting the group membership.

Wilks's Lambda (Λ)

The Wilks's lambda is used to test if the discriminant function is significant (i.e., how accurate the prediction will be). The Wilks's lambda values vary between 0 and 1. Values that are near zero denote high discrimination. The significance of lambda is determined through the p-parameter, similar to the Analysis Of Variance (ANOVA) significance.

Standardized Discriminant Coefficients

The standardized discriminant coefficient indicates the relative importance of the quantitative variables in predicting the qualitative attribute. The relative importance is obtained by examining the magnitude of the standardized coefficients (ignoring the sign). The larger the magnitude, the greater is that variable's contribution.

The use of the classification function coefficients Equation (6), Wilks's lambda (Λ) test for significance p and the standardized discriminant coefficients are exemplified in the remainder of the Section.

Results of Discriminant Analysis for QoE in P2P TV Wilks's Lambda Test for Model Significance

Table 12 shows the Wilks's lambda test of significance for the discriminant function. The p-value are highly significant ($p < 0.001$), indicating that the levels of the packet loss parameters differentiated among the two qualitative responses.

The limitation to having only one quantitative variable is that it provides no insight into how variables interact with each other in the prediction.

THE CLASSIFICATION FUNCTION COEFFICIENTS

Equation (8) shows the derived classification functions that can be used to predict user perception based on a particular packet loss parameter.

$$h_{(Acceptable)} = -1.491 + 0.157 \times Packet\ loss$$
$$h_{(Unacceptable)} = -3.495 + 0.294 \times Packet\ loss \quad (8)$$

The prediction model of user perception correctly classified 72.9% of the responses in our original set of test cases. To assess the predictive accuracy of this model in a new sample, it was

*Table 13. **Predicted user ratings for P2P packet loss parameter**. The point where the packet loss acceptability threshold is first reached is shown in bold type.*

Packet loss (%)	Classification scores		Predicted ratings
	$h_{(Acceptable)}$	$h_{(Unacceptable)}$	
0	-1.491	-3.495	Acceptable
2	-1.177	-2.907	Acceptable
5	-0.706	-2.025	Acceptable
8	-0.235	-1.143	Acceptable
10	0.079	-0.555	Acceptable
14	**0.707**	**0.621**	**Acceptable**
15	0.864	0.915	Unacceptable
18	1.335	1.797	Unacceptable
25	2.434	3.855	Unacceptable

estimated that if the leave-one-out technique were used, an equivalent proportion of cases would be accurately classified.

Summarizing Discriminant Analysis for QoE in P2P TV

The prediction model derived for QoE in P2P TV should be interpreted with caution because of the original subjective data problems. The original subjective test data had no results between packet loss rate of 0 and 14% and also, the data sample was of relatively small size. However, the aim of this Section was to illustrate the QoE framework in terms of its applicability to QoE-based management.

QoE-BASED MANAGEMENT STRATEGIES

This Section demonstrates how QoE management strategies can be deployed. These management strategies show how QoE data may be used to provide the best possible user experience together with the most efficient utilization of network resources. Network bandwidth is a key metric for network management and planning processes (Farrera, 2005, p. 8). It must be efficiently managed so as to avoid resource over-provisioning.

The implementation of certain mechanisms in the Joost system made it difficult to measure correlations between the user's perception and some of the network parameters. For instance, the buffer implementation mitigated the negative effects of the delay parameters. The Joost P2P system was sensitive to only packet loss parameter. Hence packet loss was the only network parameter used in the subjective assessments and in deriving the prediction model.

Music Video Content Model

To demonstrate the use of the packet loss prediction model (represented by the classification functions in Equation (9)), its classification functions were evaluated for a range of packet loss parameters. A case was predicted as belonging to the classification group for which it has the highest classification score. The full results are shown in Table 13.

$$\begin{aligned} h_{(Acceptable)} &= -1.491 + 0.157 \times Packet\ loss \\ h_{(Unacceptable)} &= -3.495 + 0.294 \times Packet\ loss \end{aligned} \tag{9}$$

Figure 20. Bandwidth utilization by the Action movie content type (without QoE management)

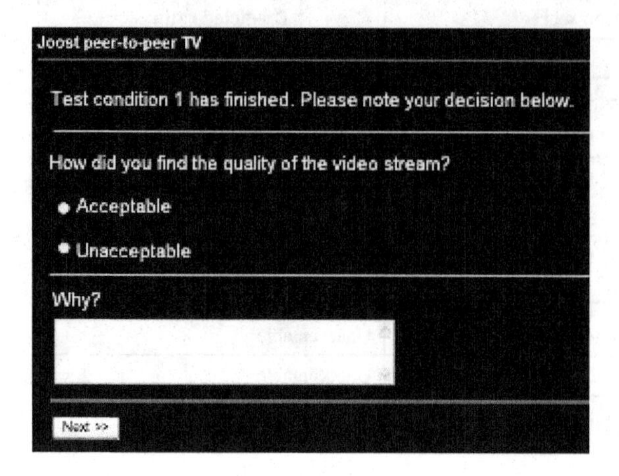

From Table 13 the model predicts that beyond 15% packet loss rate, the user perception of quality is likely to be unacceptable. Only 23 subjects provided the original subjective test data. The data was also recorded at huge steps of packet loss rates – 0%, 14%, 18%, 21% and 25%. As an improvement to this study, the range between the first two packet loss rates, need to be subdivided. This will allow a more fine-grained assessment of the relationship between network quality and user's experience.

The acceptability regions obtained in this Section could also be expressed by means of a three-dimensional graph. However, it should be noted that when illustrated graphically, the results from the graphs will be sensitive to extreme independent variable values. Since the classification function procedure assigns a case as belonging to the group for which it has the highest classification score.

Bandwidth Management Using QoE

An action movie clip was encoded as follows without any QoE considerations: frame rate at 25 fps, audio bit rate at 12.2 kbps and video bit rate at 384 kbps. The bandwidth that was consumed by this video clip over time is as shown in Figure 20. The corresponding quantization parameter[5] (Qp) values over the same time range is shown in Figure 21.

Assuming a network LAN of 10 Mbps and a T1 connection (1.55 Mbps), about 24% of the T1 connection was utilized by the video clip. However, if the QoE management approach were employed, the network's bandwidth utilization could be reduced whilst still providing an acceptable service quality. To minimize the use of the network's bandwidth, the values of the encoding parameters were strategically degraded using knowledge about the degree of influence each of them had on the end user perception. For example, for this action movie content played on a mobile phone terminal, it was found that the typical user

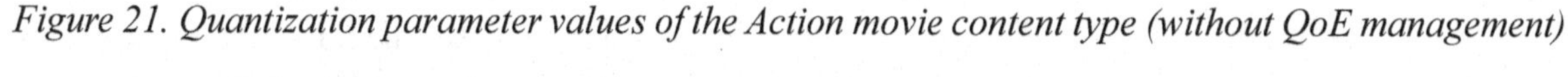

Figure 21. Quantization parameter values of the Action movie content type (without QoE management)

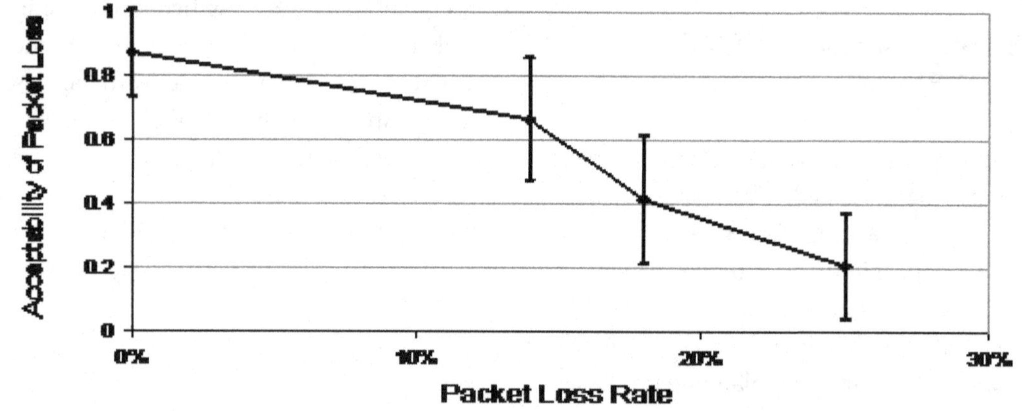

Figure 22. Comparison of bandwidth utilization by the action movie content type (with and without QoE optimization)

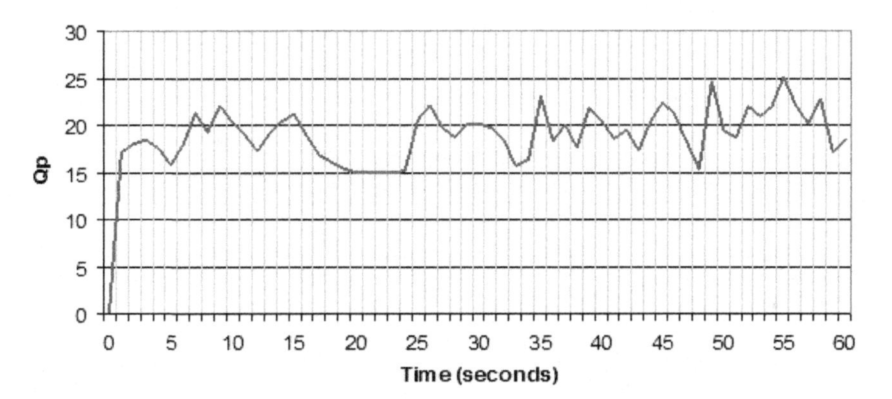

was more sensitive to reductions in the frame rate. Guided by this information, the video encoding bit rate for the action movie content was reduced from 384 kbps to 256 kbps, and the frame rate reduced from 25 fps to an acceptability limit of 18 fps. The bandwidth now being utilized by the action movie was reduced (see Figure 22), whilst still maintaining the same picture quality as is shown by the similar QP values for the "with and without" QoE scenarios in Figure 23.

With QoE considerations employed in encoding the video clip, only 15% of the T1 connection was used, thus achieving approximately a 38% savings on network resources. When further decreases of bandwidth were necessary, the video encoding bit rate parameter was degraded first, since the users are less sensitive to it, as opposed to the frame rate. Further decrements in the frame rate were only made if the user's perception of quality was not adversely affected. With this type of control mechanism, the bandwidth usage could be decreased whilst still preserving the best possible video experience.

Figure 23. Comparison of the quantization parameter values for both the action movie (with and without QoE optimization)

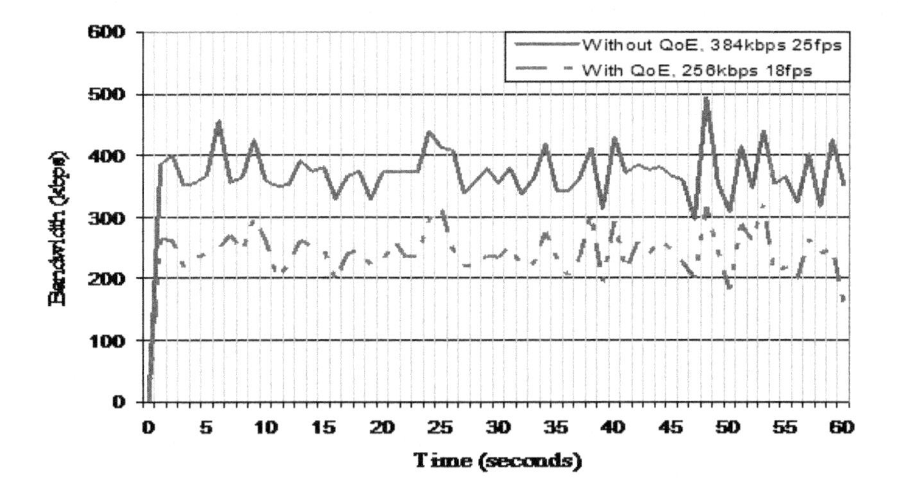

Figure 24. Bandwidth utilized by the news content type (without QoE management)

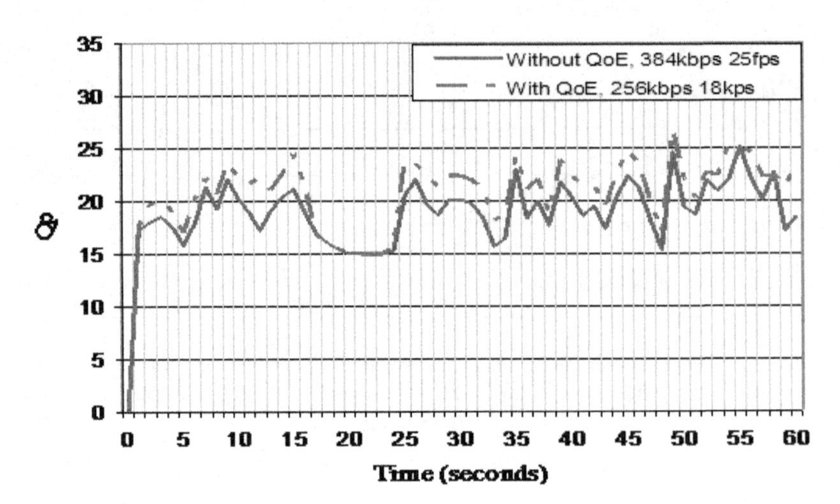

Provision of Specific Service Levels Using QoE

QoE management was also used in providing different levels of service quality (policy-based charging schemes). Consider a news content type that was encoded at a frame rate of 25 fps, with audio and video bit rate of 12.2 kbps and 384 kbps respectively, without any QoE considerations. Figure 24 (Figure 25) shows the bandwidth utilization (quantization parameter) value over a sixty-second time period.

Assuming a network LAN of 10 Mbps, having a T1 connection (1.55 Mbps), about 24% of the T1 connection was used by this news video clip. For the news content type, it was found that the typical user was most sensitive to reductions in the video encoding bit rate. In this scenario, the frame rate of the news content was significantly reduced without affecting the user's experience. It was gradually reduced to a limit of 8 fps. This limit preserved the number of bits per frame thereby preserving picture quality. The video encoding bit rate parameter was reduced to an acceptability threshold of 160 kbps. The network bandwidth was now significantly reduced as depicted in Figure 26.

Figure 25. Quantization parameter values for the news content type (without QoE management)

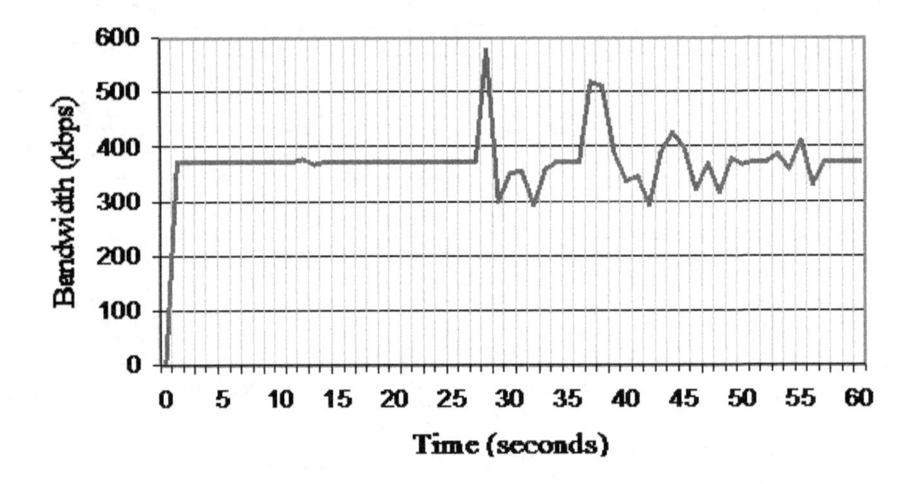

Figure 26. Comparison of the bandwidth utilized by the news content type (with and without QoE optimization)

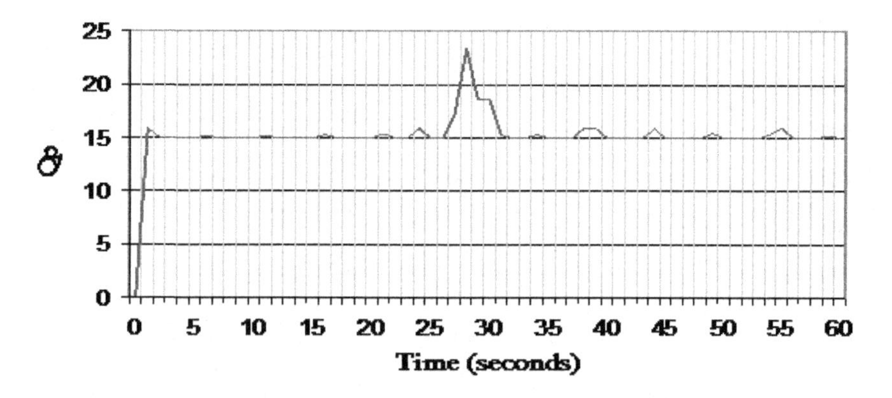

Figure 27. Comparison of the quantization parameter values for the news content type (with and without QoE optimization)

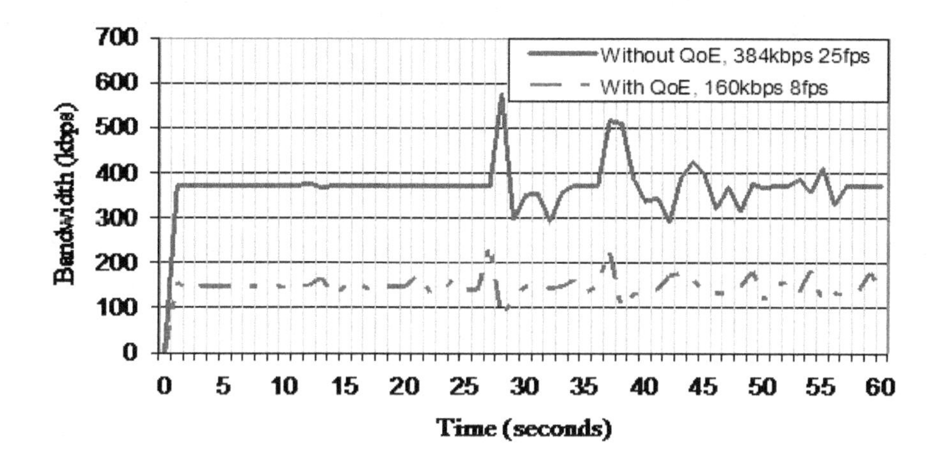

Figure 28. Comparisons of bandwidth utilized by the news content type for the case (without QoE optimization, and two cases with QoE optimization)

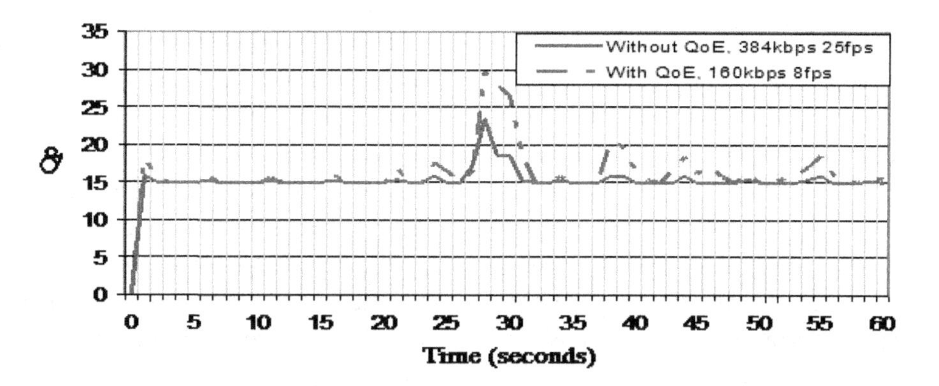

Figure 29. Comparison of the quantization parameter values for the news content type for the case (without QoE optimization, and two cases with QoE optimization)

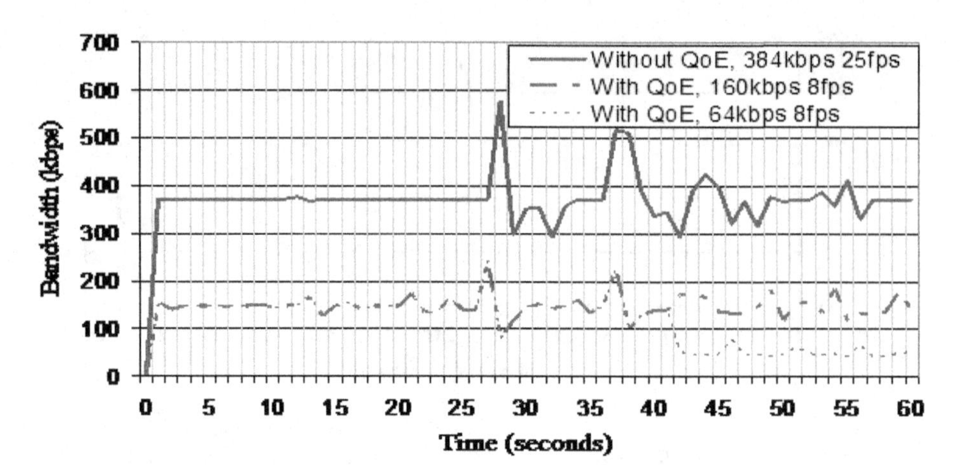

In Figure 27, it can be seen that when the frame rate was decreased, the number of bits in each frame increased thus improving the sharpness of the video clip. A comparison of the quantization parameter values for the "without QoE" and "with QoE" (Figure 28) cases, yielded more or less the same picture quality.

With QoE considerations, the news video clip used only 10% of the T1 connection, thus achieving approximately a 58% savings on network resources. When further decreases in bandwidth were required, the encoding video bit rate parameter was further degraded to a threshold of 64 kbps (see Figure 28). In this way, the video quality (Figure 29) was kept as high as possible within the available bandwidth whilst the user's QoE was hardly affected.

In this QoE approach, pricing could change depending on how QoE was maintained. For example, if a user were willing to accept greater degradations in QoE, the service would be cheaper or vice versa, thereby enabling user-centric service level agreements (SLAs). More importantly, policies and SLAs will directly target the user's experience, rather than the raw network parameters. Another advantage of the QoE approach is that service providers can selectively deliver specific levels of quality to specific market segments in such a way as to maximise their revenue.

FUTURE RESEARCH DIRECTIONS

The work in this Chapter has contributed towards the relatively new but growing discipline of QoE management in content delivery systems. It has done so by examining several issues such as users' requirements for video services on mobile terminals, and P2P streaming on PCs. The research has also led for the development of QoE prediction models. However, certain areas have been identified were more research is required.

Whilst investigating the users' requirements for mobile video services, the primary focus was on the measurement and quantification of the acceptability thresholds. The thresholds were specific to the codecs used in encoding the video contents. However, many codecs exists, which have various implementation algorithms. So the verification and estimation of these thresholds for other codec implementations are needed. The acceptability thresholds were affected by only the artefacts generated by the encoding process. As such, they did not account for the effects of the artefacts introduced during transmission. Further experiments could be performed in which the network parameters could be included, manipulated and mapped to the QoS/QoE layered model.

The coding method used in this Chapter was based on generic parameters such as video bit rates and frame rates. Further study could be carried out to investigate the user's sensitivity to quantization parameters and frame rates.

The processes in carrying out the QoE management framework in order to formulate the prediction models were manual and time consuming. It is vital in future studies to investigate the automation of the processes. For example, machine learning techniques seem the appropriate way to go, since they can enable dynamic rule-based restoration of QoE levels, using information from the real-time network conditions.

Future research could be extended to define classification algorithms for a video content's characteristics (complexity). Such an algorithm could start by examining the video's spatial and temporal domains. In the spatial domain, discrete wavelet transform (DWT) could be used to determine the amount of the spatial energy intensity that is contained in a video. Different videos will have different spatial energy intensities. This enables them to be categorised into groups such as low (L), medium (M) or high (H). In the temporal domain, motion vectors could be calculated to estimate the motion activity in a video. The values of the motion can also be categorised as low (L), medium (M) or high (H). The results obtained from the spatial and temporal algorithms could then be used to produce a two-dimensional classification of the video content characteristics. In this way, the model derived for a particular content genre will most probably improve.

Our initial study on QoE analysis for P2P TV streaming system was limited to music video content. It will be interesting to investigate the user experiences of different content types on the P2P streaming platform.

CONCLUSION

QoE analysis for P2P TV streaming system is an apt form of assessment for this up and coming method of multimedia delivery. In general, QoE analysis is a superset of QoS analysis that more comprehensively captures users' requirements. What is clear from this Chapter is that the same criteria for selecting content cannot be applied to mobile devices. The type of video shots must be tailored to smaller screens if not lower resolutions and motion should be restricted. News clips are tolerated because audio is more important in them than visual content. These are examples of the disconcerting results that the methods psychophysical testing and discriminant analysis reveal. This is a wide open area of which the Chapter has only initiated the study.

REFERENCES

Aldridge, R. P. (1996). *Continuous quality assessment of digitally-coded television pictures.* PhD Thesis, University of Essex, Colchester.

Apteker, R. T., Fisher, J. A., Kisimov, V. S., & Neishlos, H. (1995). Video acceptability and frame rate. *IEEE Transactions on Multimedia, 3*(3), 32–40.

Balaouras, P., & Stavrakakis, I. (2005). Multimedia transport protocols for wireless networks . In Salkintzis, A. S., & Passas, N. (Eds.), *Emerging wireless multimedia services and technologies.* John Wiley & Sons, Ltd.

Bouch, A., Sasse, M. A., & DeMeer, H. (2000). *Of packets and people: A user-centred approach to quality of service*. In Eighth International Workshop on Quality of Service.

Brun, P., Hauske, G., & Stockhammer, T. (2004). Subjective assessment of H.264/AVC video for low-bitrate multimedia messaging services. In *IEEE International Conference on Image Processing* (pp. 1145-1148).

Claypool, M., & Tanner, J. (1999). *The effects of jitter on the perceptual quality of video.* In 7th ACM International Conference on Multimedia.

comScore. (2007). *comScore study reveals that mobile TV currently most popular among males and younger age segments*. Press Release.

Detenber, B. H., & Reeves, B. (1996). A bio-informational theory of emotion: Motion and image size effects on viewers. *The Journal of Communication, 46*(3), 66–84. doi:10.1111/j.1460-2466.1996.tb01489.x

Diot, C., Levine, B. N., Lyles, B., Kassem, H., & Balensiefen, D. (2000). Deployment issues for the IP multicast service and architecture. *IEEE Network, 14*(1), 78–88. doi:10.1109/65.819174

Farrera, M. P. (2005). *Packet-by-packet analysis of video traffic dynamics on IP networks.* PhD Thesis, University of Essex, UK, Department of Electronic Systems Engineering.

Fechner, G. T. (1966). *Elements of psychophysics* (Adler, H. E., Trans.). Holt, Rinehart and Winston, Inc.

Fredendall, G. L., & Behrend, W. L. (1960). Picture quality - Procedures for evaluating subjective effects of interference. *SMPTE Journal, 48*, 1030–1034.

Gale, A. G. (1997). Human response to visual stimuli . In Hendee, W. R., & Wells, P. N. T. (Eds.), *The perception of visual information* (pp. 127–147). Berlin, Germany: Springer Verlag. doi:10.1007/978-1-4612-1836-4_5

Gescheider, G. A. (1997). *Psychophysics: The fundamentals*. Lawrence Erlbaum Associates, Inc.

Ghanbari, M. (2003). *Standard codecs: Image compression to advanced video coding*. Stevenage, UK: Institution of Electrical Engineers

Ghinea, G. Thomas, J. P., & Fish, R. S. (1999). *Multimedia, network protocols and users - Bridging the gap.* In ACM Multimedia.

Gilliam, D. (2006). *The appeal of mobile video: Reading between the lines*. Retrieved from http://www.tdgresearch.com/tdg_opinions_the_appeal_of_mobile_ video.htm

Goodchild, J. (2005). Integrating voice, data and video . In *IP video implementation and planning guide*. United States Telecom Association.

Hands, D. (1997). *Mental processes in the evaluation of digitally-coded television pictures.* PhD Thesis. University of Essex, Colchester.

Hands, D. S. (2004). A basic multimedia quality model. *IEEE Transactions on Multimedia, 6*(6), 806–816. doi:10.1109/TMM.2004.837233

Hefeeda, M., Habib, A., Botev, B., Xu, D., & Bhargava, B. (2003). *PROMISE: Peer-to-peer media streaming using CollectCast*. In 11th ACM International Conference on Multimedia.

Hei, X., Liang, C., Liang, J., Liu, Y., & Ross, K. W. (2007). A measurement study of a large-scale P2P IPTV system. *IEEE Transactions on Multimedia, 9*(8), 1672–1687. doi:10.1109/TMM.2007.907451

Hong, D. W., & Hong, C. S. (2003). A QoS management framework for distributed multimedia systems. *International Journal of Network Management, 13*(2), 115–127. doi:10.1002/nem.465

ISO 9241-11. (1998). *Guidance on usability. Ergonomic requirements for office work with visual display terminals (VDTs) - Part 11.*

Ito, R. M., Tasaka, S., & Fukuta, Y. (2004). *Psychometric analysis of the effect of end-to-end delay on user-level QoS in live audio-video transmission.* In IEEE International Conference on Communications.

ITU-R Recommendation BS.1116. (1997). *Methods for the subjective assessment of small impairments in audio systems including multichannel sound systems.*

ITU-R Recommendation BS.1679. (2004). *Subjective assessment of the quality of audio in large screen digital imagery applications intended for presentation in a theatrical environment.*

ITU-T Recommendation E.800. (1994). *Terms and definitions related to quality of service and network performance including dependability.*

ITU-T Recommendation P.911. (1998). *Subjective audiovisual quality assessment methods for multimedia applications.*

ITU-T Recommendation P.910. (1999). *Subjective video quality assessment methods for multimedia applications.*

ITU-T Study Group 12. (2007). *Definition of quality of experience (QoE).*

Joly, A., Nathalie, M., & Marcel, B. (2001). *Audio-visual quality and interactions between television audio and video.* In International Symposium on Signal Processing and its Applications.

Jones, B. L., & McManus, P. R. (1986). Graphic scaling of qualitative terms. *SMPTE Journal, 95*(11), 1166–1171. doi:10.5594/J04083

Jumisko-Pyykkö, S., Vadakital, V. K. M., Liinasuo, M., & Hannuksela, M. M. (2006). *Acceptance of audiovisual quality in erroneous television sequences over a DVB-H channel.* In Workshop in Video Processing and Quality Metrics for Consumer Electronics.

Kato, S., Boon, C. S., Fujibayashi, A., Hangai, S., & Hamamoto, T. (2005). *Perceptual quality of motion of video sequences on mobile terminals.* In the 7th IASTED International Conference on Signal and Image Processing.

Klecka, W. R. (1980). *Discriminant analysis.* Beverley Hills, CA: Sage Publications.

Knoche, H., McCarthy, J., & Sasse, M. (2006). How low can you go? The effect of low resolutions on shot types in mobile TV. *Journal of Multimedia Tools and Applications Series, 36*(1-2), 145–166. doi:10.1007/s11042-006-0076-5

Knoche, H., & McCarthy, J. D. (2005). *Good news for mobile TV.* In the Wireless World Research Forum 14.

Knoche, H., McCarthy, J. D., & Sasse, M. A. (2005). *Can small be beautiful? Assessing image size requirements for mobile TV.* In the 13th ACM International Conference on Multimedia.

Knoche, H., & Sasse, M. A. (2008). Getting the big picture on small screens: Quality of experience in mobile TV . In Ahmad, A. M. A., & Ibrahim, I. K. (Eds.), *Multimedia transcoding in mobile and wireless networks* (pp. 31–46). Information Science Reference. doi:10.4018/978-1-59904-984-7.ch003

Kouvelas, I., Hardman, V., & Watson, A. (1996). *Lip synchronisation for use over the Internet: Analysis and implementation.* In IEEE Global Telecommunications Conference.

Kuang, T., & Williamson, C. (2001). *RealMedia streaming performance on an IEEE 802.11b wireless LAN.* In IASTED Wireless and Optical Communications.

Levine, B. N., & Shefner, J. M. (2000). *Fundamentals of sensation and perception.* Oxford University Press.

Li, B., & Yin, H. (2007). Peer-to-peer live streaming on the Internet: Issues, existing approaches, and challenges. *IEEE Communications Magazine, 45*(6), 94–99. doi:10.1109/MCOM.2007.374425

Lloyd, E., Maclean, R., & Stirling, A. (2006). *Mobile TV- Results from the BT Movio DAB-IP pilot in London.* EBU Technical Review. Retrieved from http://www.ebu.ch/en/technical/trev/trev_frameset-index.html

Lombard, M., Ditton, T. B., Grabe, M. E., & Reich, R. D. (1997). The role of screen size in viewer responses to television fare. *Communication Reports*, *10*(1), 95–106. doi:10.1080/08934219709367663

López, D., Gonźalez, F., Bellido, L., & Alonso, A. (2006). Adaptive multimedia streaming over IP based on customer oriented metrics. In *IEEE International Symposium on Computer Networks* (pp. 185-191).

Lund, A. (1993). The influence of video image size and resolution on viewing-distance preferences. *SMPTE Journal*, *102*, 406–415. doi:10.5594/J15915

Maccarthaigh, M. (2007). *Joost network architecture*. (PowerPoint slides available online.)

Marfia, G., Pau, G., Rico, P., & Gerla, M. (2007). P2P streaming systems: A survey and experiments. *ST Journal of Research.*

McCarthy, J., Sasse, M. A., & Miras, D. (2004). *Sharp or smooth? Comparing the effects of quantization vs. frame rate for streamed video.* In SIGCHI Conference on Human Factors in Computing Systems.

Moore, M. S., Mitra, S. K., & Foley, J. M. (2002). *Defect visibility and content importance implications for the design of an objective video fidelity metric.* In International Conference on Image Processing.

Moreira, J., Antonello, R., Fernandes, S., Kamienski, C., & Sadok, D. (2008). A step towards understanding Joost IPTV. In *Networks Operations and Management Operations* (pp. 211-214).

MPEG-4 AVC Verification Tests. (2011). Report on the formal verification tests on AVC (ISO/IEC 14496-10 | ITU-T Rec. H.264). Retrieved from http://www.chiariglione.org/mpeg/quality_tests.htm

Murphy, S., Searles, M., Rambeau, C., & Murphy, L. (2004). *Evaluating the impact of network performance on video streaming quality for categorised video content.* In International Packet Video Workshop.

Narita, N. (1993). Graphic scaling and validity of Japanese descriptive terms used in subjective-evaluation tests. *SMPTE Journal*, *102*, 616–622. doi:10.5594/J03770

Nokia. (2004). *Quality of experience (QoE) of mobile services: Can it be measured and improved?* Whitepaper.

Nokia. (2005). *Finnish mobile TV: Pilot results.* Retrieved from http://www.mobiletv.nokia.com/download_counter.php?file=/onAir/finland/files/RI_Press.pdf

Olista. (2007). *Live trials by Olista with European mobile operators demonstrate common barriers for mobile data services.* Press Release 120207-1.

Patrick, A. S., Singer, J., Corrie, B., Nöel, S., Khatib, K., Emond, B., et al. (2004). A QoE sensitive architecture for advanced collaborative environments. In the *First International Conference on Quality of Service in Heterogeneous Wired/Wireless Networks* (pp. 319-322).

Qiu, D., & Srikant, R. (2004). Modeling and performance analysis of BitTorrent-like peer-to-peer networks. *SIGCOMM*, *34*(4), 367–378. doi:10.1145/1030194.1015508

Raatikainen, P. (2005). On developing networking technologies and pervasive services. *Wireless Personal Communications*, *33*, 261–269. doi:10.1007/s11277-005-0571-4

Rainey, S., Petty, G., & Cutten, M. W. (2005). Content acquisition challenges . In *IP Video Implementation and Planning Guide* (pp. 43–61). United States Telecom Association.

ITU-R Recommendation BT500-11. (2002). *Methodology for the subjective assessment of the quality of television pictures.*

Reeves, B., & Nass, C. (1998). *The media equation: How people treat computers, television, and new media like real people and places*. Chicago, IL: University of Chicago Press.

Robert, E. M. (1986). On judging quality by price: Price dependent expectations, not price dependent preferences. *Southern Economic Journal, 52*(3), 665–672. doi:10.2307/1059265

Sadka, A. H. (2002). *Compressed video communications*. Chichester, UK: John Wiley & Sons, Ltd. doi:10.1002/0470846712

Satamoto, K., Aoyama, S., Asahara, S., Yamashita, K., & Okada, A. (2008). Lecture Notes in Computer Science: *Vol. 5068. Relationship between viewing distance and visual fatigue in relation to feeling of involvement* (pp. 232–239). Berlin, Germany: Springer.

Siller, M. (2006). *An agent-based platform to map quality of service to experience in active and conventional networks.* Ph.D Thesis, University of Essex, Colchester, U.K.

Sodergard, C. (2003). *Mobile television - Technology and user experiences report on the mobile-TV project.* VTT Information Technology.

Soldani, D., Li, M., & Cuny, R. (2006). *QoS and QoE management in UMTS cellular systems*. Chichester, UK: Wiley and Sons. doi:10.1002/9780470034057

Steinmetz, R. (1996). Human perception of jitter and media synchronization. *IEEE Journal on Selected Areas in Communications, 14*(1), 61–72. doi:10.1109/49.481694

Sugawara, M., Mitani, K., Kanazawa, M., Okano, F., & Nishida, Y. (2006). Future prospects of HDTV-Technical trends towards 1080p. *SMPTE Motion Imaging Journal, 115*(1), 10–15. doi:10.5594/J11496

Tang, Y., Luo, J. G., Zhang, Q., Zhang, M., & Yang, S. Q. (2007). Deploying P2P networks for large-scale live video-streaming service. *IEEE Communications Magazine, 45*(6), 100–106. doi:10.1109/MCOM.2007.374426

Teunissen, K. (1996). The validity of CCIR quality indicators along a graphical scale. *SMPTE Journal, 105*(1), 144–149. doi:10.5594/J04650

UMTS Forum report 11. (2000). *Enabling UMTS third generation services and application.*

VQEG Final Report from the Video Quality Experts Group on the Validation of Objective Models of Video Quality Assessment. Retrieved from http://www.vqeg.org

Watson, A., & Sasse, M. A. (1998). *Measuring perceived quality of speech and video in multimedia conferencing applications.* In ACM Multimedia Conference.

Webster, A. A., Jones, C. T., Pinson, M. H., Voran, S. D., & Wolf, S. (1993). An objective video quality assessment system based on human perception . In *SPIE Human Vision.* Visual Processing and Digital Display. doi:10.1117/12.152700

Winkler, S., & Dufaux, F. (2003). *Video quality evaluation for mobile applications.* In SPIE Visual Communications and Image Processing.

Winkler, S., & Faller, C. (2005). Maximizing audiovisual quality at low bitrates. In *Proceedings of Workshop on Video Processing and Quality Metrics.*

Zapater, M. N., & Bressan, N. (2007). *A proposed approach for quality of experience assurance for IPTV.* In First International Conference on Digital Society.

Zhang, X., Liu, J., Li, B., & Yum, T. P. (2005). CoolStreaming/DONet: A data-driven overlay networks for peer-to-peer live media streaming. In *IEEE INFOCOM* (pp. 2102-2111).

ADDITIONAL READING

Agboma, F., & Liotta, A. (2007). Addressing user expectations in mobile content delivery. *Journal of Mobile Information Systems*, *3*, 153–163.

Armitage, G. (2000). *Quality of service in IP networks: Foundations for a multi-service internet*. Macmillan Technical Publishing.

Bai, Y., & Ito, R. M. (2004). QoS control for video and audio communication in conventional and active networks: Approaches and comparison. *IEEE Communications Surveys*, *6*(1), 42–49. doi:10.1109/COMST.2004.5342233

Benham, D. (2005). Video service to network linkages . In *IP Implementation and Planning Guide*. United States Telecom Association USA.

Bharrathsingh, K. (2005). Quality of experience as an integral part of network engineering. *Nortel Technical Journal,* 31-35.

Callet, L. P., Viard-Gaudin, C., & Péchard, S. (2006). No reference and reduced reference video quality metrics for end to end QoS monitoring. *IEICE Transactions on Communications*, *89*(3), 289–296. doi:10.1093/ietcom/e89-b.2.289

De Vleeschauwer, B., Simoens, P., Van de Meersssche, W., Latré, S., De Turck, F., Dhoedt, B., et al. (2007). *Autonomic QoE optimization in the access node knowledge plane*. Paper presented at the Broadband Europe Conference.

Jurca, D., Chakareski, J., Wagner, J. P., & Frossard, P. (2007). Enabling adaptive video streaming in P2P systems. *IEEE Communications Magazine*, *45*(6), 108–114. doi:10.1109/MCOM.2007.374427

Lombard, M. (1995). Direct responses to people on the screen: Television and personal space. *Communication Research*, *22*(3), 288–324. doi:10.1177/009365095022003002

Perkis, A., Munkeby, S., & Hillestad, O. I. (2006). A model for measuring quality of experience. In *Proceedings of the 7th Nordic Signal Processing Symposium*.

Robins, M. (2006). *Delivering optimal quality of experience (QoE) for IPTV services*. Spirent Communications, Whitepaper.

Sun, H., Lin, Y., & Shu, L. (2008). The impact of varying frame rates and bit rates on perceived quality of low/high motion sequences with smooth/complex texture. *Journal of Multimedia Systems*, *14*(1), 1–13. doi:10.1007/s00530-007-0101-1

KEY TERMS AND DEFINITIONS

Method of Limits: Is a technique used in psychophysics, in which either a sensation is increased or decreased until it becomes just perceptible. Psychophysics is a branch of psychology that uses such quantitative methods to determine sensations.

MPEG-4 AVC: Otherwise known as H.264/AVC (Advanced Video Coding) standard is the most recent video codec standard, produced jointly by MPEG and the ITU-T (hence the alternative names). MPEG-4 AVC is approximately 50% more efficient in coding Standard Definition TV, which it mainly achieves by the variety of different motion compensation modes available, together with enhanced sub-pixel motion detection.

Quality of Experience: Is a subjective measure of the user experience. In video streaming it has been found that Quality of Service, which is an objective measure of packet loss and latencies, may be a poor indicator of a user's experience. Factors that can influence Quality of Experience (QoE or QoX) are display device type, quality expectations, mobility, and effect of visual artifacts.

Sobel Filter: Is a method in image processing of detecting edges by convolving a small spatial filter with an image. By applying the filter in separable fashion in the vertical and horizontal directions, computational savings are made. The filter works by finding the variation of gradient within an image, thus indicating where abrupt changes in image intensity (edges) have occurred.

VQEG: Video Quality Experts Group is a long established (since 1997) organization that has developed methods for assessing video quality.

ENDNOTES

1. One-time fee is described in Sodergard (2003) as a direct fee that is payable to the service provider when assessing a new service for the first time or allows a lifetime access to the service. A one-time fee can also be an indirect fee hidden in the price for new hardware and, thus, sometimes invisible to the customer.
2. In Ad-based (Sodergard, 2003), the cost of using the service is purely financed by sponsors or advertisers, if the user is willingly to receive advertisements on their mobile devices.
3. The actual value of which in a real scenario with varying quality will lie between the bounds.
4. Since Joost is the originator of its content, the control of the AQoS parameters is not permitted.
5. Low Qp values result in high quality for each encoded frame, but also in high bit rates and vice versa (Brun, *et al.*, 2004).

Chapter 10
Reliable Multicast Streaming

Javier Gálvez Guerrero
i2CAT Foundation, Spain

ABSTRACT

Video streaming is becoming one of the most important services deployed over telecommunication networks such as the Internet and triple-play operators' networks. This service differs from the rest in being loss sensitive and highly delay sensitive and requiring a considerable amount of bandwidth in order to offer a smooth transmission of packets through the network. While upgrading network elements with quality of service and multicast transmission capabilities becomes prohibitive for most network operators, peer-to-peer (P2P) architectures appear to be smart and efficient solutions to the previous issues. Many different P2P systems have been proposed and deployed to offer reliable video streaming services. These approaches address issues such as multicast transmission, quality of service enablement, mobility robustness, and video distribution according to network and user device capabilities. This chapter gives an overview of the different issues related to performance and reliability in multicast streaming over wireless networks and presents several alternatives facing them, including amendments to the already existing multicast mechanism of data distribution, video scalability and how peer-to-peer networking can provide a cost-effective solution to such problems.

INTRODUCTION

Media streaming services over IP networks such as videoconference and video-on-demand have enormously increased in popularity in the last decade. Providers of such services find multicast an enormously advantageous system for distributing video and audio data to their customers as it can increase the bandwidth occupation efficiency, although its deployment implies great investments due to the need for updates to legacy hardware.

DOI: 10.4018/978-1-4666-1613-4.ch010

In addition, IEEE 802.11 and 3G networks have spread and evolved in such a way that they can offer access to these services through the widely available wireless infrastructure.

However, multicast transmissions are prone to considerably diminish the quality-of-experience of such services when deployed in wireless environments. It is widely known that wireless networks are characterized by signal strength fluctuation and vulnerability to interference. Moreover, in multicast sessions, the lack of effective proactive and reactive mechanisms intended to avoid transmission errors causes a noticeably poor performance in media streaming applications, which are quite sensitive to packet loss. Additionally, currently available communication devices have many different capabilities regarding network connectivity, display size and resolution, power and battery durability, data storage and processing features, thus, producing a highly heterogeneous network.

In order to alleviate such problems and provide a suitable *quality-of-service (QoS)* to all users, several video and audio compression algorithms and network protocols improvements have been developed. Thus, QoS in such services can be achieved by considering many factors, such as bandwidth management, multicast traffic acknowledgment, video scalability and peer-to-peer networking, amongst others.

Basic solutions to overcome multicast performance issues include packet duplication, usage of different media codecs and copying multicast traffic into unicast streams targeted to client devices. However, such solutions do not scale well or do not perform as expected in wireless environments, so specific approaches have to be considered. Such solutions include implementations of leader-based acknowledgment protocols (Li & Herfet, 2008) (Ding et al., 2004) (Miroll et al., 2010), NACK-based protocols such as *NACK Oriented Reliable Multicast (NORM)* (NORM) and more robust transmissions of data with additional *Forward Error Correction (FEC)* protection like those included in the *File Delivery over Unidirectional Transport (FLUTE) protocol* (Paila et al., 2004).

Furthermore, scalable video provides a number of benefits in terms of reliability, these mainly related to the heterogeneity of networks and devices where video services are being deployed. With *Scalable Video Coding* (SVC) and *Multiple Description Coding* (MDC), content can be offered to a huge amount of receivers with different computational, display and network access capabilities, while these can choose which enhancement layers receive and process them in order to play the video. Thus, SVC and MDC can be used to deploy new QoS-enabled video services to a many different devices.

As previously stated, multicast distribution of media is considered a scalable technique when compared with unicast and simulcast methods. Despite this, current networks do no fully support multicast transport given how expensive the network devices upgrade is. This situation has made network operators search for a cheaper, more scalable and reliable network transmission architecture without the drawbacks of multicast systems. As a result, the first considered approach was *Content Delivery Networking (CDN)*, which reduces the probability of network congestion in streaming services by distributing contents through different servers along the network, then placing the source servers closer to the end users. However, this solution faces a number of problems. CDNs are not scalable since a large number of simultaneous connection requests can lead to servers' overload. Additionally, CDNs are based in the client-server model, so if the link between the two end points is broken there is no alternative to receive the video streaming data. Thus, deploying a reliable CDN platform for video streaming services is still prohibitive for most network operators and does not provide with important advantages when compared with multicast techniques.

On the other hand, *peer-to-peer* (P2P) networks can be implemented to build an application level overlay network where user devices (peers) behave both as client, server and routing elements, then relying to a great extent in user device capabilities instead of network ones to stream data. In order to provide end devices with different video quality and requirement alternatives, P2P networks can build different multicast trees, each transporting a SVC layer or a MDC description. Otherwise, the overlay network can be built without defining any structure, so video data is streamed when requested by peers. Both techniques allow saving resources with regard to unicast and simulcast transmissions without the need of upgrading network devices with multicast capabilities. As a result, merging video splitting coding techniques such as SVC and MDC with multicast P2P overlay networking, video streaming services could be offered in heterogeneous networks while guaranteeing reliability, device adaptability and scalability.

Then, this chapter aims at analyzing the main advantages and drawbacks of multicast streaming in wireless scenarios and introduces some approaches considered when trying to provide reliability, focusing on multicast distribution of scalable video over peer-to-peer wireless networks. The next section focuses on spotting all performance issues related to multimedia streaming when the service is offered over a wireless network. After giving some examples of coarse solutions, approaches specifically designed to overcome multicast reliability problems are introduced. Next, coding, transport and signaling techniques for scalable video are presented as an alternative to fulfill reliable transport of multicast video streaming requirements. Then, peer-to-peer networking combined with scalable video is shown to be the best solution for current architectures, introducing some state-of-the-art implementations. Finally, the future of reliable multicast streaming is tried to be depicted bearing in mind the foundations analyzed through the chapter.

PERFORMANCE ISSUES

Network Conditions

IEEE 802.11 devices have spread worldwide thanks to their simplicity and usefulness. Different standard specifications of the wireless access technology give many options of capacity, range and reliability. While IEEE 802.11 a and g specifications can provide up to 54 Mbps capacity at 2,4 and 5 Ghz frequency bands respectively with a maximum range of 100 meters using *Orthogonal Frequency Division Multiplexing (OFDM)* modulation, IEEE 802.11 b specification reaches 11 Mbps capacity at 2,4 frequency band with *Complement Code Keying (CCK)* modulations, less robust than OFDM..

However, these maximum rates and ranges are reduced in practice due to many causes. Wireless access technologies are known to be vulnerable to interference, signal fluctuation and multipath propagation, arising data loss and transmission errors (Rappaport, 2002). These losses can be characterized by two different patterns: burst and random errors, which have different effects in service performance. IEEE 802.11 a and g specifications include FEC mechanisms, aimed at reducing performance degradation caused by random errors. Despite that, these proactive techniques are based upon information redundancy and, along with packet encapsulation headers and the radio channel management, causes the maximum user application level capacity to drop to 32 Mbps.

The recent IEEE 802.11 n specification includes some enhancements like *Multiple Input Multiple Output (MIMO)* technology, which benefits from multi-path propagation, a higher number of subcarriers to be used, beamforming and traffic prioritization thanks to an IEEE 802.11e based extension. These improvements offer a wireless access technology with up to 100 Mbps capacity, more reliability and a wider range for connected devices.

In addition, differences between unicast and multicast transmissions must be mentioned as they are quiet relevant in streaming services. While unicast sessions are established between an origin node and a destination node, multicast ones imply one origin node creating a single stream that reaches all nodes in a multicast destination group. Packets sent in unicast sessions are acknowledged, but those sent in multicast sessions are not, so acknowledgment flooding and 'crying baby' effects are avoided. Then, in IEEE 802.11 networks, where there may appear many transmission errors, lost multicast packets are never retransmitted. Thus, backoff intervals due to collided packets are never implemented in such sessions, which may cause channel access monopolization and parallel unicast sessions to be indefinitely blocked.

Digital Video Compression

Video digitalization diminishes the transmission error vulnerability. However, digital compression algorithms need to be used in order to reduce the network capacity requirements for video data. These techniques are based on visual redundancy and imperceptible elements suppression, reaching efficient compression levels. Thus, different video frame types are defined: I-type frames completely define a video frame while P- and B-type ones only contain differential information regarding previous and future frames, respectively, thus requiring less capacity.

Despite the advantages of digital video compression, these techniques introduce frame dependency, which may imply different problems in case of frame loss. While differential frame loss leads to spurious pixel blocks in the image recovery process, loss of reference I-type frames affects the whole *Group of Pictures (GOP)*, implying an important impact on video quality. The widely known MPEG-1 video codec establishes that, when using the common IBBPBBPBBPBB GOP scheme, a 1% *Packet Error Rate (PER)* can lead to 10% of corrupt frames. Newer codecs, such as MPEG-4 reduce loss impact in video quality, although they do not completely eradicate it.

These losses are directly related to the patterns previously mentioned. As it is shown in Figure 1, random errors have a higher probability of affecting a greater number of frames, while burst errors affect consecutive packets, thus fewer frames. As a consequence of this and due to the fact that MPEG-4 algorithms try to reduce the effects of burst errors, random ones are more likely to cause video quality degradation problems in this scenario.

Multimedia Streaming

Services such as web browsing and file transport are offered over IP networks using TCP. This protocol guarantees an error-free packet transmission by means of data retransmission and congestion

Figure 1. Error patterns in wireless channels

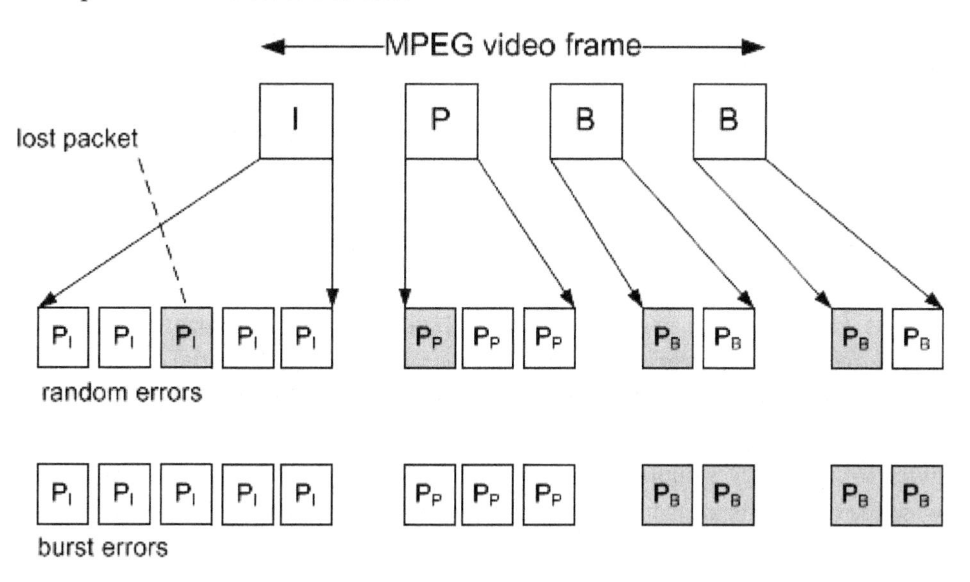

control algorithms. However, retransmissions lead to delays in packet reception, especially in wireless networks where errors are more common. Streaming services prioritize the transmission continuity against packet loss, so the UDP protocol is widely used instead of TCP for such applications.

Additionally, in order to properly process the UDP packets stream, *Realtime Transport Protocol (RTP)* (Schulzrinne et al., 2003) and *Realtime Transport Control Protocol (RTCP)* (Schulzrinne et al., 2003) protocols are used. While RTP eases the packet reordering through timestamps and sequence numbers, RTCP establishes a streaming session control information exchange between the origin and destination nodes. The information periodically exchanged includes number of lost packets, jitter and other network condition parameters through which the server nodes can adapt the streaming session for specific quality of service thresholds. In unicast transmissions, *Real Time Streaming Protocol (RTSP)* (Schulzrinne et al., 1998) is also used to establish video streaming sessions and remote server control.

Coarse Solutions

As said before, multicast data distribution faces random and burst errors in wireless environments. In order to reduce the impact that such loss of data can produce, a very basic approach is to deploy packet duplication. As the name suggests, this proposal consists in create an exact copy of each packet to be streamed and perform its transmission to destination. Combinations of packet ordering and transmission delay at the sender side can be implemented depending on the network conditions, thus adapting to the requirements of specific scenarios. However, such an approach is designed to only face random errors, being unable to protect the streaming session from burst errors. Moreover, this solution produces a huge overhead, given every packet is duplicated, thus requiring twice the original bandwidth.

Regarding media codecs, they have been always mainly focused on enhancing the compression efficiency while preserving original fidelity as high as possible. However, the newest techniques such as those included in the MPEG-4 codec take into account resiliency with different design approaches. First, such a codec introduces the new concept of object coding, which disrupts the general vision of having a reference frame and many differential frames which include just the relative changes. With object coding the different elements of a video sequence can be split into their respective differential vectors. The first advantage of such mechanism is that while in standard coding losing a frame means losing all information regarding it, with object coding only information regarding a specific object of the frame may be lost. As a result, instead of having artifacts in the frame or in specific macroblocks, with object coding these artifacts can be localized while all other objects can be properly decoded. However, the overall video quality depends on how much data has been lost.

Aside of such dependency reduction, MPEG-4 explicitly increases the robustness in error prone environments by means of three techniques: resynchronization, data recovery and error concealment. Classic resynchronization, based on spatial resynchronization, is performed by inserting markers at specific places in the bit stream to tag group of blocks. Such an approach allows resynchronizing the bit stream in case of data loss when these markers are received. However, as the markers are included at the beginning of groups of blocks it is dependent on the variable encoding rate, thus being unevenly spread through the stream. MPEG-4 performs a uniform distribution of resynchronization markers along the stream. Additionally, the packet header information included by the MPEG-4 codec helps in the task of recovering the stream synchronization and increases the probabilities of data concealment. Data recovery can be performed once the stream has been resynchronized. MPEG-4 does it by means of *Reversible Variable Length Codes (RVLC)* (MPEG, 2004), which allow recovering loss data by reading variable length codewords in forward and reverse directions, thus increasing the robustness of the bit stream. Finally, error concealment includes different techniques to recover from located errors, pointed to by the resynchronization markers and the data included in the packet headers. Some of these techniques include simple block replication from previous or current frames, while splitting texture, motion data and vectors.

All these improvements of resilient streaming of video have been proved not to be enough for wireless environments when transmitting multicast contents. While unicast does not really need the mechanisms presented by MPEG-4 due to the availability of retransmissions, data loss recovery in multicast streaming is not fully accomplished. The main cause for such failure is the unpredictable behavior of wireless network conditions. Neither the combination of MPEG-4 with adaptive 802.11coverage can completely eradicate such an issue.

Another classic approach when trying to provide robust streaming appears with bandwidth management. Again, this is not a solution especially targeted to multicast distribution of data, but it is included as a use case. Traffic balancing and resource reservation are the most widely spread alternatives. However, although continuous updates on multicast trees of traffic balancing approaches and proper resource reservation through the *Resource Reservation Protocol (RSVP)* (Braden et al., 1997) can help in increasing the bandwidth efficiency and avoiding network congestion, they have also shown no remarkable results in wireless multicast streaming when a high level of robustness becomes a requirement. The reason is quite simple: such techniques are not applicable to end devices and the local area network where they are connected to, but are deployed in the backbone and are specific for increasing the speed and guarantee the success of transmission on such core wired network.

As just noted, multicast streaming does not suffer from considerable losses in a distribution network if it is wired and a number of protocols are deployed to provide reliability. It is on the wireless access network where the problems arise. Given that packet duplication and usage of advanced media codecs do not perform as expected, another basic approach has been implemented by some vendors in their own devices, thus corresponding to private solutions, although sharing the same concept. Such a concept relies on a "translation" of multicast streams into multiple unicast streams in the wireless network, one per end device waiting for the stream to be received. This "translation" allows the multicast traffic being acknowledged due to the inherent unicast management in wireless networks, so it behaves in the same way as having multiple independent unicast sessions. Obviously, such approach is not scalable and only works in scenarios where a small amount of clients are expected to concurrently be receiving multicast traffic.

MULTICAST RELIABILITY

Overview

Current standard MAC layer protocols do not provide multicast and broadcast transport of data with any reliability mechanism. As a result different approaches have been proposed at different levels, considering link level and application layer amendments. In the application layer, the classic *Automatic Repeat Request (ARQ)* has been tested along with FEC error control schemes. However, such mechanisms usually introduce delays that real-time services cannot cope with.

MAC level proposals to provide reliable multicast and broadcast communication have always been focused on the idea of acknowledging received data as unicast sessions do. In order to avoid the effect of acknowledgement packets flooding and 'crying baby' effects, the specific approach for multicast sessions is based on the selection of a leader of the group, which is committed to the task of managing retransmission queries when required.

Finally, as neither of the previous approaches have finally become a standard or been taken into account in any known application or service deployment, the IEEE organization recently created a task group in charge of defining an amendment to the IEEE 802.11 standard which should consider reliability and robust transmission of real-time data in multicast scenarios. Such amendment corresponds to IEEE 802.11aa.

FEC-Based Solutions at the Application Layer

FEC mechanisms are based in providing redundancy to data before being transmitted. With such an approach, an (n,k) block erasure code, using Reed-Solomon coding for example, is produced, which means that the data of k source packets is mingled into n packets (being n > k, and n – k the so-called parity packets) where redundancy has been introduced. While all n packets are transmitted, receiving whichever k of them guarantees the correct recovery of the original data, thanks to the spread redundancy.

The ARQ mechanism has been tested in combination with FEC. Such approach is known as *Hybrid Error Correction (HEC)*. The behavior of HEC is very simple: when the number of received FEC-coded data packets is not enough to recover the original data, the ARQ component issues a retransmission request to the sender so parity packets can be sent again. The amount of packets retransmitted can be configured and setup by the protocol itselfThe combination of FEC and ARQ mechanisms in HEC approaches has been shown to increase the efficiency of retransmissions and can provide a certain level of reliability at the cost of a higher overhead (Li & Herfet, 2009).

FEC redundancy can be really useful in wireless scenarios where a high percentage of lost packets is due to isolated random errors, given parity can be used to recover the lost data. However, higher bandwidth is required, as more data has to be sent and such overhead increases with the reliability the service has to be provided with. Despite this, the *Multimedia Broadcast/Multicast Service (MBMS)* (3GPP, 2010), where low quality video and audio services are provided through 3G/3.5G networks, uses a robust and efficient FEC protection both in multimedia download (by means of the FLUTE protocol) and in multimedia streaming, achieving satisfactory results. However, providing high quality multimedia with the same mechanism over such networks would require more bandwidth because of the high redundancy introduced in order to provide a reliable service.

Leader-Based Acknowledgement Protocols

As previously stated, when packets are lost in multicast transmissions they cannot be recovered by native link layer procedures, as unicast sessions do in the same scenario. In the previous section it has been shown how ARQ has been combined with FEC coding, thus providing application level retransmissions when required. However, retransmissions at such a level may cause the data to get delayed so much that it can become useless when eventually received. Link level retransmissions would be far more ideal.

In order to manage multicast reliability at the link level different proposals based on the selection of a manager node in the multicast group have emerged. The basic idea is quite simple: when data are lost, instead of every node generating a retransmission request, only the manager passively claims the packet to be resent. Different variations of this main idea exist.

The so-called *Leader-Based Protocol (LBP)* (Kuri & Kaseri, 1999) is the simplest approach, where the leader sends an ACK packet to acknowledge the successful reception of each data packet. If any error occurs, no ACK packet is sent by the leader, and the sender, after waiting a timeout period, retries by sending again the last data packet. Non-leader nodes, when successfully receiving data packets do not perform any action, but when they receive a packet with errors send a negative ACK (NACK) packet which is intended to collide with the ACK packet issued by the leader, in case it received the data packet successfully. The ACK/NACK jam is known as JACK and produces the retransmission of the data packet, as the sender is not able to decipher any ACK message. This mechanism has two problems: firstly, if the entire data packet is lost, the non-leaders cannot issue a NACK as they are unable to guess when to do it in order to guarantee a collision with the leader's ACK in case it exists. Secondly, a packet will not be considered as successfully sent until every node in the multicast group receives it at the same time. This means that, in case some non-leaders properly receive the data packet but some others do not, the retransmission will be claimed, and both nodes, those which did not receive correctly the data packet and those that did receive it will be expecting the retransmitted packet to be error-free again, which can significantly delay the transmission of the next packet in the sequence.

In order to avoid the previous issues of the LBP protocol, it was enhanced with a simple upgrade and it was renamed as *Beacon-driven Leader Based Protocol (BLBP)* (Li & Herfet, 2008). The performed upgrade was the addition of a beacon frame issued prior to the data transmission which included control data used for ACK/NACK synchronization and avoiding claiming for retransmission of already received data. As a result, BLBP adds reliability to multicast and broadcast sessions where there are a small group of receivers, so managing the data reception becomes lightweight. If the group increases its size, as this proposal is not scalable, considering single or multiple leaders for all receivers would not perform properly, due to problems on distributed synchronization and, probably, massive control traffic flooding.

It must be noted that using negative acknowledgements was previously used by another proposal called NORM transport protocol, an initiative of the Naval Research Laboratory of the United States. Such a protocol was specifically designed to provide reliable end-to-end transport of data in heterogeneous IP networks. Although the proposal has several options and is moderately complex, its basis is to avoid acknowledging every received packet in a multicast session and just sending NACKs for those that are received with errors or not received at all, somewhat resembling the LBP approach. Different options in the protocol definition allow avoiding control messages flooding and a bunch of more efficient alternatives exist in which responsibilities are widely distributed among the different members of the multicast group. In the end, it can be concluded that the different proposals of LBP-based protocols are just particular cases of the NORM protocol, which has been considered by the IETF to be periodically revised and supported.

IEEE 802.11aa Initiative

The previous approaches reflect the efforts of several research activities performed by students and professors with experience in multicast reliability. With the exception of the NORM protocol, no other standard has emerged from such proposals. As a result, the IEEE institution has considered the need to define a task group to standardize the requirements in order to implement an efficient and reliable protocol for providing multicast transport for real-time services, such as video and audio streams. Such task group has been named IEEE 802.11aa and is officially committed with the mission of producing an amendment to the IEEE 802.11 standard focused on the reliability and adaptability of the multicast service while keeping full compatibility with legacy devices.

Although the amendment is still in the draft status, some features are already known. Such features include the definition of the *More Reliable*

Group (MRG), which claims for a higher level of reliability, four different acknowledgement policies depending on the degree of reliability to be provided, a new power management mode called *MRG Service Period (MRG-SP)*, and a stream prioritization mechanism based on *Enhanced Distributed Channel Access (EDCA)* and simple parameter selection.

Further information regarding this amendment will appear in the near future as it is an ongoing effort and has produced considerable interest in the research and industry communities, as it is clearly aimed to solve the wireless multicast performance issues not from an experimental point of view.

SCALABLE VIDEO

Overview

Scalability in video coding and transmission has been a research topic for many years (Fallah et al., 2008) (Li et al., 2010) (Sun et al., 2002), being the main goal to enable the decoding of a complex video bit stream on both powerful and limited devices. However, the scalable approaches included in previous standards have not been successful due to different reasons, including legacy video transmission systems, increased coding complexity and significant loss in compression efficiency. Moreover, simulcast emerged as a straight-forward alternative when trying to provide different quality streams to a heterogeneous group of devices, although requiring a higher bandwidth, thus being less efficient than the original scalable proposal. Indeed another competitor appeared, especially in 3G networks: media transcoding, which by means of adaptive network systems deployed QoS-enabled videoconference and multimedia streaming services. Neither of such approaches reached the expected efficiency of the original and more general scalable video proposal, but their ease of deployment filled the gap.

A video stream is known to be scalable as soon as some parts of the stream can be somehow removed while still receiving a valid stream with enough decodable information, such that it represents the source content with less quality but still acceptable. There are different types of scalability, although the most important are temporal, spatial and quality scalability. In temporal scalability, each substream represents the source content with a different frame rate, while in spatial scalability, each of the available substreams provide a different picture size. Quality scalability, also known as SNR scalability, preserves spatio-temporal resolution, but with a lower SNR ratio. Combinations of these types of scalability are allowed in order to produce a rich multitude of representations of the source content.

Scalable Video Coding (SVC) provides many benefits which can be useful in current multimedia services. As mentioned before, video applications have been spread in a way that people can access them from different types of devices, including mobile devices with small-sized screens and limited energy resources, and high resolution panels with high speed network connections. It makes no sense for the first device to receive a high quality video stream if the hardware does not allow for rendering and displaying such content at this level of detail. Moreover, bandwidth and energy consumption required for such task can be overwhelming. On the other side, having a low quality stream in a high definition display with high speed wired network connection may be disappointing for the user willing to have a great user experience after installing a device with outstanding features. SVC allows each device to receive the substream or substreams they are interested in according to their features or the current status of different parameters, such as network congestion or battery level. Moreover, scalability can be very useful when managing bandwidth occupancy and congestion, so, in case it is necessary, a part of the scalable stream (i. e. a substream or some substreams of the whole scalable stream) can be discarded, thus not being further forwarded. This can alleviate the network congestion while still guaranteeing the content reception at the price of not achieving the highest picture size, frame rate or image fidelity. Thus, SVC can be used to deploy QoS-enabled video services in heterogeneous networks.

So, once the benefits were clear and after some attempts at providing the MPEG technology framework with video scalability, it was in 2003 when the *Joint Video Team* (JVT), consisting of ITU-T and MPEG groups members, started conscientiously to develop the SVC extension to the H.264/AVC standard (Schwarz et al., 2007) of the MPEG-4 framework, which has acquired a wide acceptance both in research and commercial applications. Some essential requirements were considered in its definition stage: including a similar coding efficiency compared to single-layer coding; also little increase in decoding complexity when compared to single-layer coding; support of the three types of scalability; and provision of a backward compatible base layer for H.264/AVC decoding. As a result, this amendment relied on representing a video signal by one base layer and one or more enhancement layers, where these layers may increase the temporal resolution, the spatial resolution, or the SNR ratio of the video.

It must be noted that the general SVC concept is quite similar to the one in the so-called *Multiple Description Coding (MDC)* (Goyal, 2001), which basically proposes splitting a bit stream in different independent substreams. However, while the approved SVC model relies on a layered scheme where the substreams are dependent in a hierarchical way, MDC just results in a higher quality of the merged video as more streams are received. Given that SVC has been standardized and properly recognized as a mechanism to provide scalable video coding, this section will focus on such an alternative.

This section describes the fundamentals of scalable video coding and how the different substreams may be signaled and transported using already existing standards. Although SVC and MDC can be thought of as stand-alone solutions to the performance issues of multicast streaming, they cannot guarantee a reliable transport of such data. However, in the next section it will be shown how such techniques can be used in combination with P2P networking to offer the required reliability.

Coding

From the previous subsection it can be concluded that video scalability in SVC refers to the removal of parts of the video bit stream in order to adapt it to the various needs or preferences of end users as well as to varying terminal capabilities or network conditions. The main goal of the SVC standard is to enable the encoding of high-quality video bit steams into one or more subset bit streams that can be themselves be decoded with a complexity and reconstruction quality similar to that achieved in the H.264/AVC framework, while being backward compatible with this standard. In order to achieve this, SVC was standardized as an extension of H.264/AVC, thus taking advantage of the features of its main building blocks: the *Video Coding Layer (VCL)*, which defines the basis of coding relying on the classic bloc-based approach but including new features that increase compression efficiency, flexibility and adaptability; and the *Network Abstraction Layer (NAL)*, in charge of bundling the VLC output into NAL units. However, some specific features have been added in order to support scalable video coding.

In H.264/AVC the coding and display order of pictures is completely decoupled. As a result, any picture can be marked as a reference picture and used for motion-compensated prediction of subsequent pictures independently of the corresponding slice coding types (intra, predictive or bi-predictive). This feature allows the coding of picture sequences with arbitrary temporal dependencies.

In order to generate temporal scalable bit streams hierarchical prediction structures are used.

Figure 2. Hierarchical prediction structures in temporal scalability

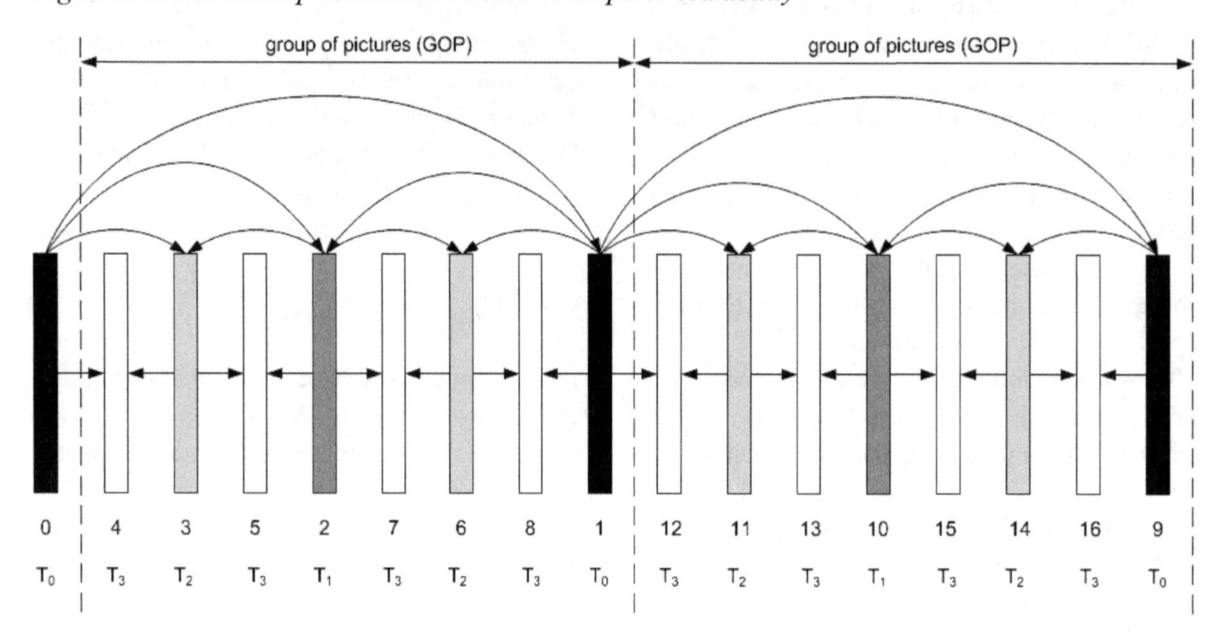

These structures allow coding the so-called key pictures in regular intervals by using only previous key pictures as references. The pictures between two key pictures are hierarchically predicted as shown in Figure 2, thus providing different enhancement layers (T_1, T_2 and T_3). While the sequence of key pictures represents the coarsest temporal resolution (i.e. the base layer, T_0 in Figure 2), hierarchically-coded layers support subsequent enhancement streams which add pictures following the temporal prediction levels. This feature is usually characterized by an increasing frame rate in the pictures decoding process.

The hierarchical prediction structures also provide an improved coding efficiency compared to classical IBBP coding. On the other hand, the encoding and decoding delay is increased as well. It should be noted that this delay can be controlled by restricting the motion-compensated prediction from pictures of the future, thus using different dependency schemes and dyadic and non-dyadic hierarchic structures.

Spatial scalability uses the conventional approach of multilayer coding, where each spatial layer corresponds to a supported spatial resolution, i.e. the frame size. In each of these layers, motion compensated prediction and intra-prediction are employed as for single-layer coding. In addition, inter-layer prediction mechanisms are incorporated as shown in Figure 3. These prediction mechanisms have been made switchable so that an encoder can freely choose which base layer information should be processed for an efficient enhancement layer coding. It must be noted that the temporal prediction structures of the spatial layers should be temporally aligned for an efficient use of the inter-layer prediction.

Spatial scalability coding efficiency strongly depends on the video sequence characteristics and the prediction structure, achieving in some tests similar results to single-layer coding but decreasing effectiveness when using prediction schemes like IPPP coding. Efficiency can be improved by adding multiple-loop decoding structures and hierarchical prediction schemes but it also significantly increases decoder complexity.

Quality scalability, also known as SNR scalability, is focused on providing a base layer with a low SNR ratio while preserving the frame size and rate. Thus, each enhancement layer provides

Figure 3. Multilayer structure for spatial scalable coding

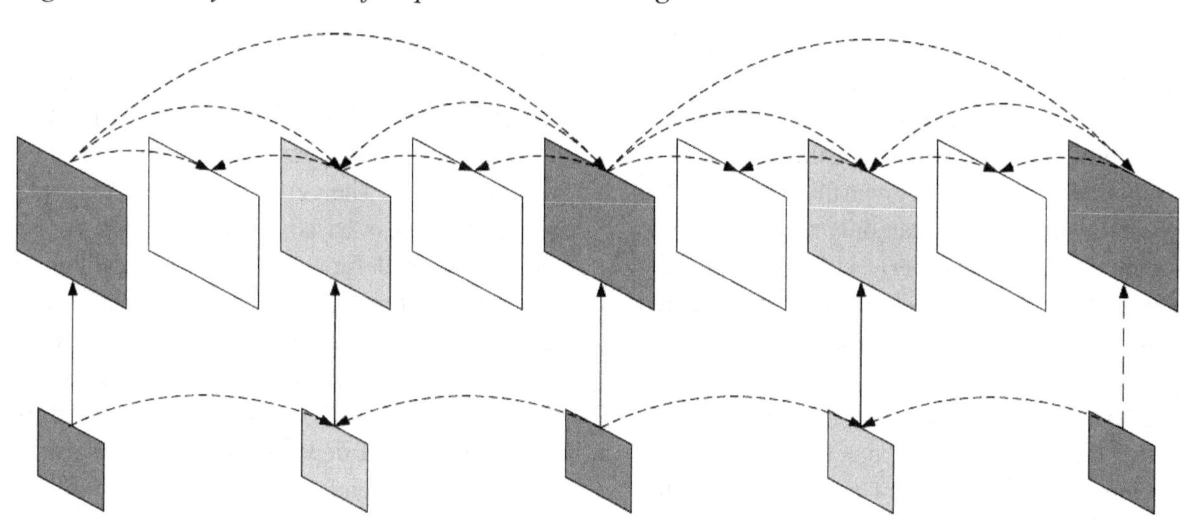

the required information in order to increase the image fidelity.

Transport and Signaling

Once the base and the different enhancement layers are coded using SVC, some mechanisms to transport, signal and synchronize them are needed in order to let the scalable video coding framework be fully supported in the network where related services are going to be deployed, generally IP networks.

Current H.264/AVC video services are usually based on RTP packetization and transport and *Session Description Protocol (SDP)* (Handley et al., 2006) multimedia descriptions. Some proposals based on these two standards of the *Internet Engineering Task Force (IETF)* have been addressed in order to provide SVC streams with packetization, adaptation and cross-layer synchronization when layers are sent in multiple RTP sessions, and representation of layer dependencies and attributes through signaling with SDP descriptions.

The original RTP payload, defined in RFC 3984, has been slightly modified in order to support the SVC amendment yet providing backward compatibility. Thus, although the payload structure and fragmentation, aggregation and packetization methods remain the same, the semantics of some fields like the *Decoding Order Number (DON)* change subtly.

RFC 3984 supports encapsulating a single NAL unit, more than one NAL unit or a fragment of a NAL unit into one RTP packet. Moreover, four types of aggregation NAL units are specified: STAP-A and STAP-B single-time and MTAP16 and MTAP24 multiple-time aggregation packet types. While the STAP-based aggregation allows encapsulating more than one NAL unit into one RTP packet with the same RTP timestamp (i.e. defining the same picture), the MTAP-based aggregation can be used to aggregate NAL units from different pictures into one RTP packet. Additionally, RFC 3984 also defines two types of fragmentation units, FU-A and FU-B, which enable fragmentation of one NAL unit into multiple RTP packets. Three packetization modes are also defined in the standard: single NAL unit, non-interleaved and interleaved. The latter, which allows using STAP-B, MTAP16, MTAP24 and FU-B, is needed to support SVC transport.

The DON field has been redefined in the SVC extension of RFC 3984, given that in the interleaved packetization mode different layers are transported in more than one RTP session. Thus, in order to support proper decoding management, explicit signaling of NAL units order through derived DON values in the interleaved packet stream is used. It must be noted that all these procedures also provide packet header overhead reduction and improved error resilience, as well as backward compatibility, given that a receiver that does not support SVC can only receive the base layer, when this is properly identified. Figure 4 depicts how different SVC layers are transported and reordered with the previous cross-layer synchronization system, which takes advantage of previous fragmentation, aggregation and packetization methods. The base layer 0 is transported in stream S_1; the enhancement layer 1 is transported in stream S_2 and enhancements layers 2 and 3 are both transmitted over S_3.

In order to enable easy identification of scalability dependencies within the transported streams and fast and efficient manipulation of these, a new NAL unit type known as PAyload Content Scalability Information (PACSI) has been defined. The PACSI NAL unit, if used, must be the first NAL unit in an aggregation packet and must not be present in other types of packets. This special NAL unit contains header scalability characteristics that are common for all the remaining NAL units in the payload, thus making it easier for network devices such as *Media Aware Network Elements (MANEs)* to decide whether to forward or discard the packets, thus contributing to network performance and efficiency.

SDP is a description language used to define multimedia sessions and the media to be transferred. It is widely used in session setup protocols such as *Session Initiation Protocol (SIP)* and RTSP. SDP implements two different description levels. While the session level describes the session itself with parameters such as the session's name or the originator contact details, the media level description defines each media stream in the session, e.g. video and audio streams, each identifying one RTP session. On both session and media levels, attributes can be used to add details about the session or media descriptions. The SDP specification defines some attributes itself and permits extending the definition with user-generated attributes, like those regarding SVC.

In order to support SVC signaling with SDP, there is a need to group the different media sessions (i.e. layers) and describe their relationship with each other. The currently proposed signaling mechanism is based on an extension of the SDP grouping defined in RFC 3388, adding a new grouping type called *Decoding DePendency* (DDP), which indicates the dependency relationship between the different RTP sessions of an SDP group. Additional attributes are used to define the type of dependency, thus supporting both layered dependency and multi-description decoding dependency.

In order to guarantee support of legacy devices, i.e. devices that implement non-scalable H.264/AVC and RFC 3984 standards but not SVC amendments, two approaches have been defined. On the one side, the RTP session containing the base layer is described by two media descriptions, one announcing the stream as non-scalable H.264/AVC and the other announcing it as SVC's base layer. On the other side, in point-to-multipoint scenarios, it would be easier to separate H.264/AVC and SVC NAL units in two different RTP sessions.

The following example in Box 1 depicts the signaling of dependency relationships between layered media streams within a SDP description.

This example shows how different media streams sessions are separated in their transport. An H.264/AVC base layer is transported in its own RTP session indicated by a media identifier of "mid:1". Three additional SVC enhancement layers are defined in three different RTP sessions, while their relation is indicated by the "depend" attribute. It can be noted that the layer tagged with "mid:2" may be a quality enhancement layer; "mid:3" may be a further temporal enhancement layer and "mid:4" may be a spatial enhancement layer. This can be guessed from the different parameters and arguments included in the media descriptions.

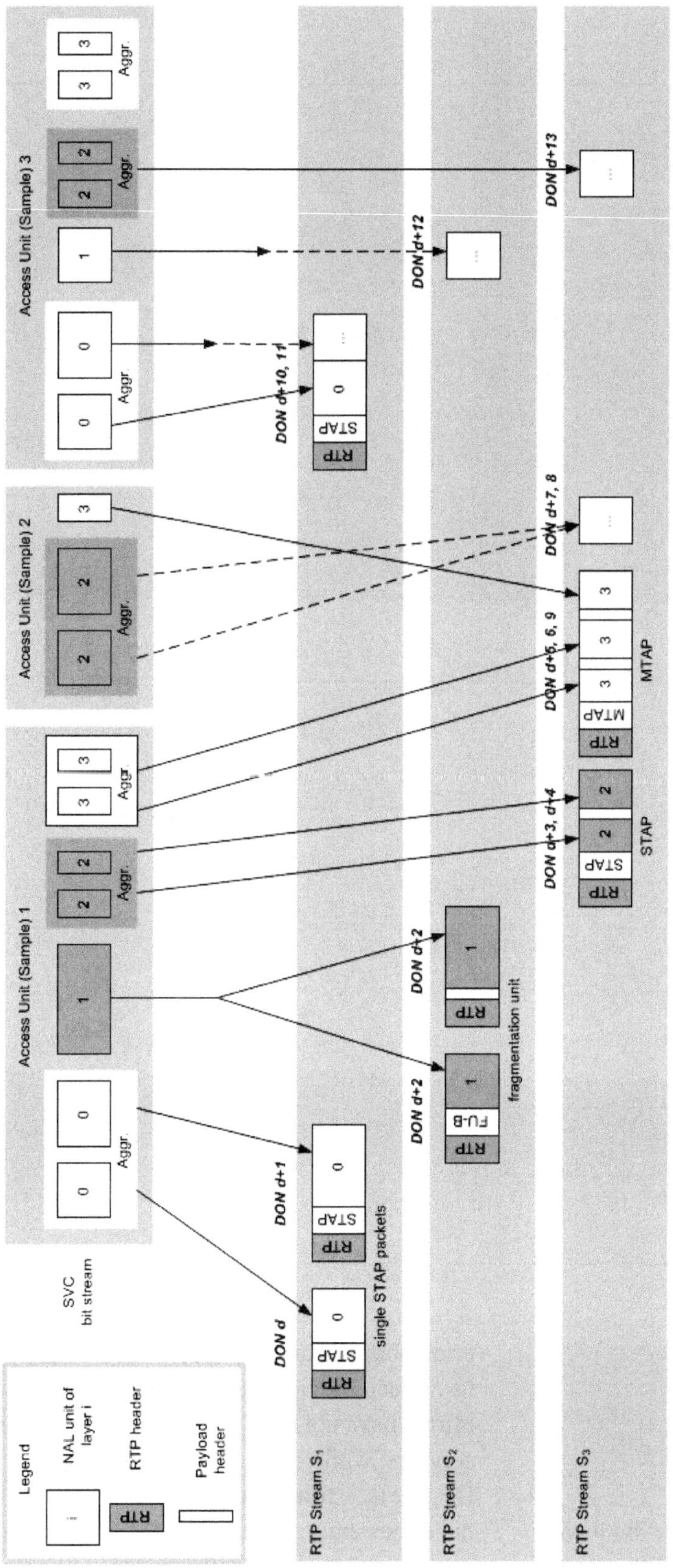

Figure 4. Cross-layer synchronization with packetization in interleaved mode.

Box 1.

```
v=0
o=svcsrv 289083124 289083124 IN IP4
host.example.com
s=LAYERED VIDEO SIGNALING Seminar
t=0 0
c=IN IP4 224.2.17.12/127
a=group:DDP 1 2 3 4
m=video 40000 RTP/AVP 94
b=AS:96
a=framerate:15
a=rtpmap:94 H264/90000
a=mid:1
a=depend:lay
m=video 40002 RTP/AVP 95
b=AS:96
a=framerate:15
a=rtpmap:95 SVC/90000
a=mid:2
a=depend:lay 1
m=video 40004 RTP/AVP 96
b=AS:64
a=framerate:30
a=rtpmap:96 SVC/90000
a=mid:3
a=depend:lay 1 2
m=video 40004 RTP/SAVP 100
c=IN IP4 224.2.17.13/127
b=AS:512
k=uri:conditional-access.example.com
a=framerate:15
a=rtpmap:100 SVC/90000
a=mid:4
a=depend:lay 1 2
```

MULTICAST PEER-TO-PEER OVERLAY NETWORKING

Overview

Different approaches to provide the multicast transmission mechanisms with reliability, considering both link and application level alternatives, have been previously explained. However, such efforts have not been deployed in real streaming services, while being kept as research activities. Despite this, further refinements of original ideas have been emerging since their first proposal and some ideas have eventually produced standardization initiatives, like the definition of the NORM protocol or the IEEE 802.11aa task group committed to create a standard for reliable multicast streaming.

On the other hand, multicast P2P overlay networking is based on the creation of multicast groups linked by application level edges where unicast sessions are established between each pair of nodes (peers). As a result, all the issues introduced by the inherent link-level multicast features are avoided, mainly by providing data transmission acknowledgements. Peer-to-peer networking, in combination with scalable video coding, appears as an easy-to-deploy alternative to all previous research activities. As a result, some real video streaming services based on such mechanisms have been offered in the Internet during the last few years. As depicted in the next examples, grouped under the different architectures peer-to-peer networks can be based on, all of them were initially conceived as research activities, but some of them eventually ended up as Internet services or have been used as the basis for them.

Although P2P can offer a proper video streaming networking architecture, some important issues must be considered. These include billing, digital rights management and peers mobility, which must be analyzed in order to design a potentially commercial video streaming platform relying in P2P networking. Once these issues are solved, merging video splitting coding techniques such as SVC and MDC with P2P overlay networking, video streaming services could be offered in heterogeneous networks while guaranteeing reliability, device adaptability and scalability.

Tree-Based Architectures

DONet/CoolStreaming

DONet is a Data-driven Overlay Network (Zhang et al., 20005) for live media streaming, meaning this that nodes periodically exchange data availability information with a set of partners and retrieve unavailable data from one or more partners while supply available data to other partners. This represents an unstructured and data-centric design of a streaming overlay with no father or children peers defined, thus avoiding node dynamics drawbacks to a certain extent.

A DONet node has three key modules: a membership manager, a partnership manager, and a scheduler. Additionally, these nodes have a unique identifier and a membership cache (mCache) containing a partial list of the identifiers of the active nodes in the DONet.

For each segment of a video stream, a DONet node can be either a supplier, a receiver or both, depending dynamically on the segment's availability information, except the so-called origin node, which is always a supplier, e.g. a dedicated video server. This node receives all joining requests and redirects them to a deputy node randomly selected from its mCache. The joining node receives a list of candidate nodes from the deputy node and processes it in order to establish its partners in the overlay. The *Scalable Gossip Membership (SCAM)* protocol is used to distribute the membership messages (i.e. mCache information) among DONet nodes, being scalable and lightweight, essential features for an algorithm to be used in a P2P system. All this behavior is controlled by the membership manager.

As it relies on an unstructured overlay network design, neither the partnerships nor the data transmission directions are fixed in a DONet. Besides that, the availability of the stream segments in the buffer of a node is represented by a buffer map. The partnership manager is in charge of continuously exchanging the node's buffer map with its partners, while the scheduler manages the retrieved information and programs which segment is to be fetched from which partner accordingly. The scheduling algorithm is designed in order to meet two constraints: the playback deadline for each segment; and the available streaming bandwidth from the partners. To do so, the algorithm first calculates the number of potential suppliers for each segment. Then, since a segment with less potential suppliers is considered to be more difficult to meet the deadline constraints, the algorithm tries to schedule first the segments with fewer suppliers and waits for those with more. When there are multiple potential suppliers, the one with the highest bandwidth and enough available time is selected, thus meeting the second requirement.

Regarding the overlay network prune process, when a peer leaves gracefully, it sends a departure message similar to a joining request message. When there is a node failure, a partner that detects the failure issues the departure message on behalf of the failed node. In order to avoid duplicated failure messages in this case, only the first received message is gossiped. Additionally, each node periodically establishes new partnerships with nodes randomly selected from its mCache, then maintaining a stable number of partners to strengthen the overlay in the presence of node departures while exploring it to look for better quality connections.

Based on DONet mechanisms, its developers implemented a public Internet video streaming service called CoolStreaming, which was accessible worldwide during one year and achieved a reasonable success in the home user environment. From their experience and received feedback, DONet/CoolStreaming developers concluded that the current Internet had enough available bandwidth to support TV-quality streaming and the larger the data-driven overlay is, the better the streaming quality it delivers. Apart from these practically obvious conclusions, they faced some real-environment issues such as *Network Address Translation (NAT)* traversing problems and lack of stream remote control, which were to be considered as future work.

HotStreaming

HotStreaming (Wu et al., 2006) is basically a revision of CoolStreaming developed by a different group of researchers, which included some improvements to the original P2P overlay implementation. These improvements are essentially the addition of two new policies to the partnership formation process and the usage of MDC descriptions.

HotStreaming uses the SCAM protocol as a partnership protocol just as CoolStreaming does. Although it achieves a good load balance level, the randomness of the process can lead to some stability problems in dynamic environments where some peers can become isolated. To overcome this, HotStreaming uses two policies called Preferential Random Forwarding and Preferential Random Selection. The former increase the probability of partners with larger partnerships to be chosen as the targets of subscription messages while the subsequent forwarding of the subscription messages will be sent to partners with smaller partnership; in the latter, a peer with a small partnership size selects all nodes from all subscriptions it has received until reaching a default partnership size limit.

HotStreaming also implements the MDC approach while defining a new coding scheme called MDC-STHI (Spatial Temporal Hybrid Interpolation), consisting of four streams containing even or odd and full-size or down-sampled combinations streams respectively. All streams can be combined taking into account the devices' network and display capabilities while guaranteeing video playback continuity with a single stream being received, given MDC coding properties.

So, HotStreaming implements a MDC approach, where descriptions are requested using algorithms defined in CoolStreaming but with some new policies aimed to enhance the stability of the platform and then the streaming service.

P2P-Leveraged Mobile Live Streaming

P2P-Leveraged Mobile Live Streaming (P2PMLS) (Jinfeng et al., 2007) is a mobile video streaming system which uses the multiple network interfaces of mobile hosts to build different kinds of connections. Thus, P2PMLS aims to use WiFi or Bluetooth technologies to form a P2P overlay

Figure 5. P2PMLS system architecture

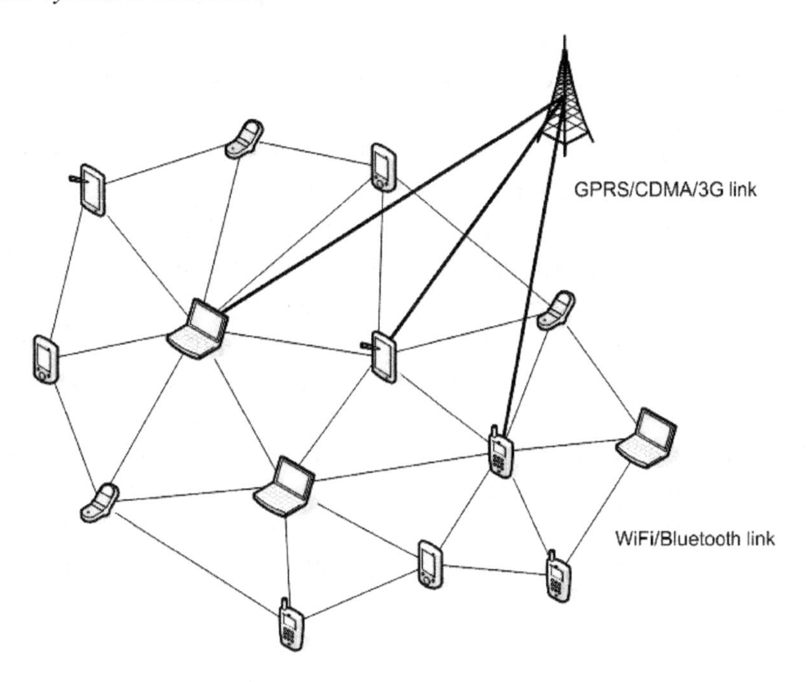

network, while cellular data networks such as UMTS or GPRS are intended for seed peers to receive the streaming data from a server.

In the system architecture of P2PMLS, depicted in Figure 5, mobile hosts have at least two wireless interfaces, one is used to communicate in the wide area cellular network, and the other is used to communicate in the local area wireless network and then build a P2P overlay network. Some peers in this overlay network behave as seed nodes, which directly receive the streaming data from the streaming server through the cellular network and propagate it to peers who request it.

The server side is comprised of a streaming server and a management server. The former streams the different contents, which are encoded in sub-streams through MDC techniques, to users, while the latter manages the users in the network and the content information.

On the other side, a mobile peer node includes four modules. The endpoint communication module provides the functions to communicate to mobile hosts over a wireless link, both the overlay network and the cellular network. The mobility support layer module manages the power state of the mobile host and collects its neighbors' energy states, detects the mobility pattern of the mobile host in terms of direction and speed and identifies mobile hosts in the overlay network with the same mobility pattern and periodically monitors and probes the network bandwidth and exchanges this information with neighbor peers. The peer layer module manages member and partnership processes and, finally, the streaming protocol layer module implements the different standard streaming protocols to receive and send video data.

In order to join the P2PMLS platform, a joining node first gets the content list from the management server and requests the information related to a specific program, including the seed nodes and the different MDC descriptions location. If there are no seed nodes for the selected descriptions, the new peer requests one or more sub-streams from the streaming server, according to the node resources and capabilities, and becomes a seed node for the received descriptions. Otherwise, the membership manager chooses a subset of neighbor peers for every description according to the previously commented algorithms and starts receiving the video data. If the new peer's partnership requests are rejected, the peer asks the streaming server for it and becomes a seed node.

Peers can leave the system when the mobile host's source power is exhausted, moves out of the communication scope or the user exits from the application. In the two former cases, given that they can be predicted in advance, the leaving mobile peers broadcast BYE messages in the overlay network. The unpredicted leaving peers are detected by the manager server through probing messages, then issuing recovery alerts to the affected peers. Additionally, depending on the information provided by the mobility support layer, the membership manager periodically updates the list of neighbors, so the peer adapts the partnership in order to increase the reliability of service.

Unstructured Architectures

KQstream

Kindred-based QoS-aware Live Media Streaming (KQStream) (He et al., 2007) is a streaming system aimed at working in heterogeneous P2P environments using MDC through a multi-tree structure in order to deliver content in a granular mode. The two pillars of this implementation are the kindred-based topology construction and QoS-aware streaming process.

KQStream defines a KindredSet structure containing information about a peer, its parents, siblings and children, all sharing a common MDC description. This structure is used in order to exchange information about the available description and peers' resources with neighbor peers. In order to join a multicast tree, joining nodes will obtain this information and make a selection taking into account a user's preferences, available bandwidth and buffer size regarding the joining peer itself and other parameters based on candidate peers such as link delay, contribution degree or tree depth. After this process, the joining node will be attached to a specific description and will start receiving streaming media content through the joined multicast tree. Peer leaving is managed through QUIT messages to all kindreds and with periodical probing to detect a leaving peer under unexpected failures in peer communication.

In order to improve the overlay performance, three optional policies are included in the topology management process. These are load balancing to avoid parent-demand overload in nodes with better network conditions; shorten tree depth, which causes descriptions with the shortest depth to be requested first; and shorter delay preference to guarantee a better general connectivity.

The QoS-aware mechanism in KQStream is based on the individual streaming state of peers, allowing users to configure their playback requirements while not requesting media quality over their available downstream bandwidth. With re-buffering techniques, KQStream allows peers to manage buffer occupancy and the number of requested descriptions according to the received quality and buffer size. Additionally, a peer-attachment regulating technique to face loss rate fluctuations and peer dynamics is implemented, so attachment attempts are limited in order to guarantee streaming stability and topology construction efficiency.

SplitStream

SplitStream (Castro et al., 2003) designers claim and it is true, that conventional tree-based multicast P2P overlay networks are not well matched to a cooperative environment, given that forwarding multicast traffic is carried by the subset of peers that are interior nodes in the tree, while the majority of peers are leaf nodes and contribute no resources. The problem is further aggravated in high-bandwidth consuming applications, such as video streaming, where many peers may not be able to become an interior node because of the resources and network requirements. SplitStream

Figure 6. SplitStream multicast trees

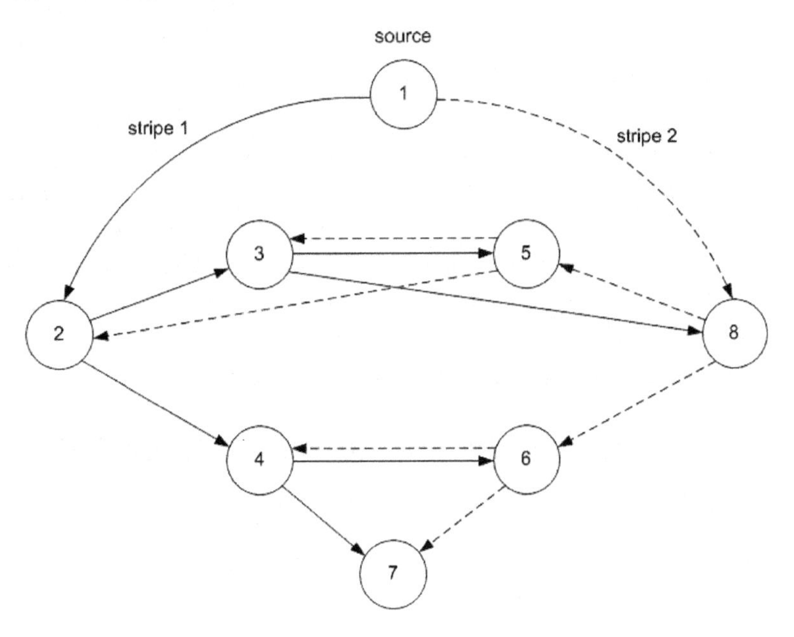

addresses these issues by splitting the content into stripes (MDC descriptions or SVC layers, although non-video traffic can also be transported over this system) and multicast each stripe using a separate tree. Thus, peers join as many trees as stripes they wish to receive and they specify an upper bound on the number of stripes that they are willing to forward according to their capabilities. As depicted in Figure 6, the group of multicast trees is built such that an interior node in one tree is a leaf node in all the remaining trees (i.e. a disjoint interior node), then building a cooperative multicast P2P overlay system where the forwarding load can be spread across all participating peers.

This approach is based on Pastry and Scribe, which rely on distributed hash tables in order to guarantee a fast peer location. Essentially, SplitStream uses Pastry routing and Scribe hash key assignations in order to create disjoint interior nodes and distinguish different multicast trees. It must be noted that a joining peer is always accepted by a SplitStream node, regardless of whether it has reached its maximum number of children. However, the parent node rejects one of its children after submitting them to a decision algorithm, which takes into account if children belong to another group and the node IDs and group keys similarity to the parent's one (the more similar, the physically closer they are). The orphan node starts searching for a new parent with similar IDs. SplitStream also defines a spare capacity group, which contains all peers with fewer children that their forwarding capacity limit and that can serve children for other stripes.

FUTURE RESEARCH DEVELOPMENTS

The future of multicast reliability, especially when considering multimedia streaming services, will probably rely on the results of the IEEE 802.11aa initiative and P2P overlay networking with scalable video, which have been proved to be the alternatives which more clearly target real-time video and audio streaming solutions for a wide deployment in both the Internet and private IP networks.

Many research efforts will continue to produce mechanisms to provide pure multicast reliability by means of the techniques introduced in this chapter, refining previous algorithms and implementations and reducing network overhead, amongst other improvements. However, IEEE 802.11aa is expected to gather all the proper algorithms and define a standardized basis for future implementation of such mechanisms. Maybe an extremely efficient algorithm can be defined in the next years, but probably it will be darkened by the IEEE standard. Hopefully the new discoveries in FEC-based and leader-based solutions can be amended to future revisions of the 802.11aa standard to enhance its implementation results.

On the other side, P2P overlay networks where scalable video is deployed has emerged as a more direct solution for the current situation. The success of P2P networks has become a door for real implementations of adaptive video streaming services, as seen through some examples introduced. Despite this, not many commercial applications have been really successful, so no impact has been detected on the video streaming over the Internet business. Once P2P networks are conceived as a really useful network architecture instead of a piracy vehicle, end clients, service providers and network companies will really take into consideration this approach.

CONCLUSION

In this chapter the performance issues introduced by the multicast transport mechanism have been explained in order to set the scenario for the requirements when designing solutions for reliable multicast video and audio streaming. Next, different alternatives to provide such reliability have been shown, including basic approaches that did not fulfill the expected results, improvements on the link-level multicast definition, amendments designed to add robustness to transmissions by means of FEC protection, usage of video scalability techniques and its combination with P2P networking, amongst others. All mentioned approaches emerged from research activities initiated to overcome the known pitfalls of multicast transport of real-time data. Some of them acquired enough recognition, being reflected in the creation of standards like the NORM protocol and the IEEE 802.11aa task group or in the deployment of real streaming services, like CoolStreaming. Finally, according to what has been stated through the chapter, the future research directions have been briefly commented.

ACKNOWLEDGMENT

This work has been supported by the Spanish Government, MICINN, under research grant TIN2010-20136-C03.

REFERENCES

Castro, M., Druschel, P., Kermarrec, A.-M., Nandi, A., Rowstron, A., & Singh, A. (2003). SplitStream: High-bandwidth multicast in a cooperative environment. *Proceedings of the Nineteenth ACM Symposium on Operating Systems Principles.*

Ding, P., Holliday, J., & Celik, A. (2004). A leader based priority ring reliable multicast in WLANs. *IASTED International Conference on Communications Systems and Networks*, (pp. 70–75).

Fallah, Y. P., Mansour, H., Khan, S., Nasiopoulos, P., & Alnuweiri, H. M. (2008). A link adaptation scheme for efficient transmission of H.264 Scalable video over multirate WLANs. *IEEE Transactions on Circuits and Systems for Video Technology*, *18*(7), 875–887. doi:10.1109/TCSVT.2008.920745

Goyal, V. K. (2001). Multiple description coding: Compression meets the network. *IEEE Signal Processing Magazine*, *18*(5), 74–94. doi:10.1109/79.952806

3GPP TS 26.346 V9.4.0. (2010). *Technical specification: 3rd generation partnership project; technical specification group services and system aspects; multimedia broadcast/multicast service (MBMS); Protocols and codecs (Release 9).* Braden, R., Zhang, L., Berson, S., Herzog, S., & Jamin, S. (1997). Resource ReSerVation Protocol (RSVP) – Version 1 functional specification. *IETF, RFC 2205.*

Handley, M., Jacobson, V., & Perkins, C. (2006). SDP: Session description protocol. *IETF, RFC 4566.*

He, Y., Gu, T., Guo, J., & Dai, J. (2007). *KQStream: Kindred-based QoS-aware live media streaming in heterogeneous peer-to-peer environments* (pp. 56–61). Parallel Processing Workshops.

Jinfeng, Z., Jianwei, N., Rui, H., Jianping, H., & Limin, S. (2007). P2P-leveraged mobile live streaming. *Advanced Information Networking and Applications Workshops*, (pp. 195 – 200).

Kuri, J., & Kasera, S. K. (1999). Reliable multicast in multi-access wireless LANs. *Proceedings - IEEE INFOCOM*, 760–767.

Li, M., Chen, Z., & Tan, Y.-P. (2010). Scalable video transmission over multiuser MIMO-OFDM systems. *CHINACOM, 2010*, 1–8.

Li, Z., & Herfet, T. (2008). Beacon-driven leader based protocol over a GE channel for MAC layer multicast error control. *International Journal of Communications, Network and System Science, 1*(20, 144-153.

Li, Z., & Herfet, T. (2009). MAC layer multicast error control for IPTV in wireless LANs. *IEEE Transactions on Broadcasting, 55*(2), 353–362. doi:10.1109/TBC.2009.2016502

Miroll, J., Zhao, L., & Herfet, T. (2010). *Wireless feedback cancellation for leader-based MAC layer multicast protocols: Measurement and simulation results on the feasibility of leader-based MAC protocols using feedback cancellation on the 802.11aa wireless multicast network.* IEEE 14th International Symposium on Consumer Electronics.

MPEG-4 Overview – (V.21 – Jeju version). (2002). *International Organization for Standardization, ISO/IEC JTC1/SC29/WG11, Coding of moving pictures and audio.*

NORM. (n.d.). Retrieved from http://cs.itd.nrl.navy.mil/work/norm/

Paila, T., Luby, M., Lehtonen, R., Roca, V., & Walsh, R. (2004). FLUTE: File delivery over unidirectional transport. *IETF, RFC 3926.*

Rappaport, T. S. (2002). *Wireless communications: Principles and practice*. Upper Saddle River, NJ: Prentice Hall.

Schulzrinne, H., Casner, S., Frederick, R., & Jacobson, V. (2003). RTP: A transport protocol for real-time applications. *IETF*, RFC 3550.

Schulzrinne, H., Rao, A., & Lanphier, R. (1998). Real time streaming protocol (RTSP). *IETF, RFC 2326.*

Schwarz, H., Marpe, D., & Wiegand, T. (2007). Overview of the scalable video coding extension of the H.264/AVC standard. *IEEE Transactions on Circuits and Systems for Video Technology, 17*(9), 1103–1120. doi:10.1109/TCSVT.2007.905532

Sun, X., Wu, F., Li, S., Gao, W., & Zhang, Y.-Q. (2002). Seamless switching of scalable video bitstreams for efficient streaming. *ISCAS, 2002*, 385–388.

Wu, J.-C., Peng, K.-J., Lu, M.-T., Lin, C.-K., Cheng, Y.-H., & Huang, P. ... Chen, H. H. (2006). *HotStreaming: Enabling scalable and quality IPTV services.* Retrieved April 11th from http://nslab.ee.ntu.edu.tw/publication/conf/hotstreaming-iptv06.pdf

Zhang, X., Liu, J., Li, B., & Yum, Y.-S. P. (2005). CoolStreaming/DONet: A data driven overlay network for efficient live media streaming. *Proceedings of the IEEE INFOCOM,* (pp. 2102-2111).

ADDITIONAL READING

Alay, Ö., Korakis, T., & Wang, Y. (2010). Dynamic rate and FEC adaptation for video multicast in multi-rate wireless networks. *Mobile Networks and Applications Journal, 15*(3), 425–434. doi:10.1007/s11036-009-0202-5

Bikfalvi, A., & Nozzilla (2008). *A novel peer-to-peer architecture for video streaming* (Master Thesis), Director: García, J. Madrid.

Camarillo, G., Holler, J., & Schulzrinne, H. (2002). Grouping of media lines in the session description protocol (SDP). *IETF, RFC 3388.*

Dixit, S., & Wu, T. (2004). *Content networking in the mobile internet.* Chichester, UK: Wiley. doi:10.1002/047147827X

Dujovne, D., & Turletti, T. (2003). Multicast in 802.11 WLANs: An experimental study. *9th ACM International Symposium on Modeling Analysis and Simulation of Wireless and Mobile Systems,* (pp. 130-138).

González, J. D., & Paradells, J. (2007). Design and implementation of a personalized electronic service guide (ESG) system able to offer digital contents to portable devices over IP unicast/multicast platforms.

Handley, M., Jacobson, V., & Perkins, C. (2006). SDP: Session description protocol. *IETF, RFC 4566.*

IEEE. 802.11. (2007). *Standard part 11: Wireless LAN medium access control (MAC) and physical layer (PHY) specifications.* IEEE Computer Society, June 2007.

Jun, J., Peddabachagari, P., & Sichitiu, M. (2003). Theoretical maximum through-put of IEEE 802.11 and its applications. *Second IEEE International Symposium on Network Computing and Applications,* (pp. 249 – 256).

Koucheryavy, Y., Moltchanov, D., & Harju, J. (2003). Performance evaluation of live video streaming service in 802.11b WLAN environment under different load conditions. *Interactive Media on Next Generation Networks, 2889,* 30–41. doi:10.1007/978-3-540-40012-7_3

Lin, C. H., Ke, C. H., Shieh, C. K., & Chilamkurti, N. K. (2006). The packet loss effect on MPEG video transmission in wireless networks. *Proceedings of the 20th International Conference on Advanced Information Networking and Applications,* (pp. 565–572).

Miroll, J., Li, Z., & Herfet, T. (2010). *Wireless feedback cancellation for leader-based MAC layer multicast protocols: Measurement and simulation results on the feasibility of leader-based MAC protocols using feedback cancellation on the 802.11aa wireless multicast network.* IEEE 14th International Symposium on Consumer Electronics.

Mushtaq, M., & Ahmed, T. (2008). P2P-based mobile IPTV: Challenges and opportunities. *Computer Systems and Applications, AICCSA, 2008,* 975–980.

Rieckh, J. (2008). *Scalable video for peer-to-peer streaming* (Master Thesis), Director: Superiori L., Rupp, M., Warner, W. Vienna, 2008.

Rowstron, A., & Druschel, P. (2001). Pastry: Scalable, distributed object location and routing for large-scale peer-to-peer systems. *IPFiP/ACM International Conference on Distributed Systems Platforms (Middleware)*, (pp. 329 – 350).

Weber, J., & Newberry, T. (2007). *IPTV crash course*. Boston, MA: McGraw Hill.

Wenger, S., Hannuksela, M. M., Stockhammer, T., Westerlund, M., & Singer, D. (2005). RTP payload format for H.264 video. *IETF RFC 3984*.

Wenger, S., Wang, Y.-K., & Schierl, T. (2007). Transport and signaling of SVC in IP networks. *IEEE Transactions on Circuits and Systems for Video Technology*, *17*(9), 1164–1173. doi:10.1109/TCSVT.2007.905523

Wenger, S., Wang, Y.-K., & Schierl, T. (2011). *RTP payload format for SVC video*. IETF Internet Draft draft-ietf-avt-rtp-svc-27.txt.

KEY TERMS AND DEFINITIONS

Content Delivery Networking (CDN): Network architecture approach where contents are spread all over the infrastructure by replication in order to reduce transport delay and other related issues.

Scalable Video Coding (SVC): Coding technique which splits a video content in different dependent layers. Such technique allows video streaming quality to be adapted to network conditions by reducing the required bandwidth and the content quality in terms of spatial resolution, temporal resolution or SNR resolution.

Multiple Description Coding (MDC): Alternative approach to SVC where the layers (here known as descriptors) are independent, so the more are received the better the quality, no matter which ones.

Simulcast: Network data distribution scheme where different packets streams are sent in parallel to multiple destinations at the same time, generally following a multicast tree. It can be though as different multicast flows transmitted over the same multicast tree at the same time.

'Crying baby' effect: When more resources are dedicated to a specific node in a network to solve a related issue, other nodes can start experiencing the same issue, thus clambering for attention. That is why it is called a 'crying baby' effect; when one baby starts feeling unattended, every other baby ends up claiming for attention, being impossible to satisfy all of them.

Section 3
Network Perspectives

Chapter 11
An Open Service Platform Based on an IP Metropolitan Mesh Network

Raúl Aquino-Santos
University of Colima, México

Arthur Edwards-Block
University of Colima, México

Víctor Rangel-Licea
National Autonomous University of México, Mexico

ABSTRACT

This chapter proposes an open-service platform based on an IP metropolitan mesh network suitable for multimedia services in an all-IP network environment. To guarantee mobile applications in the metropolitan mesh network simulated, the authors evaluated the five most prominent mobile ad hoc network (MANET) routing algorithms: Ad hoc On Demand Distance Vector (AODV), Dynamic Source Routing protocol for mobile ad hoc networks (DSR), Optimized Link State Routing Protocol (OLSR), Temporally-Ordered Routing Algorithm (TORA), and Geographic Routing Protocol (GRP). The metropolitan mesh network architecture is based on the IEEE 802.16-2004 Standard that supports the IP protocol and the interaction with MANET protocols. The MANET routing protocols are evaluated in terms of delivery ratio, MANET delay, routing overhead, overhead, WiMAX delay, WiMAX load, and WiMAX throughput. Results show that proactive routing algorithms are more efficient than the reactive routing algorithms for the IP metropolitan-mesh network simulated.

INTRODUCTION

Mobile ad-hoc networks have attracted considerable attention and interest from the commercial sector as well as the standards community. The introduction of new technologies such as IEEE 802.11g and IEEE 802.16 greatly facilitate the deployment of ad-hoc technology beyond the military domain (Santos et al., 2010). However, to the best of our knowledge, mobile ad hoc routing algorithms based on IP IEEE 802.16-mesh networks, have been insufficiently unexplored.

DOI: 10.4018/978-1-4666-1613-4.ch011

The IEEE 802.16-2004 standard includes a mesh topology which can be used efficiently in Mobile Ad-Hoc Networks or MANETs.

WiMAX (Worldwide Interoperability for Microwave Access) is a wireless communication system that can provide high data rate communications in IP *metropolitan area networks (MANs)*. Over the years, the IEEE 802.16 workgroup has developed a number of standards for WiMAX. The first standard was published in 2001 to support communications in the 10-66 GHz frequency band. In 2003, IEEE 802.16a was introduced to provide additional physical layer specifications for the 2-11 GHz frequency band. These two standards were further revised by the IEEE 802.16-2004 (Chite & Daigle, 2003). Subsequently, IEEE 802.16e was also approved as the official standard for mobile applications. IEEE 802.16m (Ahmadi, 2011) is the current standard for WiMAX, aimed at fulfilling the 4G specification.

In the *physical (PHY)* layer, IEEE 802.16 supports four PHY specifications for the licensed bands. These four specifications are Wireless-MAN-SC (single carrier), -SCa, -OFDM (orthogonal frequency – division multiplexing), and *OFDMA (orthogonal frequency –division multiple access)*. In addition, the standard also supports different PHY specifications (-SCa, -OFDM, and –OFDMA) for the unlicensed bands, including wireless high-speed unlicensed MAN (WirelessHUMAN). Most PHY specifications are designed for *non-line-of-sight (NLOS)* operation in frequency bands below 11 GHz, except –SC, which operates in the 10-66 GHz frequency band. To support multiple subscribers, IEEE 802.16 supports *both time-division duplex (TDD)* and *frequency-division duplex (FDD)* operations.

The mobile version of IEEE 80.16 also supports the following features to enhance the performance of the wireless system: 1) *multiple input, multiple output (MIMO)* techniques such as transmit/receive diversity multiplexing and 2) *adaptive modulation and coding (AMC)* is used to better match instantaneous channel and interference conditions. Furthermore, multiple antenna schemes can also be used to improve performance by increasing the transmitted data rates through spatial multiplexing and, importantly, in the *medium access control (MAC)* layer, IEEE 802.16-2004 supports two modes: *point-to-multipoint (PMP)* and mesh. The former organizes nodes into a cellular-like structure consisting of a *base station (BS)* and *subscriber stations (SSs)*. The channels are divided into uplink (from SS to BS) and downlink (from BS to SS), and both uplink and downlink channels are shared among the SSs. PMP mode requires all SSs to be within the transmission range and clear *line of sight (LOS)* of the BS. On the other hand, in mesh mode, an ad hoc network can be formed with all nodes acting as relay routers in addition to their sender and receiver roles, although there may still be nodes that serve as BSs and provide backhaul connectivity.

This chapter proposes a simulation model for an IP metropolitan IEEE 802.16 mesh network which can transmit all-IP services and whose proposed architecture can be deployed in emergency scenarios because it does not require a base station.

The remainder of this chapter is organized as follows: the second section describes the classification of mobile ad hoc routing algorithms. The following section analyzes the state of the art of routing algorithms for wireless mesh networks. There follows an explanation of the simulation model for the IP metropolitan-mesh network. The chapter then discusses the scenario simulated and results obtained, followed by the concluding sections, which summarize our work and propose future research.

MOBILE AD HOC ROUTING ALGORITHMS

Mobile ad hoc algorithms can be categorized into two different categories: non-positional algorithms and positional algorithms. Non-positional algorithms can discover the network topology without relying on specific geographical information; on the other hand positional algorithms make use of geographical information which is derived from a GPS receiver or by implementing an alternative approach such as trilateration or any other estimation technique.

Non-positional algorithms can be further classified as proactive (table-driven), reactive (on-demand), or hybrid. Proactive, or table-driven algorithms, periodically update the topology information of their nodes, making routes immediately available when needed. The disadvantage of these algorithms, however, is that they require additional bandwidth to periodically transmit topology control traffic, resulting in significant network congestion because each individual node must maintain the necessary routing information and is responsible for propagating topology updates in response to instantaneous changes in network connectivity (Perkins, 2000). Important examples of non-positional proactive protocols include *Optimized Link State Routing (OSLR)* (Clausen et al. 2003) and *Topology Dissemination Based on Reverse Path Forwarding (TBRPF)* (Ogier et al., 2004). These two protocols record the routes for all of the destinations in the ad hoc network, resulting in minimal initial delay (latency) when communicating with arbitrary destinations. Such protocols are also called proactive because they store route information before it is actually needed and are table driven because the information is available in well-maintained tables.

On the other hand, on-demand, or reactive protocols, acquire routing information only as needed. Reactive routing protocols often use less bandwidth for maintaining route tables. The disadvantage of these protocols, however, is that the *Route Discovery (RD)* latency for many applications can substantially increase. Most applications may suffer delay when they start because a destination route must be acquired before communication can begin. On-demand protocols must realize a route discovery process before the first data packet can be sent, resulting in reduced control traffic overhead at the cost of increased latency in finding the destination route (Zou et al., 2002). Examples of reactive, or on-demand protocols, include *Ad-Hoc On-Demand Distance Vector (AODV)* routing (Perkins, Belding-Royer & Das, 2003), and *Dynamic source Routing (DSR)* algorithms (Johnson, Maltz, & Hu, 2007).

A protocol that combines both proactive and reactive approaches is called a hybrid (Schaumann, 2002). The most popular protocol in this category is the *Zone Routing Protocol (ZRP)*. In ZRP, the network is divided into overlapping routing zones that can use independent protocols within and between each zone. ZRP is considered a hybrid protocol because it combines proactive and reactive approaches to maintain valid routing tables without causing excessive overhead. Communication within a specific zone is realized by the *Intrazone Routing Protocol (IARP)*, which provides effective direct neighbor discovery (proactive routing). On the other hand, communication between different zones is realized by the *Inter-zone Routing Protocol (IERP)*, which provides routing capabilities among nodes that must communicate between zones (reactive routing).

Scalability represents the principal disadvantage of purely proactive and reactive routing algorithms in highly mobile environments. A second disadvantage is their very low communication throughput, which sometimes results from a potentially large number of retransmissions (Mauve et al., 2001). To overcome these limitations, however, several newer types of routing algorithms that employ geographic position information have been developed, including: *Location-Aided Routing (LAR)* (Ko, & Vaidya, 1998), *Distance Routing Effect Algorithm for Mobility (DREAM)* (Basagni et al., 1998), *Grid Location Service (GLS)* (Li, 2000), *Greedy Perimeter Stateless Routing for Wireless Networks (GPSR)* (Karp, & Kung, 2000), and *Location Routing Algorithm with Cluster-Based Flooding* (Santos et al. 2005).

STATE OF THE ART OF ROUTING ALGORITHMS FOR WIRELESS MESH NETWORKS

Several wireless mesh routing protocols have been reported in the literature. Grace (2000) describes the *Mobile Mesh Border Discovery Protocol (MMBDP)*, which is a robust, scalable, and efficient mobile ad hoc routing protocol based on the "link state" approach. A node periodically broadcasts its own *Link State Packet (LSP)* to each neighbor participating in the protocol. LSPs are relayed by nodes, thus allowing each node to have full topology information for the entire ad hoc network. From its topology database, a node is able to compute the lowest cost of unicast transmission routes to all other nodes in the mobile ad hoc network. In (Ogier, Templin & Lewis, 2004), the authors propose the *Topology Dissemination Based on Reverse-Path Forwarding (TBRPF)* protocol, which is a proactive, link-state routing protocol designed for mobile ad hoc networks. TBRPF provides hop-by-hop routing along the shortest paths to each destination. Each node running TBRPF computes a source tree, based on partial topology information stored in its topology table, using a modification of Dijkstra´s algorithm (Dijkstra, 1959). To minimize overhead, each node reports only part of its source tree to neighbors. TBRPF uses a combination of periodic and differential updates to keep all neighbors informed of the reported part of its source tree. Each node also has the option of reporting additional topology information to provide improved robustness in highly mobile networks. *Dynamic Source Routing (DSR-MP)* is also described in (Ogie et al., 2004). In the multi-path version of the DSR protocol, each ROUTE_REQUEST packet received by the destination is responded to with an independent ROUTE_REPLY packet. In (Pirzada et al., 2006), AOMDV, a well-known ad hoc routing algorithm and variant of AODV, is described. AOMDV provides loop-free and disjoint alternate paths. During route discovery, the source node broadcasts a ROUTE_REQUEST packet that is flooded throughout the network. In contrast to AODV, each recipient node creates multiple reverse routes while processing the ROUTE_REQUEST packets that are received from multiple neighbors. The authors in (Pirzada et al., 2007) present the *Ad-hoc On-demand Distance Vector Hybrid Mesh (AODV-HM)* Protocol. The aim of AODV-HM is to maximize the involvement of mesh routers into the routing process without significantly lengthening the paths. In addition, the objective of the authors is to maximize channel diversity in the selected path. To implement these features, they make two changes to the RREQ header. First, they add a 4-bit counter (MR-Count) to indicate the number of mesh routers encountered on the path taken by the RREQ. They further add a 7-bit field (Rec-Chan) to advertise the optimal channel to be used for the Reverse Route.

The weakness of the previous mesh routing protocols is that they are implemented with local area network technology (IEEE 802.11x), which is most suitable for short-range wireless area networks.

Relatively little work has been done that is related to ad hoc routing protocols for IEEE 802.16 Mesh Networks. In Zhou et al., (2009), AODV was implemented in an IEEE 802.16 Mesh Network, where the authors propose a scheduling scheme to piggyback AODV routing messages to the WiMAX mesh MAC protocol. They do this to speed up the routing discovery process and reduce overhead in control slots. Hao et al. (2010) only compare Ad Hoc On Demand Distance Vector (AODV) Routing and Optimized Link State Routing (OLSR) and Delay Tolerant Network (DTN) routing protocols such as Epidemic Routing and Spray and Wait Routing in terms of "delay tolerance" in a maritime communications environment.

The main contribution of this work is to simulate our model for IP metropolitan mesh networks, which is based on the IEEE 802.16-2004 Standard. The simulation we propose using the *(AODV, DSR, OLSR, TORA and GRP)* mobile ad hoc routing algorithm to present a complete performance comparison in terms of "*delivery ratio*", "*MANET delay*", "*routing overhead*", "*overhead*", "*WiMAX delay*", "*WiMAX load*" and "*WiMAX throughput*".

SIMULATION MODEL FOR THE IP METROPOLITAN MESH NETWORK

This section presents the OPNET simulation model (v.14) we employed for the IP metropolitan mesh network.

The initialization process at the mesh node consists of three principal phases: (1) the creation of the broadcast flow, (2) the neighbor discovery process and finally, (3) the establishment of individual unicast flows for each of the neighbors, as shown in Figure 1.

The generation of two broadcast flows, each with its corresponding *CID (connection identifier)*, is one of the first processes occurring at the instant mesh nodes begin to interact. One of these flows is utilized to transmit the broadcast information generated in the upper layers of the mesh node (data), while the second flow is the control broadcast that sends information to the MAC layer. The discovery process and neighbor node localization is modeled by the simulator (i.e. software), which registers objects based on the existing distances between each one of the nodes.

Figure 1. Initialization process of the mesh mode.

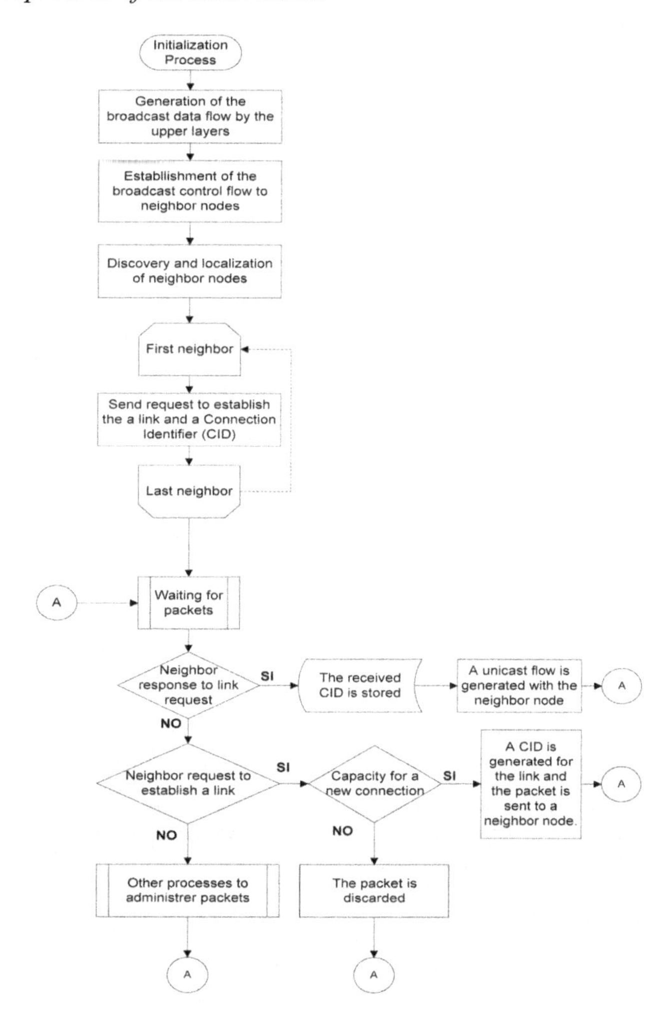

Figure 2. Simulation model for IP WiMAX-mesh architecture.

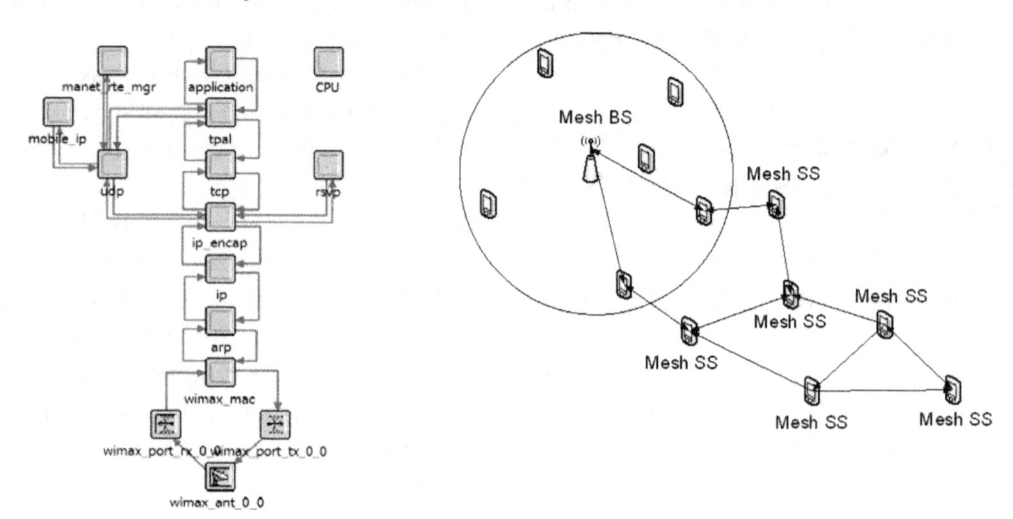

A unicast Hello message is sent to the upper data layers for each neighbor node that is discovered. This is done to administer the connection, accept the request, forward a response to the corresponding CID and establish one link for the unicast flow connection. If the node responds affirmatively, it stores the request and establishes the other link point identified by a connection (CID), thus establishing both endpoints of the data flow with their immediate neighbors. In this way, both endpoints of the data flow are established. It is important to mention that establishing connections to the mesh node itself may be necessary at some point. For this reason, mesh nodes should generate both the identification and the connection, forwarding them, if possible, to the requesting node.

Our simulation is implemented in OPNET Modeler (https://enterprise1.opnet.com/support) as shown in Figure 2. OPNET Modeler is an important network simulator that can be used to design and study communication networks, devices, protocols and applications. This chapter, therefore, evaluates five routing algorithms used for mesh topologies; however the WiMAX point-to-multipoint architecture can also be simulated with only minor modifications.

SCENARIO SIMULATED AND RESULTS OBTAINED BY SIMULATIONS

Figure 3 shows the scenario simulated in OPNET. Fifteen wireless mesh nodes are distributed in an area of eight square kilometers. Table 1, describes the main parameters utilized in the simulation model.

Tables 2, 3, 4, 5, and 6 show the simulation parameters for *Ad hoc On Demand Distance Vector (AODV)*, *Dynamic Source Routing protocol for mobile ad hoc networks (DSR)*, *Optimized Link State Routing Protocol (OLSR)*, *Temporally-Ordered Routing Algorithm (TORA)* and *Geographic Routing Protocol (GRP)*, respectively.

The scenario shown in Figure 3 has been simulated extensively in OPNET, and has been evaluated for 300 s using five mobile routing algorithms: AODV, DSR, OLSR, TORA and GRP. Fifteen fixed nodes were placed in an area of 8 square kilometers and one node was placed in one of corners. This corner node acted as the source node, which sent data packets each second to the node acting as the destination node, which was located in the opposite corner of the rectangular area. The data packet size was of 1024 bits and

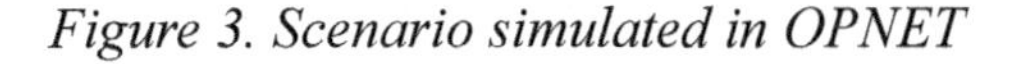
Figure 3. Scenario simulated in OPNET

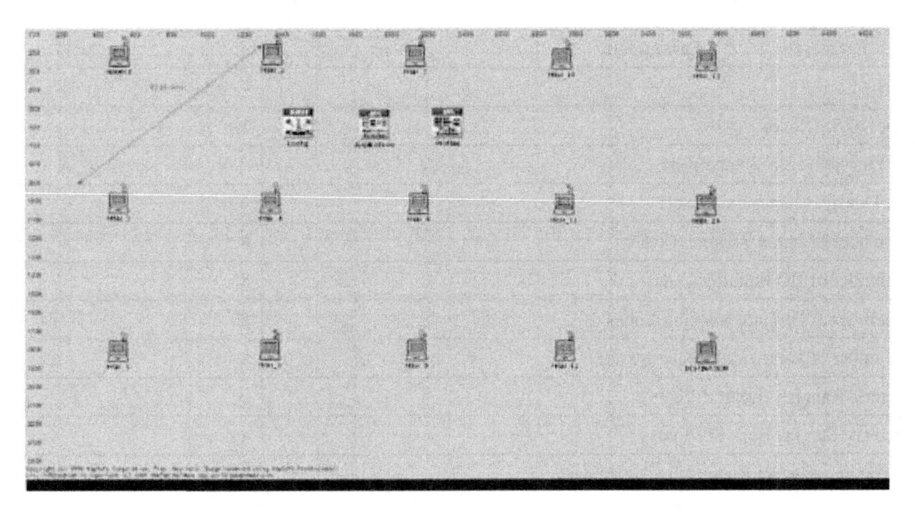

sent at a *constant bit rate (CBR)* of one second. Each node had a configured transmission range of 1 kilometer so that the packets sent by the source node had to traverse several intermediate nodes to reach the destination node.

Figure 4 shows the delivery ratio simulated in OPNET for the following wireless routing algorithms: AODV, DSR, OLSR, TORA/IMEP and GRP. AODV and DSR are reactive routing algorithms while OLSR and GRP are proactive routing algorithms. However, TORA/IMEP can be simulated as proactive or reactive. For this work, we implemented the reactive feature of TORA/IMEP. DSR presents the best performance in terms of delivery ratio. DSR does not require any neighbor sensing mechanism, which is very efficient when nodes are static. On the other hand, the worst behavior in terms of delivery ratio is obtained by GRP. GRP floods the network with HELLO packets to obtain the position information of all the network nodes which may cause packet loss.

MANET delay simulation results are presented in Figure 5. The MANET delay metric is defined as the sum of all of the possible delays caused by buffering during route discovery, queuing, re-transmission delays at the MAC layer, and propagation and transfer times. Simulation results show that GRP and OLSR are the routing algorithms with the least delay. GRP's routing strategy depends on geographic position information and is based on the source sending a message to the destination's geographic location instead of using its network address. On the other hand, OLSR employs the neighbor sensing mechanism which provides information on up to two-hop neighbors. This mechanism allows OLSR to reduce the transmission delay between the source-destination pair. AODV has the worst behavior because it employs the expanding ring search technique. In this technique, the originating node initially uses a Time to Live (TTL) equal to TTL_START in the route request IP header and sets the timeout for receiving a route reply to RING_TRAVERSAL_TIME seconds. In case of a timeout without a route reply, the originator broadcasts the request again with TTL incremented by TTL_INCREMENT. This continues until TTL reaches TTL_THRESHOLD, beyond which TTL is set to NET_DIAMETER (network-wide broadcast) for each attempt. There is a limit on maximum attempts, determined by the Route Request Retries attribute, as set in the Route Discovery Parameters. This technique is not efficient when the members of the source-destina-

Table 1. Simulation parameters for the IP WiMAX-mesh model

Parameters	Value
Antenna Gain (dBi)	-1 dB
Maximum Number of SS Nodes	10
Minimum Power Density (dBm/subchannel)	-90
Maximum Power Density (dBm/subchannel)	-60
CDMA Codes: Number of Initial Ranging Codes	8
CDMA Codes: Number of HO Ranging Codes	8
CDMA Codes: Number of Periodic Ranging Codes	8
CDMA Codes: Number of Bandwidth Request Codes	8
Back Off Parameters: Ranging Back Off Start	2
Back Off Parameters: Ranging Back Off End	4
Back Off Parameters: Bandwidth Request Back Off Start	2
Back Off Parameters: Bandwidth Request Back Off End	4
Neighbor Advertisement Interval (frames)	10
Neighborhood ID	0
Scanning Interval Definitions: Scanning Threshold (dB)	0.0
Scanning Interval Definitions: Scan Duration (N) (Frames)	5
Scanning Interval Definitions: Interleaving Interval (P) (Frames)	240
Scanning Interval Definitions: Scan Interaction (T)	10
Scanning Interval Definitions: Start Frame (M) (Frames)	5
Handover Parameters: Resource Retain Time (100 milliseconds)	2 (200 milliseconds)
Channel Quality Averaging Parameter	4/16
Neighbor Distance	1000 mts.
MAC Address	Auto Assigned
Maximum Transmission Power (W)	0.01
Mesh Role	Uncoordinated
PHY Profile	Wireless OFDMA 20 MHz
PHY Profile Type	OFDM
Multipath Chanel Model	ITU Pedestrian A
Path Loss Model	Free Space
Terrain Type (Suburban Fixed)	Terrain Type A
Shadow Fading Standard Deviation	Disable Shadow fading
Ranging Power Step (mW)	0.25
Timers: T3 (ms)	50
Timers: T4 (ms)	10
Contention Ranging Retries	16

Table 2. Simulation parameters for the ad hoc on demand distance vector (AODV) routing algorithm (Perkins, 2000)

Feature	Description
Ad Hoc On Demand Distance Vector (AODV)	
Route discovery	The AODV model implements the complete set of route discovery mechanisms, including broadcasting route requests (RREQ), establishing a reverse path while forwarding RREQ, backforwarding route replies (RREP) by intermediate nodes or the destination node, and creating a forward path while forwarding RREP. Rate limit, maximum number of retries, and exponential backoff are also implemented between each retry during the route discovery.
Maintaining sequence numbers	The AODV model maintains sequence numbers to avoid loops during route formation or update. Each node maintains a sequence number that is updated accordingly while receiving AODV control packets.
Hello messages	A node offers link connectivity information by broadcasting local Hello messages. Only the nodes that are part of an active route use Hello message. At HELLO_INTERVAL seconds, each node checks whether it has sent a broadcast within the last HELLO_INTERVAL seconds. If not, it broadcasts a Hello message.
Maintaining local connectivity	Each forwarding node maintains its continued connectivity to its active next hops. If a node does not receive a packet (Hello or otherwise) from a neighbor within ALLOWED_HELLO_LOSS × HELLO_INTERVAL seconds, it assumes that the link to the neighbor is lost.
Route maintenance	Each AODV routing table entry is associated with a route expiry timer. This timer is calculated during the route discovery process. The timer is refreshed with each packet using the specific route entry. If a route is not used for more than ACTIVE_ROUTE_TIME seconds, it is marked INVALID and cannot be used for forwarding packets. An active route can also be marked INVALID after detecting a link break (next hop connection failure). A route is deleted from the routing table only DELETE_PERIOD seconds after it has been marked invalid.
Route reply by intermediate node	During route discovery, an intermediate node having a "sufficiently fresh" route to the destination can send a reply to the source. This feature can be switched off by enabling the DESTINATION ONLY flag. With this flag set, only the destination node can reply to a route request.
Gratuitous route reply	In order for a destination to discover routes to the originating node, the originating node sets a "gratuitous route reply" (G) flag in the route request. If an intermediate node replies to a route request having the G flag set, it also unicasts a gratuitous route reply to the destination node.
Expanding ring search	To prevent unnecessary network-wide dissemination of route requests, originating nodes use the expanding ring search technique. In this technique, the originating node initially uses a TTL equal to TTL_START in the route request IP header and sets the timeout for receiving a route reply to RING_TRAVERSAL_TIME seconds. In case of timeout without a route reply, the originator broadcasts the request again with TTL incremented by TTL_INCREMENT. This continues until TTL reaches TTL_THRESHOLD, beyond which TTL is set to NET_DIAMETER (network-wide broadcast) for each attempt. There is a limit on maximum attempts, determined by the Route Request Retries attribute, as set in the Route Discovery Parameters.
Local repair	When a link break occurs in an active route, the node upstream of the break can attempt to repair the link locally if local repair is enabled on that node. During the local recovery, the node buffers the currently undeliverable packet and starts a route discovery for the destination node. Local repair attempts are often invisible to the originating node.
Route Discovery Parameters	
Route Request Retries	5
Route Request Rate Limit (pkts/sec)	10
Gratuitous Route Reply Flag	disabled
Destination Only Flag	disabled
Acknowledgement Required	disabled

continued on following page

Table 2. Continued

Feature	Description
Active Route Timeout (seconds)	10
Hello Interval (seconds)	uniform (1,11)
Allowed Hello Loss	2
Net Diameter	8
Node Traversal Time (seconds)	0.04
Route Error Rate Limit (pkts/sec)	10
Timeout Buffer	2
TTL Parameters	
TTL Start	1
TTL Increment	2
TTL Threshold	7
Local Add TTL	2
Packet Queue Size (packets)	infinity
Local Repair	enabled
Addressing Mode	IPv4

Table 3. Simulation parameters for the dynamic source routing (DSR) algorithm (Perkins, 2000)

Feature	Description
Dynamic Source Routing (DSR)	
Route discovery	The DSR model implements the complete set of route discovery mechanisms sending broadcasting route requests to find a route and receives route replies with a specific route to the destination.
Route maintenance	As DSR is designed for mobile networks, route maintenance is done to verify whether the next hop along the source route is reachable. The complete set of route maintenance mechanisms consisting of transmitting acknowledgement requests and receiving acknowledgements is implemented.
Route Cache	Each node maintains a route cache consisting of network routing information. The route cache is implemented as a "path cache," where the node maintains a set of paths to each destination. The node selects the route with the fewest number of hops to the destination.
Replying to Route Requests using Cached Routes	A node can reply to a route request even if it is not the destination. It can do this by searching its own route cache for a route to the destination of the route request.
Non-propagating route request	DSR implements a "non-propagating" route request mechanism where the route request is broadcast only to the immediate neighbors of the node performing route discovery. This request is not re-broadcasted by the neighbors. If one of the neighbors is the destination or has a route to the destination of the request, it sends a reply. If the node performing route discovery does not receive a reply within a specified period, it times out and sends out a "propagating" route request which is broadcast throughout the entire network.
Packet Salvaging	When an intermediate node forwarding a packet detects that the next hop for the packet is broken, if the node has an alternate route to the destination of the packet, the node "salvages" the packet by sending it along this alternate route.

continued on following page

Table 3. Continued

Feature	Description
Automatic Route Shortening	The route used by a packet may be automatically shortened if one or more of the intermediate nodes in the route become no longer necessary. This may happen when a node operating in promiscuous mode receives a packet in which it is not the next hop, but is named in the unexpended portion of the route. A specific node can then remove the nodes that are no longer needed.
Initial Request Period (seconds)	0.5
Non Propagating Request Timer (seconds)	0.03
Gratuitous Route Reply Timer (seconds)	1
Route Maintenance Parameters	
DSR Routes Export	Do Not Export
Route Replies using Cached Routes	Enabled
Packet Salvaging	Enabled
Non Propagating Request	Disabled
Broadcast Jitter (seconds)	uniform(0, 0.01)
Route Cache Parameters	
Max Cached Routes	infinity
Route Expiry Timer (seconds)	300
Send Buffer Parameters	
Max Buffer Size (packets)	infinity
Expiration Timer (seconds)	30
Route Discovery Parameters	
Request Table Size (nodes)	64
Maximum Request Table Identifiers (identifiers)	16
Maximum Request Retransmissions (retransmissions)	16
Maximum Request Period (seconds)	10

Table 4. Simulation parameters for the optimized link state routing (OLSR) algorithm (Perkins, 2000)

Feature	Description
Optimized Link State Routing (OLSR)	
MPR flooding mechanism	The OLSR model implements the MPR (Multi Point Relay) flooding mechanism to broadcast and flood Topology Control (TC) messages in the network. The algorithm is implemented as suggested in OLSR RFC 3626. This mechanism takes advantage of controlled flooding by allowing only selected nodes (MPR nodes) to flood the TC message. Each node selects an MPR to reach its two-hop neighbors.
Neighbor sensing mechanism	The OLSR model implements the neighbor sensing mechanism through periodic broadcast of Hello messages. These Hello messages are one-hop broadcasts (never forwarded) that carry neighbor type and neighbor quality information. The neighbor sensing mechanism provides information on up to two-hop neighbors. Generation and processing of the Hello messages are implemented as suggested in the OLSR RFC.
Topology discovery/ diffusion mechanism	Periodic and triggered Topology Control (TC) messages implement the topology discovery/diffusion mechanism in the OLSR model. TC messages are generated by MPR nodes and carry information about MPR selector nodes. These messages are diffused throughout the network using controlled flooding, thus helping to form a topology of <reachable nodes, previous hop> on each node.
Willingness	Willingness default
Hello Interval (seconds)	2.0
TC interval (seconds)	5.0
Neighbor Hold Time (seconds)	6.0
Topology Hold Time (seconds)	15.0
Duplicate Message Hold Time (seconds)	30.0
Addressing Mode	IPv4

Table 5. Simulation parameters for the temporally ordered routing algorithm (TORA/IMEP) routing algorithm (Perkins, 2000)

Feature	Description
Temporally Ordered Routing Algorithm (TORA)	
Route Creation	The model supports the route discovery to other TORA-enabled nodes in the network through the use of QRY and UPD packets.
Route Elimination	The model supports removal of routes to neighbors that have lost connectivity.
Routing mode of operation	The model supports the "On-Demand" and "Proactive" modes of operation for route discovery.
IMEP neighbor discovery	The model supports the discovery of neighbors through the full exchange of Beacon, Echo, ACK and New-Color objects.
IMEP route injection	The model supports the injection of routes for one-hop neighbors into the common route table o the node.
IMEP upper-layer packet segmentation and reassembly	The model supports the segmentation and reassembly of upper layer packets.
Mode Operation	On-Demand

continued on following page

Table 5. Continued

Feature	Description
OPT Transmit Interval (seconds)	300
IP Packet Discard Timeout (seconds)	10
Beacon Period (seconds)	20
Max Beacon Timer (seconds)	60
Max Retries (number of attempts)	3
Max IMEP Packet Length (bytes)	1,500
Route Injection	Disabled

Table 6. Simulation parameters for the geographic routing protocol (GRP) routing algorithm (Perkins, 2000)

Feature	Description
Geographic Routing Protocol (GRP)	
Hello messages	A node maintains its list of neighbor nodes by periodically broadcasting Hello messages. A configurable parameter allows users to specify the interval between Hello messages.
Maintaining local connectivity	If a node does not receive a Hello message from a neighboring node for a period exceeding the specified "Neighbor Expiry Time," it assumes the link to the neighbor is lost.
Initial flooding	To bootstrap the network, all nodes initiate full flooding throughout the network. The number of initial floods sent out by each node is specified as the "Number of Initial Floods" attribute.
Node positions	It is assumed that each node can determine its own position using GPS. The position of other nodes is determined through flooding.
Position updates	When a node moves more than a specified distance, it sends out a flooding message with its new position. One can specify the how far the node can move before it must transmit its new position. Flooding messages are also sent out when a node crosses a quadrant boundary.
Fuzzy routing	To reduce the overhead caused by flooding updates, the scope of the flooding is limited. This is known as fuzzy routing. In fuzzy routing, when a node sends a position update, only nodes that "need to know" about the change receive the flood.
Neighborhoods (quadrants)	The entire network is divided into quadrants (neighborhoods) for the purposes of optimized flooding. The size of the neighborhood can be specified using the global attribute. The neighborhoods are organized in a hierarchical manner. Each higher level neighborhood is partitioned into 4 smaller lower- level neighborhoods. To illustrate, consider a system with top level neighborhoods A and B. Within A, we define four more neighborhoods Aa, Ab, Ac and Ad. Within B, we define Ba, Bb, Bc and Bd. Similarly, within Aa we define Aa1, Aa2, Aa3 and Aa4.
Information management	When a node receives a flooding message from another node, it maintains the position information of the node based on the quadrant the node lies in. If the destination node lies in the same quadrant as the source node, it maintains the exact location of the node. However, if the destination node lies in a neighboring quadrant, the source node only maintains summarized information about the location of the destination node. For example, if a source node in quadrant A receives a flood from a destination node in quadrant B, it maintains the location of the destination node as located in quadrant B (instead of the exact position).
Neighborhood flooding	Each node knows the neighborhood that it belongs to using its location information and the knowledge of the neighborhood definitions. When a node needs to send out a position update due to distance moved within its own quadrant, it only sends out the flooding message to nodes within its own quadrant.

continued on following page

Table 6. Continued

Feature	Description
Flooding based on neighborhood transitions	When a node moves to a new quadrant, it determines the highest level where the neighborhood crossing occurred. If the node has moved from neighborhood Aa1 to Aa2, it will generate a flood scoped to neighborhood Aa. Similarly, if a node moves from neighborhood Aa1 to Ab1, the node will initiate a flood with scope equal to neighborhood A.
Forwarding	GRP is based on the shortest distance to destination. Each node that receives the data packet considers which of its neighbor nodes is closest to the destination and picks that neighbor to forward the packet to. To avoid loops, neighbor nodes that have already been traversed are omitted.
Backtracking	Sometimes, forwarding packets based on shortest distance can lead to blocked routes where there are no new nodes to which to forward the packet. Backtracking is a mechanism where the packet is returned to the previous hop where a new next hop selection can be made.

Figure 4. Delivery ratio for AODV, DSR, TORA, OLSR and GRP

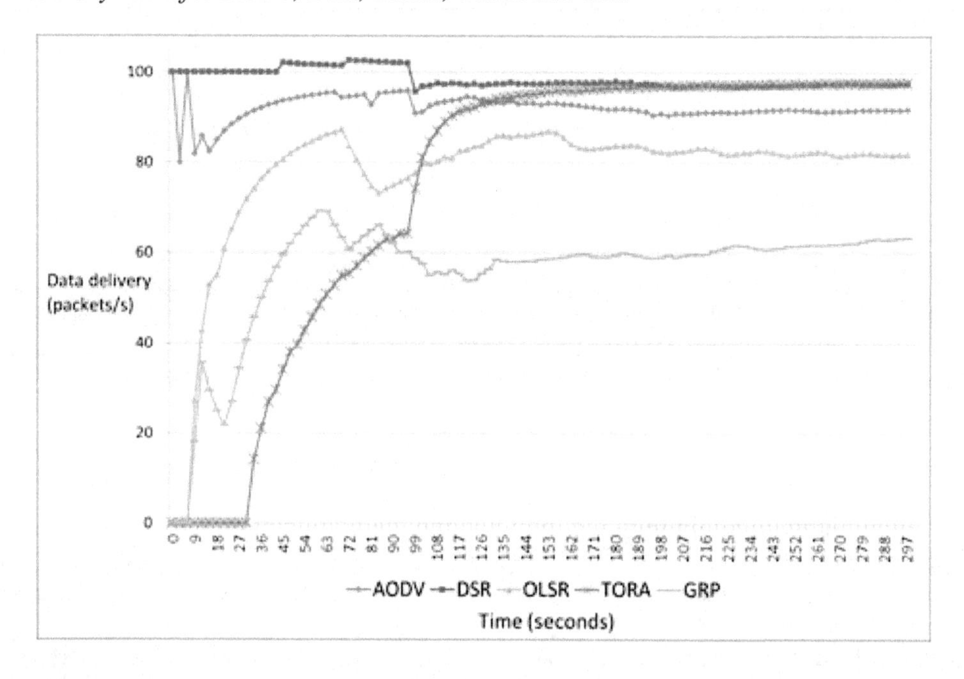

tion pair are far away from each other as in our scenario.

Figure 6 compares the routing overhead of the five routing algorithms considered in this study. The routing overhead is defined as the total number of routing packets transmitted during simulation. GRP and TORA/IMEP are the routing algorithms with the lowest routing overhead. AODV and DSR perform the worst because of their routing strategy. AODV employs its expanding ring search technique and DSR uses a multi-path reply mechanism which increases the routing overhead.

Figure 7 shows the overhead for the five routing algorithms evaluated. Overhead is the total number of routing packets that are generated, divided by the total number of data packets transmitted, plus the total number of routing packets. GRP and TORA/IMEP have the lowest overhead due to their routing mechanisms. GRP's routing strategy depends on geographic position information and is based on the source sending a message to the destination's geographic location instead of using its network address and TORA/IMEP

Figure 5. MANET delay for AODV, DSR, TORA, OLSR and GRP

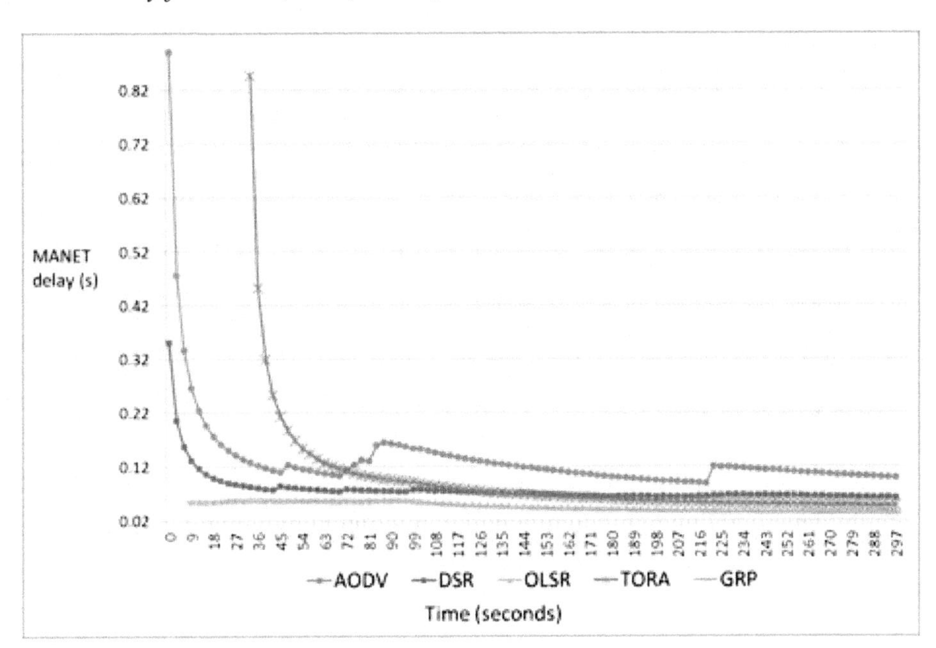

Figure 6. Routing overhead for AODV, DSR, TORA, OLSR and GRP

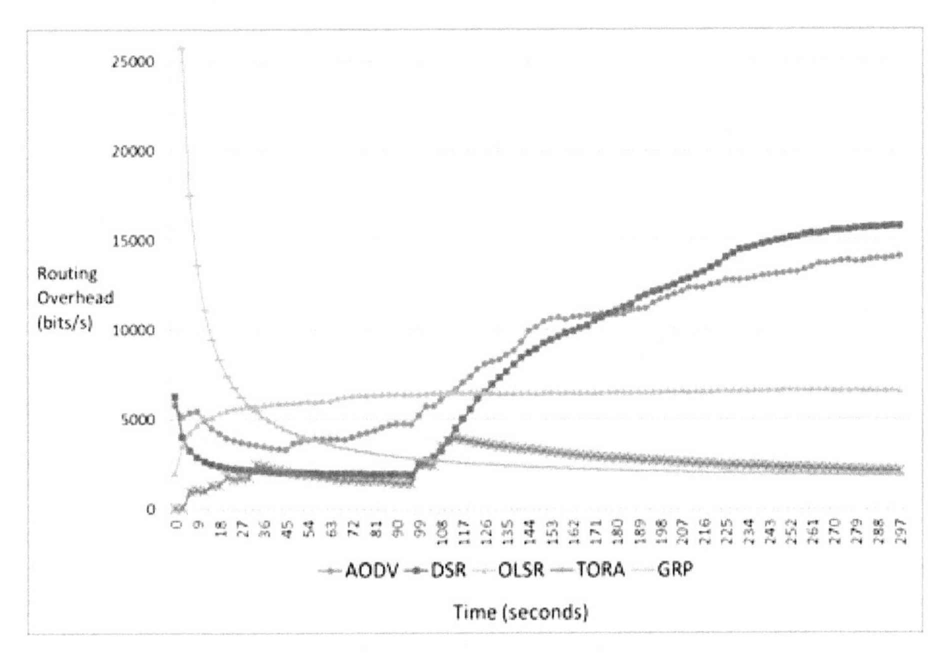

builds and maintains a Directed Acyclic Graph rooted at a destination.

AODV performs the worst in terms of overhead due to its expanding ring search technique.

Figure 8 shows the WiMAX delay. This delay is measured at the MAC layer, which is different from the MANET delay, which is measured at the network layer. Both delays, however, are peer-to-peer. OLSR has the shortest delay, due to its routing strategy which provides information on up to two-hop neighbors. DSR has the longest delay because of its routing strategy, which increases

Figure 7. Overhead for AODV, DSR, TORA, OLSR and GRP

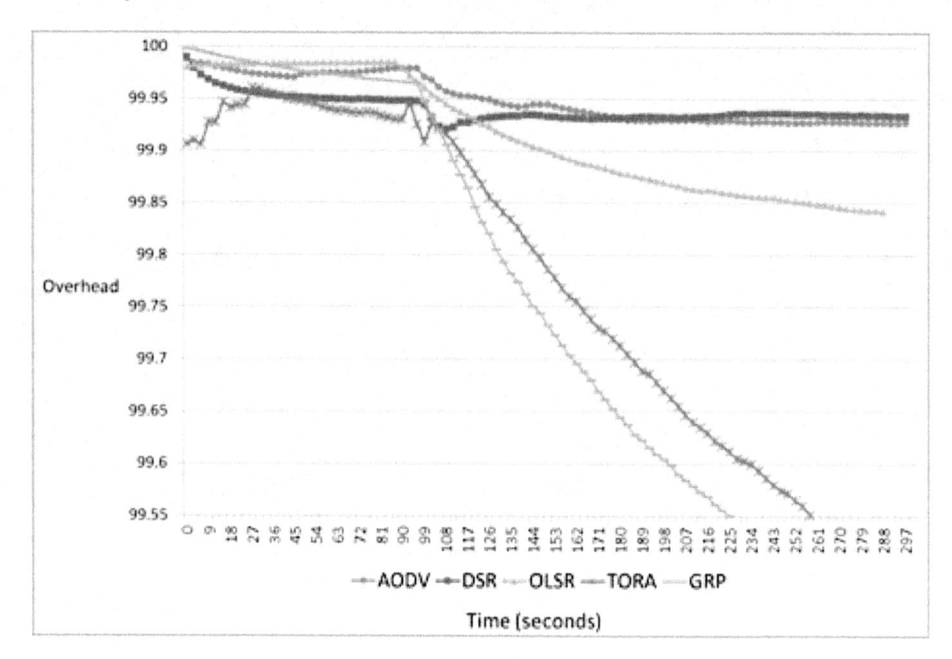

Figure 8. WiMAX delay for AODV, DSR, TORA, OLSR and GRP

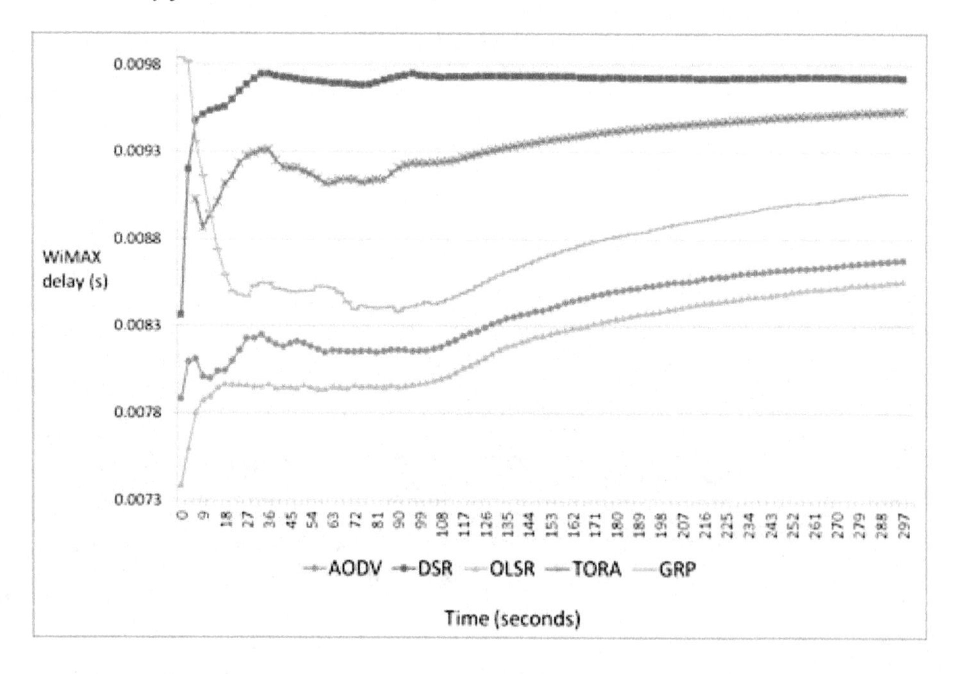

the packet header size according to the distance between the source-destination pair.

WIMAX load is compared in Figure 9; which is defined as the total load (in bits/sec) submitted to the WiMAX layer by all higher layers in all WiMAX nodes of the network. GRP shows the lowest load at the MAC layer. GRP divides the network into quadrants, which avoids the dissemination of control packets, reducing the network load. However, DSR increases its packet header size according to the distance between

Figure 9. WiMAX load for AODV, DSR, TORA, OLSR and GRP

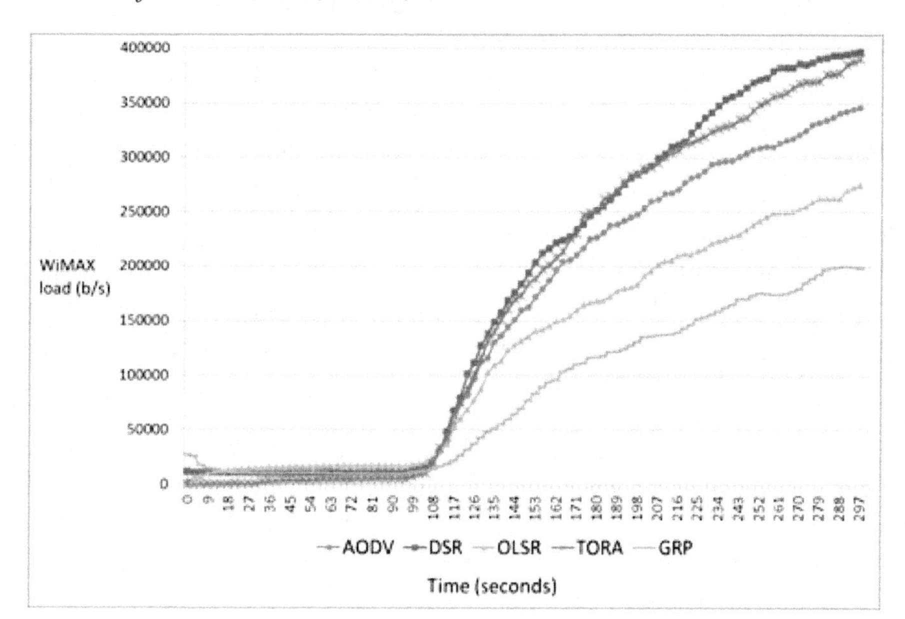

Figure 10. WiMAX throughput for AODV, DSR, TORA, OLSR and GRP

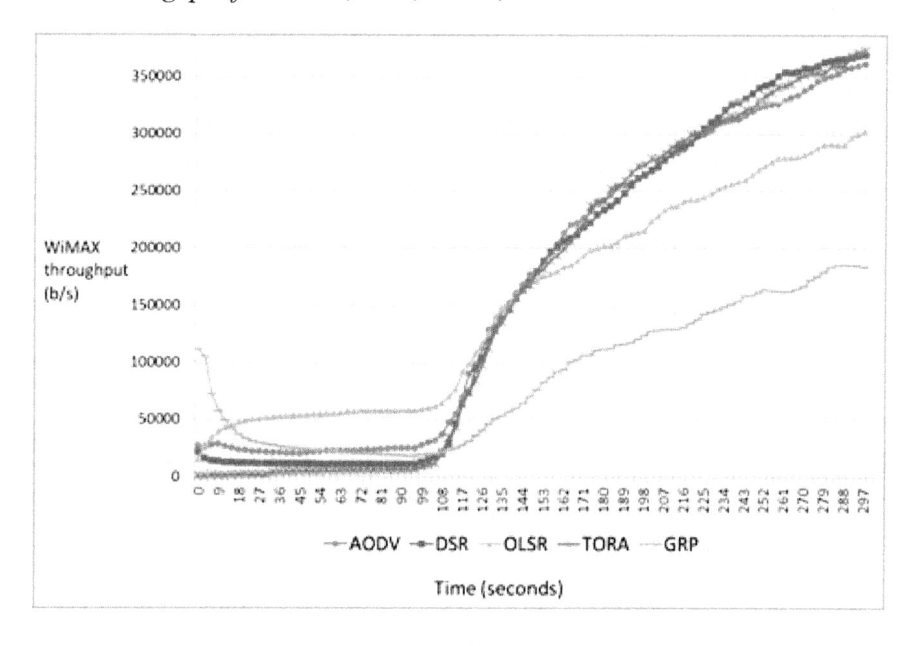

source-destination pair, thus increasing the network load.

Figure 10 shows the WiMAX throughput for the five wireless routing algorithms evaluated. AODV, DSR and TORA/IMEP are the routing algorithm with the highest throughput and GRP and OLSR with the lowest throughput.

FUTURE DEVELOPMENTS

Our future work in routing algorithms for wireless mesh networks will be developed in vehicular ad hoc networks. An emergency systems is being developed in a vehicles to avoid possible collisions, additionally, a videoconference platform based on IEEE 802.16.4 is being also developed to help enterprises to be technologically updated.

CONCLUSION

This chapter has evaluated five wireless routing algorithms: AODV, DSR, OLSR, TORA/IMEP and GRP over an IP metropolitan mesh network, which was evaluated in terms of delivery ratio, MANET delay, routing overhead, overhead, WiMAX delay, WiMAX load and WiMAX throughput.

Results show that AODV performs well with WIMAX in terms of throughput. DSR functions excellently in terms of delivery ratio and WiMAX throughput. OLSR behaves well in terms of MANET and WiMAX delay. TORA/IMEP performs best in terms of routing overhead, overhead and WiMAX throughput. Lastly, but equally important, GRP performs well in terms of MANET delay, routing overhead, overhead and WiMAX load.

In sum, proactive algorithms perform better in terms of MANET, WiMAX delay and load, because they do not require flooding the network with messages requesting the position of the destination node. However, proactive algorithms require a greater number of control packets to maintain their routing table, substantially increasing packet loss which results in low throughput.

ACKNOWLEDGMENT

We would like to thank to the National Council of Science and Technology in Mexico (CONACYT) for its support under project numbers 143582, 105279, and the National Autonomous University of Mexico-DGAPA for its support of the project PAPIIT IN108910.

REFERENCES

Ahmadi, S. (2011). *Mobile WiMAX: A systems approach to understanding IEEE 802.16m radio access technology*. Amsterdam, The Netherlands: Academic Press.

Basagni, S., Chalamtac, I., & Syrotiuk, V. (1998). A distance routing effect algorithm for mobility (DREAM). *4th Annual ACM/IEEE International Conference on Mobile Computing and Networking,* (pp. 76-84).

Chite, V. A., & Daigle, J. N. (2003). Performance of IP-based services over GPRS. *IEEE Transactions on Computers, 52*(6), 727–741. doi:10.1109/TC.2003.1204829

Clausen, T., Jacket, P., Laouiti, A., Minet, P., Muhlethaler, P., Qayyum, A., & Viennot, L. (2003). Optimized link state routing protocol (OLSR). Retrieved March 8, 2011, from http://www.ietf.org/rfc/rfc3626.txt

Dijkstra, E. W. (1959). A note on two problems in connection with graphs. *Numerische Mathematik, 1*, 269–271. doi:10.1007/BF01386390

Grace, K. (2000). *Mobile mesh border discovery protocol*. Work in Progress. (Internet Draft). Retrieved March 8, 2011, from http://www.mitre.org/work/tech_transfer/mobilemesh/draft-grace-manet-mmrp-00.txt

Hao-Min, L., Ge, Y., Pang, A. C., & Pathmasuntharam, J. S. (2010). Performance study on delay tolerant networks in maritime communication environments. *OCEANS 2010 IEEE – Sydney*, (pp. 1-6). doi:10.1109/OCEANSSYD.2010.5603627

Johnson, D., Maltz, D., & Hu, Y. (2007). *The dynamic source routing protocol for mobile ad hoc networks* (DSR). Retrieved March 8, 2011, from http://www.ietf.org/rfc/rfc4728.txt

Karp, B., & Kung, H. (2000). GPSR: Greedy perimeter stateless routing for wireless networks. *Proceedings of the 6th Annual ACM/IEEE International Conference on Mobile Computing and Networking (MobiCom 2000)*, (pp. 243-254).

Ko, Y., & Vaidya, N. (1998). Location-aided routing (LAR) in mobile ad hoc networks. *Proceedings of the 4th Annual ACM/IEEE International Conference on Mobile Computing and Networking*, (pp 66-75).

Li, J., Jannotti, J., De Couto, D., Karger, D., & Morris, R. (2000). *A scalable location service for geographic ad hoc routing* (pp. 120–130). ACM Mobicom.

Mauve, M., Widmer, J., & Hartenstein, H. (2001). A survey on position-based routing in mobile ad-hoc networks. *IEEE Network Magazine, 15*(6), 30–39. doi:10.1109/65.967595

Ogier, R., Lewis, M., & Templin, F. (2004). *Topology dissemination based on reverse-path forwarding (TBRPF)*. Retrieved March 8, 2011, from http://www.ietf.org/rfc/rfc3684.txt

Ogier, R., Templin, F., & Lewis, M. (2004). *Topology dissemination based on reverse-path forwarding (TBRPF)*. Work in Progress. Retrieved March 8, 2011, from http://www.faqs.org/rfcs/rfc3684.html

Perkins, C. (2000). *Ad hoc networking*. Boston, MA: Addison Wesley.

Perkins, C., Belding-Royer, E., & Das, S. (2003). *Ad hoc on-demand distance vector (AODV) routing*. Retrieved March 8, 2011, from http://www.ietf.org/rfc/rfc3561.txt

Pirzada, A., Portmann, M., & Indulska, J. (2006). Performance comparison of multi-path AODV and DSR protocols in hybrid mesh networks. *14th IEEE International Conference on Networks*, (pp. 1-6).

Pirzada, A., Portmann, M., & Indulska, J. (2007). Hybrid mesh ad-hoc on-demand distance vector. *Proceeding of the Thirtieth Australasian Conference on Computer Science*, (pp. 49-58).

Santos, R., Edwards, A., Edwards, R., & Seed, N. (2005). Performance evaluation of routing protocols in vehicular ad-hoc networks. *International Journal of Ad Hoc and Ubiquitous Computing, 1*(1/2), 80–91. doi:10.1504/IJAHUC.2005.008022

Santos, R., Rangel, V., Mendez, A., García-Ruiz, M. A., & Edwards-Block, A. (2010). Analyzing IEEE 802.11g and IEEE 802.16e technologies for single-hop inter-vehicular communications. In M. Watfa (Ed.), Advances in vehicular ad-hoc networks: Developments and challenges (pp. 120-149). Hershey, PA: IGI Global. Retrieved from http://www.igi-global.com/bookstore/chapter.aspx?titleid=43168

Schaumann, J. (2002). Analysis of the zone routing protocol. Retrieved March 8, 2011, from http://www.netmeister.org/misc/zrp/zrp.pdf

Zhou, M.-T., Harada, H., Kong, P.-Y., Ang, C.-W., Ge, Y., & Pathmasuntharama, J. S. (2009). A method to deliver AODV routing messages using WiMAX mesh MAC control messages in maritime wireless networks. *Proceedings of IEEE International Symposium on Personal, Indoor and Mobile Radio Communications*, (pp. 1537-1541). doi:10.1109/PIMRC.2009.5449888

Zou, X., Ramamurthy, B., & Magliveras, S. (2002). Routing techniques in wireless ad hoc networks –Classification and comparison. *Proceedings of the Sixth World Multiconference on Systemics, Cybernetics and Informatics*, (pp. 1-6).

APPENDIX: ABBREVIATIONS

AMC: adaptive modulation and coding
AODV: Ad hoc On Demand Distance Vector
AODV-HM: Ad-hoc On-demand Distance Vector Hybrid Mesh
AOMDV: Ad hoc On-demand Multipath Distance Vector
BS: Base Station
CBR: Constant Bit Rate
CDMA: Code Division Multiple Access
CID: Connection Identifier
DREAM: Distance Routing Effect Algorithm for Mobility
DSR: Dynamic Source Routing
DSR-MP: Multi Path Dynamic Source Routing
DTN: Delay Tolerant Network
FDD: Frequency-division duplex
GLS: Grid Location Service
GPS: Global Positioning System
GPSR: Greedy Perimeter Stateless Routing for Wireless Networks
GRP: Geographic Routing Protocol
IARP: Intrazone Routing Protocol
IERP: Inter-zone Routing Protocol
IP: Internet Protocol
LAR: Location-Aided Routing
LOS: Line Of Sight
LSP: Link State Packet
MAC: Medium Access Control
MAN: Metropolitan Area Network
MANET: Mobile Ad hoc Network
MIMO: Multiple Input, Multiple Output
MMBDP: Mobile Mesh Border Discovery Protocol
NLOS: Non-Line-Of-Sight
OFDM: Orthogonal Frequency Division Multiplexing
OLSR: Optimized Link State Routing Protocol
OPNET: Optimized Network Engineering Tools
PMP: Point-to-Multipoint
RD: Route Discovery
RREQ: Route Requests
RREP: Route Replies
SS: Subscriber Station
TBRPF: Topology Dissemination Based on Reverse Path Forwarding
TC: Topology Control
TDD: time-division duplex
TORA: Temporally-Ordered Routing Algorithm

TTL: Time To Live
UDP: User Datagram Protocol
WiMAX: Worldwide Interoperability for Microwave Access
Wireless-MAN-SC: Wireless Metropolitan Area Network - Single Carrier
ZRP: Zone Routing Protocol

Chapter 12
Optimal Route Selection Algorithm for Multi-Homed Mobile Network

Sulata Mitra
Bengal Engineering and Science University, India

ABSTRACT

This chapter develops the concept of route optimization in a multi-homed mobile network. In a future wireless network a user may have multiple mobile devices, each having multiple network interfaces and needing interconnection with each other as well as with other networks to form a mobile network. Such mobile networks may be multi-homed i.e. having multiple points of attachment to the Internet. It forwards packets of mobile network nodes inside it to Internet using suitable routes. But there may be multiple routes in a mobile network for forwarding packets of mobile network node. Moreover, the mobile network nodes inside a mobile network may have packets of different service types. So the optimal route selection inside a mobile network depending upon the service type of mobile network node is an important research issue. Two different route optimization schemes to create point to point network among mobile network nodes are elaborated in this chapter. This chapter is aimed at the researchers and the policy makers making them aware of the different means of efficient route selection in a multi-homed mobile network as well as understanding the problem areas that need further vigorous research.

INTRODUCTION

The users' in future wireless networks expect to be connected to the Internet from "anywhere" at "anytime," in fixed wireless locations or while on the move. A user may roam over a series of networks during his global travel. Internet browsing "on-the-move," video conferencing and file transfer are some of the new expected services in near future. In future wireless network a user may have more than one mobile devices say a mobile phone, a laptop, and a personal digital assistant. Each of these devices is likely to have multiple network interfaces that enable them to interconnect with each other as well as with other networks. These devices moving with the user

DOI: 10.4018/978-1-4666-1613-4.ch012

together are an example of a small scale *mobile network (MN)*. The access networks deployed on public transportations such as ships, trains, buses and aircrafts are examples of MNs at a larger scale. The introduction of security and privacy enhancing mechanism in future wireless networks is required without which antisocial and criminal behaviour jeopardizes the benefit of the system deployment. The hosts are identified with an IP address in such an environment. The Internetwork assures that the IP packets are indeed sent to and received from the authentic hosts. So it is required to verify the authentication of a host's identity by the routing infrastructure implicitly. A host in this environment claims to own an address that some other node is currently using, with the intention of launching a masquerade, man-in-the-middle or denial-of-service attack against the owner of the given address. Figure 1 shows the architecture of an MN. The *mobile routers (MRs)* in a MN act as a gateway between the entire MN and the rest of the Internet. A number of *mobile network nodes (MNNs)* are attached behind a MR as shown in Figure 1. The MRs which are directly connected to the Internet are the root MRs, and the MRs which are directly connected to MNNs are the leaf MRs. For example, MN1 in Figure 1 has MR1, MR3, MR5 as root MRs and MR2, MR4, MR6 as leaf MRs. A MN may change its point of attachment to the Internet. Such a MN is identified within the Internet topology by *mobile network prefix (MNP)* which is a bit string that consists of some number of initial bits of an IP address. The network to which a MN is usually connected is called its home network. The *home agent (HA)* in the home network keeps track of the location of MN. When the MN moves out of its home network and enters into a foreign network, it maintains its connectivity with its home agent through an MR-HA tunnel. The home agent tunnels all subsequent packets for MN to MR using this tunnel. So MNNs in different MNs or in the same MN create a point-to-point network among them through MR-HA tunnel. For example, MNN1 in MN1 creates a point-to-point network with MNN4 in MN2 through MR-HA tunnel as shown by firm arrow in Figure 1. MNN2 in MN1 creates a point to point network with MNN3 in MN1 through MR-HA tunnel as shown by the dotted line in Figure 1.

Each MR has one or more egress interface and one or more ingress interface. The packets forwarded from the MN to the rest of the Internet are transmitted through one of the MR's egress interfaces and packets forwarded to the MN are transmitted through one of the MR's ingress in-

Figure 1. Architecture of MN

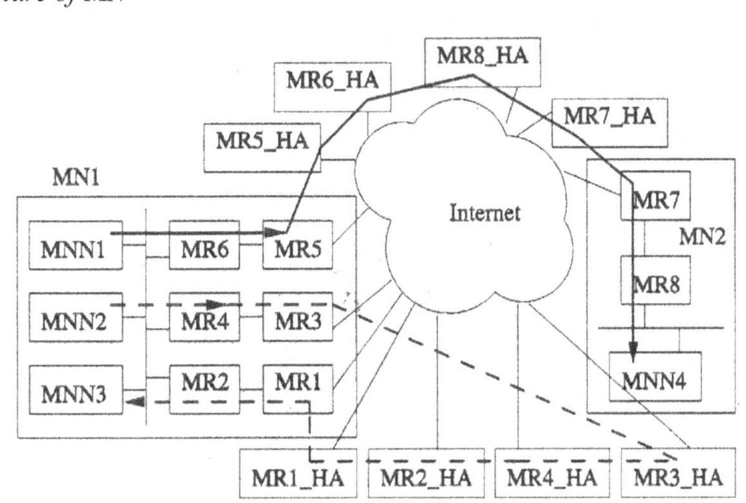

terfaces. But there may be multiple routes inside a MN for forwarding packets of MNN to the Internet. For example, in Figure1 the packets of MNN1 may be forwarded to the Internet either through the route MR6->MR5 or through the route MR4->MR3 or through the route MR2->MR1. The MNNs inside the MN may want to transmit packets of different service types for which the optimal route selection from MN to Internet is required. Moreover when an egress interface of a MR associated with an optimal route loses the connection to the global Internet, the MR can make use of its alternate egress interface if any or can utilize a route provided by an alternate MR inside MN. Two such optimal route selection schemes (Mitra 2010, Mitra & Pyne 2011) are elaborated in this chapter. Both the schemes consider a MN as shown in Figure 2. It has 6 MRs, single HA and single MNP. The proposed MN has a local fixed node (LFN) inside it. A LFN does not move w.r.t. the MN. The MN supports three different service types (data, voice, and video). The data related service is assumed as delay insensitive. The availability of bandwidth is assumed as most important to achieve fast data transfer and the packet loss should be at a minimum to achieve lossless data transfer. The voice related service is assumed as delay sensitive. So delay and packet loss should be at a minimum. The video related service is assumed as lossless and delay sensitive. Both voice and video type of service need moderate bandwidth. In (Mitra 2010) when a MNN wants to initiate a session for a particular service type it sends a route request in the form of an MNN_LFN message to the LFN inside the MN. LFN executes a dynamic route selection algorithm to select an optimal route for the desired service type of MNN from MN to Internet (Opt_Route) and sends Opt_Route to MNN in the form of LFN_MNN message. MNN sends MNN_MR message to the leaf MR associated with Opt_Route for initiating the session. MNN sends the first packet of the session corresponding to the desired service type to the leaf MR associated with Opt_Route. The leaf MR along with the other MRs associated with Opt_Route execute egress interface selection algorithm to select the best egress interface for the desired service type of MNN (Best_Eg) and delivers the packet of MNN to the next hop of Opt_Route using Best_Eg till it reaches to the Internet. But an attacker MNN may send a lot of MNN_LFN messages for route selection which may fills up the LFN queue and prevents the other authentic MNNs from accessing the service of the LFN. This results in the obvious consequence of the LFN service failure which in turn increases session loss and reduces throughput of the network. The solution of this problem is addressed in (Mitra & Pyne,2010).

Such dynamic route selection algorithm must be capable of making a decision based on incomplete information and in a region of uncertainty. Fuzzy logic deals with reasoning that is approxi-

Figure 2. Proposed mobile network

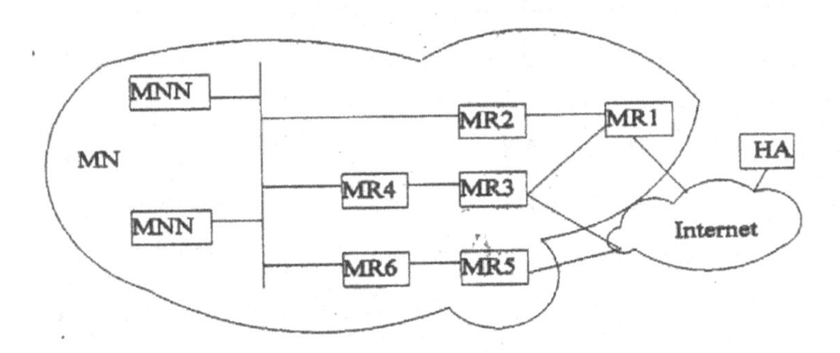

mate rather than fixed and exact. In traditional logic theory, binary sets have two-valued logic: true or false. But the fuzzy logic variables may have a truth value that ranges in degree between 0 and 1. Fuzzy logic can be viewed as a theory for dealing with uncertainty about complex systems and as an approximation theory. The route selection algorithm in (Mitra, 2010) is implemented using fuzzy logic in (Mitra & Pyne, 2011).

The existing routing protocols like linked state routing and distance vector routing consume a lot of bandwidth in message exchange among the neighbor routers. The authors in (Munasinghe & Jamalipour) proposed the integration of NEMO support for enabling group mobility management between multiple heterogeneous networks and route optimization for supporting nested mobility. A two-level intelligent route control scheme is proposed in Alshaer & Elmirghani (2009) which enables multi-homed NEMOs to optimally distribute traffic among their mobile routers' egress interfaces based on a specific mobile routing policy, traffic connection scheduling and routing path performance metrics. The nested level of NEMO is limited to one in both of these two schemes. However the NEMO in Figure 1 is more complex in terms of the number of routes from MNN to Internet and its nested level is two.

The chapter is organized as follows: Some of the published work related to the route selection in mobile network is presented in the next section. Then the route optimization schemes are elaborated. . The final section concludes the chapter.

BACKGROUND

Mobile IP (or IP mobility) is an Internet Engineering Task Force standatd communication protocol. It allows mobile device users to move from one network to another while maintaining a permanent IP address. It allows location independent routing of IP datagrams on the Internet.Initial protocol used to support mobility in network layer was Internet Protocol version 4 (IPv4). IPv4 addressing was based on 32-bit addresses. As Internet experienced a mushroom growth, the IPv4 addressing scheme became insufficient to provide the necessary support. Hence, a new version of IPv6 was introduced. IPv6 addresses are 128-bit long and is considered sufficient for years to come. The existing Mobile IP (MIP) protocols (Perkins, 2002;Johnson, Perkins & Arkko, 2004] can not support the network mobility (NEMO) as the mobility service should be provided transparently to every node inside the network. A NEMO basic support protocol has been proposed (Devarapalli et al., 2005) to support this kind of network. It is an extension of MIPv6 (Johnson et al., 2004). In Cho et al., (2005). Cho et al. (2005) proposed a home agent based (HA-based) dynamic load sharing mechanism for multihomed mobile networks. A dynamic neighbor MR authentication and registration mechanism using the Return Routerability procedure of MIPv6 is considered in this work. The proposed scheme measures tunnel latency using periodic BU/BACK messages and the HAHA protocol (Wakikawa, Devarapalli & Thubert, 2004). The HA can share traffic load with the neighbor MR-HA tunnel depending upon the measured tunnel latency.

Shima et al. (2006) proposed two operational experiments of NEMO. The first experiment is based on NEMO basic support in a real environment. The real environment was the WIDE 2005 autumn camp meeting (Shima et al., 2006). At the meeting a wireless network was provided to the attendees. The MR of the proposed mobile network had two network interfaces, one was for external connectivity and the other was used to provide the mobile network. But the result of this experiment shows a serious service disruption problem during handover. The second network mobility experiment uses the WIDE 2006 spring meeting environment (Shima et al., 2006). The multiple Care of Addresses (CoA) registration mechanism (Wakikawa et al., 2006) is used in this experiment which helps to use multiple network interfaces concurrently. The MR was equipped with three network interfaces. It can connect to a new network before leaving an old network. The multiple CoA mechanism is useful for seamless handover of a mobile network and the mobile network is practically usable as a moving network.

In Adeniji et al., (2008), the authors proposed a policy based routing protocol. It extends the *prefix scope binding update (PSBU)* message to carry sufficient topology information about nested mobile network to HA. The binding associates the network prefix with the mobile router's CoA and a sequence of intermediated mobile router's CoAs. The mobile network prefix identifies the home link within the Internet topology. The same IP prefix is used by all the mobile network nodes. The CoAs of the MRs are the addresses of the intermediated hops during packet routing into the mobile network. The MR will send a PSBU message with a chain of CoAs to register with the HA and the core network. The HA and core network build a binding entry in their binding cache after receiving this message. The HA and core network send packets to mobile networks using an optimal routing path. The proposed routing protocol helps to achieve high throughput. But it has a considerable binding update message overhead which in turn increases the traffic congestion in the home link.

ROUTE OPTIMIZATION SCHEMES

Two route optimization schemes are elaborated on in the following subsections. The first scheme is named as Scheme_I (Mitra 2010) and the second scheme is named as Scheme_II (Mitra & Pyne 2011). The main objective of Scheme_I and Scheme_II is to select Opt_Route for the MNN from MN to Internet for their desired service type.

The proposed MN (Figure 2) has 6 MRs (NO_OF_MR=6). MR1, MR3 and MR5 are the root MRs and MR2, MR4, MR6 are the leaf MRs. The number of egress interface (NO_OF_EI) of each MR is assumed as 4 (NO_OF_EI=4). Both Scheme_I and Scheme_II consider three sets as X=(E1,E2,E3,E4)=(X1,X2,X3,X4), Y=(Delay, Unused bandwidth, Packet loss, Cost)=(Y1,Y2,Y3,Y4) and Z=(data,voice,video)=(Z1,Z2,Z3). The set X indicates 4 egress interfaces of each MR and $1 \leq X \leq NO\_OF\_EI$, the set Y indicates 4 parameter values (NO_OF_P=4) to determine the status of each egress interface and $1 \leq Y \leq NO\_OF\_P$, the set Z indicates 3 different service types (NO_OF_ST=3) that are supported by the MN and $1 \leq Z \leq NO\_OF\_ST$. Each MR determines the status of each of its egress interfaces depending upon the values of the 4 parameters such as delay, unused bandwidth, packet loss and costThe parameter values per egress interface is maintained by each MR in the form of g(X,Y). For example, g(X1,Y1), g(X1,Y2), g(X1,Y3) and g(X1,Y4) elements of g(X,Y) are $Delay_{E1}$ (delay), un_BW_{E1} (unused bandwidth), PL_{E1} (packet loss) and $Cost_{E1}$ (cost) respectively of egress interface E1; g(X2,Y1), g(X2,Y2), g(X2,Y3) and g(X2,Y4) elements of g(X,Y) are $Delay_{E2}$ (delay), un_BW_{E2} (unused bandwidth), PL_{E2} (packet loss) and $Cost_{E2}$ (cost) respectively of egress interface E2; g(X3,Y1), g(X3,Y2), g(X3,Y3) and g(X3,Y4) elements of g(X,Y) are $Delay_{E3}$ (delay), un_BW_{E3} (unused bandwidth), PL_{E3} (packet loss) and $Cost_{E3}$ (cost) respectively of egress interface E3; g(X4,Y1), g(X4,Y2), g(X4,Y3) and g(X4,Y4) elements of g(X,Y) are $Delay_{E4}$ (delay), un_BW_{E4} (unused bandwidth), PL_{E4} (packet loss) and $Cost_{E4}$ (cost) respectively of egress interface E4;.

Each MR executes egress interface selection algorithm to determine Best_Eg from its 4 egress interfaces depending upon their parameter values. It also determines its own status depending upon the status of its egress interfaces and sends its status to the LFN using MR_LFN message.

There are 4 possible routes (NO_OF_R=4) of transmission from MN (Figure 2) to the Internet as shown in Table 1. The selection of route r1 as Opt_Route depends upon the combined status of

Table 1. Possible routes in proposed MN

Route	Path	Route length in hops
r1	MNN->MR2->MR1->Internet	3
r2	MNN->MR4->MR3->Internet	3
r3	MNN->MR6->MR5->Internet	3
r4	MNN->MR4->MR3->MR1->Internet	4

MR2 and MR1, selection of route r2 as Opt_Route depends upon the combined status of MR4 and MR3, selection of route r3 as Opt_Route depends upon the combined status of MR6 and MR5, selection of route r4 as Opt_Route depends upon the combined status of MR4, MR3 and MR1.

Route optimization Scheme_I: The route optimization Scheme_I is elaborated in this section.

The MNN_LFN message contains 2 components as MNN identification (MNN_id) of the session and Service type (S_type). In case of 100000 MNN, the number of bits required to represent MNN_id is 17. In case of 3 different service types supported by MN, the number of bits required to represent the service type is 2. So the length of this message is 19 bits. The LFN_MNN message contains the identification of Opt_Route, the identification of the leaf MR associated with Opt_Route and session identification (Session_id). In case of 4 routes, the number of bits require to represent the identification of the selected route is 2. In case of 6 MRs, the number of bits required to represent the identification of the leaf MR is 3.

Each session has a unique Session_id as assigned by the LFN after selecting Opt_Route for that session. The LFN maintains one counter (session_count) to count the number of active session. The counter value increases by 1 after selecting each route per session. The number of bits require to represent Session_id is $\log_2$(session_count). So the length of LFN_MNN message is 5+$\log_2$(session_count) bits. The MNN_MR message has 3 different formats (Figure 3). The format as shown in Figure 3(a) is used as the header of the first packet. Its length (len_first) is 26+$\log_2$(S_no)+$\log_2$(P_no)+$\log_2$(session_count) bits where S_no indicates the sequence number of each packet and P_no indicates the number of packets in the corresponding session. The format as shown in Figure 3(b) is used as the header of the last packet and its length (len_last) is 2+$\log_2$(S_no)+$\log_2$(session_count) bits. The format as shown in Figure 3(c) is used as the header for all the intermediate packets and its length (len_int) is $\log_2$(session_count)+$\log_2$(S_no) bits. The S_flag and F_flag in Figure 3(a) and Figure 3(b) indicate the start flag and finish flag respectively. S_flag is set in the first packet of the session to indicate the start of the session and F_flag is set in the last packet of the session to indicate the end of the session.

Figure 3. Packet header as MNN_MR message (a) For first packet (b) For last packet (c) For intermediate packet

(a)

MNN_id	LFN_MNN message	S_type	S_no	S_flag	F_flag	P_no

(b)

Session_id	S_no	S_flag	F_flag

(c)

Session_id	S_no

Function of each MR: The function of each MR for the selection of Opt_Route is considered for discussion in this section.

Maintenance of Routing Table: Each MR maintains a routing table to keep the record of various sessions in the form (MNN_id, Session_id, P_no). The LFN_MNN message part of the header which is available with the first packet of the corresponding session has the value of the Session_id attribute. It is also used by all the MRs associated with Opt_Route to insert the value of MNN_id and P_no attributes in the routing table. The leaf MR associated with Opt_Route receives packet from MNN and the other MRs associated with the same route receive packet from their predecessor MR. Each MR associated with Opt_Route inserts a record in the routing table after receiving the first packet of that session and deletes the record from the routing table after receiving the last packet of that session. When a MR receives a packet, it searches the routing table using Session_id as the searching key to retrieve the corresponding record. If found, it verifies the MNN_id and transmits the packet to the next hop of Opt_Route using Best_Eg. A route remains idle for a long time if the corresponding MNN becomes out of order or stops transmission or go out of the coverage area of the MN. A route becomes out of order in case of failure of the link(s) associated with it. The MRs associated with such route delete the corresponding record from the routing table and makes the resources associated with such route free which helps to improve the resource utilization within the MN.

Computation of the parameter values of Ej: Each MR computes the element values of g(X,Y). For p^{th} MR (MRp), where $1 \leq p \leq 6$, g(X,Y) is identified as gp(X,Y). The element values of gp(X,Y) are $Delay_{E1\_p}$, $un\_BW_{E1\_p}$, PL_{E1_p}, $Cost_{E1\_p}$ for egress interface E1 of MRp; $Delay_{E2\_p}$, $un\_BW_{E2\_p}$, PL_{E2_p}, $Cost_{E2\_p}$ for egress interface E2 of MRp; $Delay_{E3\_p}$, $un\_BW_{E3\_p}$, PL_{E3_p}, $Cost_{E3\_p}$ for egress interface E3 of MRp; $Delay_{E4\_p}$, $un\_BW_{E4\_p}$, PL_{E4_p}, $Cost_{E4\_p}$ for egress interface E4 of MRp. Let MRp selects its jth egress interface (Ej where $1 \leq j \leq NO\_OF\_EI$) as Best_Eg for the desired service type of ith MNN (MNN_i). In this section the computation of the 4 parameter values of Ej by MRp is considered for discussion.

The predecessor MR of MRp associated with Opt_Route for the session of MNN_i includes the current time stamp in the header of each packet of MNN_i before sending it to MRp. MRp measures the time stamp after transmitting the same packet using Ej to the next hop of Opt_Route. The difference of the two time stamp (δ_{ij}) is considered as the delay per packet of MNN_i using Ej. The initial value of delay at Ej of MRp ($Delay_{Ej\_p}$) is assumed as 0.0 msec. So $Delay_{Ej\_p}$ is increased by δ_{ij} after transmitting a single packet of MNN_i with computation complexity O(1).

In case the MRs are in the WiFi network, the available bandwidth per egress interface of the MR can be assumed as the bandwidth of WiFi network. The initial value of the unused bandwidth at Ej of MRp ($un\_BW_{Ej\_p}$) is assumed as the available bandwidth at Ej (av_BW_{Ej}) and $desire\_BW_{ij}$ indicates the bandwidth which is required for the service type of MNN_i. So after receiving the first packet from MNN_i, $un\_BW_{Ej\_p}$ is reduced by $desire\_BW_{ij}$ and after receiving the last packet from MNN_i, $un\_BW_{Ej\_p}$ is increased by $desire\_BW_{ij}$ with computation complexity O(1). It has been assumed that each MR of MN knows the desired bandwidth per service type.

The packet loss at any egress interface is the sum of the packet loss due to time out and buffer overflow. A counter is maintained at each egress interface to count the number of loss of packets. The initial value of packet loss counter at Ej of MRp (PL_{Ej_p}) is assumed as 0. Each MR searches all the packets in the buffer at Ej for time out and increases PL_{Ej_p} by 1 after removing a packet from the buffer at Ej due to time out with computation complexity O(number of packets in egress buffer). PL_{Ej_p} is also increased by 1 after removing a packet from the buffer at Ej due to buffer overflow with computation complexity O(1). The packet loss at Ej is computed in % as (PL_{Ej_p}/total packet at Ej)*100.

The cost per egress interface is the sum of cost of all the MNNs using that particular egress interface. The cost of each MNN is the sum of route selection cost and transmission cost. The route selection cost depends upon the overhead due to message exchange for the selection of Opt_Route. Now the overhead due to message exchange is the sum of bits in MNN_LFN message, LFN_MNN message and MNN_MR message.

The transmission cost is the product of the amount of data in bits and cost/bit. Now the amount of data in bits is the product of the number of packet and size of packet in bits. The initial value of cost at Ej of MRp ($Cost_{Ej\_p}$) is assumed as 0. Let $Cost_{ij}$ indicates the cost for the service type of MNN_i using Ej, where

$Cost_{ij}$=[52+(P_no+1)*$\log_2$(session_count)+P_no*$\log_2$(S_no)+$\log_2$(P_no)+ (P_no*P_sz)]*cost/bit, P_sz indicates the size of packet for the desired service type. P_sz per service type is fixed and is known to each MR of MN. The average packet size for data, voice and video service type is assumed as 8000, 640 and 720 bits respectively. After receiving the first packet from MNN_i. $Cost_{Ej\_p}$ is increased by $Cost_{ij}$ with computation complexity O(1).

Each MR performs the same computation to calculate the 4 parameter values (Delay, Unused bandwidth, Packet loss, Cost) of its 4 egress interfaces. Each MR also maintains a table (Table 2) to keep the parameter values of 4 egress interfaces in the form (Egress, Delay, Unused bandwidth, Packet loss, Cost). The parameter values of Ej at MRp are shown in Table 2.

Computation of status parameter values: Each MR computes its own status parameter values at the end of transmission of each session. MRp computes $Delay_p$, un_BW_p, PL_p and $Cost_p$ as its own status parameters using the element values of gp(X,Y).

$$Delay_p = (Delay_{E1\_p} \wedge Delay_{E2\_p} \wedge Delay_{E3\_p} \wedge Delay_{E4\_p})$$

$$un\_BW_p = (un\_BW_{E1\_p} \vee un\_BW_{E2\_p} \vee un\_BW_{E3\_p} \vee un\_BW_{E4\_p})$$

$$PL_p = (PL_{E1\_p} \wedge PL_{E2\_p} \wedge PL_{E3\_p} \wedge PL_{E4\_p})$$

$$Cost_p = (Cost_{E1\_p} \wedge Cost_{E2\_p} \wedge Cost_{E3\_p} \wedge Cost_{E4\_p})$$

Egress Interface Selection Algorithm per Service Type: Each MR executes this algorithm after receiving the first packet of a session. Ej is Best_Eg of MRp for data service if $un\_BW_p = un\_BW_{Ej\_p}$, for voice service if $Delay_p = Delay_{Ej\_p}$, and for video service if $PL_p = PL_{Ej\_p}$. For example if $Delay_p = Delay_{E3\_p}$, E3 is the Best_Eg of MRp for voice service.

In the best case one egress interface of MRp is determined as Best_Eg for the desired service type of MNN_i by this algorithm. In the worst case multiple egress interfaces of MRp are determined as Best_Eg for the desired service type of MNN_i. In such a case the algorithm uses best possible egress interface selection function to determine the best egress interface. Let the qth and n^{th} egress interface (Xq and Xn) of MRp are selected as Best_Eg for Z^{th} service type of MNN_i where ($1 \le q \le NO\_OF\_EI$, $1 \le n \le NO\_OF\_EI$ and $q \ne n$). $Delay_{Xq\_p}$, $un\_BW_{Xq\_p}$, PL_{Xq_p}, $Cost_{Xq\_p}$ are the 4 parameter values of Xq and $Delay_{Xn\_p}$, $un\_BW_{Xn\_p}$, PL_{Xn_p}, $Cost_{Xn\_p}$ are the 4 parameter values of Xn.

Table 2. Parameter values of Ej at MRp

Egress	Delay	Unused bandwidth	Packet Loss	Cost
Ej	$Delay_{Ej}$	un_BW_{Ej}	PL_{Ej}	$Cost_{Ej}$

The flow chart of the best possible egress interface selection function for data class of traffic (Flow_data_E) is shown in Figure 4(i), for voice and video class of traffic (Flow_vo_vi_E) is shown in Fig.4(ii). Both Flow_data_E and Flow_vo_vi_E use func Y2 and para func. The flow chart of func Y2 and para func are shown in Figure 4(iii).

Function of each LFN: LFN computes the values of the parameters such as Delay, Unused bandwidth, Packet loss and Cost of the 4 routes

Figure 4. (i). Flow_data_E (ii). Flow_vo_vi_E (iii). Flow chart for (a) func Y2 (b) para func for para=Y1 or Y3 or Y4.

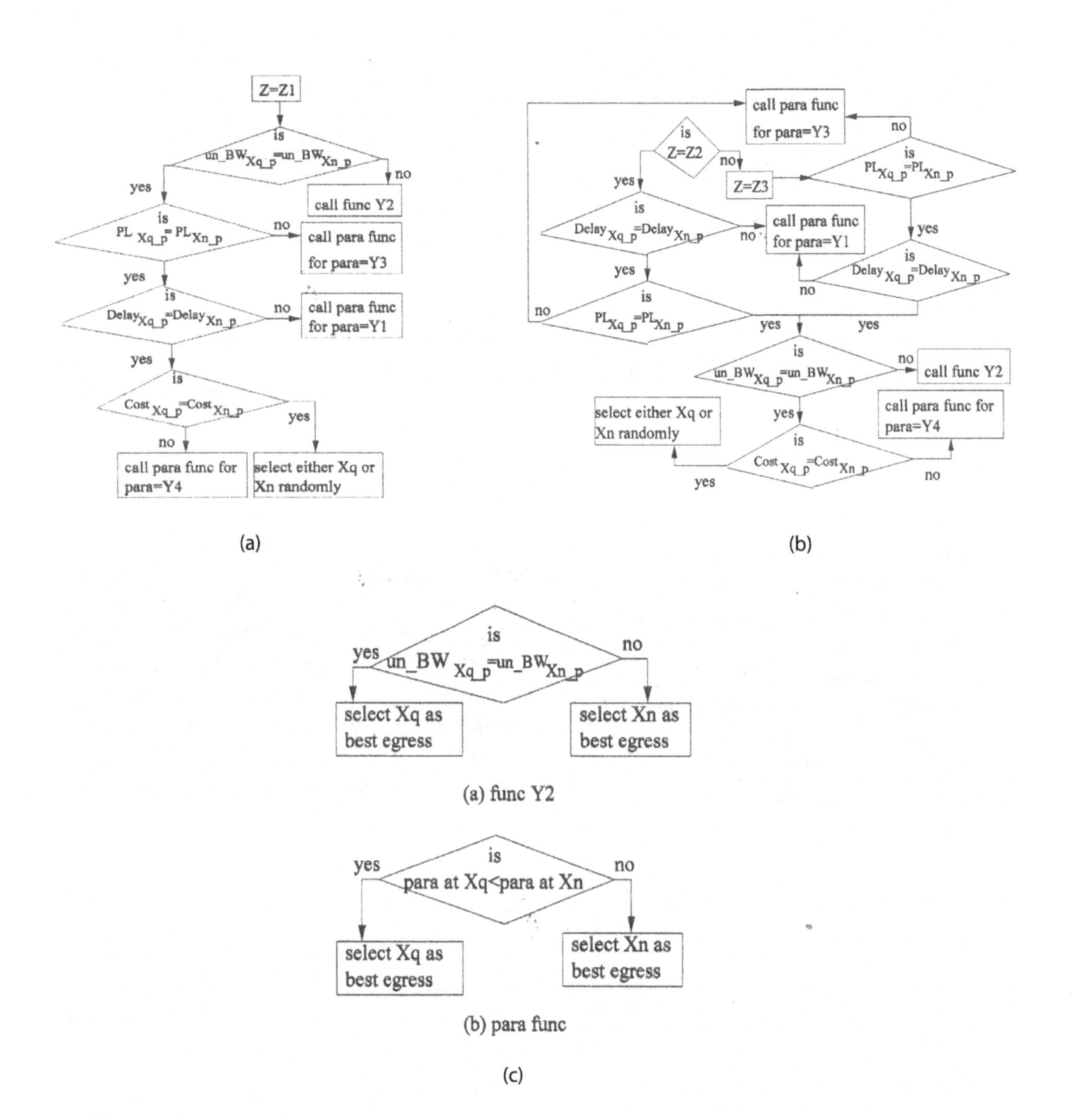

from MN to Internet after receiving MR_LFN message. The LFN also executes route selection algorithm after receiving MNN_LFN message. The function of the LFN is considered for discussion in this section.

Computation of parameter values for 4 routes: The MRs MR2 and MR1 are associated with the route r1. Delay_r1, un_BW_r1, PL_r1 and Cost_r1 are the 4 parameter values of r1.

$$\text{Delay\_r1}=(\text{Delay}_{MR2} \vee \text{Delay}_{MR1})$$

$$\text{un\_BW\_r1}=(\text{un\_BW}_{MR2} \wedge \text{un\_BW}_{MR1})$$

$$\text{PL\_r1}=(\text{PL}_{MR2} \vee \text{PL}_{MR1})$$

$$\text{Cost\_r1}=(\text{Cost}_{MR2} \vee \text{Cost}_{MR1})$$

The MRs MR4 and MR3 are associated with r2. Delay_r2, un_BW_r2, PL_r2 and Cost_r2 are the 4 parameter values of r2.

$$\text{Delay\_r2}=(\text{Delay}_{MR4} \vee \text{Delay}_{MR3})$$

$$\text{un\_BW\_r2}=(\text{un\_BW}_{MR4} \wedge \text{un\_BW}_{MR3})$$

$$\text{PL\_r2}=(\text{PL}_{MR4} \vee \text{PL}_{MR3})$$

$$\text{Cost\_r2}=(\text{Cost}_{MR4} \vee \text{Cost}_{MR3})$$

The MRs MR6 and MR5 are associated with r3. Delay_r3, un_BW_r3, PL_r3 and Cost_r3 are the 4 parameter values of r3.

$$\text{Delay\_r3}=(\text{Delay}_{MR6} \vee \text{Delay}_{MR5})$$

$$\text{un\_BW\_r3}=(\text{un\_BW}_{MR6} \wedge \text{un\_BW}_{MR5})$$

$$\text{PL\_r3}=(\text{PL}_{MR6} \vee \text{PL}_{MR5})$$

$$\text{Cost\_r3}=(\text{Cost}_{MR6} \vee \text{Cost}_{MR5})$$

The MRs MR4, MR3 and MR1 are associated with r4. Delay_r4, un_BW_r4, PL_r4 and Cost_r4 are the 4 parameter values of r4.

$$\text{Delay\_r4}=(\text{Delay}_{MR4} \vee \text{Delay}_{MR3} \vee \text{Delay}_{MR1})$$

$$\text{un\_BW\_r4}=(\text{un\_BW}_{MR4} \wedge \text{un\_BW}_{MR3} \wedge \text{un\_BW}_{MR1})$$

$$\text{PL\_r4}=(\text{PL}_{MR4} \vee \text{PL}_{MR3} \vee \text{PL}_{MR1})$$

$$\text{Cost\_r4}=(\text{Cost}_{MR4} \vee \text{Cost}_{MR3} \vee \text{Cost}_{MR1})$$

Route selection algorithm: The optimal route selection algorithm for different service types in an MN are considered for discussion in this section.

Route selection for data service: The proposed algorithm selects the route having maximum unused bandwidth for data service. LFN computes a parameter value r_data as

$$\text{r\_data}=(\text{un\_BW\_r1} \vee \text{un\_BW\_r2} \vee \text{un\_BW\_r3} \vee \text{un\_BW\_r4}).$$

The algorithm selects route r1 if r_data=un_BW_r1, route r2 if r_data=un_BW_r2, route r3 if r_data = un_BW_r3 and route r4 if r_data=un_BW_r4.

Route selection for voice service: The proposed algorithm selects the route having minimum delay for voice service. LFN computes a parameter value r_voice as

r_voice=(Delay_r1∧Delay_r2∧Delay_r3∧Delay_r4). The algorithm selects route r1 if r_voice=Delay_r1, route r2 if r_voice=Delay_r2, route r3 if r_voice=Delay_r3 and route r4 if r_voice=Delay_r4.

Route selection for video service: The proposed algorithm selects the route having minimum packet loss for video service. LFN computes a parameter value r_video as

r_video=(PL_r1∧PL_r2∧PL_r3∧PL_r4). The algorithm selects route r1 if r_video=PL_r1, route r2 if r_video=PL_r2, route r3 if r_video=PL_r3 and route r4 if r_video=PL_r4.

In the best case, one route from MN to Internet is selected as Opt_Route for Z^{th} service type of MNN_i by the algorithm. In the worst case, let the route rd and rt be selected as Opt_Route for Z^{th}

service type of MNN_i, where $(1 \leq d \leq NO\_OF\_RT$, $1 \leq t \leq NO\_OF\_RT$ and $d \neq t)$. In such a case the algorithm uses the best possible route selection function to determine the best route from the MN to the Internet for the Zth service type of MNN_i.

Delay_rd, un_BW_rd, PL_rd, Cost_rd are the 4 parameter values of route rd and Delay_rt, un_BW_rt, PL_rt, Cost_rt are the 4 parameter values of route rt. The flow chart of the best possible route selection function for data class of traffic (Flow_data_R_I) is shown in Figure 5(i), for voice and video class of traffic (Flow_vo_vi_R_II) is shown in Figure 5(ii). Both Flow_data_R_I and Flow_vo_vi_R_II use unused bandwidth func and para func. The flow chart of unused bandwidth func and para func are shown in Figure 5(iii).

Figure 5. (i). Flow_data_R_I. (ii). Flow_vo_vi_R_I. (iii). Flow chart for (a) Unused Bandwidth func (b) Para func.

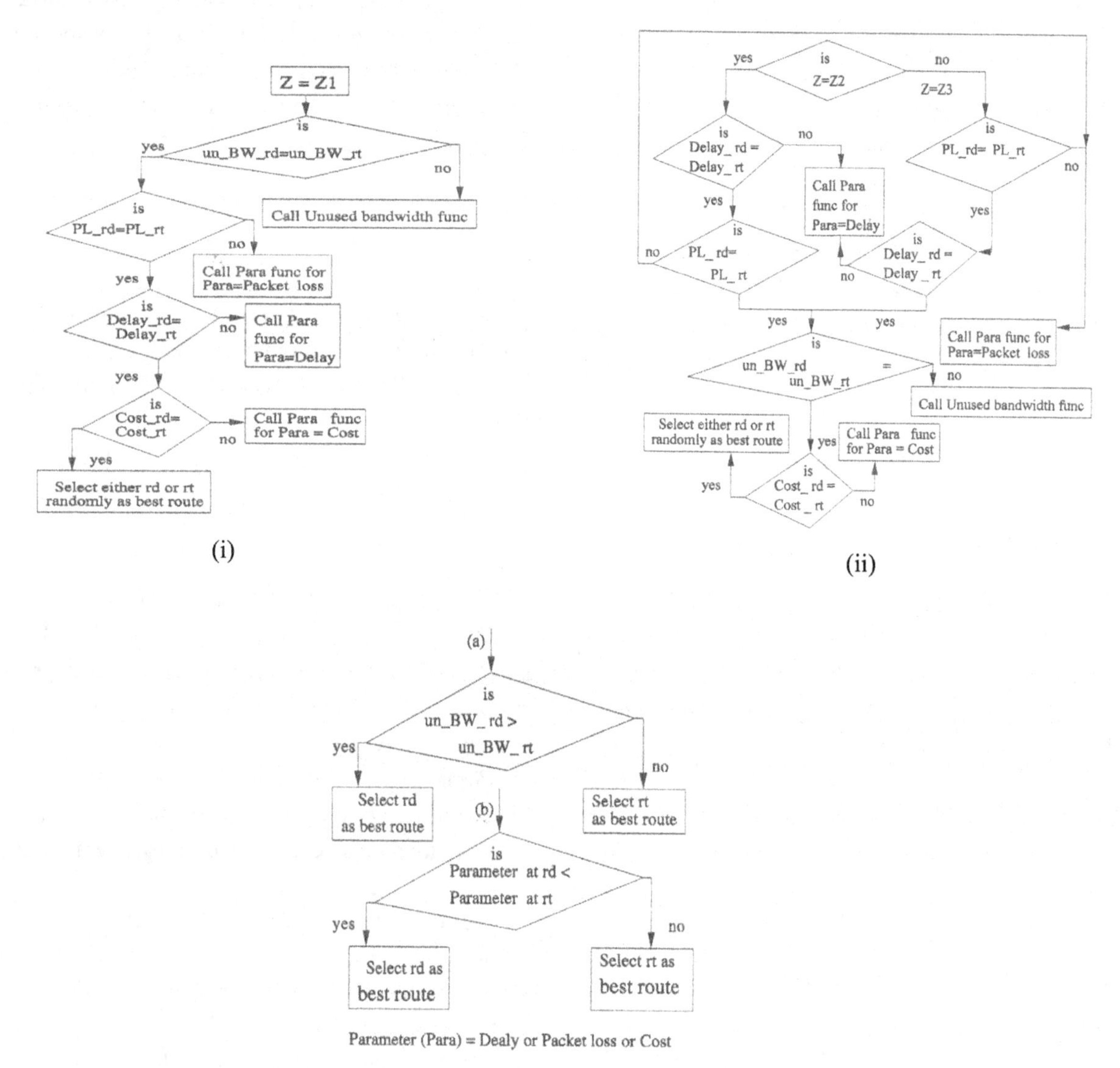

NEMO_SIM Simulator

Scheme_I is simulated using NEMO_SIM simulator. It is an application based object oriented simulator. When a user gives a complete MN as input to NEMO_SIM, the NEMO_SIM automatically creates an environment of a MN where communication can take place. The NEMO_SIM is implemented using Java, because of platform free usage of the executable Java programs and also for further extension of the simulator to be accessed online. Java has a good set of Application Program Interfaces that largely benefits the development of complex simulation software. NEMO_SIM can be a part of NS2 simulation environment by using AgentJ (Taylor et al., 2006), which is a Java Virtual Machine for NS2. NEMO_SIM can also act as an extended part of JNS 1.7, Java Network Simulator [//jns.sourceforge.net].

The MN in the proposed scheme (Figure 2) is the combination of some interconnected processing units such as MNN, LFN, MR. The processing units are treated as threads and the whole MN is considered as a complex producer-consumer problem in a large scale. Java provides a facility for using multiple threads and thread synchronization, which is the main ingredient for building NEMO_SIM. The processing units and the corresponding threads are shown in Figure 6. The function of all the threads is discussed below:

MNN_REQ (T_1) Thread: It sends MNN_LFN message to LFN. A MNN has only one T_1 thread.

LFN_MNN (T_2) Thread: It receives MNN_LFN message request from MNN, runs the route selection algorithm and sends LFN_MNN message to MNN.

MNN_SERVICE_START (T_3) Thread: It receives LFN_MNN response message and starts a new session. A MNN has only one T_3 thread.

MNN_SERVICE (T_4) Thread: It creates a new session for the desired application, transmits packet corresponding to the desired application to the ingress interface of the leaf MR corresponding to the optimal route. After transmitting all the packets successfully T_4 thread dies. A MNN has zero or more T_4 thread depending upon how many sessions are still alive.

MR_ROUTE_UPDATION (T_5) Thread: It sends MR_LFN message.

LFN_MR (T_6) Thread: It receives MR_LFN message.

MR_PACKET_RECEIVE_FORWARD (T_7) Thread: It receives a packet from the ingress queue and forwards it to the best egress interface as selected by the egress interface selection algorithm.

MR_egress (T_8) Thread: It receives a packet from the best egress queue and forwards it to the ingress queue of the next hop. It also computes packet loss due to the overflow at the egress queue.

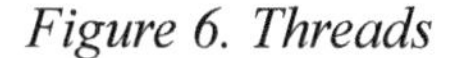

Figure 6. Threads

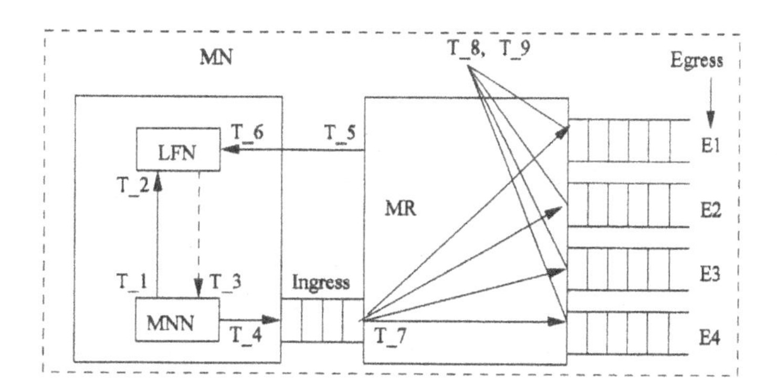

MR_EGRESS_PACKET_LOSS(T_9) Thread: It discards the packets from the egress queue due to time out.

Simulation of Scheme_I: The simulation experiment of Scheme_I is carried out considering the internal network of MN (Figure 2) as WiFi (IEEE 802.11a). The size of LFN buffer, MR egress as well as ingress buffer and MNN buffer are assumed as 1000, 10^5 and 1000 respectively.

Figure 7, Figure 8 and Figure 9 show in the plots of throughput, session loss and route selection time vs. traffic load of the proposed MN (Figure 2). The traffic load is computed as the ratio of arrival rate and departure rate of session request by the LFN. The route selection time is constant

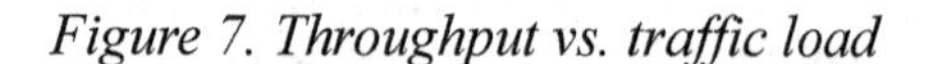

Figure 7. Throughput vs. traffic load

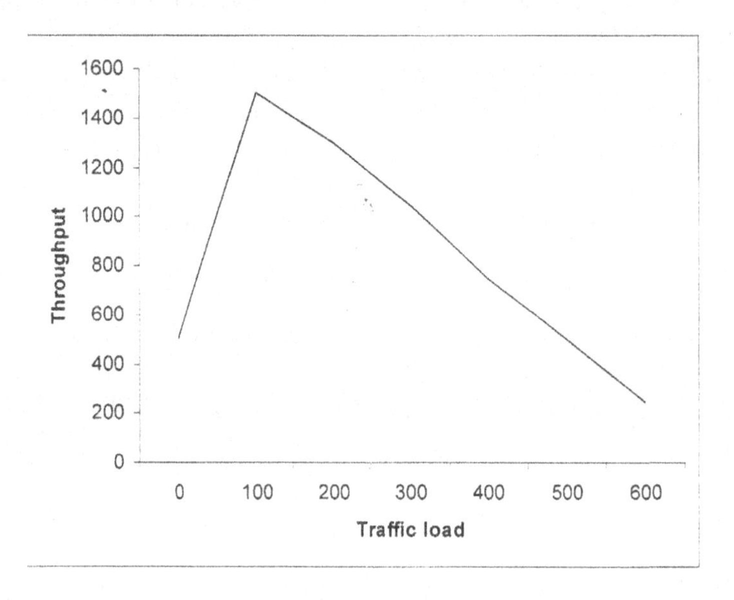

Figure 8. Session loss vs. traffic load

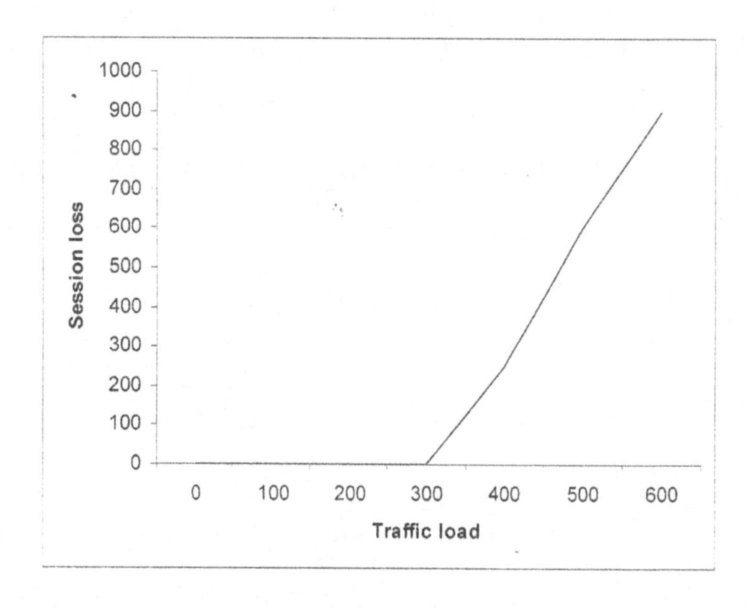

Figure 9. Route selection time vs. traffic load

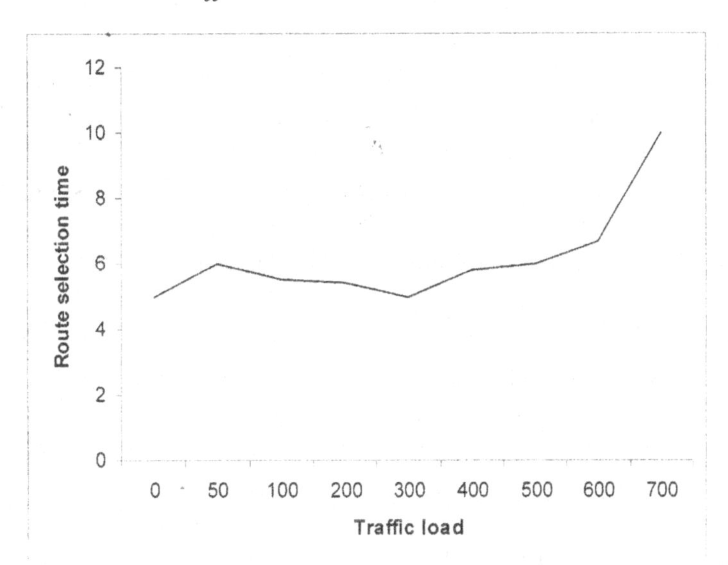

up to traffic load 300 and then it increases slowly, which causes a decrease in throughput with traffic load. Session loss is zero up to traffic load 350 and then it increases slowly.

But in Scheme_I MNN sends MNN_LFN message as route request message to the LFN for its session initiation. An attacker MNN may send a lot of such route request message to the LFN intentionally to flood the LFN queue. But the LFN is a fixed local infrastructure inside MN. It is a very smart device, capable of handling a large number of requests and possessing a large computational power. So it would be extremely difficult for a single MNN to make the LFN inactive by sending requests.

Thus, in reality the mechanism of *Distributed Denial of Service (DDoS)* attack is considered. The main culprit MNN makes a number of MNNs obey its orders by taking them under its control. Then it manipulates all the MNNs to act as traps for other MNNs and so on. The MNNs which are converted to slave machines are termed as Zombies (Gibson). After a certain time a huge number of MNNs become Zombies. All these MNNs send route request message to the LFN which in turn fills up the LFN queue and prevents the other authentic MNNs from accessing the service of the LFN. This results in the obvious consequence of the LFN service failure which in turn increases session loss and reduces throughput of MN.

The solution of such security issue for Scheme_I is addressed in (Mitra & Pyne, 2010). It uses a priority queue based solution in which LFN assigns a priority value to each route request of MNN. If the number of route requests from a particular MNN increase, the priority value assign to each such route request reduces. As a result when a new route request arrives from a new MNN, LFN assigns a higher priority to it. But this method is useful only if the LFN can recognize the MNN identification distinctly. In case of a DDoS attack, the route requests come from different MNNs making it practically impossible to stop this attack by this method. A non preemptive roun- robin scheduling algorithm is also used in (Mitra & Pyne, 2010) to schedule the route requests at the LFN queue. This process of implementation provides an equal amount of time to each MNN by using time-slicing technique. Both the priority queue-based algorithm and non-preemptive round-robin scheduling algorithm give an opportunity to the authentic MNNs for accessing the service of LFN.

The performance of Scheme_I is evaluated with and without incorporating the solution of the security issue using NEMO_SIM simulator in (Mitra & Pyne, 2010). Two threads are added in NEMO_SIM simulator during the simulation of (Mitra & Pyne, 2010), MNN_service_stop thread and False_MNN_service thread. The attacker MNN sends a lot of MNN_LFN message for route selection to the LFN. The LFN sends LFN_MNN messages in response to each MNN_LFN message to the attacker MNN. The MNN_service_stop thread accepts the first LFN_MNN message and discards the other LFN_MNN message. The False_MNN_service thread at MNN generates spurious packets and transmits them using Opt_Route, as specified in the first LFN_MNN message which unnecessarily utilize a lot of resources available in the MN. The simulation experiment is conducted to observe the variation of throughput and session loss vs. simulation time in the presence of an attacker MNN.

The simulation experiment is conducted in the same simulation environment as in Scheme_I. It considers three different cases as follows:

- No attacker MNN is present
- Only one MNN is an attacker MNN
- Three MNNs are attacker MNN

Figure 10 shows the plot of throughput vs. simulation time for all the 3 cases. The throughput is maximum in CASE I and minimum in CASE III. The throughput in CASE II is less than CASE I and higher than CASE III. Figure 11 shows the plot of session loss vs. simulation time for all the 3 cases. The session loss is minimum in CASE I and maximum in CASE III. The session loss in CASE II is less than CASE III and higher than CASE I. Figure 12 shows the plot of throughput vs. simulation time of MN after incorporating priority queue based solution. It can be observed from Figure 12 that throughput increases with simulation time. Moreover it is higher than the throughput corresponding to CASE III in Figure 10. Figure 13 shows the plot of throughput vs. simulation time of MN after incorporating round-robin scheduling. It can be observed from Figure 13 that throughput increases with simulation time. Moreover it is higher than the throughput corresponding to CASE III in Figure 10.

Figure 14 shows the plot of route selection time vs. session count when the scheduling of the LFN queue is round robin type (CASE I), when the LFN queue is implemented as priority queue (CASE II) and the LFN queue is implemented as a first in first out queue (CASE III). In CASE III the time complexity of placing a request in the queue and of removing a request from the queue

Figure 10. Throughput vs. simulation time for LFN flooding attack

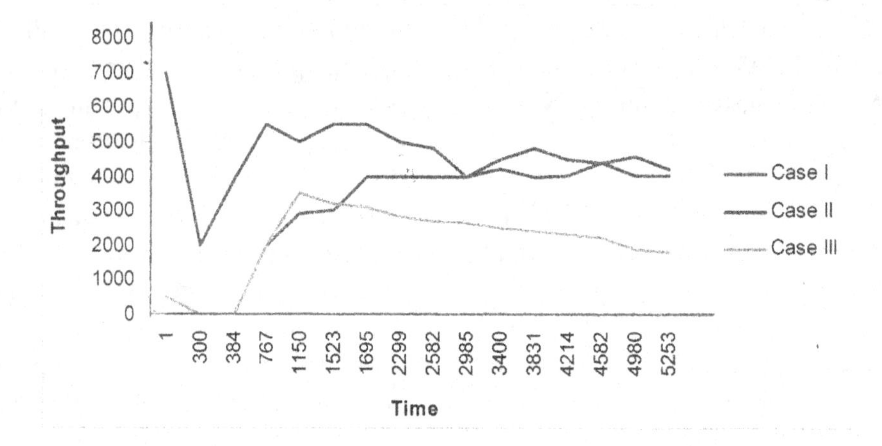

Figure 11. Session loss vs. simulation time for LFN flooding attack

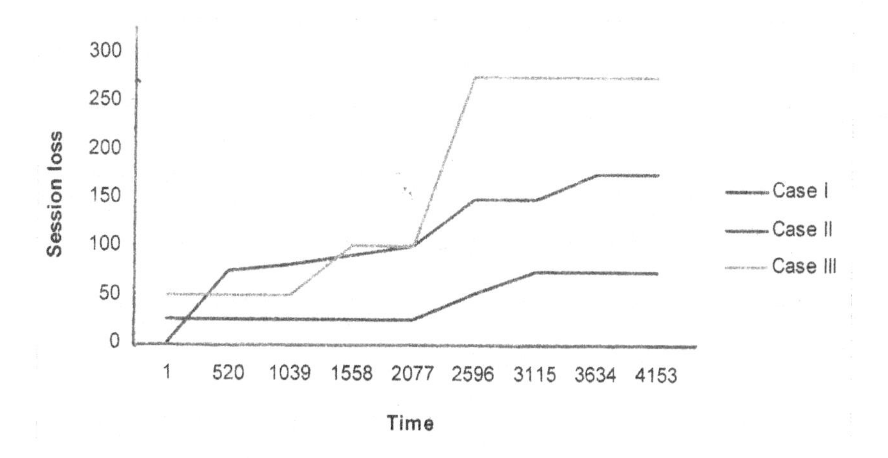

Figure 12. Throughput vs. simulation time with priority based solution

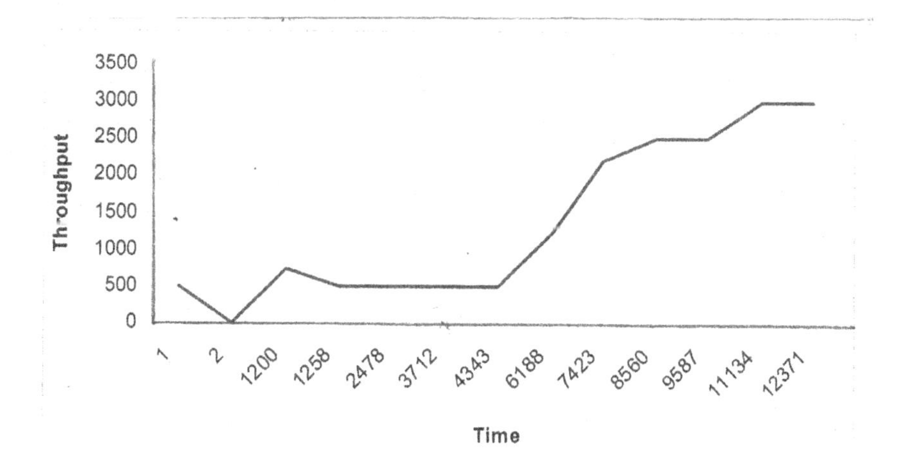

Figure 13. Throughput vs. simulation time with round robin solution

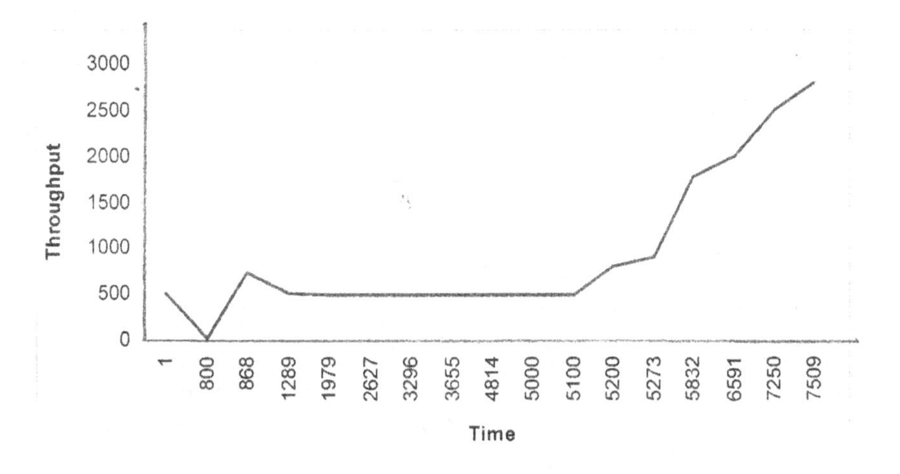

Figure 14. Route selection time vs. session count with and without solution

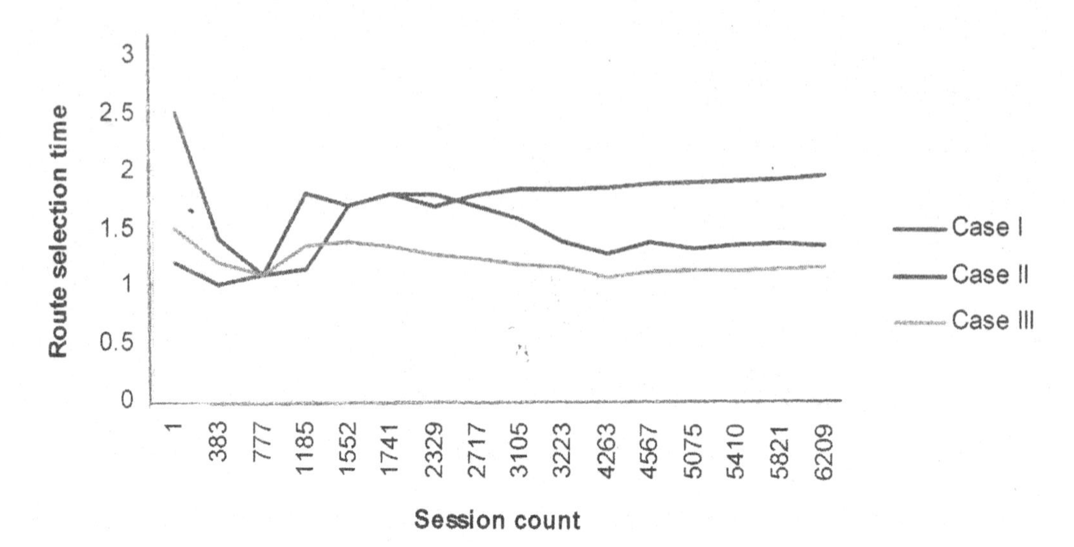

is O (1). In CASE I the time complexity of placing a request in the queue is O(1) but the time complexity of removing a request from the queue on an average case is O(n) where n is the number of requests present in the LFN queue. In CASE II the priority queue is implemented as a heap. So the total time complexity of placing a request in the queue and of removing a request from the queue is O ($nlog_2n$), which is greater than both O (1) and O (n). From the plot it can be observed that the route selection time is maximum in CASE II and minimum in CASE III. The route selection time in CASE I is less than CASE II and higher than CASE III.

Route optimization Scheme_II: The route optimization Scheme_II is elaborated in this section.

The MNN_LFN message has 3 components as MNN_id, S_type, and request number. If the total number of requests per MNN is req_no, the number of bits require to represent request number is $\log_2$(req_no). So the maximum length of MNN_LFN message is assumed as 19+$\log_2$(req_no) bits. LFN_MNN message contains Session_id, l_i-1 number of (MR_id,E_id) pairs where l_i is the length of ri^{th} route in number of hops and $1 \leq i \leq$NO_OF_R. For example let the route r1 is selected as Opt_Route. MR1 and MR2 are the MR identification (MR_id) of the MRs, which are associated with the route r1. So this message contains two (MR_ id, E_id) pair as (MR2,E_id of suitable egress interface of MR2) and (MR1,E_id of suitable egress interface of MR1). The E_id of suitable egress interface of MR1 and E_id of suitable egress interface of MR2 are the egress identification (E_id) of Best_Eg as determined by MR1 and MR2 respectively using best egress determination algorithm. In case of 6 MRs, the number of bits require to represent MR_id is 3. In case of 4 egress interfaces of each MR, the number of bits require to represent E_id is 2. So the number of bits require to represent a single (MR_id,E_id) pair is 5. In case of the routes r1, r2, r3 LFN_MNN message contains 2 such pair as 2 MRs are associated with these routes. So the length of LFN_MNN message is the sum of $\log_2$(session_count) bits and 10 bits. In case of route r4 LFN_MNN message contains 3 such pair as 3 MRs are associated with r4. So the length of LFN_MNN message is the sum of $\log_2$(session_count) bits and 15 bits. LFN generates Session_id and l_i-1 number of (MR_id,E_id) pairs of this message with computation complexity O(l_i).

The MNN_MR message of Scheme_II is identical to the MNN_MR message of Scheme_I. Each MR generates MR_LFN message after executing best egress determination algorithm. The format of this message which is generated by MRp is shown in Figure 15. This message has 3 parts. The part1 of this message (Figure 15(a)) contains 3 services Z1, Z2 and Z3 along with their identification number (S_id) 0, 1 and 2 respectively. S_id for service type Z1 is 0. So the 0th element of the variable array (variable_array[0]) in part3 of this message contains the 4 parameter values (i.e. NO_OF_P) (Y1, Y2, Y3, Y4) of Best_Eg which is determined by MRp for service type Z1 in the form gp(X,Y). Similarly the 1th element (variable_array [1]) and the 2th element (variable_array[2]) of the variable array in part3 of the message contain 4 parameter values of Best_Eg which is determined by MRp for service type Z2 and Z3 respectively. For 3 services the numbers of bits require to represent S_id is 2. So the maximum length of part1 is assumed as 6 bits.

The part2 of this message (Figure 15(b)) contains fuzzy values corresponding to Best_Eg of MRp per service type. Let the egress interfaces X1, X2 and X3 of MRp are selected as Best_Eg for service type Z1, Z2 and Z3 respectively. So the part2 of this message (Figure 15(b)) indicates X1 is Best_Eg of MRp for service type Z1, X2 is Best_Eg of MRp for service type Z2 and X3 is Best_Eg of MRp for service type Z3. MRp(X1,Z1), MRp(X2,Z2) and MRp(X3,Z3) are the 3 fuzzy values as determined by MRp during the execution of the best egress determination algorithm. The size of each fuzzy value is assumed as 32 bits. So the maximum length of part2 is assumed as 3x32=96 bits.

The part3 of the MR_LFN message (Figure 15(c)) is an array having variable number of elements and so this array is known as variable array. Each element of this array indicates Best_Eg in the form X and its 4 parameter values (i.e. NO_OF_P) in the form gp(X,Y) for MRp per service type. So each element contains (NO_OF_P+1) number

Figure 15(a). Part1 of MR_LFN message (b). Part2 of MR_LFN message (c). Part3 of MR_LFN message

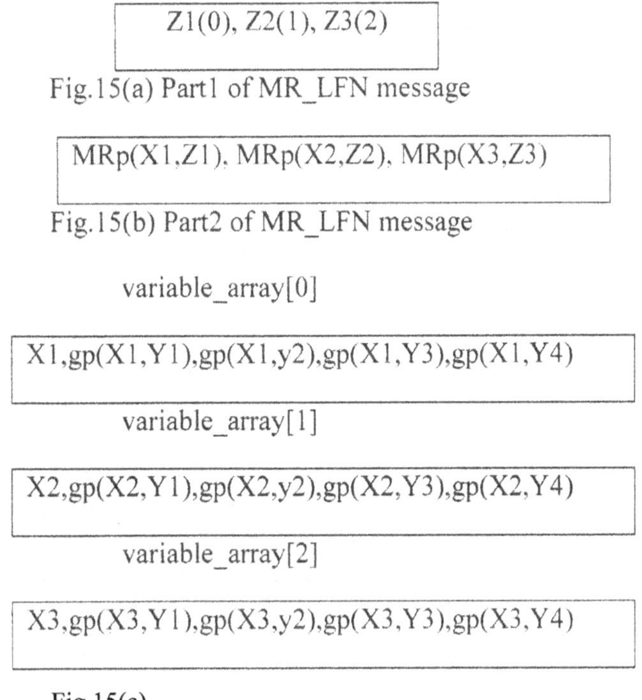

of components. The number of elements in this array is decided depending upon 3 different cases. In the worst case Best_Eg of 3 different service types as determined by MRp are different i.e. X1≠X2≠X3. In the average case Best_Eg of any 2 different service types as determined by MRp are identical i.e. ((X1=X2 and X1≠X3 and X2≠X3) or (X1=X3 and X1≠X2 and X2≠X3) or (X2=X3 and X1≠X3 and X1≠X2).

In the best case, Best_Eg of 3 different service types as determined by MRp are identical i.e. (X1=X2=X3). So the number of elements in the variable array for the worst case is 3 (i.e. NO_OF_ST) one per service type, for the average case is 2 and for the best case is 1. For example, the 0th element of variable_array (variable_array[0]) in Figure 15(c) contains X1 as E_id of Best_Eg of MRp for service type Z1. It also contains gp(X1,Y1), gp(X1,Y2), gp(X1,Y3) and gp(X1,Y4) as the 4 parameter values of the egress interface X1. Each MR has 4 egress interfaces and the number of bits require to represent each egress identification number is 2. The size of each gp(X,Y) is assumed as 32 bits. So the maximum length of part3 in the worst case is assumed as 3*(2+4*32)=390 bits. Hence the maximum length of the MR_LFN message is 6+96+3(2+4*32)=492 bits.

The maximum length of part3 in the average case is assumed as 2*(2+4*32)=260 bits. Hence, the maximum length of the MR_LFN message is 6+96+2(2+4*32)=362 bits. The maximum length of part3 in the best case is assumed as 1*(2+4*32)=130 bits. Hence, the maximum length of the MR_LFN message is 6+96+1*(2+4*32)=232 bits. The computation complexity to generate MR_LFN message depends upon the computation complexity of generating part3 of this message due to its variable length. MR generates a single element having (NO_OF_P+1) number of components in the best case with computation complexity O(NO_OF_P) and (NO_OF_ST) number of elements each having (NO_OF_P+1) number of components in the worst case with computation complexity O(NO_OF_ST*NO_OF_P).

Function of each MR: The function of each MR for the selection of Opt_Route is considered for discussion in this section.

Maintenance of routing table: Each MR maintains a routing table (Table 3) to keep the records of various sessions in the form (MNN_id, E_id, Session_id, desire_BW, Next_hop). One record is maintained for each session of a MNN. Each MR associated with Opt_Route inserts a record in the routing table after receiving the first packet of that session and deletes the record from the routing table after receiving the last packet of that session. The value of the attributes MNN_id and Session_id of each record are obtained from the header available with the first packet of the corresponding session (Figure 3(a)) as discussed in Scheme_I.

The desire_BW attribute of each record indicates the required bandwidth to maintain the desired service type which is known to each MR within MN. The Next_hop attribute of each record indicates the next node associated with Opt_Route. The Next_hop attribute is Internet in the routing table which is maintained by the root MR associated with Opt_Route. It is MR in the routing table which is maintained by the leaf MR and by all the intermediate MRs associated with Opt_Route. The LFN_MNN message part of the header avail-

Table 3. Routing table

MNN_id	E_id	Session_id	desire_BW	Next_hop
MNN_i	Ej	S_s	BW_z	MR/Internet

able with the first packet (Fig.3(a)) of the corresponding session is used by all the MRs associated with Opt_Route to insert the value of the attributes E_id and Next_hop in a record. For example, if Opt_Route is r1, MR2 is the leaf MR and MR1 is the root MR (Figure 2). The LFN_MNN message part in the header of the first packet contains (MR2,E_id of suitable egress interface of MR2),(MR1,E_id of suitable egress interface of MR1). MR2 inserts "E_id of suitable egress interface of MR2" as the E_id attribute and MR1 as the Next_hop attribute in the corresponding record. MR1 inserts "E_id of suitable egress interface of MR1" as the E_id attribute and Internet as the Next_hop attribute in the corresponding record. When a MR receives packet from a MNN, it searches the routing table using Session_id as the searching key to retrieve the record of MNN with computation complexity O(1). The MR retrieves the value of the attribute E_id from the record of MNN for delivering the packet using Bet_Eg and the value of the attribute Next_hop from the record of MNN for delivering the packet to the next hop of Opt_Route.

The record for MNN_i using Ej is shown in Table 3.

Maintenance of fizzifiers F1 and F2: Each MR maintains 2 fuzzifiers F1 and F2. The function of F1 and F2 are considered for discussion in this section.

Operation of F1: It considers the sets X and Y as in Scheme_I. The element values of g(X,Y) are computed by each MR in the same way as discussed in Scheme_I. F1 transforms g(X,Y) into the binary fuzzy relation P(X,Y) on fuzzy sets X and Y if at least one element of g(X,Y) changes. P(X,Y) is defined as

$$\begin{bmatrix} F1\_De_{E1} & F1\_BW_{E1} & F1\_PI_{E1} & F1\_\text{cost}_{E1} \\ F1\_De_{E2} & F1\_BW_{E2} & F1\_PI_{E2} & F1\_\text{cost}_{E2} \\ F1\_De_{E3} & F1\_BW_{E3} & F1\_PI_{E3} & F1\_\text{cost}_{E3} \\ F1\_De_{E4} & F1\_BW_{E4} & F1\_PI_{E4} & F1\_\text{cost}_{E4} \end{bmatrix}$$

where each element in P(X,Y) is the fuzzy value of the corresponding element in g(X,Y). For example, $F1\_De_{E1}$, $F1\_BW_{E1}$, $F1\_PL_{E1}$, and $F1\_cost_{E1}$ are the fuzzy values corresponding to $Delay_{E1}$, un_BW_{E1}, PL_{E1}, and $Cost_{E1}$ respectively.

Computation of P(X,Y): The output parameters of fuzzifier F1 are the 4 fuzzy values of 4 parameters per egress interface which is maintained by each MR in the form P(X,Y). The computation of the 4 fuzzy values at the output of F1 for Ej using the appropriate fuzzy membership function is discussed below.

$$F1\_De_{Ej} = 1 - (Delay_{Ej} \,/\, \sum_{j=1}^{NO\_OF\_EI} (Delay_{Ej}))$$

if $\sum_{j=1}^{NO\_OF\_EI} (Delay_{Ej}) \neq 0$

$F1\_De_{Ej}$ is assumed as 1 if

$$\sum_{j=1}^{NO\_OF\_EI} (Delay_{Ej}) = 0$$

$Delay_{Ej} \,/\, \sum_{j=1}^{NO\_OF\_EI} (Delay_{Ej})$ indicates delay at Ej w.r.t. NO_OF_EI number of egress interfaces. If $Delay_{Ej}$ is very high, $F1\_De_{Ej}$ becomes very low which indicates the status of Ej is not good for the parameter delay.

$F1\_De_{Ej}$ is computed using (NO_OF_EI-1) number of addition, 1 division and 1 subtraction with computation complexity O(NO_O_EI). The F1 fuzzifier at MRp computes $F1\_De_{E1}$, $F1\_De_{E2}$, $F1\_De_{E3}$ and $F1\_De_{E4}$ of its 4 egress interfaces E1, E2, E3 and E4 respectively.

$F1\_BW_{Ej}$=(un_BW_{Ej})/(available bandwidth at Ej)

If un_BW_{Ej} is very high $F1\_BW_{Ej}$ is very close to 1 which indicates the status of Ej is good for the parameter bandwidth.

$F1\_BW_{Ej}$ is computed with computation complexity O(1). The F1 fuzzifier at MRp computes

$F1\_BW_{E1}$, $F1\_BW_{E2}$, $F1\_BW_{E3}$ and $F1\_BW_{E4}$ of its 4 egress interfaces E1, E2, E3 and E4 respectively.

$F1\_PL_{Ej}$=1-(PL_{Ej})/(total packet at Ej)

$F1\_PL_{Ej}$ is assumed as 1 if (total packet at Ej) is zero.

(PL_{Ej})/(total packet at Ej) indicates the probability of packet loss at Ej. If PL_{Ej} is very high, $F1\_PL_{Ej}$ is very low which indicates the status of Ej is not good for the parameter packet loss.

$F1\_PL_{Ej}$ is computed with computation complexity O(1). The F1 fuzzifier at MRp computes $F1\_PL_{E1}$, $F1\_PL_{E2}$, $F1\_PL_{E3}$ and $F1\_PL_{E4}$ of its 4 egress interfaces E1, E2, E3 and E4 respectively.

$$F1\_cost_{Ej} = 1 - (Cost_{Ej} / \sum_{j=1}^{NO\_OF\_EI} \left(Cost_{Ej}\right))$$

if $\sum_{j=1}^{NO\_OF\_EI} \left(Cost_{Ej}\right) \neq 0$

$F1\_cost_{Ej}$ is assumed as 1 if

$$\sum_{j=1}^{NO\_OF\_EI} \left(Cost_{Ej}\right) = 0$$

$Cost_{Ej} / \sum_{j=1}^{NO\_OF\_EI} \left(Cost_{Ej}\right)$ indicates cost at Ej w.r.t. NO_OF_EI number of egress interfaces. If $Cost_{Ej}$ is very high, $F1\_cost_{Ej}$ becomes very low which indicates the status of Ej is not good for the parameter cost.

$F1\_cost_{Ej}$ is computed using (NO_OF_EI-1) number of addition, 1 division and 1 subtraction with computation complexity O(NO_OF_EI). The F1 fuzzifier at MRp computes $F1\_cost_{E1}$, $F1\_cost_{E2}$, $F1\_cost_{E3}$ and $F1\_cost_{E4}$ of its 4 egress interfaces E1, E2, E3 and E4 respectively.

Computation of F2: This considers set Y and set Z as in Scheme_I. The parameter value Y1 is in msec, Y2 is in kbps, Y3 is in % and Y4 is in unit. The minimum allowable delay among the allowable delays of all data, voice and video related services (www.3gpp.org, 1999) is 250 msec, 150 msec and 150 msec respectively. The maximum required bandwidth to maintain data, voice and video related services (www.3gpp.org, 1999) is 28.8 kbps, 21 kbps and 32 kbps respectively. All the services must be lossless in the best case. So the minimum packet loss for data, voice and video related services are assumed as 0%, 1% and 1%. The minimum cost of each service type is assumed as the product of average packet size for any service type as considered in Scheme_I and cost/bit. The cost/bit is assumed as 1 unit. So the minimum cost for data, voice and video service types are 8000, 640 and 720 unit respectively.

Each MR maintains the minimum allowable values of Delay, Packet loss, Cost and maximum allowable values of Desired bandwidth corresponding to the three different service types in the form of h(Y,Z). It is defined as

$$\begin{bmatrix} h(Y1,Z1) & h(Y1,Z2) & h(Y1,Z3) \\ h(Y2,Z1) & h(Y2,Z2) & h(Y2,Z3) \\ h(Y3,Z1) & h(Y3,Z2) & h(Y3,Z3) \\ h(Y4,Z1) & h(Y4,Z2) & h(Y4,Z3) \end{bmatrix}$$

$$= \begin{bmatrix} 250 & 150 & 150 \\ 28.8 & 21 & 32 \\ 0 & 1 & 1 \\ 8000 & 640 & 720 \end{bmatrix}$$

F2 transforms h(Y,Z) into the binary fuzzy relation Q(Y,Z) on fuzzy sets Y and Z only once when the system starts functioning as the value of all the elements in h(Y,Z) is constant. Q(Y,Z) is defined as

$$\begin{bmatrix} F2\_De_{Z1} & F2\_De_{Z2} & F2\_De_{Z3} \\ F2\_BW_{Z1} & F2\_BW_{Z2} & F2\_BW_{Z3} \\ F2\_PL_{Z1} & F2\_PL_{Z2} & F2\_PL_{Z3} \\ F2\_\mathrm{cost}_{Z1} & F2\_\mathrm{cost}_{Z2} & F2\_\mathrm{cost}_{Z3} \end{bmatrix}$$

where each element in Q(Y,Z) is the fuzzy value of the corresponding element in h(Y,Z). For example, $F2\_De_{E1}$, $F2\_BW_{E1}$, $F2\_PL_{E1}$, and $F2\_cost_{E1}$ are the fuzzy values corresponding to h(Y1,Z1), h(Y2,Z1), h(Y3,Z1) and h(Y4,Z1) respectively.

Computation of Q(Y,Z): The output parameters of fuzzifier F2 are the 4 fuzzy values of 4 parameters per service type which is maintained by each MR in the form Q(Y,Z). The fuzzy values at the output of F2 indicates minimum allowable values of Delay, Packet loss, Cost and maximum allowable desired bandwidth corresponding to the 3 different service types (i.e. NO_OF_ST). The computation of 4 fuzzy values at the output of F2 for Z^{th} service type using the appropriate fuzzy membership function is discussed below.

$F2\_De_Z = a'k/h(Y1,Z)$

where a'k is computed as $h(Y1,Z1) \wedge h(Y1,Z2) \wedge h(Y1,Z3)$ and is equal to 150 msec. So a'k is lesser than h(Y1,Z) and $F2\_De_Z$ lies within 0 to 1. The F2 fuzzifier at MRp computes $F2\_De_{Z1}$, $F2\_De_{Z2}$, $F2\_De_{Z3}$ for service types Z1, Z2, Z3 respectively.

$F2\_BW_z = h(Y2,z)/b'k$

where b'k is computed as $h(Y2,Z1) \vee h(Y2,Z2) \vee h(Y2,Z3)$ and is equal to 32 Kbps. So b'k is greater than h(Y2,Z) and $F2\_BW_Z$ lies within 0 to 1. The F2 fuzzifier at MRp computes $F2\_BW_{Z1}$, $F2\_BW_{Z2}$, $F2\_BW_{Z3}$ for service types Z1, Z2, Z3 respectively.

$F2\_PL_Z = 1 - h(Y3,Z)$

h(Y3,Z) indicates minimum allowable packet loss for Z^{th} service type in percentage whereas $F2\_PL_Z$ indicates how lossless the service type Z is. The F2 fuzzifier at MRp computes $F2\_PL_{Z1}$, $F2\_PL_{Z2}$, $F2\_PL_{Z3}$ for service types Z1, Z2, Z3 respectively.

$F2\_cost_Z = d'k/h(Y4,Z)$

where d'k is computed as $h(Y4,Z1) \wedge h(Y4,Z2) \wedge h(Y4,Z3)$ and is equal to 640 unit. So d'k is lesser than h(Y4,Z) and $F2\_cost_Z$ lies within 0 to 1.

a'k, b'k and d'k are computed using (NO_OF_ST-1) number of min, max and min operation respectively with computation complexity O(NO_OF_ST). The F2 fuzzifier at MRp computes $F2\_cost_{Z1}$, $F2\_cost_{Z2}$, $F2\_cost_{Z3}$ for service types Z1, Z2, Z3 respectively. $F2\_De_Z$, $F2\_BW_Z$, $F2\_PL_Z$ and $F2\_cost_Z$ are computed with computation complexity O(1).

Computation of R(X,Z): Each MR computes the fuzzy relation R(X,Z) using the max-min composition of the fuzzy relations P(X,Y) and Q(Y,Z). The min operation is used to determine the status of Ej per parameter for Z^{th} service type in the worst case and the max operation is used to determine the status of Ej in the best case. P(X,Y) contains the fuzzy values corresponding to the 4 parameter values per egress interface and Q(Y,Z) contains the fuzzy values corresponding to the 4 parameter values per service type. R(X,Z) contains the fuzzy values to indicate the status of each egress interface per service type. So R(X,Z)=P(X,Y).Q(Y,Z). R(X,Z) is computed if at least one element of P(X,Y) changes

$$\begin{bmatrix} F1\_De_{E1} & F1\_BW_{E1} & F1\_PI_{E1} & F1\_\text{cost}_{E1} \\ F1\_De_{E2} & F1\_BW_{E2} & F1\_PI_{E2} & F1\_\text{cost}_{E2} \\ F1\_De_{E3} & F1\_BW_{E3} & F1\_PI_{E3} & F1\_\text{cost}_{E3} \\ F1\_De_{E4} & F1\_BW_{E4} & F1\_PI_{E4} & F1\_\text{cost}_{E4} \end{bmatrix} \bullet$$

$$\begin{bmatrix} F2\_De_{Z1} & F2\_De_{Z2} & F2\_De_{Z3} \\ F2\_BW_{Z1} & F2\_BW_{Z2} & F2\_BW_{Z3} \\ F2\_PL_{Z1} & F2\_PL_{Z2} & F2\_PL_{Z3} \\ F2\_\text{cost}_{Z1} & F2\_\text{cost}_{Z2} & F2\_\text{cost}_{Z3} \end{bmatrix}$$

Box 1.

$$P_{11}=(F1\_De_{E1}\wedge F2\_De_{Z1})\vee(F1\_BW_{E1}\wedge F2\_BW_{Z1})\vee(F1\_PL_{E1}\wedge F2\_PL_{Z1})\vee(F1\_cost_{E1}\wedge F2\_cost_{Z1})$$
$$P_{12}=(F1\_De_{E1}\wedge F2\_De_{Z2})\vee(F1\_BW_{E1}\wedge F2\_BW_{Z2})\vee(F1\_PL_{E1}\wedge F2\_PL_{Z2})\vee(F1\_cost_{E1}\wedge F2\_cost_{Z2})$$
$$P_{13}=(F1\_De_{E1}\wedge F2\_De_{Z3})\vee(F1\_BW_{E1}\wedge F2\_BW_{Z3})\vee(F1\_PL_{E1}\wedge F2\_PL_{Z3})\vee(F1\_cost_{E1}\wedge F2\_cost_{Z3})$$
$$P_{21}=(F1\_De_{E2}\wedge F2\_De_{Z1})\vee(F1\_BW_{E2}\wedge F2\_BW_{Z1})\ (F1\_PL_{E2}\wedge F2\_PL_{Z1})\vee(F1\_cost_{E2}\wedge F2\_cost_{Z1})$$
$$P_{22}=(F1\_De_{E2}\wedge F2\_De_{Z2})\vee(F1\_BW_{E2}\wedge F2\_BW_{Z2})\ (F1\_PL_{E2}\wedge F2\_PL_{Z2})\vee(F1\_cost_{E2}\wedge F2\_cost_{Z2})$$
$$P_{23}=(F1\_De_{E2}\wedge F2\_De_{Z3})\vee(F1\_BW_{E2}\wedge F2\_BW_{Z3})\ (F1\_PL_{E2}\wedge F2\_PL_{Z3})\vee(F1\_cost_{E2}\wedge F2\_cost_{Z3})$$
$$P_{31}=(F1\_De_{E3}\wedge F2\_De_{Z1})\vee(F1\_BW_{E3}\wedge F2\_BW_{Z1})\vee(F1\_PL_{E3}\wedge F2\_PL_{Z1})\vee(F1\_cost_{E3}\wedge F2\_cost_{Z1})$$
$$P_{32}=(F1\_De_{E3}\wedge F2\_De_{Z2})\vee(F1\_BW_{E3}\wedge F2\_BW_{Z2})\vee(F1\_PL_{E3}\wedge F2\_PL_{Z2})\vee(F1\_cost_{E3}\wedge F2\_cost_{Z2})$$
$$P_{33}=(F1\_De_{E3}\wedge F2\_De_{Z3})\vee(F1\_BW_{E3}\wedge F2\_BW_{Z3})\vee(F1\_PL_{E3}\wedge F2\_PL_{Z3})\vee(F1\_cost_{E3}\wedge F2\_cost_{Z3})$$
$$P_{41}=(F1\_De_{E4}\wedge F2\_De_{Z1})\vee(F1\_BW_{E4}\wedge F2\_BW_{Z1})\vee(F1\_PL_{E4}\wedge F2\_PL_{Z1})\vee(F1\_cost_{E4}\wedge F2\_cost_{Z1})$$
$$P_{42}=(F1\_De_{E4}\wedge F2\_De_{Z2})\vee(F1\_BW_{E4}\wedge F2\_BW_{Z2})\vee(F1\_PL_{E4}\wedge F2\_PL_{Z2})\vee(F1\_cost_{E4}\wedge F2\_cost_{Z2})$$
$$P_{43}=(F1\_De_{E4}\wedge F2\_De_{Z3})\vee(F1\_BW_{E4}\wedge F2\_BW_{Z3})\vee(F1\_PL_{E4}\wedge F2\_PL_{Z3})\vee(F1\_cost_{E4}\wedge F2\_cost_{Z3})$$

$$\begin{bmatrix} P_{11} & P_{12} & P_{13} \\ P_{21} & P_{22} & P_{23} \\ P_{31} & P_{32} & P_{33} \\ P_{41} & P_{42} & P_{43} \end{bmatrix}$$

Where as shown in Box 1:

R(X,Z) has NO_OF_ST*NO_OF_EI number of elements. The computation of each element in R(X,Z) needs (NO_OF_P) number of min operations and (NO_OF_P-1) number of max operations. So each element of R(X,Z) is computed with computation complexity O(2*NO_OF_P-1).

Best egress determination algorithm: Each MR in the MN uses this algorithm to determine Best_Eg per service type supported by the MN independently. Let MRp execute this algorithm for the Zth service type. MRp computes MRp(X,Z) as Rp(X1,Z)∨Rp(X2,Z)∨Rp(X3,Z)∨Rp(X4,Z), where Rp(X1,Z), Rp(X2,Z), Rp(X3,Z) and Rp(X4,Z) are the R(X,Z) fuzzy relation as computed by MRp for Z^{th} service type. If MRp(X,Z)=Rp(X3,Z), X3 is Best_Eg of MRp for Z^{th} service type. The computation of MRp(X,Z) needs (NO_OF_EI-1) number of max operations and so it is computed with computation complexity O(NO_OF_EI). In the best case one egress interface of MRp is determined as Best_Eg for Z^{th} service type by the algorithm with computation complexity O(NO_OF_EI). In the worst case, multiple egress interfaces of MRp are determined as Best_Eg for Z^{th} service type. Let (NO_OF_EI) number of egress interfaces of MRp is selected as Best_Eg for Z^{th} service type. The algorithm uses best possible egress interface selection function as discussed for Scheme_I for (NO_OF_EI-1) number of times to select the best egress interface of MRp from (NO_OF_EI) number of egress interfaces for Z^{th} service type depending upon the 4 parameter values (i.e. NO_OF_P), Y1, Y2, Y3, Y4 with computation complexity O(NO_OF_EI*NO_OF_P).

Function of LFN: The LFN maintains 4 data structures and generates NO_OF_R number of routing tables when the system starts functioning. It updates the data structure and routing table after receiving Best_Eg per service type information from MR in the form of MR_LFN message. LFN also uses best route selection algorithm to select Opt_Route after receiving a route request in the form of MNN_LFN message to initiate a session from MNN.

Maintenance of data structure: The data structure 1 is MR_EG_STA. Each element of MR_EG_STA[p][j][Z] is a fuzzy value to indicate the status of Ej at MRp for Z^{th} service type. So it is a 3-D array of dimension NO_OF_MR

* NO_OF_EI * NO_OF_ST. All the elements of this 3-D array are initialized by 1.0 when the system starts functioning. LFN updates the element values of MR_EG_STA data structure after receiving MR_LFN message. LFN updates the value of the 3 array elements (i.e. NO_OF_ST) one for each service type, MR_EG_STA[p][X1][Z1], MR_EG_STA[p][X2][Z2] and MR_EG_STA[p][X3][Z3] corresponding to the egress interfaces X1, X2 and X3 of MRp using MRp(X1,Z1), MRp(X2,Z2) and MRp(X3,Z3) respectively with computation complexity O(NO_OF_ST). The fuzzy values corresponding to MRp(X1,Z1), MRp(X2,Z2) and MRp(X3,Z3) are obtained from the part2 of MR_LFN message.

The data structure 2 is MR_EGRESS_INFORM. Each element of MR_EGRESS_INFORM[p][j][y] indicates the value of yth parameter at Ej of MRp. So it is a 3-D array of dimension NO_OF_MR*NO_OF_EI*NO_OF_P. For example, the value of the elements

MR_EGRESS_INFORM[p][j][Y1],

MR_EGRESS_INFORM[p][j][Y2], MR_EGRESS_INFORM[p][j][Y3] and MR_EGRESS_INFORM[p][j][Y4]

indicate Delay (Y1), Unused bandwidth (Y2), Packet loss (Y3) and Cost (Y4) at Ej of MRp. This array has 4 elements for 4 parameter values (i.e. NO_OF_P) corresponding to each egress interface of MRp. The element values of the array corresponding to Delay is initialized by 0.0 msec, the element values of the array corresponding to Unused bandwidth is initialized by the available bandwidth at Ej in kbps, the element values of the array corresponding to Packet loss is initialized by 0.0% and the element values of the array corresponding to Cost is initialized by 0.0 units when the system starts functioning. LFN updates the element values of MR_EGRESS_INFORM data structure after receiving MR_LFN message. LFN updates the value of the 4 array elements (i.e. NO_OF_P) MR_EGRESS_INFORM[p][X1][Y1], MR_EGRESS_INFORM[p][X1][Y2], MR_EGRESS_INFORM[p][X1][Y3] and MR_EGRESS_INFORM[p][X1][Y4] corresponding to the egress interface X1 of MRp using the values gp(X1,Y1), gp(X1,Y2), gp(X1,Y3) and gp(X1,Y4) respectively. The values corresponding to gp(X1,Y1), gp(X1,Y2), gp(X1,Y3) and gp(X1,Y4) are obtained from 0th element of the variable array as specified in part3 of the MR_LFN message by MRp. In the best case the number of elements in the variable array is 1 and the element has (NO_OF_P+1) number of components. So the computation complexity for the updating of data structure 2 is O(NO_OF_P). In the worst case the number of elements in the variable array is 3 (i.e. NO_OF_ST) one per service type and each element has (NO_OF_P+1) number of components. So the computation complexity for the updating of data structure 2 is O(NO_OF_ST*NO_OF_P). The LFN repeats the same steps of updating for the egress interface X2 and X3 of MRp.

The data structure 3 is BE_EG. Each element of BE_EG[p][Z] indicates Best_Eg of MRp for Z^{th} service type. So it is a 2-D array of dimension NO_OF_MR * NO_OF_ST. All the elements of this array are initialized by 1 when the system starts functioning. LFN updates the element values of BE_EG data structure after receiving MR_LFN message from MRp. LFN updates the values of the 3 array elements (i.e. NO_OF_ST) one per service type, BE_EG[p][Z1], BE_EG[p][Z2] and BE_EG[p][Z3] using X1, X2 and X3 respectively with computation complexity O(NO_OF_ST). The value corresponding to X1, X2 and X3 are obtained from 0th element, 1th element and 2th element of the variable array in part3 of the MR_LFN message.

The data structure 4 is MR_Route_relation. The value of the element MR_Route_relation[ri][MRp] is 1 if MRp is associated with ri^{th} route. So it is a 2-D array of dimension NO_OF_R *

Figure 16. MR_Route_relation data structure.

Route	MR1	MR2	MR3	MR4	MR5	MR6
r1	1	1	0	0	0	0
r2	0	0	1	1	0	0
r3	0	0	0	0	1	1
r4	1	0	1	1	0	0

NO_OF_MR. The value of the elements in this data structure is constant as shown in Figure 16.

Generation and updation of routing tables: LFN generates NO_OF_R number of routing tables for NO_OF_R number of routes when the system starts functioning. The number of rows in each routing table is equal to the number of service type (NO_OF_ST). The ri[th] routing table has one column to specify the service type, l_i-1 number of columns to specify Best_Eg of l_i-1 number of MRs associated with ri[th] route, one column to specify the value attribute for ri[th] route and Z[th] service type (VALUE$_i$(Z)). LFN generates NO_OF_R number of routing tables with computation complexity O(NO_OF_R*NO_OF_ST*l_i) as each routing table has NO_OF_ST number of entries each having l_i+1 number of attributes. LFN inserts the value of Best_Eg in the l_i-1 number of columns corresponding to l_i-1 number of MRs associated with the ri[th] route in the routing table ri after receiving MR_LFN message. Let the LFN receives MR_LFN message from MRp where 1≤p≤NO_OF_MR. LFN searches the MR_Route_relation data structure to find the route ri where 1≤i≤NO_OF_R with which MRp is associated. The LFN inserts BE_EG[MRp][Z1], BE_EG[MRp][Z2] and BE_EG[MRP][Z3] in ri[th] routing table. In the best case, MRp is associated with a single route and so insertion is required in the 3 rows (i.e. NO_OF_ST) corresponding to the attribute MRp in the ri[th] routing table with computation complexity O(NO_OF_ST). In the worst case, MRp is associated with NO_OF_R number of routes and so insertion is required in the 3 rows (i.e. NO_OF_ST) corresponding to the attribute MRp in NO_OF_R number of routing tables with computation complexity O(NO_OF_R*NO_OF_ST). For example, let LFN receives MR_LFN message from MR4. LFN searches the MR_Route_relation data structure where it finds MR_Route_relation[r2][MR4] and MR_Route_relation[r4][MR4] is 1.

So MR4 is associated with route r2 and route r4. LFN inserts BE_EG[MR4][Z1], BE_EG[MR4][Z2] and BE_EG[MR4][Z3] in Table 5 and Table 7 corresponding to the route r2 and route r4 respectively with computation complexity O(2*NO_OF_ST) as insertion is required in 2 number of routing tables.

The LFN computes the value attribute of the routing table after receiving MNN_LFN message. The computation of the value attribute for the service type Z1 and for routes r1, r2, r3, r4 are discussed below. The LFN repeats the same steps of operation to compute the value attribute for the services Z2, Z3 and for routes r1, r2, r3, r4.

Let us consider the routing table (Table 4) for route r1. The route r1 is associated with MR1 and MR2. LFN computes VALUE$_{r1}$(Z1) which is the value attribute for route r1 and service type Z1 using the min operation between MR_

Table 4. Routing table for route r1

Services	MR1	MR2	$VALUE_{r1}(Z)$
Z1	BE_EG[MR1][Z1]	BE_EG[MR2][Z1]	$VALUE_{r1}(Z1)$
Z2	BE_EG[MR1][Z2]	BE_EG[MR2][Z2]	$VALUE_{r1}(Z2)$
Z3	BE_EG[MR1][Z3]	BE_EG[MR2][Z3]	$VALUE_{r1}(Z3)$

Table 5. Routing table for route r2

Services	MR3	MR4	$VALUE_{r2}(Z)$
Z1	BE_EG[MR3][Z1]	BE_EG[MR4][Z1]	$VALUE_{r2}(Z1)$
Z2	BE_EG[MR3][Z2]	BE_EG[MR4][Z2]	$VALUE_{r2}(Z2)$
Z3	BE_EG[MR3][Z3]	BE_EG[MR4][Z3]	$VALUE_{r2}(Z3)$

EG_STA[MR1][BE_EG[MR1][Z1]][Z1] and MR_EG_STA[MR2][BE_EG[MR2][Z1]][Z1].

Let us consider the routing table (Table 5) for route r2. The route r2 is associated with MR3 and MR4. LFN computes $VALUE_{r2}(Z1)$ which is the value attribute for route r2 and service type Z1 using the min operation between MR_EG_STA[MR3][BE_EG[MR3][Z1]][Z1] and MR_EG_STA[MR4][BE_EG[MR4][Z1]][Z1].

Let us consider the routing (Table 6) for route r3. The route r3 is associated with MR5 and MR6. LFN computes $VALUE_{r3}(Z1)$ which is the value attribute for route r3 and service type Z1 using the min operation between MR_EG_STA[MR5][BE_EG[MR5][Z1]][Z1] and MR_EG_STA[MR6][BE_EG[MR6][Z1]][Z1].

Let us consider the routing table (Table 7) for route r4. The route r4 is associated with MR1, MR3 and MR4. LFN computes $VALUE_{r4}(Z1)$ which is the value attribute for route r4 and service type Z1 using the min operation between

MR_EG_STA[MR1][BE_EG[MR1][Z1]][Z1],
MR_EG_STA[MR3][BE_EG[MR3][Z1]][Z1],
MR_EG_STA[MR4][BE_EG[MR4][Z1]][Z1].

Table 6. Routing table for route r3

Services	MR5	MR6	$VALUE_{r3}(Z1)$
Z1	BE_EG[MR5][Z1]	BE_EG[MR6][Z1]	$VALUE_{r3}(Z1)$
Z2	BE_EG[MR5][Z2]	BE_EG[MR6][Z2]	$VALUE_{r3}(Z2)$
Z3	BE_EG[MR5][Z3]	BE_EG[MR6][Z3]	$VALUE_{r3}(Z3)$

Table 7. Routing table for route r4

Services	MR1	MR3	MR4	$VALUE_{r4}(Z)$
Z1	BE_EG[MR1][Z1]	BE_EG[MR3][Z1]	BE_EG[MR4][Z1]	$VALUE_{r4}(Z1)$
Z2	BE_EG[MR1][Z2]	BE_EG[MR3][Z2]	BE_EG[MR4][Z2]	$VALUE_{r4}(Z2)$
Z3	BE_EG[MR1][Z3]	BE_EG[MR3][Z3]	BE_EG[MR4][Z3]	$VALUE_{r4}(Z3)$

The computation of each value attribute needs (l_i-1) number of min operation and so has computation complexity O(l_i-1).

Best route selection algorithm: LFN in the MN uses this algorithm to determine Opt_Route using the value attribute information available in the routing table. The algorithm selects route r1 for the service type Z1

if $VALUE_{r1}(Z1) \vee VALUE_{r2}(Z1) \vee VALUE_{r3}(Z1) \vee VALUE_{r4}(Z1) = VALUE_{r1}(Z1)$,

selects route r2 for the service type Z1

if $VALUE_{r1}(Z1) \vee VALUE_{r2}(Z1) \vee VALUE_{r3}(Z1) \vee VALUE_{r4}(Z1) = VALUE_{r2}(Z1)$,

selects route r3 for the service type Z1

if $VALUE_{r1}(Z1) \vee VALUE_{r2}(Z1) \vee VALUE_{r3}(Z1) \vee VALUE_{r4}(Z1) = VALUE_{r3}(Z1)$ and

selects route r4 for the service type Z1

if $VALUE_{r1}(Z1) \vee VALUE_{r2}(Z1) \vee VALUE_{r3}(Z1) \vee VALUE_{r4}(Z1) = VALUE_{r4}(Z1)$

Each route selection needs (NO_OF_R-1) number of max operation and so has computation complexity O(NO_OF_R).

If the two different routes are selected as Opt_Route for Z^{th} service type, the algorithm uses "Best possible route selection function" to select the best route.

Best possible route selection function: Let the route rd and rt are selected as Opt_Route for Z^{th} service type where ($1 \le d \le NO\_OF\_R$, $1 \le t \le NO\_OF\_R$ and $d \ne t$). LFN computes the average delay, average unused bandwidth, average packet loss and average cost for rd and rt of MRp to select the best route for Z^{th} service type (BestRoute(Z)). The expression to compute the average delay (avg_delay_ri), average unused bandwidth (avg_unused_BW_ri), average packet loss (avg_packet_loss_ri) and average cost (avg_cost_ri) of ri^{th} route are as given below:

$$avg\_delay\_ri = \sum_{p=1}^{li-1} MR\_EGRESS\_INFORM[p][BE\_EG[p][Z]][Y1] / (l_i - 1)$$

$$avg\_unused\_BW\_ri = \sum_{p=1}^{li-1} MR\_EGRESS\_INFORM[p][BE\_EG[p][Z]][Y2] / (l_i - 1)$$

$$avg\_packet\_loss\_ri = \sum_{p=1}^{li-1} MR\_EGRESS\_INFORM[p][BE\_EG[p][Z]][Y3] / (l_i - 1)$$

$$avg\_cost\_ri = \sum_{p=1}^{li-1} MR\_EGRESS\_INFORM[p][BE\_EG[p][Z]][Y4] / (l_i - 1)$$

Each expression is evaluated using (l_i-2) number of addition and 1 division with computation complexity O(l_i).

The best possible route selection function uses the expression of avg_delay_ri, avg_unused_BW_ri, avg_packet_loss_ri and avg_cost_ri to compute the average delay, average unused bandwidth, average packet loss and average cost for route rd (i=d) and for route rt (i=t).

In the worst case all the NO_OF_R number of routes is selected as Opt_Route. In such a case the algorithm uses best possible route selection function for (NO_OF_R-1) number of times to select the best route for Z^{th} service type with computation complexity O(NO_OF_R*l_i). The avg_delay_rd, avg_unused_BW_rd, avg_packet_loss_rd and avg_cost_rd are the average delay, average unused bandwidth, average packet loss and average cost of route rd respectively. The avg_delay_rt, avg_unused_BW_rt, avg_packet_loss_rt and avg_cost_rt are the average delay, average unused bandwidth, average packet loss and average cost of route rt respectively.

The flow chart of the best possible route selection function for data class of traffic (Flow_data_R_II) is shown in Figure 17(i), for voice and video class of traffic (Flow_vo_vi_R_II) is shown in Figure 17(ii). Both Flow_data_R_I and Flow_vo_vi_R_II use func1 Y2 and para func1. The flow chart of func1 Y2 and para func1 are shown in Figure 17(iii).

LFN uses the same procedure to select the optimal route for the service types Z2 and Z3.

Computation complexity of the proposed algorithm: In this section the computation complexity of the best egress determination algorithm to select Best_Eg and best route selection algorithm to select a Opt_Route for Z^{th} service type is considered for discussion.

Computation complexity of the best egress determination algorithm: It is the sum of the computation complexity of g(X,Y) calculation, computation complexity of F1 and F2 fuzzifier, computation complexity of R(X,Z) calculation, computation complexity of executing best egress determination algorithm and computation complexity of generating MR_LFN message.

The computation complexity to compute g(j,Y1), g(j,Y2), g(j,Y3) and g(j,Y4) for Ej are O(1), O(1), O(number of packets in egress buffer) and O(1) respectively. So the computation complexity of g(X,Y) calculation per egress interface of a MR is O(number of packets in egress buffer) and for NO_OF_EI number of egress interfaces of a MR are O(NO_OF_EI*number of packets in egress buffer).

The computation complexity to compute $F1\_de_{Ej}$, $F1\_BW_{Ej}$, $F1\_PL_{Ej}$ and $F1\_cost_{Ej}$ for Ej by F1 fuzzifier are O(NO_OF_EI), O(1), O(1) and O(NO_OF_EI) respectively. So the computation complexity of F1 fuzzifier per egress interface of a MR is O(NO_OF_EI) and for NO_OF_EI number of egress interfaces of a MR are O(NO_OF_EI*NO_OF_EI).

The computation complexity to compute $F2\_De_Z$, $F2\_BW_Z$, $F2\_PL_Z$ and $F2\_cost_Z$ for Z^{th} service type by F2 fuzzifier is O(1).

Each MR computes R(X,Z) by using max-min composition among the fuzzy relations P(X,Y) and Q(Y,Z). Each element of R(X,Z) is computed with computation complexity O(2*NO_OF_P-1). R(X,Z) has (NO_OF_EI*NO_OF_ST) number of elements. So all the elements of R(X,Z) are computed with computation complexity O(NO_OF_EI*NO_OF_ST*NO_OF_P).

The best egress determination algorithm of a MR has computation complexity O(NO_OF_EI) to select a single egress interface for a service type. If multiple egress interfaces are selected as Best_Eg for a service type, the algorithm uses the best possible egress interface selection function with computation complexity O(number of best egress * NO_OF_P) to select the best egress interface.

The MR_LFN message generation has computation complexity O(NO_OF_P) in the best case and O(NO_OF_ST * NO_OF_P) in the worst case.

So the computation complexity of the best egress determination algorithm is O(NO_OF_EI*NO_OF_EI) + O(NO_OF_EI*NO_OF_ST*NO_OF_P) + O(NO_OF_EI*number of packets in egress buffer) + O(number of best egress*NO_OF_P)+O(NO_OF_ST*NO_OF_P).

Computation complexity of the best route selection algorithm: It is the sum of the computation complexity of data structure updation, computation complexity of the routing table updating, computation complexity of executing best route selection algorithm and computation complexity of generating LFN_MNN message.

The LFN updates data structure 1 and data structure 3 with computation complexity O(NO_OF_ST) whereas updates data structure 2 with computation complexity O(NO_OF_P) in the best case and O(NO_OF_ST*NO_OF_P) in the worst case. The computation complexity to search an element from data structure 4 is O(1).

The LFN inserts the value of Best_Eg in the ri^{th} routing table corresponding to the ri^{th} route after receiving MR_LFN message with computation complexity $O((l_i-1)*NO\_OF\_ST)$. So the com-

Figure 17. (i). Flow_data_R_II. (ii). Flow_vo_vi_R_II. (iii). Flow chart for (a) func1 Y2 (b) para func1.

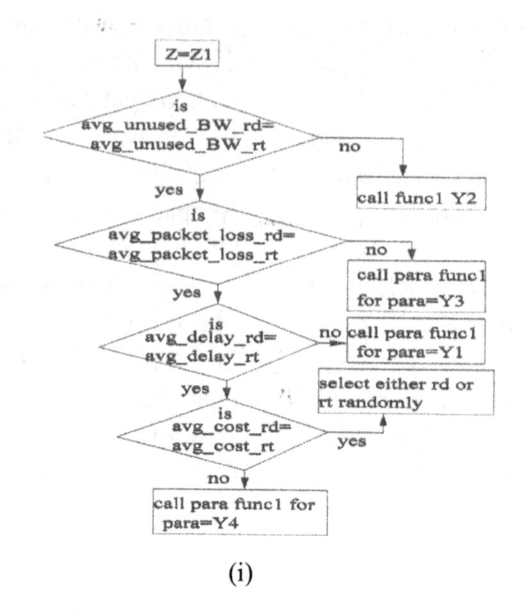

(i)

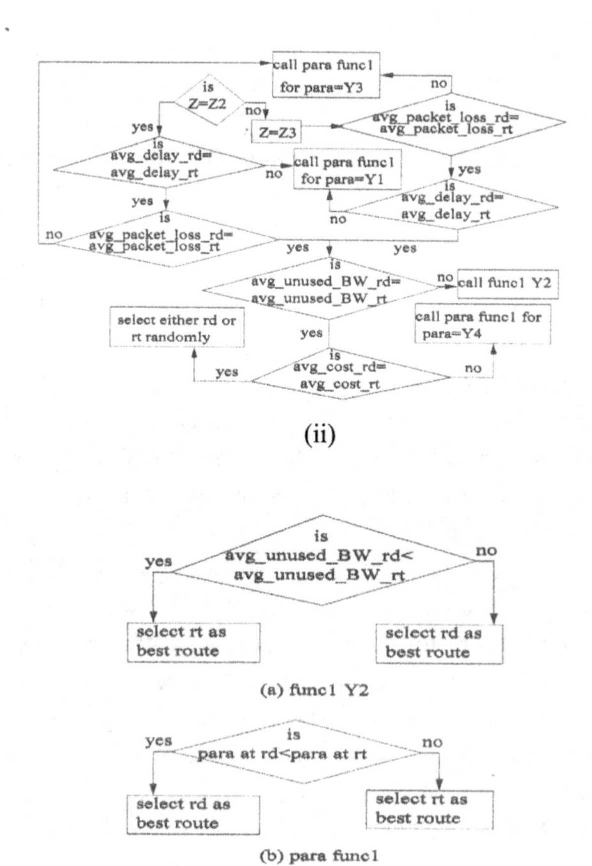

(ii)

(iii)

putation complexity for insertion in NO_OF_R number of routing tables is

$$\sum_{i=1}^{NO\_OF\_R} O((l_i-1)*NO\_OF\_ST).$$

LFN computes NO_OF_R number of value attributes for each service type after receiving MNN_LFN message with computation complexity $O(NO\_OF\_R*l_i)$.

The best route selection algorithm has computation complexity O(NO_OF_R) to select a single route. If multiple routes are selected as Opt_Route for a service type, the algorithm uses best possible route selection function with computation complexity O(number of best route*l_i) to select the best route.

The LFN generates LFN_MNN message with computation complexity $O(l_i)$.

So the computation complexity of the best route selection algorithm is $O(l_i * NO\_OF\_ST * NO\_OF\_R)+O(NO\_OF\_R*l_i)+O(NO\_OF\_ST * NO\_OF\_P)$ + O(number of best route * l_i).

Simulation of Scheme_II: The performance of Scheme_II is studied using NEMO_SIM simulator. It uses the same set of threads as in Scheme_I during simulation. The simulation experiment is carried out for 3 different cases considering the internal network of MN (Figure 2) as WiFi (IEEE 802.11a). Each case of experiment considers different size of LFN buffer, MR egress and ingress buffer and MNN buffer as mentioned below:

CASE I: LFN buffer size 1000, MR egress and ingress interface buffer size 10^5, MNN buffer 1000

CASE II: LFN buffer size 1500, MR egress and ingress interface buffer size 150000, MNN buffer 1500

CASE III: LFN buffer size 500, MR egress and ingress interface buffer size 50000, MNN buffer 500

Figure 18, Figure 19 and Figure 20 show the plot of throughput, route selection time and session loss vs. traffic load of the proposed MN (Figure 2) for 3 cases. The traffic load is computed as the ratio of arrival rate and departure rate of session request from MNN by LFN. In CASE I route selection time is constant up to traffic load 300 and then it increases slowly which causes decrease in throughput with traffic load. Session loss is zero up to traffic load 350 and then it increases slowly. In CASE II route selection time is constant up to traffic load 550 and then it increases slowly which causes the decrease in throughput with traffic load.

Figure 18. Throughput vs. Traffic load for MN in Figure 2.

Figure 19. Route selection time vs. Traffic load for MN in Figure 2.

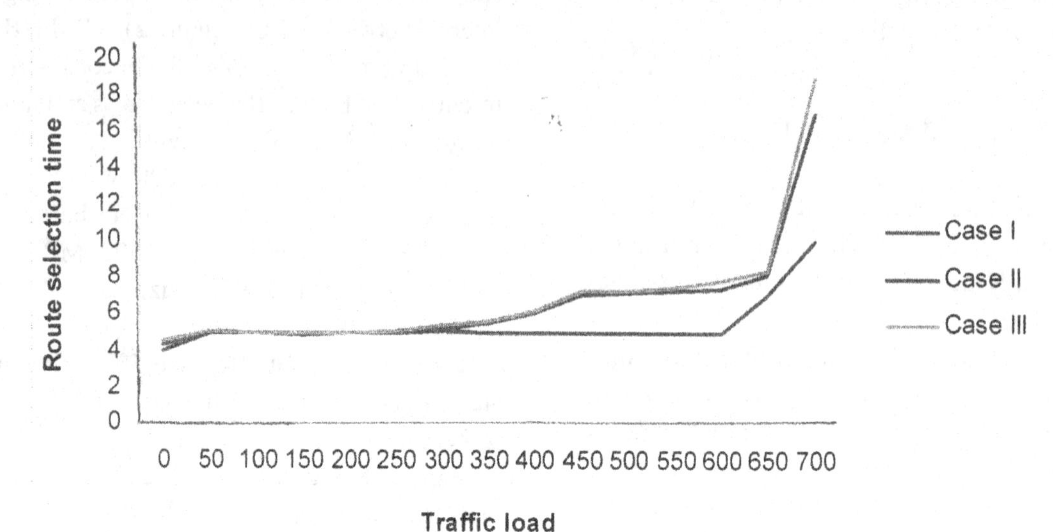

Figure 20. Session loss vs. Traffic load for MN in Figure 2.

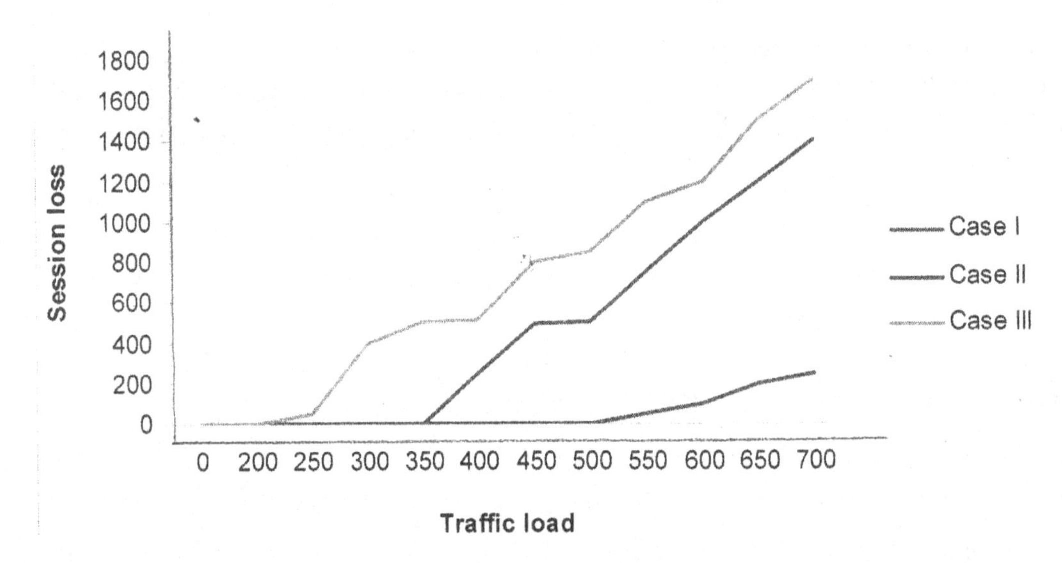

Session loss is zero up to traffic load 550 then it increases slowly with traffic load. In CASE III route selection time is constant up to traffic load 200 and then it increases slowly which causes the decrease in throughput with traffic load. Session loss is zero up to traffic load 350 then it increases slowly with traffic load.

The route selection time is the minimum, throughput is the maximum and session loss is the minimum in CASE II due to the maximum buffer. The route selection time and session loss is the maximum in CASE III due to the minimum buffer. The route selection time and session loss in CASE I is higher than CASE II but lower than CASE III due to moderate buffer. The route selection time in CASE III is lesser than the route selection time in CASE I up to traffic load 200 which causes higher throughput in CASE III than in CASE I. The route selection time in CASE III is higher than the route selection time in CASE I from traffic load 200 which causes lesser throughput in CASE III than in CASE I.

FUTURE RESEARCH DEVELOPMENTS

The growth of the Mobile Internet continues apace. Mobile handsets and laptops will soon have several wireless interfaces as the newest wireless technologies being developed and deployed. The multihomed mobile node should be able to select the best interface for a particular traffic. This is a challenging task especially in heterogeneous networks environments where quality of service parameters change their values rapidly. The switching among interfaces should cause minimal disruption to on-going communication sessions. Such switching mechanism needs to be very fast as real-time application can only tolerate delays in the range of milliseconds. Finally the implementation of such schemes should not change the existing Internet infrastructure and network applications. A clean architecture should be designed where multi-homing and mobility are separated so that changes in one should not affect the other. If a mobile node is not multihomed, it should still be able to use mobility support protocol. Such multi-homing architecture must be able to detect any failure quickly for providing reliable communication.

CONCLUSION

In mobility, addresses change sequentially when mobile node moves across networks whereas multihoming is a process of attaching a subscriber to more than a single access point in the network. This subscriber can be a single host, a router or a whole network site. The multihomed mobile node frequently changes its location in the Internet. So such a node is configured with multiple addresses at a given time and it can choose the communication path according to some rules/performances. In traditional Internet environment, IP address plays the role of both identifying a node and representing its location. This is sufficient when Internet nodes are static and single homed. But in case of multihomed mobile nodes having more than one globally routable address there is a need to decouple the identification of nodes from their location. Each address represents location of node in Internet. However, for maintaining sessions during communication a single identity must be chosen. In both mobility and multi-homing scenarios, switching takes place between different address. However, the motivations are different for each. The mobile node roaming in a heterogeneous network environment should also be able to take advantage of being multi-homed.

In this chapter two route optimization schemes in multi-homed MN have been considered. Scheme_I is a dynamic route optimization scheme whereas Scheme_II uses fuzzy logic to implement dynamic route optimization in a mult-homed MN. To conclude this can be stated that both Scheme_I and Scheme_II can be extended to achieve bi-directional communication between MNN in MN and any correspondent node (CN) in other fixed or mobile network. In case of high network mobility, communication between MNN and CN takes place through a home agent whereas direct communication between MNN and CN is possible in case of lower network mobility to achieve route optimization. Moreover, the proposed route selection algorithm can be extended by incorporating the fault tolerant capabilities in case of route failure in mobile network.

REFERENCES

Adeniji, S. D., Khatun, S., Raja, R. S. A., & Borhan, M. A. (2008). Design and analysis of resource management support software for multihoming in vehicle of IPv6 network. *AsiaCSN*, (pp. 13-17).

Alshaer, H., & Elmirghani, J. M. H. (2009). An intelligent route control scheme for multihomed mobile networks. *IEEE Vehicular Technology Conference*, (pp. 1-5).

Cho, S., Na, J., & Kim, C. (2005). A dynamic load sharing mechanism in multihomed mobile networks. *IEEE International Conference on Communications*, (pp. 1459-1463).

Devarapalli, V., Wakikawa, R., Petrescu, A., & Thubert, P. (2005). *Network mobility basic support protocol*. IETF RFC 3963.

Gibson, S. (2009). *Distributed reflection denial of service*. Gibson Research Corporation. Retrieved from http://grc.com/dos/drdos.htm.

Java Network Simulator. (n.d.). Retrieved from http://jns.sourceforge.net

Johnson, D., Perkins, C., & Arkko, J.(2004). *Mobility support in IPv6*. IETF RFC 3775.

Mitra, S. (2010). Dynamic route optimization in a multihomed mobile network. *International Journal of Computational Vision and Robotics*, *2*(1), 121–135. doi:10.1504/IJCVR.2010.036076

Mitra, S., & Pyne, S. (2010). Security issue of a route selection algorithm in multihomed mobile networks. *International Conference on VLSI Design and Communication Systems*, (pp. 31-37).

Mitra, S., & Pyne, S. (2011). Fuzzy logic based route optimization in a multihomed mobile networks. *Wireless Networks*, *1*(17), 213–229. doi:10.1007/s11276-010-0274-y

Munasinghe, K. S., & Jamalipour, A. (2010). Route optimization for roaming heterogeneous multi-homed mobile networks. *International Conference on Signal Processing and Communication Systems*, (pp. 1-7).

Perkins, C. (2002). *IP mobility support for IPv4*. IETF RFC 3344.

Shima, K., Uo, Y., Ogashiwa, N., & Uda, S. (2006). Operational experiment of seamless handover of a mobile router using multiple care-of address registration. *Journal of Networks*, *3*(1), 23–30.

Taylor, I., Downard, I., Adamson, B., & Macker, J. (2006). Agentj: Enabling Java NS-2 simulations for large scale distributed multimedia applications. *Second International Conference on Distributed Frameworks for Multimedia*, (pp. 1-7).

Wakikawa, R., Devarapalli, V., & Thubert, P. (2004). *Inter home agents protocol (HAHA)*. IETF Internet Draft, draft-wakikawa-mip6-nemo-haha-01.

Wakikawa, R., Ernst, T., & Nagami, K. (2006). *Multiple care-of addresses registration. IETF, draft-wakikawa-mobileip-multiplecoa-05, RFC 5648*. Retrieved from.

KEY TERMS AND DEFINITIONS

DDoS Attack: A *distributed denial-of-service (DDoS)* attack is one in which a multitude of compromised systems attack a single target, thereby causing denial of service for users of the targeted system. The flood of incoming messages to the target system essentially forces it to shut down, thereby denying service to the system to legitimate users.

Heterogeneous Network Environment: Future heterogeneous wireless environments will be characterized by the coexistence of a large variety of wireless access technologies, with different protocol stacks. They support a number of applications and services with different quality of service requirements.

Home agent: In *Mobile Internet Protocol (Mobile IP)*, a home agent is a router on the home network of a mobile node that maintains its current location information as identified in its care-of address. A home agent may work in conjunction with a foreign agent, which is a router on the visited network. The home agent delivers the Internet traffic to the node through the foreign agent.

Mobile Network Prefix: A bit string that consists of some number of initial bits of an IP address which identifies the entire mobile network within the Internet topology. All nodes in a mobile network necessarily have an address containing this prefix.

Mobile Network: An entire network, moving as a unit, which dynamically changes its point of attachment to the Internet.

Mobile Router: A router capable of changing its point of attachment to the Internet, moving from one link to another link.

Chapter 13
Clean-Slate Information-Centric Publish Subscribe Networks

Laura Carrea
University of Essex, UK

Raul Almeida
University of Essex, UK

ABSTRACT

The Internet architecture of today does not seem suited to the current Internet usage, as the application layer is more and more content-centric, while the network layer is ossified around the IP concept. In this chapter, the authors explore a redefinition of the whole Internet architecture where nothing is taken for granted, especially IP addresses. The review focuses on the forwarding and topology components of the EU FP7 PSIRP architecture and on a few of the problematic issues and the ongoing discussions around a pioneering clean-slate design of the way to organize networks.

INTRODUCTION

The Internet architecture of today is a packet-based internetworking architecture that was created as an efficient multiplexed utilization of heterogeneous interconnected networks. The introduction of the Internet Protocol suite (TCP/IP) opened up the growth of the Internet and slowly the backbone was privatized and became distributed. Because a central coordinating element is missing, it has been difficult to apply major architectural changes to the Internet and the Internet architecture (i.e. hierarchical routing, TCP/IP, *Domain Name System (DNS)*) has remained the same since the 1980s when it was created (Jacobson, 2006a). Since then, only incremental improvements have been introduced to supply new services reducing the management costs of the network: the *Classless Inter-Domain Routing (CIDR)* (Fuller et al., 1993) was proposed to slow the growth of routing tables on routers across the Internet, and to slow the rapid exhaustion of IPv4 addresses; the *Border Gateway Protocol (BGP)* (Rekhter, 1995) was introduced to mirror the business relationship between providers and later extended for large scale deployments; the *Multi-Protocol Label Switching (MPLS)* (Rosen et al., 2001) was introduced to improve the performances of

DOI: 10.4018/978-1-4666-1613-4.ch013

IP routers. Later, *Virtual Private Networks (VPN)* and, recently Carrier Ethernet have appeared as new data services. Moreover, other solutions have been deployed for issues of Internet design in an open commercial environment such as *Network Address Translation (NAT)* boxes, which offer limited protection for unwanted traffic fracturing network connectivity and which have extended address spaces (Touch, 2002), (Handley, 2006) and Mobile IP (Perkins, 2002), which offers mobility to the host, using network indirection points.

However, despite all those efforts, Internet is considered as *ossified* (for example (Handley, 2006), (Anderson et al., 2005)), as all the solutions are considered as patching approaches based on ad-hoc extensions and overlays.

The original design of the Internet is centered on *best-effort* delivery between network-attached devices, forming the base of the concept of the *Internet Protocol (IP)* address. Every host has a unique IP address which acts as a location (where) utilized for routing purposes and at the same time as an identifier (who) of the host. The fact that the IP address has these two different functions is considered the root of many of the limitations of today's Internet architecture (Meyer et al., 2007) and a *split* is considered necessary.

In the past and still today, efforts for the future Internet have been based on revisiting single concepts, such as the IP locator/identifier, to improve end-host reachability, end-to-end security, mobility and routing issues without questioning the host centricity of the communication. Still, the Internet moves a datagram in a best effort manner independently of the semantics and the purpose of the data transport.

The first person who envisioned the necessity of a true network revolution was Van Jacobson with a talk at Stanford (Jacobson, 2006a). He mentioned the evolution of networking from telephony which was about connecting wires to other wires, to TCP/IP which is about machines connected to the wires, to the next generation which should be about interconnecting information. This new idea implies the necessity of rethinking the fundamentals of the communication and it is based on the observation that the large-scale usage of the Internet is not an end-to-end communication but for data dissemination. This shift towards a networking which is based on information has started to happen already but as an overlay (Rothenberg et al., 2008). In fact, *Service Oriented Architectures (SOA)*, XML routers, *Deep Packet Inspection (DPI)*, *Content Delivery Networks (CDN)* and peer-to-peer overlay technologies are a clear effort from the 'top' to move towards a networking which is focused on information rather than the host. In these types of technologies a big issue is to be able to move large amounts of labeled data having, at the network level, to reach a particular host. A mismatch is clearly observable: user/applications are related to *what* they want to access while networks work with *who* wants to access data. The mapping between the models requires a lot of conventions and configurations (middleware). In fact, acquiring data is not a conversation but it is, in fact, a dissemination (Jacobson, 2006feb). In this case, only the data matters but not necessarily the source of the data, as long as the integrity and the authenticity of the data (and not necessarily the source) are guaranteed (Koponen et al., 2007). Dissemination can be obtained with a conversation but this creates a lot of the problems and results in the limitations which are present in the Internet today (Jacobson, 2006b). As Van Jacobson pointed out, the problem is not that the Internet does not work. It works in fact really well but the usage of the Internet has changed. The Internet has created a lot of content that it was not designed for.

After Van Jacobson's talk, a few research activities started to focus on the investigation of this new way to do networking based purely on a content/information/data (throughout this chapter content, information or data are considered synonyms) centric paradigm, completely redesigning the Internet architecture from scratch. Pioneering work has been carried out in the projects EU FP7 PSIRP (PSIRP) and PARC CCN (Jacobson et al., 2009).

Van Jacobson himself proposed Content-Centric Networks (Jacobson et al., 2009). The proposal focuses on the content that becomes explicitly addressable in the whole architecture from the lower levels where packets (small pieces of content) are distinctly identified with names and they have attributes. The matching between the content and the interest is based on their names and the packet is forwarded to the requester.

The other pioneering work on full content/data/information-centric networks are the PSIRP/PURSUIT projects (PSIRP), (PURSUIT) which follow an approach that is different from data naming. The philosophy behind content identification is similar to the approach proposed in DONA, based on flat labels (Koponen et al., 2007). This chapter focuses on the efforts made within the framework of the PSIRP/PURSUIT projects.

We would like to emphasize at this point that the efforts we are talking about go beyond the philosophy of content delivery networks or peer-to-peer networks. The work aims to redesign the core of the Internet, the Internet Protocol suite (TCP/IP) which is inherently unfair and inefficient for data dissemination purposes (e.g. multiple flows of peer-to-peer applications and redundancy of information (Anand et al., 2008)) and which is considered insecure. For example, *Denial-of-Service (DOS)* attacks produce costly services outages and BGP insecure routing introduces worries and expenses. These insecurities originate from an imbalance of power in the Internet of today: the network delivers best-effort packets to a destination even if the receiver does not want to receive it.

Focusing on dissemination and information-centric design, the redesign of the Internet architecture grants a way to straightforwardly deal with the following, just to mention a few:

- *Security*: security is applied to the content itself rather than to the communication channel so that each type of content can be differently secured
- *Mobility*: content and user identifier are not bound to a specific location
- *Content-aware resource mapping* through metadata
- *Reduced content replication* as content has identifiers in the network
- *Spam protection*: a more receiver-driven model.

This chapter is an introduction to the work performed in the PSIRP/PURSUIT projects and after an overview of the structure of the architecture, it focuses on a review of the work done on the routing and forwarding functions.

A NEW ARCHITECTURE: THE PSIRP ARCHITECTURE

Many networking applications of today are publish/subscribe in nature. For example, *RDF Site Summary or Really Simple Syndication (RSS)* feeds, instant messaging, presence services, many web site and most middleware systems are either based on the publish/subscribe information paradigm or internally implement a publish/subscribe system. The new architecture, introduced in the PSIRP project (PSIRP), takes into consideration the needs of the application layer and tries to achieve data dissemination through the primitives of the architecture implementation. The fundamental ingredient of the architecture has to become the data itself, in order to make data dissemination the fundamental way of networking throughout the whole system. For these reasons, the new architecture is based on information throughout the layers and it has to be designed with the purpose of interconnecting information rather than end-points. Consequently, the primary function of the network is to locate and deliver information rather than locate hosts and arrange a communication between them.

The main ideas which are the fundaments of the new architecture and characterize the radical changes with respect to the current architecture are drawn from different sources. They can be summarised as in the following (Trossen, 2009):

- The *basic primitives* used in the implementation of the system: they are publish and subscribe instead of the traditional send and receive for all the possible layers and protocols in the architecture. This means that when Alice wants to make some data available to Bob, instead of sending them, she publishes the data presuming that Bob has subscribed or will subscribe to it. Creation and consumption of information are decoupled in space and/or in time and synchronization (Eugster et al., 2003). For this reason, there are typically multiple subscribers simultaneously. Publish/subscribe is a data dissemination method that is well known and it is mainly implemented as an overlay. The main characteristics of the architecture are (Trossen, 2009):
- *Data handling*. Data becomes the first class citizen of the architecture. Any piece of data can be considered as a separate publication. For example, a single transport segment sent over a wireless device is a short-lived local publication. But on the other hand, a library can be seen as a publication which consists of millions of sub-publications, even if at the moment the exact semantic of subscribing to a very large data set is not yet clear.
- *Efficient caching* which loses the coupling of the publishers and the subscribers in time.
- *–Multicast* which efficiently helps to disseminate data in space as packets are not replicated along the delivery tree. This contributes to reducing bandwidth consumption.

We can notice that the operation of publishing is controlled by the information producer and the operation of subscribing is controlled by the information consumer. In this way, the communication can take place only if each of the two parties has taken action. The architecture is not anymore sender-centered but there is a balance of power between publisher and subscribers.

It is important to emphasize that the primitives (publish and subscribe) are the basis of communication for *all the layers and all the protocols* (with the exception of compatibility mode for legacy applications especially during the evolutionary transition) from the physical level up to the application level.

- The *organization of the architecture*: it is layer-less or recursive rather than layered. This means that the same architecture is applied in a recursive manner on top of itself whereupon each higher recursive layer utilizes the functions offered by the lower recursive layers. The functions within this layer-less architecture should be seen as really functional elements rather than layers. At the bottom of the architecture the recursion terminates when the bits are transmitted over the medium (Nikander, 2010). This kind of architecture has been inspired by the work of John Day (Day, 2008) and the work of Touch (Touch et al., 2006), (Touch et al., 2008).
- The *first class citizen* of the architecture: data and contexts rather than location address (Nikander et al., 2010), (Trossen, 2009). The focus of the architecture becomes efficient data delivery instead of connecting different hosts. Data has no meaning without a context. As data are identified with a flat label, it is necessary to know the context or contexts for a given topic. The context can be defined as a collection of data that somehow belong together. It is quite natural that each given datum may belong to a number of contexts at any given point in time. Context primar-

ily means the 'neighborhoods' of a piece of data, namely other pieces of data that a given datum is semantically associated with. However, the semantic to define the context is not yet established.

The functional elements which will be used recursively throughout all the architecture are mainly three (Trossen, 2009):

• *Rendezvous:* This is the function which is responsible for matching publications and subscriptions. There is a variety of publish/subscribe networking approaches (Eugster, 2003). The topic-based mechanism seems to offer the best scalability. In this scheme, participants publish events and subscribe to topics which are identified with a keyword. In our case, the keyword is a randomly looking long bit flat label (Caesar et al., 2006) called a RId (*Rendezvous Identifier* - see next section). The concept of topic is very similar to the concept of group. Subscribing to a topic T can be viewed as becoming a member of the group T and publishing an event on the topic T can be viewed as broadcasting that event among the members of the group T. The topic in this context is viewed as a particular event service identified with an identifier (Eugster et al., 2003). Having flat labels, the use of hierarchies in organizing topics proposed in PSIRP appears to be a key improvement to the original scheme. However, this concept is still an object of discussion. As the opposite of groups where the addressing is flat, the hierarchy offers the possibility to organize topics according to relationships. A subscription made to some node in the hierarchy implies subscription to all the sub-topics of the node. The role of the rendezvous function can be summarized in this way: an application first subscribes to a local rendezvous service (which means it publishes its own interest in the Rendezvous). If the application has any data that it wants to make available to other applications, it publishes a list of the publications through the rendezvous service. Similarly, if the application wants to receive data that has been (or will be) published by some other applications it subscribes those publications to the rendezvous service (which means publishing a subscription list). In this way, the rendezvous service maintains a database containing all the publications labeled with the flat labels (randomly looking labels) called RIds. There will probably be a large number of rendezvous systems corresponding for example to various communities, networks, interest groups and managing domains. Typically, the local rendezvous has to establish subscriptions to other rendezvous systems that the applications have expressed interest in. Whenever the rendezvous identifies a publication which has a publisher and one or more subscribers, it requests the topology function to construct the delivery tree connecting the publisher and the subscribers. In this way, rendezvous functions and topology functions constitute the *control plane* in the architecture.

- *Topology Management*: The topology manager function is responsible for:
- Constructing the delivery tree connecting the publisher and all the subscribers,
- Providing the publisher with the suitable forwarding information contained in the delivery tree which will be encapsulated by the publisher in the packet header.
- Installing state (which in this context means forwarding infomation) in the nodes for 'stateful' routing if necessary.

As the forwarding fabric used in PSIRP allows dynamic switching between source-routing-like forwarding mechanisms and stateful routing (as described in the forwarding function section), the topology manager has the new important task to make decisions on where to store forwarding information: in case of a source-routing-like mechanism, the forwarding information will be stored in the packet header, while in the case of stateful routing the forwarding information will be stored in the forwarding nodes.

In order to perform the tasks, the topology function has to keep track of the physical and logical topology of the network. It creates the graph of the network using the connectivity information (link state) which will be labeled like all the other publications in the network.

The topology manager function can typically be reduced to trivial physical connectivity or to more complex mechanisms like spanning trees, shortest path or loop-free distance-vector mechanisms. The challenges of constructing an efficient topology manager function are especially related to the multicast nature of publish/subscribe traffic, and to the multilayer nature of the network.

- *Packet Forwarding:* The packet forwarding mechanism is based on Bloom filters to encode source-routing-like forwarding information in the packet header, enabling a native multicast forwarding mechanism without dependency on end-to-end addresses. Multicast is implemented in the typical fashion of copying a single packet to multiple output queues.

In this layer-less architecture, the *data* and *control* plane functions will work together utilizing each other in a component wheel (Trossen, 2009), similar to the way Haggle managers (Scott et al., 2006) are organized. Moreover, the Rendezvous, the Topology, the Forwarding functions are applied in a recursive fashion: all the functions are made available to the next layer upwards of the recursion, forming three different planes. They get repeated in a recursive manner and the way they work recursively can be described as in the following.

The rendezvous at a certain layer of the recursion needs to know which ones of the lower layer rendezvous services know about a specific topic. A request to publish or to subscribe to a specific topic is forwarded to the relevant rendezvous service going higher up. The topology completely hides the topology from the higher layers. As the architecture is data driven, there are not long living stable node identifiers but only transient time and topology dependent identifiers. In this way, the edge identifier of one layer can be completely opaque to the other layers. Only inside the layer itself the identifiers need to have meaning so that the in-layer rendezvous and topology functions can construct the needed forwarding trees. Table 1 summarizes some concepts of information centric networks versus the same concept in the old Internet design as in Rothenberg et al. (2008).

INFORMATION: RID, SID

Information is the core of the architecture. 'Everything is information, and information is everything' is a kind of motto of the PSIRP project

Table 1. Summary of new information-centric network concepts

Today's networks	Information centric networks
sender/receiver	content producer/consumer
client/server communication	publish/subscribe (sender and receiver are decoupled)
host-to-host	information retrieval
unicast	unified uni-, multi- and any-cast
explicit destination	implicit destination
end-to-end	end-to-data
host name (look-up)	data name (search)
secure channel	data integrity

(Trossen, 2009). All the network operation shall be based on information being the primary named entity across the layers. In this way, each piece of information must have a unique identifier (in particular the identifier is chosen to be a statistically unique identifier). However, since the amount of information will be huge, there is a necessity to make this vast amount of information manageable: information has to be organized. The information to have meaning needs a sort of context. No data is really meaningful without a context. The context defines the rule for interpreting data allowing it to be processed in a meaningful way. However, information items may be used as a reference to other items providing information such as data size, information owner, permission composition elements, and so on.

These ideas have generated the following concepts (Trossen, 2009):

- *The information item* - This is the simplest unit transmitted in the network and it is identified by an identifier called rendezvous identifier and indicated as RId. It is a statistically *globally* unique identifier and it can be created through a cryptographic hash of higher level identifiers. Information may also need to be grouped to create information collections. For example, a video stream can be identified as a collection of all the segments or even of all the individual images. They would form the information collection. Collections of information are information items that belong semantically together. A way to group information can be by algorithmically tying together all the identifiers. This means creating a common algorithm that generates the algorithmic rendezvous identifiers.
- *The metadata* - This enables establishing the linking of information items and it can be used for defining access control policies, quality of service parameters, data size, information owner, and anything which describes the publication being made.
- *The information scope* - Information exists in the context of scope which support grouping of information, reducing the space to be searched for the RId. Moreover, scoping limits the 'reachability' of information to parties having access to a particular mechanism (for example rendezvous) that implements the scoping (Tarkoma, 2008). Moreover, security permissions and routing policies associated with an RId may also be combined with publish/subscribe operation to generate the final information dissemination. Of course, a single publication has not necessarily to be restricted to a single scope. The same information can be within different scopes. The scope concept together with the possibility of introducing multiple levels of scoping (hierarchical scoping) can be seen as crucial for the scalability of information dissemination across the network. Scoping defines a very powerful concept that can be used to generate social relations between the consumers and the providers of the information, and the information itself. Each scope can have as well attached a governance policy as for example authentication information for the receiver of the information. From the application perspective, scopes group related data together. From the network perspective, scopes denote the party being responsible for locating a copy of the data.

In summary, each item of information is identified with the pair (RId, SId) and the metadata.

THE TOPOLOGY AND FORWARDING FUNCTIONS

As already anticipated in the introduction, the focus of this chapter will be the forwarding and the routing (topology manager) functions of the architecture. As the architecture is centered on information, the forwarding and routing functions have to deal with this concept. The task of the functions is to deliver packets in a network without stable end-to-end addresses (IP addresses). The forwarding function has been defined in Jokela et al. (2009) and further expanded in Rothenberg et al. (2009), Särelä et al. (2010a), Zahemszky et al. (2009a), Zahemszky et al. (2009b), Särelä et al. (2011), and Rothenberg et al. (2011).

The topology manager detailed in this chapter is for wired networks (multi-layer networks). However, the work (not only on the topology manager function but on the whole architecture) is still very much in progress. This chapter is intended to be an overview of the ideas developed within the PSIRP project.

THE FORWARDING FUNCTION

In an architecture without any end-to-end addresses, designing routing and forwarding mechanisms is a challenging problem. First of all, we can notice that in an information-centric network where everything is information the forwarding tree (or path) is information as well and it can be encoded with an identifier like all the other information in the network. It is generally called *Forwarding Identifier (FId)* in order to distinguish it from other identifiers. Such FIds can be used by the nodes to forward publications. The forwarding mechanism proposed in PSIRP (Jokela et al., 2009) is Bloom-filter-based and offers an approach that combines elements of source routing-like (as default case) and stateful routing in a flexible way. Traditionally, in the source routing approach all the forwarding information namely the packets' route, is stored link by link in the packet header, while the forwarding nodes contain only information about the neighbors. In this way, the packet may increase very much in size and as the whole path is exposed in the network, security may become a big issue (Bellovin, 1989). On the other hand, the size of the forwarding tables is very small. Conversely, in stateful routing the forwarding information are stored in the forwarding nodes and the packet header contains only the source and destination addresses (or location). The addresses need to be aggregated in order to make the approach scalable. In this way, the packet header is very light while the forwarding tables become quite heavy.

The Bloom filter approach allows implementation of stateless routing, avoiding the problems related to packet header size and security and on the other hand allows switching dynamically from stateless to stateful routing in a smooth and flexible way. Moreover, it allows easy generation of a more efficient multicast routing (which is fundamental, especially for a publish/subscribe network (Eugster et al., 2003)) in contrast to the more traditional unicast, when a packet which has to be simultaneously delivered to multiple destinations, has to be replicated for the number of destinations from the sender to the receiver along the whole delivery tree as schematically shown in Figure 1.

Bloom filters were first proposed by Bloom in 1970 (Bloom, 1970) as a space/time trade-off for hash tables: they reduce the space requirements for hash tables at the cost of introducing a small probability of error which is a false positive that a tested element is represented erroneously in the set. Since then, Bloom filters have been widely used for many applications as in spell checkers, distributed databases, distributed caching (Fan et al., 2000) and many other areas. Surveys of network applications of Bloom filters are illustrated in Broder et al. (2004) and Tarkoma et al. (2012). However, all the network applications of the Bloom filter consider maintaining Bloom filters

Figure 1. Multicast in contrast with unicast addressing. In multicast simultaneous delivery of information to a group of destinations is realised using the most efficient strategy to deliver the messages over each link of the network only once, creating copies only when the links to the multiple destinations split. The shades code the destinations and the number of links between nodes codes the number of packets flowing along that link.

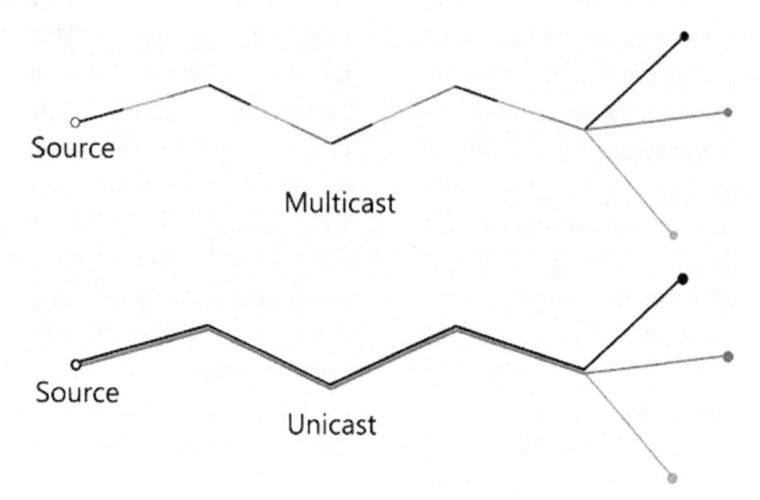

at the network nodes and checking if the incoming packets are included in the Bloom filter. Instead, this approach considers inverting the Bloom filter thinking by inserting the Bloom filter into the packet header rather than in the nodes (Jokela et al., 2009).

The Bloom Filter as a Forwarding Identifier

In the PSIRP architecture, source routing is considered always the default case in which each forwarding identifier FId is a statistically unique identifier that encodes the delivery tree for the publication and is placed in the packet header. In order to be able to construct a FId, instead of labeling the nodes for forwarding purposes, each point-to-point link is identified with a statistically unique identifier called Link-Ids. In particular, for each point-to-point link two identifiers, $\overrightarrow{AB}$ and $\overleftarrow{AB}$ one for each direction are assigned (see Figure 2). In case of a wireless link, each pair of node can be simply connected with a separate link.

Each Link-Id can be defined as an m-bits long vector $\mathbf{v}$. Using k hash functions, up to k bits are set to one in the m-bits vector, which generates the Link-Id L_0. The FId will encode the whole

Figure 2. Identifiers of unidirectional links and the composition of the name for a multicast tree

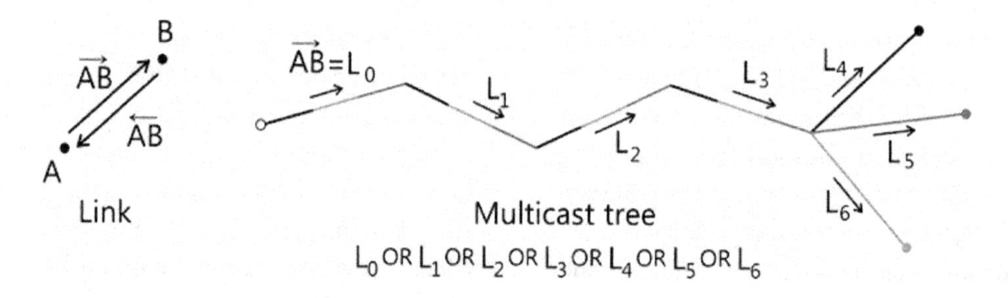

Figure 3. Example of a forwarding table in the PSIRP architecture

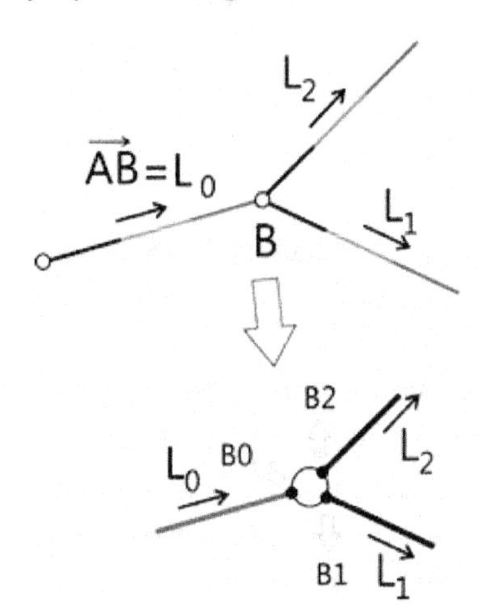

Interface	Link-Id
B1in	L_0
B2out	L_2
B1out	L_1
.	.
.	.
.	.

delivery tree as a Bloom filter (Bloom, 1970) built by OR-ing all the Link-Ids $(L_0, L_1, L_2, ...)$ composing the delivery tree as shown in Figure 2,

$$FId = L_0 \vee L_1 \vee L_2 \vee .... \quad (1)$$

We can notice that it is straightforward to generate a FId for a multicast tree. The forwarding table at each node will contain, in the default case of stateless routing approach, only the Link-Id of the links connected to that node as shown in Figure 3.

Once the packet reaches a node, the FId is AND-ed with the Link-Ids of all the outgoing links of the node (except the one which the packet was coming from). If the result of the AND operation is the Link-Id itself

$$FId \wedge L_i = L_i, \quad (2)$$

the packet is forwarded along that link. Of course, the match can result along more than one link per node. In this way, multicast turns out to be the basic addressing mechanism while unicast will be just a special case of multi-cast. There are some ways to decide whether the packet has reached the subscriber, but the most intuitive is to add the identifier of the node (NId) in the packet header and AND it with the FId in each node. It is important to notice that in an IP network, there is only one identifier: the IP address. This identifier is used to identify the source and for routing purposes. This means that as soon as the node is identified it is possible to establish a connection to it.

The operations on which the forwarding decisions are based are very simple logical operations: AND and comparison. The comparison operations can be simply parallelized since there is no memory or shared resources bottlenecks. These operations can be very easily implemented in hardware (Jokela et al., 2009).

This kind of routing is completely stateless and there are some similarities with *MultiProtocol Label Switching (MPLS)* as in both cases routing and forwarding are separated. MPLS allowed the concept of path in IP networks, introducing an identifier, a label, for it. However, as in MPLS label aggregation is not easy, multicast communication becomes quite complex (Zahemszky et al., 2010).

The Bloom filter is a space-efficient probabilistic data structure that is normally used to test whether an element is a member of a set (Mitzenmacher et al., 2005). Moreover, the Bloom filter has the advantage of offering fast searches and a small representation. For our application, fast search implies low processing time, and small representation means small packet header. In this respect, the Bloom filters turn out to be convenient tools to implement: 1) the stateful routing mechanism, where one of the limitations was high processing time, and 2) a source routing-like mechanism, where one of the limitations was heavy headers. However, Bloom filters allow a probability of error. In fact, due to their probabilistic nature, a query may give a false positive but never a false negative. Elements can be added to the set but not in general removed.

More formally, a Bloom filter is a vector **v** of m bits initially set to zero that encodes the membership of a set $A = \{a_1, a_2, .., a_n\}$ of n elements of a universe U of N elements. To add an element of the set to the Bloom filter, a set $H = \{h_1, h_2, .., h_k\}$ of k independent uniform hash functions (uniform means that the hashes are chosen at random within the set $\{1, 2, .., m\}$), each with range $\{1, 2, .., m\}$ is required and for each element $a \in A$ the bits at the positions $h_1(a), h_2(a), .., h_k(a)$ are set to one:

$$\mathbf{v}[h_1(a)] = 1, \ \mathbf{v}[h_2(a)] = 1, \dots \ \mathbf{v}[h_k(a)] = 1 \,. \tag{3}$$

In order to perform a query to check if the element b belongs to the set A encoded with the Bloom filter, all the k bits at the positions $h_1(a), h_2(a), .., h_k(a)$ are set to one. Now, if at least one bit is set to zero than necessarily the element b does not belong to the set A. Instead, if all the bits are set to one, than the element b may belong to A with a certain probability. In fact, if the element b is stored in the filter than by definition

$$b \in \mathbf{v} \ \Rightarrow \ \mathbf{v}[h_1(b)] = \mathbf{v}[h_2(b)] = \dots \mathbf{v}[h_k(b)] = 1 \tag{4}$$

but the converse

$$\mathbf{v}[h_1(b)] = \mathbf{v}[h_2(b)] = \dots \mathbf{v}[h_k(b)] = 1 \ \Rightarrow \ b \in \mathbf{v} \tag{5}$$

is not necessarily true. Namely, there is always a probability that b does not belong to A, which is called false positive probability. However, the Bloom filter is false negative free. A false negative would mean that the query over an element belonging to A gives a negative answer. For our application, it is very important that Bloom filters are false negative free, because otherwise this would mean that a message is not forwarded over a link where it should.

The False Positive Rate

In order to be able to evaluate false positive, given the universe U of N elements and given the set A of elements encoded in the Bloom filter, we denote F_P the set of false positives the elements belonging to the set $U - A$ which gives positive answer. The *false positive proportion* f_p is defined as the ratio of the number of elements in $U - A$ which give positive answer to the total number of elements in $U - A$ (Donnet et al., 2006)

$$f_p = \frac{|F_P|}{|U - A|} \,. \tag{6}$$

Alternatively we can define the following quantities:

The a-priori false positive estimate f_p. Given the number of elements n to encode in the Bloom filter, the number m of bits in the Bloom filter, and the number k of hash functions, then the *a-priori* false positive estimate is the expected false positive probability. Namely, the false positive rate is the probability for a given element that does not belong to the set A to be erroneously claimed to be present in the set. If the probability exists (the ergodic hypothesis is applied there) then it has the same value as the false positive proportion f_p. In order to calculate the false positive probability, most discussions make the hypothesis that all hash functions map each element to a random number uniformly over the range 1, 2, ..., m. With this assumption, it is easy to show that the false positive probability f_p is

$$f_p(m,n,k) = \left[1-\left(1-\frac{1}{m}\right)^{kn}\right]^k \approx \left(1-e^{-\frac{kn}{m}}\right)^k, \tag{7}$$

being $e^{-x} = \lim_{n\to\infty}\left(1+\frac{x}{n}\right)^n$.

The false positive probability depends on the number of bits of the Bloom filter m, on the number of elements to insert in the Bloom filter n, and on the number of hash functions k. The plot of the false positive probability as function of m, n, k is shown in Figures 4, 5 and 6 keeping in turn the other two variables fixed.

The $f_p(m,n_0,k_0)$ with $n = n_0$ and $k = k_0$ constant is a monotonically decreasing function where the false positive probability decreases as the number of bits in the Bloom filter increases, as shown in Figure 6. The $f_p(m_0,n,k_0)$ is a monotonically increasing function, where the false positive probability increases as the number of inserted elements (the number of links forming the delivery path in our case) increases, as shown in Figure 5. The $f_p(m_0,n_0,k)$ instead has a minimum

$$\frac{\partial f_p}{\partial k} = 0 \Rightarrow k_{\min} = \frac{m}{n}\ln 2. \tag{8}$$

Figure 4. The false positive probability f_p as function of the number of hash functions k. Three curves are shown for different value of n, the number of links to be inserted. The number of bits in the Bloom filter is m = 256.

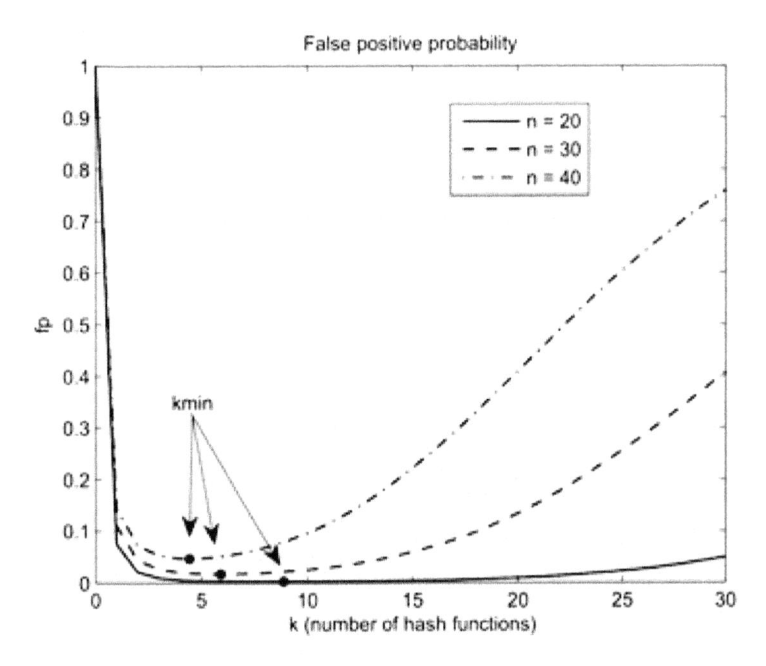

Figure 5. The false positive probability f_p as a function of the number n of links to be inserted in the Bloom filter. Four curves are shown for different value of k, the number of hash functions. The number of bits in the Bloom filter is m = 256.

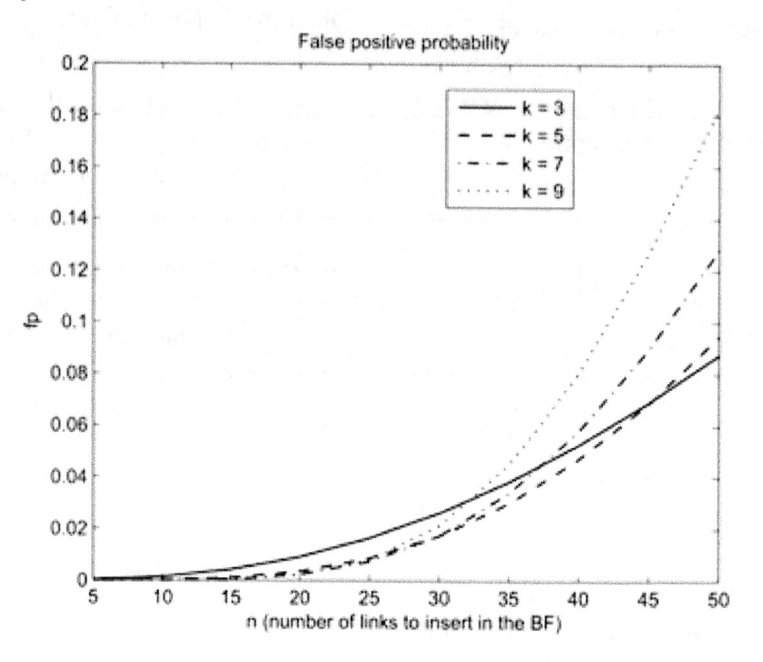

Figure 6. The false positive probability f_p as a function of the number m of bits in the Bloom filter. Three curves are shown for different value of n, the number of links to be inserted. The number of hash functions is k = 6.

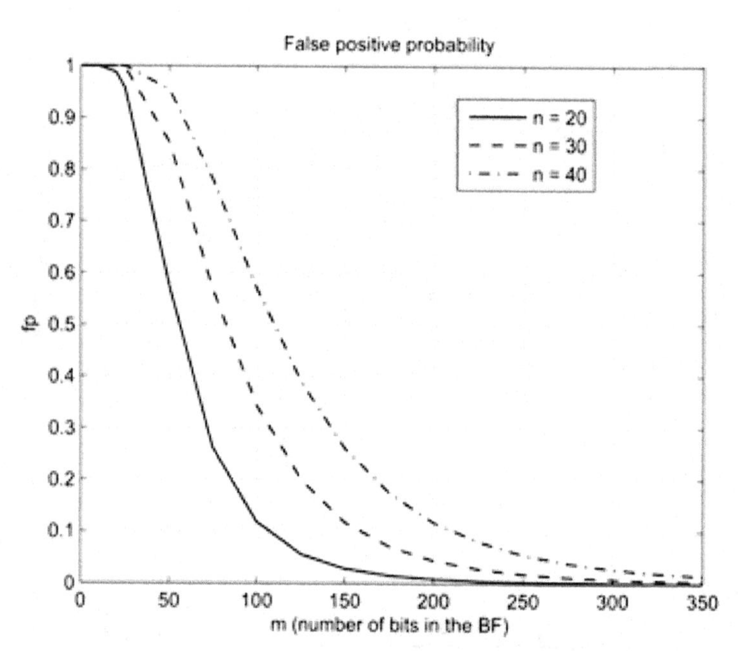

This is a global minimum but since k has to be an integer, $k_{\min}$ will be necessarily rounded to the nearest integer giving a false positive probability which will be higher than

$$f_p\left(k_{\min}\right) = 0.5^{\frac{m}{n}\ln 2}. \tag{9}$$

The rounded value of $k_{\min}$ gives the optimal value of the number of hash functions. However, the definition of f_p does not involve the number of bits which are set to one in the Bloom filter.

f_p is called *a priori false positive estimate* or *false positive estimate before hashing*. Recently, Bose (Bose et al., 2008) showed that in reality f_p as calculated in (7) is a strict lower bound on the false positive rate for any $k \geq 2$, namely for any choice of *k*, *m* and *n* (with $k \geq 2$) the false positive rate of the resulting Bloom filter is greater than f_p. Bose proposed a way to compute the false positive probability but it is not in a closed form.

The posterior false positive estimate f_{pa}. In order to be able to model the performance of the Bloom filter accurately, we define the *fill factor* ρ of the Bloom filter as:

$$\rho = \frac{s}{m}, \tag{10}$$

where s is the number of bits set to one. The *posterior false positive estimate* or *false positive estimate after hashing* f_{pa} is the expected false positive probability if the fill factor ρ of the Bloom filter is known:

$$f_{pa} = \rho^k. \tag{11}$$

The posteriori false positive estimate is a better estimator of the performance of the Bloom filter, but in practice, due to the concentration around the mean, the prior and posterior estimates are quite close as shown in Figure 7. It is clear from the right side of Figure 7 that the false

Figure 7. The posterior false positive estimate f_{pa} and the a priori false positive estimate f_p as function of the number of links n to be inserted in the Bloom filter. In this case k = 4 and m = 256. The curve f_p represents a strict lower bound of the false positive rate as we can clearly see on the right side in an enlargement of the plot on the left side.

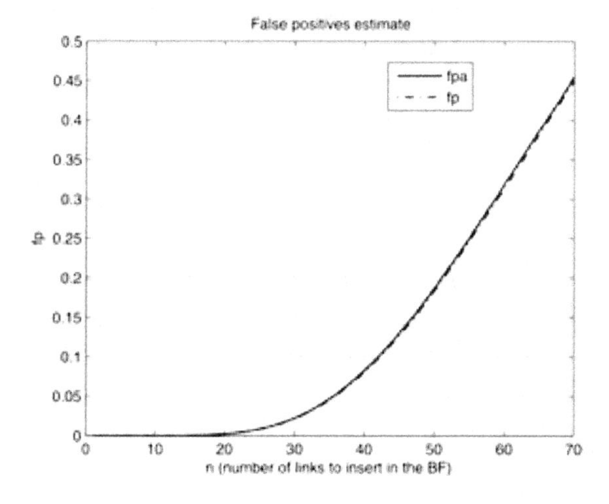

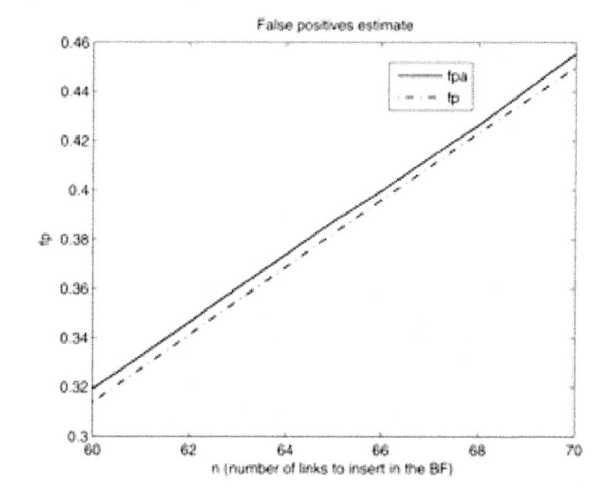

positive probability f_p as computed in (7) is a strict lower bound for the false positive rate.

The observed false positive probability f_{pr}. The actual observed false positive probability f_{pr} is the actual false positive rate that is observed when queries are made on the Bloom filter and it is defined as

$$f_{pa} = \frac{p}{q}, \quad (12)$$

where p is the number of observed false positives when a query is made on the Bloom filter and q is the number of tested elements which potentially may give false positives. Of course, it is an experimental quantity rather than a theoretical estimate like f_p.

In reality, all these quantities (in particular f_{pr}) are very interesting for our application but they describe the overall network performances only indirectly. In order to be able to describe the actual bandwidth consumption due to false positives, a new quantity f_{we} , the *forwarding efficiency,* can be introduced (Jokela et al., 2009)

$$f_{we} = \frac{N_{FId}}{N_D}, \quad (13)$$

where N_{FId} represents the number of links which form the delivery tree and N_D is the real number of links along which the packet is actually delivered. If the packet follows the path contained in the FId during the delivery than the forwarding efficiency is one; otherwise if there are false positives, the real number of links along which the packet is delivered will be higher than N_{FId} and the forwarding efficiency will drop. The quantity f_{we} can be considered as a metric to quantify the bandwidth overhead generated by sending packets over unnecessary links due to the false positives.

False Positives Reduction Method, the LITs

Different methods to reduce false positives have been produced (Tarkoma et al., 2012), (Rothenberg et al., 2010) but many of them generate false negatives, which are intolerable in a source-routing-like context. In fact, with false negatives there is a probability that packets are not forwarded through links contained in the delivery tree. Other methods such as counting Bloom filters involve the use of a bigger space. The method proposed by Lumetta (Lumetta et al., 2007) is based on the combination of two ideas from the hashing literature: the power of two choices which stems from Bloom filters (Bloom, 1970) and the work of Azar (Azar et al., 1999). The method, which consists in minimizing the number of bits set to one in the Bloom filter using two (or more) independent groups of hash functions, can yield modest reductions in the false positive probability using the same amount of space and more hashing but not introduce any false negatives. Based on the works of Lumetta (Lumetta et al., 2007) and Jimeno (Jimeno et al., 2007), Jokela (Jokela et al., 2009) proposes to take advantage of the power of choices in selecting the best of d Bloom filters during the process of generating the delivery tree. Instead of using solely one Link-Id per unidirectional link, the power of choices method suggests identifying each unidirectional link with a set of d identifiers that can vary depending on the network. Those identifiers are called LIT (Link-Id Tag). In order to obtain the maximum entropy in the bit distribution of each LIT, k hash functions are required. However, the work of Kirsch and Mitzenmacher (Kirsch et al., 2008) shows that it is sufficient to use only two random independent hash functions without any increase of asymptotic false positive probability. The two independent hash functions $h_1(x)$ and $h_2(x)$ can be used to simulate i hash functions:

$$g_i(x) = h_1(x) + i h_1(x) + i^2 \bmod m, \quad (14)$$

with m representing the size of the hashing table $\{1,..m\}$. The two hash functions $h_1(x)$ and $h_2(x)$ are system-wide parameters and just the number i is required to generate the d k hash functions necessary to compute the d LITs for any link.

Given the d identifiers for each link

$$\{\text{LIT}_1, \text{LIT}_2, \ldots \text{LIT}_d\}, \quad (15)$$

then d different FIds can be built

$$\{\text{FId}^{(1)}, \text{FId}^{(2)}, \ldots \text{FId}^{(d)}\}, \quad (16)$$

to represent the same multicast tree by OR-ing the i-th, i=1,2,…,d, LIT at a time for each link identifier of the multicast tree. For example, suppose a multicast tree composed by t links. Then:

$$\begin{aligned} \text{FId}^{(1)} &= \text{LIT}_1^{(1)} \vee \text{LIT}_2^{(1)} \vee \ldots \text{LIT}_t^{(1)} \\ \text{FId}^{(2)} &= \text{LIT}_1^{(2)} \vee \text{LIT}_2^{(2)} \vee \ldots \text{LIT}_t^{(2)} \\ &\ldots \\ \text{FId}^{(d)} &= \text{LIT}_1^{(d)} \vee \text{LIT}_2^{(d)} \vee \ldots \text{LIT}_t^{(d)} \end{aligned}, \quad (17)$$

These d FIds are all equivalent representations (identifiers) of the same multicast tree and therefore are all candidates FIds to be used to forward the packet, although only one of them, referred to as FId^{T}, will be utilized. In order to determine FId^{T} (the in-packet FId is sometimes called a *z-filter* (Jokela et al., 2009)), different conditions can be considered:

Lowest false positive after hashing: The FId^{T} will satisfy the condition of being the FId correspondent to the lowest false positive estimate after hashing f_{pa}, as defined in (11):

$$\text{FId}^T : f_{pa}^T = \min\{f_{pa}^{(1)}, \ldots, f_{pa}^{(d)}\} = \min\left\{\left(\rho^{(1)}\right)^k, \ldots, \left(\rho^{(d)}\right)^k\right\}. \quad (18)$$

Using f_{pa} to choose the actual FId^{T} is quite simple and it gives the FId corresponding to the smallest ratio of one's but, as mentioned above, this is just an estimate of the false positives rate.

Lowest observed false positive rate: The selected FId^{T} will satisfy the condition of being the FId corresponding to the lowest observed false positive rate f_{pr} defined in (12):

$$\text{FId}^T : f_{pr}^T = \min\{f_{pr}^{(1)}, \ldots, f_{pr}^{(d)}\}. \quad (19)$$

Since f_{pr} is the observed false positive probability when queries are made, the test set T_{set} of all the active links in the forwarding nodes along the delivery tree has to be used to count for false positives. In this way, the FId corresponding to the smallest number of false positives that can occur during delivery is utilized in the packet header. Using f_{pr} instead of f_{pa} produces better performance of false positives for a specific test set. However, counting false positives every time a delivery tree is formed is computationally more expensive.

- *Weighted lowest observed false positive rate:* This consists in weighting the observed false positives along the tree. The weight consents to take into account that delivering packets by mistake along some links can be worse than along others. In fact, a link may lead to a very congested area or some links have to be necessarily avoided due to routing policies or due to security reasons.

The forwarding information is stored in *d* different forwarding tables, each containing the LITs of the active links connected to the node. In order to speed up the forwarding procedure and the false positive occurrences, the index of the possible *d* correspondent to the selected FId^T can be stored in the packet header, so that, when the packet reaches the node instead of going through all the tables, it selects the table corresponding to the index in the packet header.

The price of this extension is that the forwarding table requires d-times more space than the basic solution and the packet header should also carry the d-value that will enable the forwarding node to decide which entries it should check in the FId. This price, however, is minor when compared to the benefits that are gained (remember that a false positive makes an entire packet to be unnecessarily forwarded to a link not present in the delivery tree).

Another method recently proposed to reduce false positives consists on a careful selection of *k*, the number of hash functions used to build the Link-Ids, that takes into account the network topology and the traffic profile experienced along each route (Carrea et al., 2011). The expected number of false positives F_p^T is estimated as a function of *k* and its minimum value, corresponding to the optimum value of *k* can be acquired, as shown in Figure 8. For instance, by selecting *k=5* instead of *k=9* in the LIT's formation would generate more than twice the number of false positives in the network. In fact, for

$$k = 9,\ F_p^T = 0.0158$$

while for $k = 4,\ F_p^T = 0.0371$.

Figure 8. The expected number of false positives F_p^T as a function of the number of hash functions k for a Manhattan-street network formed by 16 nodes obtained with the analytical model and the simulator. Only routes with one destination are considered. The number of bits in the Bloom filter is m = 64. The value of k corresponding to a minimum in the expected false positives is $k_{min} = 9$ and expected number of false positives $F_p^T = 0.0158$

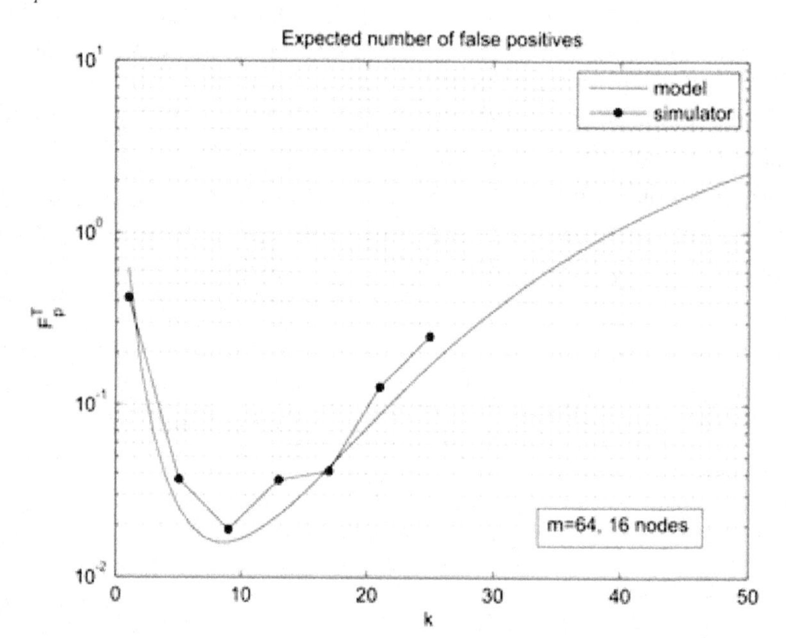

Loops

False positives can generate loops. Therefore, despite the low probability of loops it is very important to include in the functionality some kind of loop prevention strategies. A simple example of loops is when the FId^T encodes the path from the node *A* to the node *C* through the node *B*. A loop would be generated if the FId^T would match also the link connecting *C* to *A*.

The simplest strategy would be to explicitly avoid links that may generate loops in the FId selection process using the power of choice (see (19) and following). Unfortunately this method cannot guarantee a network free from loops.

Another simple method can be to include a Time To Live (TTL) field in the packet header. In case of a loop, after a certain number of hops the packet would be definitely dropped.

Another strategy proposed in (Jokela et al., 2009) which can be taken into consideration for loop prevention is related to caching the incoming Link-Id and the FId^T for a short time. Each node knows the incoming links (Link-Ids and LITs) from all the neighbors. When a packet reaches the node, it FId^T is compared with all the incoming links of the node interfaces except the one from where the packet arrives. A match means the danger of a loop. In this case, the node would cache the packet's FId^T and the Link-Id (and LITs) of the link from where the packet arrives for a short period of time. In case of a loop, the packet will come back through a different link than the cached one.

Another strategy for loop prevention is presented in Särelä et al. (2011) and consists in performing a bit permutation of the Bloom filter each time that a packet is forwarded along a link. In this case, not only an OR-ing operation is necessary when the packet reaches the node but also a permutation of the bits of the Bloom filter.

Stateful Routing and Link Recovery

As already mentioned, the Bloom filter based approach can combine elements of source routing (as default case) and stateful routing in a flexible way. A way of implementing stateful routing using Bloom filters has been proposed by associating a single LIT to a set of links, instead of all the links in the set as usual. Such set of links is commonly called *virtual link* (Jokela, et al., 2009). This aggregation is more general than the traditional one-to-one or many-to-one tunnels. It can represent any link set such as one-to-many trees, forests of partial trees as well as many-to-one concast (concurrent muticast) trees. The creation of a virtual link consists of the following steps:

- Selecting the individual links composing the virtual link;
- Assigning a Link-Id and computing the LITs;
- Updating the forwarding tables of the nodes residing along the virtual link with its Link-Id and LITs.

The individual links composing the virtual link can be contiguous and also apart, as a virtual link is actually a new entry in its node forwarding tables. Virtual links are necessarily unidirectional. In reality, a virtual link will be seen from the network as a link like any other. The virtual link represents just a way to aggregate the identifier of the links composing the delivery tree to reduce false positives and to connect edge nodes in inter-area or inter-domain scenarios. The packet with or without virtual link will always be forwarded along the same physical links.

As with any other link, the virtual link will be inserted in the FId^T. When the packet reaches a node which has links contained in the virtual link, the AND together with comparison operations will give a positive answer for the entry in the forwarding table correspondent to the virtual link.

The forwarding table will contain then the instruction to forward the packet along the right link. As with the use of virtual links the number *n* of links to be inserted in the FId^T is reduced, the probability of false positive will be consequently mitigated (see Figure 5). Unfortunately, a false positive over a virtual link will result on a false positive along all the links aggregated with it. For this reason, the virtual link identifier has to be very carefully chosen using more ones in the Link-Id than in the other links and explicitly avoiding these false positives during the selection of the best FId between the *d* FId. On the other hand, the use of virtual links adds new entries in the forwarding table increasing their size and the processing time. Moreover, having more entries in the forwarding tables means that every time a packet reaches the node, its FId is compared with more Link-Ids, causing more possible false positives occurrences. We can notice that the virtual link maintenance does not necessarily have to happen at line speed.

Stateful routing can become very important for recovery of broken links. If a link or a node suddenly fails, all the delivery trees containing that link or all the links connected to that node will be broken. Two approaches for instant rerouting have been considered in this forwarding scheme (Zahemszky et al., 2009b):

- Virtual link - This approach consists of having a separate backup path for each physical link. This separate backup path should have the same Link-Id and correspondent LITs as the physical link. In this way, the backup path would be a virtual link. This path should become active only in case of failure. As soon as a node detects a failure, it sends an activation message over the backup path using a pre computed FId containing all the links belonging to the backup path. The nodes along the backup path would then update their forwarding table by inserting a new line containing the virtual link identifier (the Link-Id and correspondent LITs of the broken link) and the interface through which the packet need to be forwarded. In this way, the packet header would not need to be modified and it would quickly flow along the backup path. The convergence time should be around zero and there should not be any service disruption.
- Pre-computed FId^T - This approach consists of having a dedicated FId for the backup path.

In this case, when a node detects a failure (a node failure or a link failure) it simply needs to OR the backup path FId with the FId^T . In this way, no additional signaling is necessary and no state needs to be included in the forwarding table. However, this method increases the false positive probability since new Link-Ids will be added to the FId^T and its fill factor ρ (defined in (11)) will increase. This method offers as well a zero convergence time and no service disruption. A hybrid approach can be used, where Pre-computed FId^T is employed just during the stage that backup path virtual link is formed.

These two methods are not only able to protect the network against a link failure but also against a node failure. In this case, it should be necessary to configure a backup tree towards all the neighbors of the failed node. The situation is equivalent to having broken all the links connected to the node. If the proposed mechanisms do not work, than a new FId^T has to be computed by the topology manager. In this case the convergence time would not be negligible anymore and some packets may be lost.

THE TOPOLOGY MANAGER FUNCTION

As the architecture is completely layer-less and recursive in essence (Touch et al., 2006) (Day, 2008), the topology function will hide the internal topology completely from upper layers (Day, 2008). At the very lowest layer, the recursion terminates. Here, it is only necessary to transmit the bits over the medium. As long as only two nodes are considered, the protocols for Rendezvous, Topology and Forwarding become trivial.

The topology function at the lowest layer consists simply of physical connectivity and at higher layers it is responsible for building spanning trees, shortest path, and loop free distance vector mechanisms. In this way, the topology function will be slightly different, depending on the recursive layer.

The topology manager is the component which generates and establishes the connection for the delivery of the information items under the request of the rendezvous function when a publication becomes active (namely when it has one publisher and at least one subscriber). The main tasks can be summarized as (Zahemsky et al., 2009a):

- Determining the delivery tree consisting of the set of links that connect the publisher to the subscribers,
- Forming the correspondent FId necessary for the actual delivery of the content,
- Deciding whether state should be installed (forwarding information) in the forwarding nodes and re-computing the FId,
- Providing the publisher with the FId.

Information for the Topology Manager

The topology component is responsible for collecting and managing network topology information from one point of view and on the other for making efficient use of the information, which is available to generate efficient data dissemination.

There are two types of information: the information about the network state and the information about the content that has to be delivered which is implicit in the publication.

- *Information about the network* - This consists of the network graph and the state of the links and they are:
    - The *nodes* in the network which can be identified by a NId. This can be obtained with a hashing process over a key or it can simply be the FId connecting the node to the local rendezvous.
    - The active *links* in the network identified by their LIds and the set of LITs. They constitute the connection available. They can be electrical or optical links, optical bypasses, and virtual links.
    - The *state* of each link and of each node (the characteristics of the link or the node) is dependent on the type of link, and is related to available bandwidth, congestion, length, delay, etc... With the state of the link the cost will be calculated. This information can be stored in the metadata of the information "link" or it can represent a publication.

If we now consider wired networks (in general multilayer networks), a node has the following components: optical node, electrical node, and controller.

We have to notice that in a multilayer network, there will be many possible links available. The topology manager needs to know which are the possible wavelengths, wavelength conversions and capabilities available in the network. The best approach seems to be that each node publish as the metadata associated to the NId, a Node

Information Vector containing all the relevant information associated to it and its adjacent links and the topology manager would run an algorithm to establish, for example, a light path having as output the light path establishment and its Link-Id.

We can notice that in this type of network we have two types of identification:

1. A semi-permanent identification which the NId that is assigned during the network attachment and it is used as the node identification,
2. A temporary identification, which is used for routing purposes and which consists of a stack of LIds. As the architecture is data driven, the topology is not described using stable network identifiers but through transient identifiers dependent on the topology and the time.

In IP networks, only one identifier is available for both the purposes, which is the IP address. This means that when the identity of a node is known, it is also known as the identifier for routing purposes. In PSIRP, the identity and the way to reach the nodes are completely decoupled.

Regarding the state of each link, this is information which needs to be somehow collected or retrieved. In an information-centric publish/subscribe network, as the primitives pub/sub are applied throughout the architecture, this kind of information has in principle to be published by the nodes. In the nodes, it has to be established what it needs to be published in order to make sure that the topology manager is able to have access to the information that is necessary. In this way, the nodes publish the links which are attached to them, together with all the characteristics and the topology manager functions subscribed to them. At least one SId is needed for grouping the information and limiting the information access to the topology manager and eventually to other nodes in the network. The state of the link is usually a very dynamic parameter. When the characteristics of the link changes, the node will release a new version of the 'publication' and the old version can be overwritten without the topology manager having continuously to subscribe to it. However, research is on-going on this topic.

- *Information about the content* The other type of information available to the topology manager is implicitly contained in the publication and in the subscriptions. There is no need for any collection mechanism. It consists of the following items:
    - Location of the publisher, for example in the form of NId, and autonomous system identifier ASId. The topology manager needs it to establish the root of the delivery tree within a domain and between domains,
    - The content identifier, the RId and the scope identifier, the SId,
    - The metadata which contains information about the content,
    - Location of the subscribers, for example in the form of a NId, and autonomous system identifier, ASId. The topology manager needs this to establish the terminals of the delivery tree within a domain and between domains,
    - The metadata associated with the subscription containing the QoS required for that requested content.

This point is very important, because if one compares with IP networks, for example, all this information needs to be explicitly collected, searched or even guessed through signaling, installation of state in the forwarding nodes, DPI, etc.

A part of the information the topology manager will need is related to the network and it will be published by the nodes. The other part of information is implicitly stored in the publication item.

The Forwarding Tree Computation

The problem of the topology manager is: given the resources available and the information about the content which has to be delivered, which is the best tree connecting the publisher and the subscribers? How can it be computed?

In order to be able to answer this question different levels of optimization and different options about where the delivery tree will be set up have to be considered. We consider first how to compute the optimal forwarding tree in a multicast scenario.

Initially the optimum tree calculation is considered. This process traditionally utilizes only the information about the network which is published by the nodes.

- *The best path between one publisher and one subscriber - Unicast* Given one publisher and one subscriber, the problem to find the shortest path given the network graph with non-negative costs is to connect two vertices of the network graph and it can be easily solved with the well-known Dijkstra's algorithm (Dijkstra, 1959). The FId is formed simply OR-ing all the links composing the shortest path.
- *The best path between one publisher and all the subscribers - Broadcast* If the set of the nodes of the minimal Steiner tree includes all the node of the network, than the problem is reduced to finding the minimal spanning tree and the two algorithms most commonly used are Prim's algorithm (Prim, 1957) and Kruskal's algorithm (Kruskal, 1956). They are greedy algorithms.
- *The best path between one publisher and many subscribers - Multicast* Since multicast is the native delivery mechanism implemented in this network. It is necessary to calculate a tree with minimal cost connecting the publisher and all the subscribers. This has been classified as a NP-complete problem (Karp, 1972) even if each link has unit cost (Garey, 1979). The problem can be formalized as *the minimum Steiner tree - MST* problem as follows: given a directed graph $G = (V, E)$ with a weight (cost) function $c : E \rightarrow \mathbb{R}^+$ on the edges, a vertex $r \in V$ called root, and a subset of vertices $k \subseteq V$ (called terminals), the aim is to find a minimum cost arborescence (the Steiner arborescence problem is the version of the Steiner tree problem where the graph is directed (Hwang et al., 1992)) rooted at r which spans all the vertices in K. The Steiner tree is a good representation for solving the routing multicast problem, but due to the computational complexity of this algorithm, heuristic algorithms are used. The literature provides a wide range of heuristics solving the problem in what it is called polynomial time (for example (Kou et al., 1981) which proposes the KMB algorithm which seems the most commonly used for data transmission.). The heuristics are mainly based on first computing shortest paths from source to each of the destinations and than using some pruning techniques or first computing the minimum spanning tree. The simplest way to build the tree is to construct the shortest path with the Dijkstra algorithm from the source to each destination. In order to build the FId of the tree, it is simply necessary to OR the FIds of the shortest (unicast) paths with minimal cost. The problem in this case is that it is not possible to really exploit the advantages of multicast delivery where the same packets are not duplicated along links in the network. This type of algorithm is likely to increase the utilization in the network.

FId Optimization

Once the optimal delivery tree has been established, the topology manager is responsible for calculating the FId which will be the identifier of the delivery tree and will be encapsulated in the packet header as described above. As in using a Bloom-filter based forwarding scheme, there is always a probability of false positive along the delivery tree. An optimization for reducing the false positives can be performed using the LITs and the optimization methods described in a previous section. The output of the optimization will be one of the d possible FIds.

It has to be noticed that the necessity for optimization due to false positives is related to the way of labeling the forwarding tree and that the false positives are dependent on the delivery tree and on the link identifiers of the links connected to the nodes along the delivery tree. Along a link there may be a false positive for a delivery tree going through the node it is connected to and not for another delivery tree.

The false positives are obviously independent of the amount of packets flowing along the links. However, this has to be taken into account that if there is a big amount of data that are flowing along a delivery tree and the number of false positives is high, those links will be unnecessarily occupied, increasing the occupancy of the link. There will be in this case a risk of congestion of those links and this can change the link state and the cost of the link causing a variation of the best delivery tree. Moreover, if a link is connected to a very congested area and along that link we will have a false positive for a certain delivery tree, and then an optimization is needed to avoid that link.

Once the false positives are reduced using the choice of the LITs, the amount of false positives may be still unacceptable. The question is related now to how to define what is unacceptable; namely it is necessary to define a threshold for that and if this threshold can be static or dynamic, depending on some criteria.

Once the threshold will be established and the false positives exceed the threshold, the FId would need to be improved. Another way to optimize the FId of the delivery tree is to add some state in the forwarding nodes establishing virtual links as discussed in a previous section. In this way, the FId of the delivery tree will be less populated by ones, because the number of links n which will form the FId will be less.

SIMULATOR AND PRELIMINARY RESULTS

A simulator has been developed at the University of Essex for the design of information centric networks with heterogeneous link layer technologies. The simulator is a C++ program developed and maintained by Dr. Raul Almeida and it is a general purpose event-driven simulator. It is the first simulator of information centric network over heterogeneous technology that has ever been developed. The simulator engine is very simple and very flexible and an example output of the simulator shows the improvement offered by using the selection mechanism of the FId corresponding to the lower value of the F_{pa}, as shown in Figure 9. We can notice that the false positives probability do not depend on the amount of traffic in the network but only on the identifiers which are assigned to the links.

SUMMARY

We have presented some ideas on the information centric networks which have been developed within the framework of the PSIRP/PURSUIT project. The chapter was mainly focused on the forwarding and topology manager functions of the architecture. Given the functionalities Rendezvous, Topology and Forwarding and given the communication primitives the networking mechanism can be summarized as in the follow-

Figure 9. The use of the best-of-d mechanism to construct the FId shows an improvement in the observed false positive f_{pr}

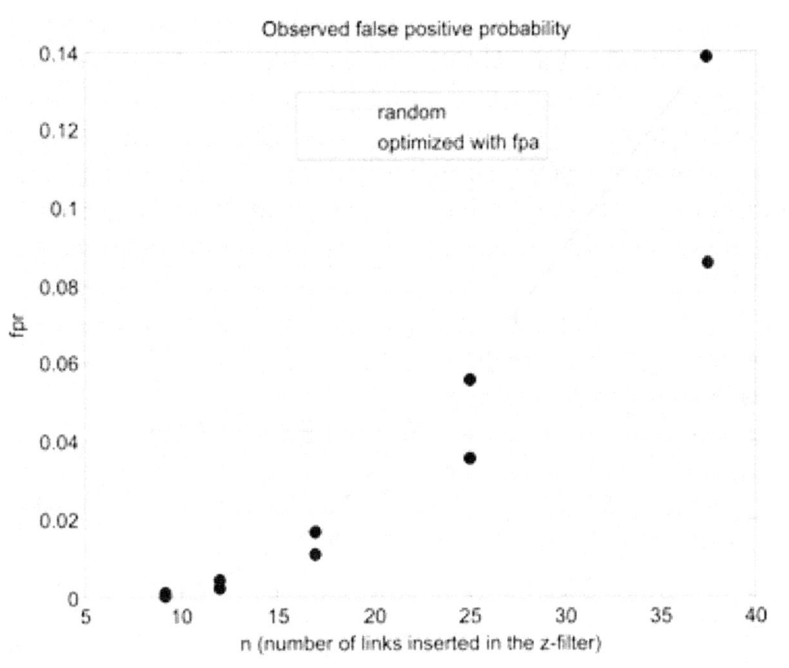

ing. A node wants to make available some data to the network. It calls the function publish(RId, SId, NId, metadata). If another node wants to access this data, it calls the function subscribe(RId, SId, NId, metadata). The rendezvous function is responsible for matching the publication and the subscriptions. Once the match is achieved, the rendezvous function calls the topology manager function to construct the FId for the actual data dissemination with the identifiers of the publisher and the subscribers, together with the metadata of the publication (carrying characteristic of the publication) and the subscription (carrying potentially the QoS required). With this information and the graph of the network the topology manager constructs the multicast tree and the forwarding identifier of the tree. It is been proposed to create the forwarding identifier as a Bloom filter. At this point, the FId is passed to the publisher node which will encapsulate the forwarding information in the packet header of each packet constituting the publication. The FId will have the same size independently of the size of the tree and, given the Fid, it is not possible to reconstruct the LinkIds which has been used to generate the FId. Once the packet is forwarded into the network, the forwarding decisions along the nodes in the tree are simple: they are based on an AND and a comparison operation. The forwarding table of each node is very short and it contains only its own links. The cost of reducing forwarding tables and simplifying the multicast delivery is some possible bandwidth wastage and potentially large packet header.

We can notice that without the agreement of the subscriber no data can reach the subscriber. In this way, the model is also receiver driven. Actually senders and receivers have the same power.

FUTURE RESEARCH DEVELOPMENTS

The pioneering work on the subject at EU FP7 PSIRP (PSIRP) and PARC CCN (Jacobson et al., 2009) is continued into the newly started projects Name Data Networking NDN (Zhang et al., 2010) and PURSUIT (PURSUIT). Many questions may have already arisen in the mind of the careful reader.

There is a long list of question for information/content/data centric networks proposals. Regarding the forwarding and topology manager functions, the reduction of the bandwidth wastage is still an open problem in particular focusing on how to distribute state to the nodes and in-packet information. Moreover, the investigation of utilizing network coding in the forwarding plane may lead to interesting results.

Research questions also focus on fast host mobility (Särelä et al., 2010b), multicast VPN services (Zahemszky et al., 2010), edge-controlled inter-domain multicast and information centric networks with optical technologies. The topology manager function has been just drafted and many questions arise, especially regarding the use of information about the content in order to improve tree formation to produce a scalable name-based routing plane. From a broader architectural point of view, a fundamental question is whether to utilize hierarchical identifiers or to keep a flat scheme for the content labels. Also fundamental concepts like caching and its implications on congestion and error control, transport functions and forwarding strategies, which are totally content-oriented, security, delay tolerant operations, traffic modeling and much more, have to be totally revisited.

CONCLUSION

Information centric networks are a promising approach at the network level to respond to the changes of the current Internet utilization which can be detected at the application layer today (or as overlays). A pure information centric architecture is far from being completed but initial architectural approaches are being evaluated and proposed. In this chapter, we have reviewed the pioneering work done under the aegis of the PSIRP project and thus this chapter is far from being a complete review of all the efforts undertaken. The Chapter is intended to be an overview in order to stimulate interest in a way of rethinking the fundamentals of our current way of constructing networks.

REFERENCES

Anand, A., Gupta, A., Akella, A., Seshan, S., & Shenker, S. (2008). Packet caches on routers: The implications of universal redundant traffic. *SIGCOMM Computer and Communications Review*, *38*(4), 219–230. doi:10.1145/1402946.1402984

Azar, Y., Broder, A., Karlin, A., & Upfal, E. (1999). Balanced allocations. *SIAM Journal on Computing*, *29*(1), 180–200. doi:10.1137/S0097539795288490

Bellovin, S. (1989). Security problems in the TCP/IP protocol suite. *SIGCOMM Computer and Communications Review*, *19*(2), 32–48. doi:10.1145/378444.378449

Bloom, B. (1970). Space/time trade-offs in hash coding with allowable errors. *Communications of the ACM*, *13*(7), 422–426. doi:10.1145/362686.362692

Bose, H., Guo, H., Kranakis, A., Maheshwari, A., Morin, P., & Morrison, J. (2008). On the false positive rates of Bloom filters. *Information Processing Letters*, *108*(4), 210–213. doi:10.1016/j.ipl.2008.05.018

Broder, A., & Mitzenmacher, M. (2004). Network applications of Bloom filters for distributed systems. *Internet Mathematics*, *1*(4), 485–509. doi:10.1080/15427951.2004.10129096

Caesar, M., Condie, T., Kannan, J., Lakshminarayanan, K., & Stoica, I. (2008). ROFL: Routing on flat labels. *SIGCOMM Computer and Communications Review*, *36*(4), 363–374. doi:10.1145/1151659.1159955

Carrea, L., Almeida, R., & Guild, K. (2011). *A qualitative method to optimise false positive occurrences for the in-packet Bloom filter forwarding mechanism. 3rd Computer Science and Electronic Engineering* (pp. 121–126). CEEC.

Day, J. (2008). *Patterns in network architecture: A return to fundamentals*. Upper Saddle River, NJ: Prentice Hall.

Dijkstra, E. (1959). A note on two problems in connexion with graphs. *Numerische Matematik*, *1*, 269–271. doi:10.1007/BF01386390

Donnet, B., Baynat, B., & Friedman, T. (2006). Retouched Bloom filters: Allowing networked applications to trade off selected false positives against false negatives. *Proceedings of the 2006 ACM CoNEXT*, (pp. 1-12).

Eugster, P., Felber, P., Guerraoui, R., & Kermarrec, A. (2003). The many faces of publish/subscribe. *ACM Computing Surveys*, *35*(2), 114–131. doi:10.1145/857076.857078

Fan, L., Cao, P., Almeida, J., & Broder, A. (2000). Summary cache: A scalable wide-area web caching sharing protocol. *IEEE/ACM Transactions on Networking*, *8*(3), 281–293. doi:10.1109/90.851975

Fuller, V., Li, T., & Varadhan, K. (1993). *Classless inter-domain routing (CIDR): An address assignment and aggregation strategy*. IETF, RFC 1519 (Proposed Standard), obsolete by RFC 4271.

Garey, M., & Johnson, D. (1979). *Computer and intractability: A guide to the theory of NP-completeness*. New York, NY: W.H. Freeman and Co.

Handley, M. (2006). Why the Internet only just works. *UBT Technology Journal*, *24*(3), 119–129. doi:10.1007/s10550-006-0084-z

Hwang, F., Richards, D., & Winter, P. (1992). *The Steiner tree problem*. Amsterdam, The Netherlands: North Holland.

Jacobson, V. (2006a). *If a clean slate is the solution what was the problem?* Stanford Clean Slate Seminar.

Jacobson, V. (2006b). *A new way to look at networking*. Google Tech Talks.

Jacobson, V., Smetters, D., Thornton, J., Plass, M., Briggs, N., & Braynard, R. (2009). Networking named content. *Proceedings of the 5th International Conference on Emerging Networking Experiments and Technology*, (pp. 1-12).

Jimeno, M., Christensen, K., & Roginsky, A. (2007). A power management proxy with a new best-of-N Bloom filter design to reduce false positives. *IEEE International Performance Computing and Communications Conference* (pp. 125-133).

Jokela, P., Zahemszky, A., Rothenberg, C. E., Arianfar, S., & Nikander, P. (2009). LIPSIN: Line speed publish/subscribe inter-networking. *SIGCOMM Computer and Communications Review*, *39*(4), 195–206. doi:10.1145/1594977.1592592

Karp, R. (1972). Reducibility among combinatorial problems. *Complexity of Computer Computations*, *40*(4), 85–103.

Kirsch, A., & Mitzenmacher, M. (2008). Less hashing, same performance: Building a better Bloom filter. *Random Structures and Algorithms*, *32*(2), 187–218. doi:10.1002/rsa.20208

Koponen, T., Chawla, M., Chun, B., Ermolinski, A., Kim, K., Shenker, S., & Stoica, I. (2007). A data orineted (and beyond) network architecture. *SIGCOMM Computer and Communications Review*, *34*(4), 181–192. doi:10.1145/1282427.1282402

Kou, L., Markowsky, G., & Berman, L. (1981). A fast algorithm for Steiner tree. *Acta Informatica*, *15*(2), 141–145. doi:10.1007/BF00288961

Lumetta, S., & Mitzenmacher, M. (2007). Using the power of two choices to improve Bloom filters. *Internet Mathematics*, *4*(1), 17–33. doi:10.1080/15427951.2007.10129136

Meyer, D., Zhang, L., & Fall, K. (2007). *Report from the IAB workshop on routing and addressing. IETF, RFC 4984*. Informational.

Nikander, P., & Tarkoma, S. (2010). *Data in context: The role of context in the RTFM network architecture.* Ideas for a paper within the PSIRP project.

Perkins, C. (2002). *IP mobility support for IPv4.* IETF, RFC 3344 (Proposed Standard), updated by RFC 4721.

Prim, R. (1957). Shortest connection networks and some generalizations. *The Bell System Technical Journal*, *36*, 1389–1401.

PSIRP: Publish Subscribe Internet Routing Paradigm. (n.d.). Retrieved from http://www.psirp.org

PURSUIT: Pursuing a publish subscribe Internet. (n.d.). Retrieved from http://www.fp7-pursuit.eu/PursuitWeb

Rekhter, Y., & Li, T. (1995). *A border gateway protocol 4 (BGP-4).* IETF, RFC 1771 (Draft Standard), obsolete by RFC 4271.

Rosen, E., Viswanathan, A., & Callon, R. (2001). *Multiprotocol label switching architecture. IETF, RFC 3031.* Proposed Standard.

Rothenberg, C. E., Jokela, P., Nikander, P., Särelä, M., & Ylitalo, J. (2009). Self-routing denial-of-service resistant capabilities using in-packet Bloom filters. *5th European Conference on Computer Network Defense (EC2ND)*, (pp. 46-51).

Rothenberg, C. E., Macapuna, C., Magalhães, F., Verdi, F., & Wiesmaier, A. (2010). In-packet Bloom filters: Design and networking applications. *Elsevier Computer Networks*, *55*(5), 1364–1378.

Rothenberg, C. E., Verdi, F., & Magalhães, M. (2008). Towards a new generation of information-oriented internetworking architectures. *Proceedings of the 2008 ACM CoNEXT First Workshop on Re-architecting the Internet (Re-Arch08)*, (pp. 1-6).

Särelä, M., Ott, J., & Ylitalo, J. (2010b). Fast inter-domain mobility with in-packet Bloom filters. *Proceedings of the 5th ACM International Workshop on Mobility in Evolving Internet Architecture (MobiArch10),* (pp. 9-14).

Särelä, M., Rothenberg, C. E., Aura, T., Zahemszky, A., & Nikander, P. (2011). Forwarding anomalies in Bloom filter based multicast. *Proceedings of the IEEE INFOCOM,* (pp. 2399-2407).

Särelä, M., Rothenberg, C. E., Zahemszky, A., Nikander, P., & Ott, J. (2010a). *BloomCast: Security in Bloom filter based multicast.* 15th Nordic Conference in Secure IT Systems (Nordsec)

Scott, J., Crowcroft, J., Hui, P., & Diot, C. (2006). Haggle: A networking architecture designed around mobile users. *Proceedings of the 3rd Annual Conference on Wireless On-demand Netwrok Systems and Services*, (pp. 78-86).

Tarkoma, S. (Ed.). (2008). *Conceptual architecture of PSIRP including subcomponent description.* Deliverable D2.2 PSIRP Project.

Tarkoma, S., Rothenberg, C. E., & Lagerspetz, E. (2012). (To appear). Theory and practice of Bloom filters for distributed systems. *IEEE Communications Survey and Tutorials*. doi:10.1109/SURV.2011.031611.00024

Touch, J. (2002). Those pesky NATs (network address translator). *IEEE Internet Computing*, *6*(4), 96. doi:10.1109/MIC.2002.1020334

Touch, J., & Pingali, V. (2008). The RNA metaprotocol. *Proceedings of the IEEE International Conference on Computer Commications*, (pp. 1-6).

Touch, J., Wang, Y., & Pingali, V. (2006). *A recursive network architecture*. ISI Technical Report 626.

Trossen, D. (Ed.). (2009). *Architecture definition, component descriptions, and requirements*. Deliverable D2.3 PSIRP Project.

Zahemszky, A., & Arianfar, S. (2009b). *Fast reroute for stateless multicast* (pp. 1–6). Ultra Modern Telecommunications & Workshops.

Zahemszky, A., Császár, A., Nikander, P., & Rothenberg, C. E. (2009a). Exploring the pubsub routing & forwarding space. *IEEE International Conference on Communications, Workshop on the Network of the Future*, (pp. 1-6).

Zhang, L., et al. (2010). *Named data networking (NDN) project.* PARC, Technical Report NDN-0001.

ADDITIONAL READING

Day, J. (2008). *Patterns in network architecture: A return to fundamentals*. Upper Saddle River, NJ: Prentice Hall.

KEY TERMS AND DEFINITIONS

Adaptive Streaming: A kind of streaming service that automatically adapts the bitrate of the sent stream according to the momentary state of the transmission channel (mainly the available bandwidth) and the client state (screen resolution, CPU load, battery level, ...)

Content Delivery Network: A cluster of servers used to deliver media contents over the network to a wide number of users. The servers are usually located in different geographical locations in order to be closer to users.

Layered Encoding: A class of media encoding techniques that permits the division of data into hierarchical "layers" of information. Starting from a base layer, that provides the lowest level of quality, each subsequent layer improves the final quality of the reconstructed media file. It is important to notice that higher levels are dependent on the lower ones in order to be correctly decoded.

Multiple Description Coding: A class of media encoding techniques that separates the original data in multiple representations (*descriptions*) of lower quality. When several descriptions are combined together, the resulting quality is enhanced compared to the quality of a single description.

Scalable Video Coding: A particular implementation of layered encoding that is an extension of the popular H.264/MPEG-4 AVC video compression standard.

Stream Switching: A technique used to change the streaming bitrate. Multiple versions of the same video are available at different bitrates, and at fixed time intervals it is possible to choose a different bitrate in order to adapt the stream to the network conditions.

Transcoding: The process of converting a media stream to a format different from the original. Transcoding can cause a decrease in the media quality.

Unequal Error Protection: When some parts of the information that needs to be transmitted are more important than others, stronger protection should be applied to these parts, in order to reduce the impact of errors during the transmission process. Unequal Error Protection codes do so by adding more redundancy to the most important parts.

Compilation of References

3 GPP TS 26.346 V9.4.0. (2010). *Technical specification: 3rd generation partnership project; technical specification group services and system aspects; multimedia broadcast/ multicast service (MBMS); Protocols and codecs (Release 9).* Braden, R., Zhang, L., Berson, S., Herzog, S., & Jamin, S. (1997). Resource ReSerVation Protocol (RSVP) – Version 1 functional specification. *IETF, RFC 2205.*

Abanoz, T. B., & Tekalp, A. M. (2009). Optimization of encoding configuration in scalable multiple description coding for rate-adaptive P2P video multicasting. *16th IEEE International Conference on Image Processing*, (pp. 3741-3744).

Abbasi, U., & Ahmed, T. (2010). SWOR: An architecture for P2P scalable video streaming using small world overlay. *IEEE Consumer Communications and Networking Conference*, (pp. 1-5).

Abbasi, U., Mushtaq, M., & Ahmed, T. (2009). Delivering scalable video coding using P2P small-world based push-pull mechanism. *Global Information Infrastructure Symposium*, (pp.1-7).

Abboud, O., Pussep, K., Kovacevic, A., & Steinmetz, R. (2009). Quality adaptive peer-to-peer streaming using scalable video coding. *Proceedings of the 12th IFIP/IEEE International Conference on Management of Multimedia and Mobile Networks and Services*, (pp. 41-54).

Abboud, O., Pussep, K., Stingl, D., & Steinmetz, R. (2011). Media-aware networking for SVC-based P2P streaming. *Proceedings of Network and Operating System Support for Digital Audio and Video (NOSSDAV)*, (pp. 15-20).

Abboud, O., Zinner, T., Pussep, K., & Steinmetz, R. (2011). On the impact of quality adaptation in SVC-based P2P video-on-demand systems. *ACM Conference on Multimedia Systems,* (pp. 223-232).

Abboud, O., Zinner, T., Pussep, K., Oechsner, S., Steinmetz, R., & Tran-Gia, P. (2010). A QoE-aware P2P streaming system using scalable video coding. *IEEE Tenth International Conference on Peer-to-Peer Computing*, (pp. 1-2).

Abboud, O., Pussep, K., Kovacevic, A., & Steinmetz, R. (2009). *Quality adaptive peer-to-peer streaming using scalable video coding* (pp. 41–54). Wired-Wireless Multimedia Networks and Services Management.

Abolhasan, M., Wysocki, T., & Dutkiewicz, E. (2003). A review of routing protocols for mobile ad hoc networks. *Ad Hoc Networks*, *2*(1), 1–22. doi:10.1016/S1570-8705(03)00043-X

Adeniji, S. D., Khatun, S., Raja, R. S. A., & Borhan, M. A. (2008). Design and analysis of resource management support software for multihoming in vehicle of IPv6 network. *AsiaCSN,* (pp. 13-17).

Adobe (2010). *HTTP dynamic streaming on the Adobe Flash platform.* Retrieved from http://www.adobe.com/products/httpdynamicstreaming/pdfs/httpdynamicstreaming_wp_ue.pdf

Agboma, F., Smy, M., & Liotta, A. (2008). QoE analysis of a peer-to-peer television system. *Proceedings of IADIS International Conference on Telecommunications, Networks and Systems*, (pp. 114-119).

Ahmad, I., Wei, X., Sun, Y., & Zhang, Y. Q. (2005). Video transcoding: an overview of various techniques and research issues. *IEEE Transactions on Multimedia*, *7*(5), 793–804. doi:10.1109/TMM.2005.854472

Ahmadi, S. (2011). *Mobile WiMAX: A systems approach to understanding IEEE 802.16m radio access technology*. Amsterdam, The Netherlands: Academic Press.

Akamai. (n.d.). *Akamai home page*. Retrieved from www.akamai.com.

Akyol, E., Tekalp, A. M., & Civanlar, M. R. (2007). A flexible multiple description coding framework for adaptive peer-to-peer video streaming. *IEEE Journal of Selected Topics in Signal Processing*, *1*(2), 231–245. doi:10.1109/JSTSP.2007.901527

Aldridge, R. P. (1996). *Continuous quality assessment of digitally-coded television pictures*. PhD Thesis, University of Essex, Colchester.

Alhaisoni, M., Ghanbari, M., & Liotta, A. (2010). Localized multi-streams for P2P streaming. *International Journal of Digital Multimedia Broadcasting*, *2010*. doi:10.1155/2010/843574

Alhaisoni, M., Liotta, A., & Ghanbari, M. (2010). Resource-awareness and trade-off optimisation in P2P video streaming. *International Journal of Advanced Media and Communication*, *4*(1), 59–77. doi:10.1504/IJAMC.2010.030006

Allavena, A., Demers, A., & Hopcroft, J. E. (2005). Correctness of a gossip based membership protocol. *Twenty-Fourth Annual ACM Symposium on Principles of Distributed Computing*, (pp. 292-301).

Alshaer, H., & Elmirghani, J. M. H. (2009). An intelligent route control scheme for multihomed mobile networks. *IEEE Vehicular Technology Conference*, (pp. 1-5).

Anand, A., Gupta, A., Akella, A., Seshan, S., & Shenker, S. (2008). Packet caches on routers: The implications of universal redundant traffic. *SIGCOMM Computer and Communications Review*, *38*(4), 219–230. doi:10.1145/1402946.1402984

Androutsellis-Theotokis, S., & Spinellis, D. (2004). A survey of peer-to-peer content distribution technologies. *ACM Computing Surveys*, *36*(4), 335–371. doi:10.1145/1041680.1041681

Apostolopoulos, J. (2001). Reliable video communication over lossy packet networks using multiple state encoding and path diversity. *Proceedings of SPIE, Visual Communications and Image Processing*, (pp. 24-26).

Apteker, R. T., Fisher, J. A., Kisimov, V. S., & Neishlos, H. (1995). Video acceptability and frame rate. *IEEE Transactions on Multimedia*, *3*(3), 32–40.

Asioli, S., Ramzan, N., & Izquierdo, E. (2010). *A novel technique for efficient peer-to-peer scalable video transmission*. 18th EURASIP European Signal Processing Conference.

Assuncao, P. A. A., & Ghanbari, M. (1996). Post-processing of MPEG2 coded video for transmission at lower bit rates. *Proceedings of the IEEE International Conference on Acoustics, Speech, and Signal Processing*, (pp. 1998-2001).

Assuncao, P., & Ghanbari, M. (1997). Transcoding of single-layer MPEG video into lower rates. *IEEE Proceedings: Vision . Image and Signal Processing*, *144*(6), 377–383. doi:10.1049/ip-vis:19971558

Azar, Y., Broder, A., Karlin, A., & Upfal, E. (1999). Balanced allocations. *SIAM Journal on Computing*, *29*(1), 180–200. doi:10.1137/S0097539795288490

Babaoglu, O., Meling, H., & Montresor, A. (2002). Anthill: A framework for the development of agent-based peer-to-peer systems. *22nd IEEE International Conference on Distributed Computing Systems*, (pp. 15-22).

Babelgum. (n.d.). *Babelgum home page*. Retrieved from www.babelgum.com

Baccichet, P., & Schierl, T. Wiegand, Thomas, & Girod, B. (2007). *Low-delay peer-to-peer streaming using scalable video coding*. Presented at Packet Video Workshop.

Baccichet, P., Shantanu, R., & Bernd, G. (2006). Systematic lossy error protection based on H. 264/AVC redundant slices and flexible macroblock ordering. *Springer Journal of Zhejiang University-Science A*, *7*(5), 900–909. doi:10.1631/jzus.2006.A0900

Bai, F., Sadagopan, N., & Helmy, A. (2003). Important: A framework to systematically analyze the impact of mobility on performance of routing protocols for ad hoc networks. *Proceedings of the IEEE INFOCOM*, (pp. 825-835).

Balaouras, P., & Stavrakakis, I. (2005). Multimedia transport protocols for wireless networks . In Salkintzis, A. S., & Passas, N. (Eds.), *Emerging wireless multimedia services and technologies*. John Wiley & Sons, Ltd.

Basagni, S., Chalamtac, I., & Syrotiuk, V. (1998). A distance routing effect algorithm for mobility (DREAM). *4th Annual ACM/IEEE International Conference on Mobile Computing and Networking,* (pp. 76-84).

Basagni, S., Chlamtac, I., Syrotiuk, V. R., & Woodward, B. A. (1998). A distance routing effect algorithm for mobility (DREAM). *Proceedings of ACM/IEEE International Conference on Mobile Computing and Networking,* (pp. 76-84).

Baset, S. A., & Schulzrinne, H. G. (2006). An analysis of the Skype peer-to-peer internet telephony protocol. *Proceedings - IEEE INFOCOM,* 1–11. doi:10.1109/INFOCOM.2006.312

Begen, A., Akgul, T., & Baugher, M. (2011). Watching video over the web: Part 1: Streaming protocols. *IEEE Internet Computing, 15*(2), 54–63. doi:10.1109/MIC.2010.155

Bellovin, S. (1989). Security problems in the TCP/IP protocol suite. *SIGCOMM Computer and Communications Review, 19*(2), 32–48. doi:10.1145/378444.378449

Berspacher, J., Schollmeier, R., Zols, S., & Kunzmann, G. (2004). Structured P2P networks in mobile and fixed environments. *International Working Conference on Performance Modeling and Evaluation of Heterogeneous Networks,* (pp. 1-25).

Biswas, S., Tatchikou, R., & Dion, F. (2006). Vehicle-to-vehicle wireless communication protocols for enhancing highway traffic safety. *IEEE Communications Magazine, 44*(1), 74–82. doi:10.1109/MCOM.2006.1580935

BitTorrent. (n.d.). *BitTorrent home page*. Retrieved from www.bittorrent.com

Bloom, B. (1970). Space/time trade-offs in hash coding with allowable errors. *Communications of the ACM, 13*(7), 422–426. doi:10.1145/362686.362692

Bluetooth Special Interest Group. (2001). *Specification of the Bluetooth System version 1.1 core, specification Volume 1.*

Blum, J. J., Eskandarian, A., & Hoffman, L. J. (2004). Challenges of inter-vehicle ad hoc networks. *IEEE Transactions on Intelligent Transportation Systems, 5*(4), 347–351. doi:10.1109/TITS.2004.838218

Bolosky, W., Douceur, J., Ely, D., & Theimer, M. (2000). Feasibility of a serverless distributed file system deployed on an existing set of desktop PCs. *ACM SIGMETRICS Performance Evaluation Review, 28*(1), 34–43. doi:10.1145/345063.339345

Bose, H., Guo, H., Kranakis, A., Maheshwari, A., Morin, P., & Morrison, J. (2008). On the false positive rates of Bloom filters. *Information Processing Letters, 108*(4), 210–213. doi:10.1016/j.ipl.2008.05.018

Bouch, A., Sasse, M. A., & DeMeer, H. (2000). *Of packets and people: A user-centred approach to quality of service*. In Eighth International Workshop on Quality of Service.

Boukerche, A. (2004). Performance evaluation of routing protocols for ad hoc wireless networks. *Mobile Networks and Applications, 9*, 333–342. doi:10.1023/B:MONE.0000031592.23792.1c

Broch, J., Maltz, D. A., Johnson, D. B., Hu, Y.-C., & Jetcheva, J. (1998). A performance comparison of multihop wireless ad hoc network routing protocols. *Proceedings of ACM/IEEE International Conference on Mobile Computing and Networking,* (pp. 85–97).

Broch, J., Johnson, D., & Maltz, D. (2003). *The dynamic source routing protocol for mobile ad hoc networks*. IETF Internet Draft.

Broder, A., & Mitzenmacher, M. (2004). Network applications of Bloom filters for distributed systems. *Internet Mathematics, 1*(4), 485–509. doi:10.1080/15427951.2004.10129096

Brosnan, A., Maitrat, T., Colhoun, A., & MacArdle, B. (2011, May 11). *Historical development*. Retrieved from http://ntrg.cs.tcd.ie/undergrad/4ba2.02-03/p1.html

Brun, P., Hauske, G., & Stockhammer, T. (2004). Subjective assessment of H.264/AVC video for low-bitrate multimedia messaging services. In *IEEE International Conference on Image Processing* (pp. 1145-1148).

Caesar, M., Condie, T., Kannan, J., Lakshminarayanan, K., & Stoica, I. (2008). ROFL: Routing on flat labels. *SIGCOMM Computer and Communications Review, 36*(4), 363–374. doi:10.1145/1151659.1159955

Cai, Y., & Hua, K. A. (2003). Sharing multicast videos using patching streams. *Multimedia Tools and Applications, 22*(2), 125–146. doi:10.1023/A:1025516608573

Camarillo, G., & Garcia-Martin, M. A. (2008). *The 3G IP multimedia subsystem (IMS): Merging the Internet and the cellular worlds*. Chichester, UK: Wiley. doi:10.1002/9780470695135

Capovilla, N., Eberhard, M., Mignanti, S., Petrocco, R., & Vehkapera, J. (2010). An architecture for distributing scalable content over peer-to-peer networks. *Second International Conference on Advances in Multimedia (MMEDIA),* (pp. 1-6).

Carrea, L., Almeida, R., & Guild, K. (2011). *A qualitative method to optimise false positive occurrences for the in-packet Bloom filter forwarding mechanism. 3rd Computer Science and Electronic Engineering* (pp. 121–126). CEEC.

Castro, M., Druschel, P., et al. (2003). SplitStream: High-bandwidth multicast in cooperative environments. *Proceedings of the Nineteenth ACM Symposium on Operating Systems Principles*, (pp. 298-313).

Castro, M., Druschel, P., Kermarrec, A.-M., Nandi, A., Rowstron, A., & Singh, A. (2003). SplitStream: High-bandwidth multicast in a cooperative environment. *Proceedings of the Nineteenth ACM Symposium on Operating Systems Principles*.

Castro, M., Druschel, Kermarrac, A. M., & Rowstron, A. I. T. (2002). SCRIBE: A large-scale and decentralized application-level multicast infrastructure. *IEEE Journal on Selected Areas in Communications, 20*(8), 1489–1499. doi:10.1109/JSAC.2002.803069

Chakareski, J., Han, S., & Girod, B. (2003). Layered coding vs. multiple descriptions for video streaming over multiple paths. *Proceedings of Internation Conference on Multimedia.*

Chen, F., & Repantis, T. (2005). Coordinated media streaming and transcoding in peer-to-peer systems. *Proceedings of the 19th IEEE International Parallel and Distributed Processing Symposium*.

Chen, C.-C., Lien, C.-N., Lee, U., Oh, S. Y., & Gerla, M. (2006, October). Codecast: A network-coding-based ad hoc multicast protocol. *IEEE Wireless Communications, 13*(5), 76–81. doi:10.1109/WC-M.2006.250362

Cheng, B., Jin, H., & Liao, X. (2007). Supporting VCR functions in P2P VoD services using ring-assisted overlays. *Proceedings of the IEEE International Conference on Communications*, (pp. 1698–1703).

Chenghao, L., Bouazizi, I., & Gabbouj, M. (2010). Advanced rate adaption for unicast streaming of scalable video. *IEEE International Conference on Communications*, (pp.1-10).

Chen, J., Chan, S., & Li, V. (2004). Multipath routing for video delivery over bandwidth-limited networks. *IEEE Journal on Selected Areas in Communications, 22*(10), 1920–1932. doi:10.1109/JSAC.2004.836000

Chiang, C. C., Wu, H. K., Liu, W., & Gerla, M. (1997). Routing in clustered multihop, mobile wireless networks with fading channel. *Proceedings of IEEE SICON*, (pp. 197-211).

Chiasserini, C. F., & Rao, R. R. (1999). Pulsed battery discharge in communication devices. *Proceedings of the 5th Annual ACM/IEEE International Conference on Mobile Computing and Networking,* (pp. 88-95).

Chitc, V. A., & Daigle, J. N. (2003). Performance of IP-based services over GPRS. *IEEE Transactions on Computers, 52*(6), 727–741. doi:10.1109/TC.2003.1204829

Chlamtac, I., & Redi, J. (1998). Mobile computing: Challenges and opportunities. In Ralston, A., Hemmendinger, E. R. D., & Reilly, E. (Eds.), *Encyclopedia of computer science* (4th ed.). Chichester, UK: Wiley & Sons.

Cho, S., Na, J., & Kim, C. (2005). A dynamic load sharing mechanism in multihomed mobile networks. *IEEE International Conference on Communications,* (pp. 1459-1463).

Choi, B. S., Suh, D. Y., Park, G. H., Kim, K., & Park, J. A. (2009). Peer-to-peer scalable video streaming using Raptor code. First International Conference on Ubiquitous and Future Networks, (pp.137-141).

Chow, C., & Ishii, H. (2007). Enhancing real-time video streaming over mobile ad hoc networks using multipoint-to-point communication. *Computer Communications, 30*(8), 1754–1764. doi:10.1016/j.comcom.2007.02.004

Chu, Y.-H., Rao, S. G., Seshan, S., & Zhang, H. (2002). A case for end system multicast. *IEEE Journal on Selected Areas in Communications*, *20*(8), 1456–1471. doi:10.1109/JSAC.2002.803066

Cicco, L. D., Mascolo, S., Bari, P., & Orabona, V. (2010). An experimental investigation of the Akamai adaptive video streaming. *Proceedings of the 6th International Conference on HCI in Work and Learning, Life and Leisure: Workgroup Human-Computer Interaction and Usability Engineering*, (pp. 447-464).

Cisco Systems. (2010). Cisco visual networking index: Forecast and methodology, 2009 – 2014. Retrieved from http://www.cisco.com/en/US/solutions/collateral/ns341/ns525/ns537/ns705/ns827/white_paper_c11-481360_ns827_Networking_Solutions_White_Paper.html

Cisco Systems. (2010b). *Cisco visual networking index: Usage*. Retrieved from http://www.cisco.com/en/US/solutions/collateral/ns341/ns525/ns537/ns705/Cisco_VNI_Usage_WP.html

Clarke, I. Sandberg, Wiley, B., & Hong, T. W. (2001). Freenet: A distributed anonymous information storage and retrieval system. *International Workshop on Designing Privacy Enhancing Technologies*, (pp. 46-66).

Clausen, T., & Jacquet, P. (2003). *Optimized link state routing protocol (OLSR).* IETF, RFC 3626.

Clausen, T., Jacket, P., Laouiti, A., Minet, P., Muhlethaler, P., Qayyum, A., & Viennot, L. (2003). Optimized link state routing protocol (OLSR). Retrieved March 8, 2011, from http://www.ietf.org/rfc/rfc3626.txt

Claypool, M., & Tanner, J. (1999). *The effects of jitter on the perceptual quality of video.* In 7th ACM International Conference on Multimedia. comScore. (2007). *comScore study reveals that mobile TV currently most popular among males and younger age segments.* Press Release.

Conti, M., Gregori, E., & Turi, G. (2004). Towards scalable P2P computing for mobile ad hoc networks. *Proceedings of the Second IEEE Annual Conference on Pervasive Computing and Communications Workshops*, (pp. 109–113).

Corson, S., & Macker, J. (1999). *Mobile ad hoc networking (MANET): Routing protocol performance issues and evaluation consideration.* IETF, RFC 2501.

Corson, M. S., Macker, J. P., & Cirnicione, G. H. (1999). Internet-based mobile ad hoc networking. *IEEE Internet Computing*, *3*(4), 63–70. doi:10.1109/4236.780962

Cox, R., Muthitacharoen, A., & Morris, R. (2002). Serving DNS using a peer-to-peer lookup service. *First International Workshop on Peer-to-Peer Systems*, (pp. 155-165).

Cramer, C., & Fuhrmann, T. (2006). Performance evaluation of chord in mobile ad hoc networks. *Proceedings of the 1st International Workshop on Decentralized Resource Sharing in Mobile Computing and Networking*, (pp. 48 – 53).

Dabek, F., & Kaashoek, M. (2001). Wide-area cooperative storage with CFS. *ACM SIGOPS Operating Systems Review*, *35*(5), 202–215. doi:10.1145/502059.502054

Dai, L., Cui, Y., & Xue, Y. (2007). Maximizing throughput in layered peer-to-peer streaming. *IEEE International Conference on Communications,* (pp.1734-1739).

Dai, M., Loguinov, D., & Radha, H. M. (2006). Rate-distortion analysis and quality control in scalable internet streaming. *IEEE Transactions on Multimedia*, *8*(6), 1135–1146. doi:10.1109/TMM.2006.884626

Das, S. R., Perkins, C. E., & Royer, E. M. (2000). Performance comparison of two on-demand routing protocols for ad hoc networks. *Proceedings of the IEEE Conference on Computer Communications*, (pp. 3-12).

Datta, A. (2003). MobiGrid: Peer-to-Peer overlay and mobile ad-hoc network rendezvous - A data management perspective. *Proceedings of the 15th Conference on Advanced Information Systems Engineering.*

Day, J. (2008). *Patterns in network architecture: A return to fundamentals.* Upper Saddle River, NJ: Prentice Hall.

de Couto, D. S. J., Aguayo, D., Bicket, J., & Morris, R. (2003). A high throughput path metric for multi-hop wireless routing. *Proceedings of the ACM Conference om Mobile Computing and Networking,* (pp. 134-146).

Detenber, B. H., & Reeves, B. (1996). A bio-informational theory of emotion: Motion and image size effects on viewers. *The Journal of Communication*, *46*(3), 66–84. doi:10.1111/j.1460-2466.1996.tb01489.x

Devarapalli, V., Wakikawa, R., Petrescu, A., & Thubert, P. (2005). *Network mobility basic support protocol*. IETF RFC 3963.

Dijkstra, E. (1959). A note on two problems in connexion with graphs. *Numerische Matematik, 1*, 269–271. doi:10.1007/BF01386390

Dijkstra, E. W. (1959). A note on two problems in connection with graphs. *Numerische Mathematik, 1*, 269–271. doi:10.1007/BF01386390

Ding, E. G., & Bhargava, B. (2004). Peer-to-peer file-sharing over mobile ad hoc networks. *Proceedings of the Second IEEE Annual Conference on Pervasive Computing and Communications Workshops (PERCOMW '04)*, (pp. 104–108).

Ding, P., Holliday, J., & Celik, A. (2004). A leader based priority ring reliable multicast in WLANs. *IASTED International Conference on Communications Systems and Networks*, (pp. 70–75).

Ding, Y., Liu, J., Wang, D., & Jiang, H. (2010). Peer-to-peer video-on-demand with scalable video coding. *Computer Communications, 33*(14), 2010. doi:10.1016/j.comcom.2010.04.025

Diot, C., Levine, B. N., Lyles, B., Kassem, H., & Balansiefen, D. (2000). Deployment issues for the IP multicast service and architecture. *IEEE Network, 14*(1), 78–88. doi:10.1109/65.819174

Djenouri, D., Nekka, E., & Soualhi, W. (2008). Simulation of mobility models in vehicular ad hoc networks. *Proceedings of the 2008 Ambi-Sys Workshop on Software Organisation and MonITоring of Ambient Systems*, (pp. 1-7).

Do, T., Hua, K., Aved, A., Liu, F., & Jiang, N. (2009). Scalable video-on-demand streaming in mobile wireless hybrid networks. *IEEE International Conference on Communications,* (pp.1-6).

Donnet, B., Baynat, B., & Friedman, T. (2006). Retouched Bloom filters: Allowing networked applications to trade off selected false positives against false negatives. *Proceedings of the 2006 ACM CoNEXT*, (pp. 1-12).

Do, T., Hua, K., Jiang, N., & Liu, F. (2009). PatchPeer: A scalable video-on-demand streaming system in hybrid wireless mobile peer-to-peer networks. *Peer-to-Peer Networking and Applications, 2*(3), 182–201. doi:10.1007/s12083-008-0027-1

Doukas, A., & Kalivas, G. (2006). Rician K factor estimation for wireless communication systems. *Proceedings of the IEEE International Conference on Wireless and Mobile Communications*, (pp. 69-73).

Draves, R., Padhye, J., & Zill, B. (2004). Routing in multi-radio, multi-hop wireless mesh networks. *Proceedings of the ACM Conference on Mobile Computing and Networking,* (pp. 114-128).

Eberspächer, J., Schollmeier, R., Zöls, S., Kunzmann, G., & Für, L. (July 2004). Structured P2P networks in mobile and fixed environments. *Proceedings of the International Working Conference on Performance Modeling and Evaluation of Heterogeneous Networks.*

Eichler, S. (2007). Performance evaluation of the IEEE 802.11 p WAVE communication standard. *Proceedings of the IEEE 66th Vehicular Technology Conference*, (pp. 2199-2203).

Emule. (n.d.). *Emule home page*. Retrieved from http://www.emule.com/

Eugster, P., Felber, P., Guerraoui, R., & Kermarrec, A. (2003). The many faces of publish/subscribe. *ACM Computing Surveys, 35*(2), 114–131. doi:10.1145/857076.857078

Fallah, Y. P., Mansour, H., Khan, S., Nasiopoulos, P., & Alnuweiri, H. M. (2008). A link adaptation scheme for efficient transmission of H.264 Scalable video over multirate WLANs. *IEEE Transactions on Circuits and Systems for Video Technology, 18*(7), 875–887. doi:10.1109/TCSVT.2008.920745

Fan, L., Cao, P., Almeida, J., & Broder, A. (2000). Summary cache: A scalable wide-area web caching sharing protocol. *IEEE/ACM Transactions on Networking, 8*(3), 281–293. doi:10.1109/90.851975

Farrera, M. P. (2005). *Packet-by-packet analysis of video traffic dynamics on IP networks.* PhD Thesis, University of Essex, UK, Department of Electronic Systems Engineering.

FastTrack. (n.d.). *FastTrack home page*. Retrieved from http://developer.berlios.de/projects/gift-fasttrack/

Fechner, G. T. (1966). *Elements of psychophysics* (Adler, H. E., Trans.). Holt, Rinehart and Winston, Inc.

Feeney, L. M. (1999). *A taxonomy for routing protocols in mobile ad hoc networks. Technical Report*. Swedish Institute of Computer Science.

Fesci-Sayit, M., Tunali, E. T., & Tekalp, A. M. (2009). Bandwidth-aware multiple multicast tree formation for P2P scalable video streaming using hierarchical clusters. *16th IEEE International Conference on Image Processing,* (pp.945-948).

Finnie, G. (2004). Fixed-mobile convergence reality check. *Heavy Reading, 2*(26).

Fiore, M., Haerri, J., Filali, F., & Bonnet, C. (2007). Vehicular mobility simulation for VANETs. *Proceedings of 40th Annual Simulation Symposium*, (pp. 301-307).

Franchi, N., Fumagalli, M., Lancini, R., & Tubaro, S. (2005). Multiple description video coding for scalable and robust transmission over IP. *IEEE Transactions on Circuits and Systems for Video Technology, 15*(3), 321–334. doi:10.1109/TCSVT.2004.842606

Franciscani, F. P., Vasconcelos, M. A., Couto, R. P., & Loureiro, A. A. F. (2003). *Peer-to-peer over ad-hoc networks: (Re)Configuration algorithms.* International Parallel and Distributed Processing Symposium.

Fredendall, G. L., & Behrend, W. L. (1960). Picture quality - Procedures for evaluating subjective effects of interference. *SMPTE Journal, 48*, 1030–1034.

Freenet. (n.d.). *Freenet Project home page*. Retrived from http://freenetproject.org/

Fuller, V., Li, T., & Varadhan, K. (1993). *Classless inter-domain routing (CIDR): An address assignment and aggregation strategy*. IETF, RFC 1519 (Proposed Standard), obsolete by RFC 4271.

Fu, X., Lei, J., & Shi, L. (2007). *An experimental analysis of Joost peer-to-peer VoD service. Technical Report*. Institute of Computer Science, University of Goettingen.

Gale, A. G. (1997). Human response to visual stimuli . In Hendee, W. R., & Wells, P. N. T. (Eds.), *The perception of visual information* (pp. 127–147). Berlin, Germany: Springer Verlag. doi:10.1007/978-1-4612-1836-4_5

Garcia, L., Arnaiz, L., Alvarez, F., Menendez, J. M., & Gruneberg, K. (2009). Protected seamless content delivery in P2P wireless and wired networks. *IEEE Wireless Communications, 16*(5), 50–57. doi:10.1109/MWC.2009.5300302

Garey, M., & Johnson, D. (1979). *Computer and intractability: A guide to the theory of NP-completeness*. New York, NY: W.H. Freeman and Co.

Gescheider, G. A. (1997). *Psychophysics: The fundamentals*. Lawrence Erlbaum Associates, Inc.

Ghanbari, M. (2003). *Standards codecs: Image compression to advanced video coding*. Stevenage, UK: IET Press. doi:10.1049/PBTE049E

Ghinea, G. Thomas, J. P., & Fish, R. S. (1999). *Multimedia, network protocols and users - Bridging the gap.* In ACM Multimedia.

Gibson, S. (2009). *Distributed reflection denial of service.* Gibson Research Corporation. Retrieved from http://grc.com/dos/drdos.htm.

Gilliam, D. (2006). *The appeal of mobile video: Reading between the lines*. Retrieved from http://www.tdgresearch.com/tdg_opinions_the_appeal_of_mobile_ video.htm

Giordano, E., Ghosh, A., Pau, G., & Gerla, M. (2008). *Experimental evaluation of peer-to-peer applications in vehicular ad-hoc networks.* First Annual International Symposium on Vehicular Computing Systems.

Giordano, S. (2001). Mobile ad-hoc networks . In Stojmenovic, I. (Ed.), *Handbook of wireless network and mobile computing*. Chichester, UK: John Wiley & Sons.

Giordano, S. (2002). Mobile ad-hoc networks . In Stojmenovic, I. (Ed.), *Handbook of wireless networks and mobile*. New York, NY: Wiley. doi:10.1002/0471224561.ch15

Gkantsidis, C., & Rodriguez, P. (2005). Network coding for large scale content distribution. *Proceedings of the IEEE INFOCOM,* (pp. 2235-2245).

Gnutella. (n.d.). *Gnutella home page*. Retrieved from www.Gnutella.com

Goldsmith, A. (2005). *Wireless communications*. Cambridge, UK: Cambridge University Press.

Goldsmith, J., & Wicker, S. (2002). Design challenges for energy-constrained ad hoc wireless networks. *IEEE Wireless Communications Magazine*, *9*(4), 8–27. doi:10.1109/MWC.2002.1028874

Goodchild, J. (2005). Integrating voice, data and video . In *IP video implementation and planning guide*. United States Telecom Association.

Goyal, V. K. (2001). Multiple description coding: Compression meets the network. *IEEE Signal Processing Magazine*, *18*(5), 74–93. doi:10.1109/79.952806

Grace, K. (2000). *Mobile mesh border discovery protocol*. Work in Progress. (Internet Draft). Retrieved March 8, 2011, from http://www.mitre.org/work/tech_transfer/mobilemesh/draft-grace-manet-mmrp-00.txt

Grois, D., & Hadar, O. (2011). Efficient adaptive bit-rate control for scalable video coding by using computational complexity-rate-distortion analysis. *IEEE International Symposium on Broadband Multimedia Systems and Broadcasting,* (pp.1-6).

Grois, D., Kaminsky, E., & Hadar, O. (2010). Adaptive bit-rate control for region-of-interest scalable video coding. *IEEE 26th Convention of Electrical and Electronics Engineers in Israel,* (pp.761-765).

Grois, D., & Hadar, O. (2011). Recent advances in region-of-interest coding . In Del Ser Lorente, J. (Ed.), *Recent advances on video coding* (pp. 49–76). Vukavar, Croatia: Intech Open Access Publisher.

Grois, D., Kaminsky, E., & Hadar, O. (2010). *ROI adaptive scalable video coding for limited bandwidth wireless networks* (pp. 1–5). IFIP Wireless Days.

Guo, H., & Lo, K.-T. (2008). Cooperative media data streaming with scalable video coding. *IEEE Transactions on Knowledge and Data Engineering*, *20*(9), 1273–1281. doi:10.1109/TKDE.2008.18

Gupta, P., & Kumar, P. R. (2000). The capacity of wireless networks. *IEEE Transactions on Information Theory*, *46*(2), 388–404. doi:10.1109/18.825799

Gurses, E., & Kim, A. (2008). Maximum utility peer selection for P2P streaming in wireless ad hoc networks. *Proceedings of the IEEE GLOBECOM*, (pp. 1-5).

Haas, Z. J., & Tabrizi, S. (1998). On some challenges and design choices in ad-hoc communications. *Proceedings of IEEE Military Communications Conference*, (pp. 187-192).

Handley, M., Jacobson, V., & Perkins, C. (2006). SDP: Session description protocol. *IETF, RFC 4566.*

Handley, M. (2006). Why the Internet only just works. *UBT Technology Journal*, *24*(3), 119–129. doi:10.1007/s10550-006-0084-z

Hands, D. (1997). *Mental processes in the evaluation of digitally-coded television pictures.* PhD Thesis. University of Essex, Colchester.

Hands, D. S. (2004). A basic multimedia quality model. *IEEE Transactions on Multimedia*, *6*(6), 806–816. doi:10.1109/TMM.2004.837233

Hao-Min, L., Ge, Y., Pang, A. C., & Pathmasuntharam, J. S. (2010). Performance study on delay tolerant networks in maritime communication environments. *OCEANS 2010 IEEE – Sydney*, (pp. 1-6). doi:10.1109/OCEANS-SYD.2010.5603627

Harmanci, O., Kanumuri, S., Kozat, U. C., Demircin, U., & Civanlar, R. (2009). Peer assisted streaming of scalable video via optimized distributed caching. *6th IEEE Consumer Communications and Networking Conference,* (pp. 1-5).

Hefeeda, M., Habib, A., Botev, B., Xu, D., & Bhargava, B. (2003). *PROMISE: Peer-to-peer media streaming using CollectCast.* In 11th ACM International Conference on Multimedia.

Hei, X., Liang, C., Liang, J., Liu, Y., & Ross, K. W. (2006). Insights into PPLive: A measurement study of a large scale P2P IPTV system. *Proceedings of the Workshop on Internet Protocol TV (IPTV) Services over the World Wide Web in conjunction with WWW2006.*

Hei, X., Liang, C., Liang, J., Liu, Y., & Ross, K. W. (2007). A measurement study of a large-scale P2P IPTV system. *IEEE Transactions on Multimedia*, *9*(8), 1672–1687. doi:10.1109/TMM.2007.907451

Hei, X., Liu, Y., & Ross, K. (2008). IPTV over P2P streaming networks: the mesh-pull approach. *IEEE Communications Magazine*, *46*(2), 86–92. doi:10.1109/MCOM.2008.4473088

Hekmat, R., & Van Mieghem, P. (2004). Interference in wireless multi-hop ad-hoc networks and its effect on network capacity. *Wireless Networks*, *10*(4), 389–399. doi:10.1023/B:WINE.0000028543.41559.ed

He, Y., Gu, T., Guo, J., & Dai, J. (2007). *KQStream: Kindred-based QoS-aware live media streaming in heterogeneous peer-to-peer environments* (pp. 56–61). Parallel Processing Workshops.

Hipp, R. (n.d.). *SQLite*. Retrieved from http://www.sqlite.org/

Hong, D. W., & Hong, C. S. (2003). A QoS management framework for distributed multimedia systems. *International Journal of Network Management*, *13*(2), 115–127. doi:10.1002/nem.465

Hossain, T., Cui, Y., & Xue, Y. (2009). Minimizing rate distortion in peer-to-peer video streaming. *6th IEEE Consumer Communications and Networking Conference,* (pp. 1-5).

Hossfeld, T., & Leibnitz, K. (2008). A qualitative measurement survey of popular Internet-based IPTV systems. *Second International Conference on Communications and Electronics*, (pp. 156-161).

Hu, Y. C., Das, S. M., & Pucha, H. (2003). Exploiting the synergy between peer-to-peer and mobile ad hoc networks. *Proceedings of HotOS IX Workshop*, (pp. 37–42).

Huang, C.-M., Lin, C.-W., Yang, C.-C., Chang, C.-H., & Ku, H.-H. (2009). An SVC-MDC video coding scheme using the multi-core parallel programming paradigm for P2P video streaming. *IEEE/ACS International Conference on Computer Systems and Applications,* (pp. 919-926).

Hu, H., Guo, Y., & Liu, Y. (2011). Peer-to-peer streaming of layered video: Efficiency, fairness and incentive. *IEEE Transactions on Circuits and Systems for Video Technology*, *21*(8), 1013–1026. doi:10.1109/TCSVT.2011.2129290

Hui, K. Y. K., Lui, J. C. S., & Yau, D. K. Y. (2006). Small-world overlay P2P networks: construction, management and handling of dynamic flash crowds. *Computer Networks: The International Journal of Computer and Telecommunications Networking*, *50*(15), 2727–2746.

Hwang, F., Richards, D., & Winter, P. (1992). *The Steiner tree problem*. Amsterdam, The Netherlands: North Holland.

IEEE 802.16e-2005. (2005). *IEEE standards for local and metropolitan area networks, part 16: Air interface for fixed and mobile broadband wireless access systems*.

Ilyas, M., & Ahson, S. A. (2008). *IP multimedia subsystem (IMS) handbook*. Boca Raton, FL: CRC Press.

iMesh. (n.d.). *iMesh home page*. Retrieved from www.imesh.com

ISO 9241-11. (1998). *Guidance on usability. Ergonomic requirements for office work with visual display terminals (VDTs) - Part 11.*

ISO/IEC 802.11 (1999). *ANSI/IEEE Std 802.11, part 11: Wireless LAN medium access control (MAC) and physical layer (PHY) specification.*

Itaya, S., Hayashibara, N., Enokido, T., & Takizawa, M. (2005). *Scalable peer-to-peer multimedia streaming model in heterogeneous networks*. Seventh IEEE International Symposium on Multimedia.

Ito, R. M., Tasaka, S., & Fukuta, Y. (2004). *Psychometric analysis of the effect of end-to-end delay on user-level QoS in live audio-video transmission.* In IEEE International Conference on Communications.

ITU-R Recommendation BS.1116. (1997). *Methods for the subjective assessment of small impairments in audio systems including multichannel sound systems*.

ITU-R Recommendation BS.1679. (2004). *Subjective assessment of the quality of audio in large screen digital imagery applications intended for presentation in a theatrical environment*.

ITU-R Recommendation BT500-11. (2002). *Methodology for the subjective assessment of the quality of television pictures*.

ITU-T Recommendation E.800. (1994). *Terms and definitions related to quality of service and network performance including dependability.*

ITU-T Recommendation P.910. (1999). *Subjective video quality assessment methods for multimedia applications.*

ITU-T Recommendation P.911. (1998). *Subjective audiovisual quality assessment methods for multimedia applications.*

ITU-T Study Group 12. (2007). *Definition of quality of experience (QoE).*

Jacobson, V., Smetters, D., Thornton, J., Plass, M., Briggs, N., & Braynard, R. (2009). Networking named content. *Proceedings of the 5th International Conference on Emerging Networking Experiments and Technology*, (pp. 1-12).

Jacobson, V. (2006a). *If a clean slate is the solution what was the problem?* Stanford Clean Slate Seminar.

Jacobson, V. (2006b). *A new way to look at networking.* Google Tech Talks.

Jahromi, N. T., Akbari, B., & Movaghar, A. (2010). A hybrid mesh-tree peer-to-peer overlay structure for layered video streaming. *5th International Symposium on Telecommunications,* (pp. 706-709).

Jarnikov, D., & Ozcelebi, T. (2010). Client intelligence for adaptive streaming solutions. *IEEE International Conference on Multimedia and Expo*, (pp. 1499-1504).

Java Network Simulator. (n.d.). Retrieved from http://jns.sourceforge.net

Jiang, D., & Delgrossi, L. (2008). IEEE 802.11 p: Towards an international standard for wireless access in vehicular environments. *Proceedings of IEEE Vehicular Technology Conference*, (pp. 2036-2044).

Jiang, H., & Garcia-Luna-Aceves, J. J. (2001). Performance comparison of three routing protocols for ad hoc networks. *Proceedings of IEEE Tenth International Conference on Computer Communications and Networks*, (pp. 547 – 554).

Jiang, X., Dong, Y., Xu, D., & Bhargava, B. (2003). GnuStream: A P2P media streaming system prototype. *International Conference on Multimedia and Expo*, (pp. 325-328).

Jiang, D., Taliwal, V., Meier, A., Holfelder, W., & Herrtwich, R. (2006). Design of 5.9 GHz DSR-based vehicular safety communication. *IEEE Wireless Communications*, *13*(5), 36–43. doi:10.1109/WC-M.2006.250356

Jiang, S., Liu, Y., Jiang, Y., & Yin, Q. (2004). Provisioning of adaptability to variable topologies for routing schemes in MANETs. *IEEE Journal on Selected Areas in Communications*, *22*, 1347–1356. doi:10.1109/JSAC.2004.829352

Jimeno, M., Christensen, K., & Roginsky, A. (2007). A power management proxy with a new best-of-N Bloom filter design to reduce false positives. *IEEE International Performance Computing and Communications Conference* (pp. 125-133).

Jin, X., Cheng, K.-L., & Chan, S.-H. G. (2006). SIM: Scalable island multicast for peer-to-peer media streaming. *IEEE International Conference on Multimedia and Expo,* (pp.913-916).

Jinfeng, Z., Jianwei, N., Rui, H., Jianping, H., & Limin, S. (2007). P2P-leveraged mobile live streaming. *Advanced Information Networking and Applications Workshops*, (pp. 195 – 200).

Joa-Ng, M., & Lu, I. (1999). A peer-to-peer zone-based two-level link state routing for mobile ad hoc networks. *IEEE Journal on Selected Areas in Communications*, *17*(8), 1415–1425. doi:10.1109/49.779923

Johansson, P., Larsson, T., & Hedman, N. (1999). Scenario-based performance analysis of routing protocols for mobile ad-hoc networks. *Proceedings of ACM/IEEE International Conference on Mobile Computing and Networking*, (pp. 195 – 206).

Johnson, D., Maltz, D., & Hu, Y. (2007). *The dynamic source routing protocol for mobile ad hoc networks* (DSR). Retrieved March 8, 2011, from http://www.ietf.org/rfc/rfc4728.txt

Johnson, D., Perkins, C., & Arkko, J.(2004). *Mobility support in IPv6*. IETF RFC 3775.

Johnson, D. B., & Maltz, D. A. (1996). Dynamic source routing (DSR) in ad hoc wireless networks . In Imielinski, K. (Ed.), *Mobile computing*. Kluwer Academic Publishers. doi:10.1007/978-0-585-29603-6_5

Johnson, D. B., Maltz, D. A., & Broch, J. (2001). DSR the dynamic source routing protocol for multihop wireless ad hoc networks . In Perkins, C. E. (Ed.), *Ad hoc networking*. Boston, MA: Addison-Wesley.

Jokela, P., Zahemszky, A., Rothenberg, C. E., Arianfar, S., & Nikander, P. (2009). LIPSIN: Line speed publish/subscribe inter-networking. *SIGCOMM Computer and Communications Review*, *39*(4), 195–206. doi:10.1145/1594977.1592592

Joly, A., Nathalie, M., & Marcel, B. (2001). *Audio-visual quality and interactions between television audio and video.* In International Symposium on Signal Processing and its Applications.

Jones, B. L., & McManus, P. R. (1986). Graphic scaling of qualitative terms. *SMPTE Journal*, *95*(11), 1166–1171. doi:10.5594/J04083

Joost. (n.d.). *Joost home page*. Retrieved from http://www.joost.com/

Jovanovic, M. (2001). *Modeling large-scale peer-to-peer networks and a case study of Gnutella*. Master's thesis, University of Cincinnati.

Jovanovic, M. A., Annexstein, F. S., & Berman, K. A. (2001). *Scalability issues in large peer-to-peer networks-A case study of Gnutella*. University of Cincinnati Technical Report.

Jubin, J., & Tornow, J. D. (1987). The DARPA packet radio network protocols. *Proceedings of the IEEE*, *75*(1), 21–32. doi:10.1109/PROC.1987.13702

Jumisko-Pyykkö, S., Vadakital, V. K. M., Liinasuo, M., & Hannuksela, M. M. (2006). *Acceptance of audiovisual quality in erroneous television sequences over a DVB-H channel*. In Workshop in Video Processing and Quality Metrics for Consumer Electronics.

Jurca, D., Chakareski, J., Wagner, J. P., & Frossard, P. (2007). Enabling adaptive video streaming in P2P systems. *IEEE Communications Magazine*, *45*(6), 108–114. doi:10.1109/MCOM.2007.374427

Kalogeraki, V., Gunopulos, D., & Zeinalipour-Yatzi, D. (2002). A local search mechanism for peer-to-peer networks. *Proceedings of the Eleventh International Conference on Information and Knowledge Management* (pp. 300-307).

Kao, J.-Y. (2010). The method of SVC over P2P in video streaming system. *International Conference on E-Business and E-Government,* (pp.3479-3482).

Karlsson, J., Li, H., & Eriksson, J. (2005). Real-time video over wireless ad-hoc networks. *14th International Conference on Computer Communications and Networks*, (pp. 596-607).

Karp, B., & Kung, H. (2000). GPSR: Greedy perimeter stateless routing for wireless networks. *Proceedings of the 6th Annual ACM/IEEE International Conference on Mobile Computing and Networking (MobiCom 2000)*, (pp. 243-254).

Karp, R. (1972). Reducibility among combinatorial problems. *Complexity of Computer Computations*, *40*(4), 85–103.

Kato, S., Boon, C. S., Fujibayashi, A., Hangai, S., & Hamamoto, T. (2005). *Perceptual quality of motion of video sequences on mobile terminals*. In the 7th IASTED International Conference on Signal and Image Processing.

Kazaa. (n.d.). *Home page of Kazaa*. Retrieved from www.kazaa.com

Kesting, A., Treiber, M., & Helbing, D. (1999). General lane-changing model MOBIL for car-following models. *Journal of the Transportation Research Board*, *1*, 86–94.

Khan, S., Schollmeier, R., & Steinbach, E. (2004). A performance comparison of multiple description video streaming in peer-to-peer and content delivery networks. *IEEE International Conference on Multimedia and Expo*, (pp. 503-506).

Kirsch, A., & Mitzenmacher, M. (2008). Less hashing, same performance: Building a better Bloom filter. *Random Structures and Algorithms*, *32*(2), 187–218. doi:10.1002/rsa.20208

Klecka, W. R. (1980). *Discriminant analysis*. Beverley Hills, CA: Sage Publications.

Klemm, A., Lindemann, C., & Waldhorst, O. (2004). Peer-to-peer computing in mobile ad hoc network . In Calzarossa, M., & Gelenbe, E. (Eds.), *Performance tools*. Berlin, Germany: Springer Verlag. doi:10.1007/978-3-540-24663-3_9

Knoche, H., & McCarthy, J. D. (2005). *Good news for mobile TV.* In the Wireless World Research Forum 14.

Knoche, H., McCarthy, J. D., & Sasse, M. A. (2005). *Can small be beautiful? Assessing image size requirements for mobile TV.* In the 13th ACM International Conference on Multimedia.

Knoche, H., McCarthy, J., & Sasse, M. (2006). How low can you go? The effect of low resolutions on shot types in mobile TV. *Journal of Multimedia Tools and Applications Series, 36*(1-2), 145–166. doi:10.1007/s11042-006-0076-5

Knoche, H., & Sasse, M. A. (2008). Getting the big picture on small screens: Quality of experience in mobile TV . In Ahmad, A. M. A., & Ibrahim, I. K. (Eds.), *Multimedia transcoding in mobile and wireless networks* (pp. 31–46). Information Science Reference. doi:10.4018/978-1-59904-984-7.ch003

Ko, Y., & Vaidya, N. (1998). Location-aided routing (LAR) in mobile ad hoc networks. *Proceedings of the 4th Annual ACM/IEEE International Conference on Mobile Computing and Networking*, (pp 66-75)

Koponen, T., Chawla, M., Chun, B., Ermolinski, A., Kim, K., Shenker, S., & Stoica, I. (2007). A data orineted (and beyond) network architecture. *SIGCOMM Computer and Communications Review, 34*(4), 181–192. doi:10.1145/1282427.1282402

Kortuem, G., & Schneider, J. (2001). *An application platform for mobile ad-hoc networks.* Workshop on Application Models and Programming Tools for Ubiquitous Computing, International Conference on Ubiquitous Computing.

Kou, L., Markowsky, G., & Berman, L. (1981). A fast algorithm for Steiner tree. *Acta Informatica, 15*(2), 141–145. doi:10.1007/BF00288961

Kouvelas, I., Hardman, V., & Watson, A. (1996). *Lip synchronisation for use over the Internet: Analysis and implementation.* In IEEE Global Telecommunications Conference.

Kovacevic, A., Heckmann, O., Liebau, N. C., & Steinmetz, R. (2008). Location awareness—Improving distributed multimedia communication. *Proceedings of the IEEE, 96*(1), 131–142. doi:10.1109/JPROC.2007.909913

Ko, Y. B., & Vaidya, N. H. (2000). Location aided routing (LAR) in mobile ad hoc networks. *Wireless Networks, 6*(4), 307–321. doi:10.1023/A:1019106118419

Kristiansen, S., Lindeberg, M., Rodríguez-Fernández, D., & Plagemann, T. (2010). *On the forwarding capability of mobile handhelds for video streaming over MANETs.* MobiHeld 2010 — The Second ACM SIGCOMM Workshop on Networking, Systems, and Applications on Mobile Handhelds.

Kuang, T., & Williamson, C. (2001). *RealMedia streaming performance on an IEEE 802.11b wireless LAN.* In IASTED Wireless and Optical Communications.

Kubiatowicz, J., Bindel, D., Chen, Y., Czerwinski, S., & Eaton, P. (2000). Oceanstore: An architecture for global-scale persistent storage. *ACM SIGARCH Computer Architecture News, 28*(5), 190–201. doi:10.1145/378995.379239

Kumar, S., Xu, L., Mandal, M., & Panchanathan, S. (2006). Error resiliency schemes in H.264/AVC standard. *Elsevier Journal of Visual Communication and Image Representation, 17*, 425–450. doi:10.1016/j.jvcir.2005.04.006

Kuri, J., & Kasera, S. K. (1999). Reliable multicast in multi-access wireless LANs. *Proceedings - IEEE INFOCOM*, 760–767.

Kurose, J., & Ross, K. (2005). *Computer networking: A top down approach featuring the internet* (3rd ed.). Boston, MA: Addison-Wesley.

Lambert, P., De Neve, W., Dhondt, Y., & Van de Walle, R. (2006). Flexible macroblock ordering in H.264/AVC. *Journal of Visual Communication and Image Representation, 17*(2), 358–375. doi:10.1016/j.jvcir.2005.05.008

Lamy-Bergot, C., & Candillon, B. Pesquet-Popescu, B., & Gadat, B. (2009). A simple, multiple description coding scheme for improved peer-to-peer video distribution over mobile links. *Proceedings of IEEE Packet Coding Symposium.*

La, R., & Anantharam, V. (2002). Optimal routing control: Repeated game approach. *IEEE Transactions on Automatic Control, 47*(3), 437–450. doi:10.1109/9.989076

Lee, J. S., Hsu, J., Hayashida, R., Gerla, M., & Bagrodia, R. (2002). Selecting a routing strategy for your ad hoc network. *Proceedings of ACM Symposium on Applied Computing,* (pp. 906 – 913).

Lee, T.-C., Liu, P.-C., Shyu, W.-L., & Wu, C.-Y. (2008). Live video streaming using P2P and SVC. *Proceedings of the 11th IFIP/IEEE International Conference on Management of Multimedia and Mobile Networks and Services: Management of Converged Multimedia Networks and Services*, (pp. 104-113).

Lee, Y.-C., Kim, J., Altunbasak, Y., & Mersereau, R. M. (2003). Performance comparisons of layered and multiple description coded video streaming over error-prone networks. *Proceedings of the IEEE International Conference on Communications,* (pp. 35–39).

Lei, J., Shi, L., & Fu, X. (2009). An experimental analysis of Joost peer-to-peer VoD service. *Peer-to-Peer Networking and Applications*, *3*(4), 351–362. doi:10.1007/s12083-009-0063-5

Leung, M.-F., Chan, S.-H., & Au, O. (2006). COSMOS: Peer-to-peer collaborative streaming among mobiles. *IEEE International Conference on Multimedia and Expo*, (pp. 865-868).

Levine, B. N., & Shefner, J. M. (2000). *Fundamentals of sensation and perception.* Oxford University Press.

Li, C., Yuan, C., & Zhong, Y. (2009). A novel substream extraction for scalable video coding over P2P networks. *11th International Conference on Advanced Communication Technology,* (pp.1611-1615).

Li, C., Yuan, C., & Zhong, Y. (2009). Robust and flexible scalable video multicast with network coding over P2P network. *2nd International Congress on Image and Signal Processing,* (pp.1-5).

Li, J., & Chan, S.-H. G. (2010). Optimizing Segment caching for mobile peer-to-peer interactive streaming. *IEEE International Conference on Communications,* (pp.1-5).

Li, X. Y. Z., Yao, P., & Huang, J. (2006). Implementation of P2P computing in design of MANET routing protocol. *Proceedings of the First International Multi-Symposiums on Computer and Computational Sciences,* (pp. 594–602).

Li, Z., & Herfet, T. (2008). Beacon-driven leader based protocol over a GE channel for MAC layer multicast error control. *International Journal of Communications, Network and System Science, 1*(20, 144-153.

Liao, R.-F., Wouhaybi, R., & Campbell, A. (2003). Wireless incentive engineering. *IEEE Journal on Selected Areas in Communications*, *21*(10), 1764–1779. doi:10.1109/JSAC.2003.815014

Li, B., & Yin, H. (2007). Peer-to-peer live streaming on the Internet: Issues, existing approaches, and challenges. *IEEE Communications Magazine*, *45*(6), 94–99. doi:10.1109/MCOM.2007.374425

Li, J., Jannotti, J., De Couto, D., Karger, D., & Morris, R. (2000). *A scalable location service for geographic ad hoc routing* (pp. 120–130). ACM Mobicom.

Li, M., Chen, Z., & Tan, Y.-P. (2010). Scalable video transmission over multiuser MIMO-OFDM systems. *CHINACOM*, *2010*, 1–8.

Limewire. (n.d.). *Limewire home page*. Retrieved from http://www.limewire.com/

Lin, K., Liu, N., & Luo, X. (2008). An optimized P2P based algorithm using SVC for media streaming. *Third International Conference on Communications and Networking in China,* (pp.569-573).

Liu, B., Lee, W. C., & Lee, D. L. (2005). Supporting complex multi-dimensional queries in P2P systems. *The 25th IEEE International Conference on Distributed Computing Systems,* (pp. 155-164).

Liu, D., Setton, E., Shen, B., & Chen, S. (2007). PAT: Peer-assisted transcoding for overlay streaming to heterogeneous devices. *Proceedings of 17th International Workshop on Network and Operating Systems Support for Digital Audio and Video.*

Liu, X., Raza, S., Chuah, C.-N., & Cheung, G. (2008). Network coding based cooperative peer-to-peer repair in wireless ad-hoc networks. *IEEE International Conference on Communications*, (pp. 2153 - 2158).

Liu, Y., Guo, Y., & Liang, C. (2008). A survey on peer-to-peer video streaming systems. *Journal of P2P Networking and Applications, 1*(1), 18-28.

Liu, Y.-S., Nuang, Y.-M., & Hsieh, M.-Y. (2006). Adaptive P2P caching for video broadcasting over wireless ad hoc networks. *Proceedings of Joint Conference on Information Sciences.*

Liu, B., Khorashadi, B., Du, H., Ghosal, D., Chuah, C., & Zhang, M. (2009). VGSim: An integrated networking and microscopic vehicular mobility simulation platform. *IEEE Communications Magazine*, *47*(5), 134–141. doi:10.1109/MCOM.2009.5277467

Liu, Y., Guo, Y., & Liang, C. (2008). A survey on peer-to-peer video streaming systems. *Peer-to-Peer Networking and Applications*, *1*(1), 18–28. doi:10.1007/s12083-007-0006-y

Liu, Z., Shen, Y., Ross, K. W., Panwar, S. S., & Wang, Y. (2009). LayerP2P: Using layered video chunks in P2P live streaming. *IEEE Transactions on Multimedia*, *11*(7), 1340–1352. doi:10.1109/TMM.2009.2030656

Li, X., & Wu, J. (2006). Searching techniques in peer-to-peer networks . In Wu, J. (Ed.), *Handbook of theoretical and algorithmic aspects of ad hoc, sensor, and peer-to-peer networks* (pp. 613–642). Boston, MA: Auerbach Publications.

Li, Z., & Herfet, T. (2009). MAC layer multicast error control for IPTV in wireless LANs. *IEEE Transactions on Broadcasting*, *55*(2), 353–362. doi:10.1109/TBC.2009.2016502

Lloyd, E., Maclean, R., & Stirling, A. (2006). *Mobile TV- Results from the BT Movio DAB-IP pilot in London.* EBU Technical Review. Retrieved from http://www.ebu.ch/en/technical/trev/trev_frameset-index.html

Lombard, M., Ditton, T. B., Grabe, M. E., & Reich, R. D. (1997). The role of screen size in viewer responses to television fare. *Communication Reports*, *10*(1), 95–106. doi:10.1080/08934219709367663

López, D., Gonźalez, F., Bellido, L., & Alonso, A. (2006). Adaptive multimedia streaming over IP based on customer oriented metrics. In *IEEE International Symposium on Computer Networks* (pp. 185-191).

Lopez-Fuentes, F. A. (2010). Adaptive mechanism for P2P video streaming using SVC and MDC. *International Conference on Complex, Intelligent and Software Intensive Systems*, (pp.457-462).

López-Fuentes, F. A., & Steinbach, E. (2008). Adaptive multisource video multicast. *Proceedings of the IEEE International Conference on Multimedia and Expo,* (pp. 457–460).

Lopez-Fuentes, F. A. (2011). P2P video streaming combining SVC and MDC. *International Journal of Appied Mathematical Computer Science*, *21*(2), 295–306. doi:10.2478/v10006-011-0022-1

Lua, E. K., Crowcroft, J., Pias, M., Sharma, R., & Lim, S. (2005). A survey and comparison of peer-to-peer overlay network schemes. *IEEE Communications Surveys & Tutorials*, *7*(2), 72–93. doi:10.1109/COMST.2005.1610546

Lumetta, S., & Mitzenmacher, M. (2007). Using the power of two choices to improve Bloom filters. *Internet Mathematics*, *4*(1), 17–33. doi:10.1080/15427951.2007.10129136

Lund, A. (1993). The influence of video image size and resolution on viewing-distance preferences. *SMPTE Journal*, *102*, 406–415. doi:10.5594/J15915

Luo, H., Ci, S., & Wu, D. (2009). A cross-layer optimized distributed scheduling algorithm for peer-to-peer video streaming over multi-hop wireless mesh networks. *6th Annual IEEE Communications Society Conference on Sensor, Mesh and Ad Hoc Communications and Networks,* (pp.1-9).

Lv, Q., Cao, P., Cohen, E., Li, K., & Schenker, S. (2002). Search and replication in unstructured peer-to-peer networks. *Proceedings of the 16th International Conference on Supercomputing,* (pp. 84-95).

Lv, Q., Ratnasamy, S., & Schenker, S. (2002). Can heterogeneity make Gnutella scalable? *1st International Workshop on Peer-to-Peer Systems,* (pp. 94-103).

Maccarthaigh, M. (2007). *Joost network architecture*. (PowerPoint slides available online.)

Magharei, N., Rejaie, R., & Guo, Y. (2007). Mesh or multiple-tree: A comparative study of live P2P streaming approaches. *IEEE INFOCOM*, (pp. 1424-1432).

Mao, S., Lin, S., Panwar, S., & Wang, Y. (2001). Reliable transmission of video over ad-hoc networks using automatic repeat request and multi-path transport. *Proceedings of IEEE Vehicular Technology Conference*, (pp. 615-619).

Marfia, G., Pau, G., Di Rico, P., & Gerla, M. (2007). P2P streaming systems: A survey and experiments. *ST Journal of Research*.

Mastronarde, N., Turaga, D. S., & van der Schaar, M. (2006). Collaborative resource management for video over wireless multi-hop mesh networks. *IEEE International Conference on Image Processing,* (pp. 1297-1300).

Mastronarde, N., Turaga, D. S., & van der Schaar, M. (2007). Collaborative resource exchanges for peer-to-peer video streaming over wireless mesh networks. *IEEE Journal on Selected Areas in Communications, 25*(1), 108–118. doi:10.1109/JSAC.2007.070111

Matolak, D. W. (2008). Channel modeling for vehicle-to-vehicle communications. *IEEE Communications Magazine, 46*(5), 76–83. doi:10.1109/MCOM.2008.4511653

Mauve, M., Widmer, J., & Hartenstein, H. (2001). A survey on position-based routing in mobile ad-hoc networks. *IEEE Network Magazine, 15*(6), 30–39. doi:10.1109/65.967595

Mayer-Patel, K. (2007). Systems challenges of media collectives supporting media collectives with adaptive MDC. *Proceedings of the 15th International Conference on Multimedia,* (pp. 625-630).

Maymounkov, P., & Mazieres, D. (2002). Kademlia: A peer-to-peer information system based on the xor metric. *First International Workshop on Peer-to-Peer Systems,* (pp. 53-65).

McCarthy, J., Sasse, M. A., & Miras, D. (2004). *Sharp or smooth? Comparing the effects of quantization vs. frame rate for streamed video.* In SIGCHI Conference on Human Factors in Computing Systems.

Meddour, D., Mushtaq, M., & Ahmed, T. (2006). Open issues in P2P multimedia. *Proceedings of, MULTICOMM2006,* 43–48.

Medjiah, S., Ahmed, T., Mykoniati, E., & Griffin, D. (2011). Scalable video streaming over P2P networks: A matter of harmony? *IEEE 16th International Workshop on Computer Aided Modeling and Design of Communication Links and Networks (CAMAD),* (pp.127-132).

Mehrotra, S., & Weidong, Z. (2009). Rate-distortion optimized client side rate control for adaptive media streaming. *IEEE International Workshop on Multimedia Signal Processing,* (pp. 1-6).

Meyer, D., Zhang, L., & Fall, K. (2007). *Report from the IAB workshop on routing and addressing. IETF, RFC 4984.* Informational.

Milojicic, D., Kalogeraki, V., et al. (2003). *Peer-to-peer computing.* HP Labs. Technical Report.

Minar, N. (2001). Distributed systems topologies. *Proceedings of the O'Reilly P2P and Web Services Conference.*

Miroll, J., Zhao, L., & Herfet, T. (2010). *Wireless feedback cancellation for leader-based MAC layer multicast protocols: Measurement and simulation results on the feasibility of leader-based MAC protocols using feedback cancellation on the 802.11aa wireless multicast network.* IEEE 14th International Symposium on Consumer Electronics.

Mirshokraie, S., & Hefeeda, M. (2010). Live peer-to-peer streaming with scalable video coding and networking coding. *Proceedings of the First Annual ACM SIGMM Conference on Multimedia Systems,* (pp. 123-132).

Mitra, S., & Pyne, S. (2010). Security issue of a route selection algorithm in multihomed mobile networks. *International Conference on VLSI Design and Communication Systems,* (pp. 31-37).

Mitra, S. (2010). Dynamic route optimization in a multihomed mobile network. *International Journal of Computational Vision and Robotics, 2*(1), 121–135. doi:10.1504/IJCVR.2010.036076

Mitra, S., & Pyne, S. (2011). Fuzzy logic based route optimization in a multihomed mobile networks. *Wireless Networks, 1*(17), 213–229. doi:10.1007/s11276-010-0274-y

Moore, M. S., Mitra, S. K., & Foley, J. M. (2002). *Defect visibility and content importance implications for the design of an objective video fidelity metric.* In International Conference on Image Processing.

Moreira, J., Antonello, R., Fernandes, S., Kamienski, C., & Sadok, D. (2008). A step towards understanding Joost IPTV. In *Networks Operations and Management Operations* (pp. 211-214).

Morpheus. (n.d.). *Morpheus home page.* Retrieved from www.morpheus.com

Moshref, M., Motamedi, R., Rabiee, H. R., & Khansari, M. (2010). LayeredCast - A hybrid peer-to-peer live layered video streaming protocol. *5th International Symposium on Telecommunications,* (pp. 663-668).

MPEG-4 AVC Verification Tests. (2011). Report on the formal verification tests on AVC (ISO/IEC 14496-10 | ITU-T Rec. H.264). Retrieved from http://www.chiariglione.org/mpeg/quality_tests.htm

MPEG-4 Overview – (V.21 – Jeju version). (2002). *International Organization for Standardization, ISO/IEC JTC1/SC29/WG11, Coding of moving pictures and audio.*

Muller, J., Magedanz, T., & Fiedler, J. (2009). NNodeTree: A scalable peer-to-peer live streaming overlay architecture for next-generation-networks. *Network Protocols and Algorithms, 1*(2).

Munasinghe, K. S., & Jamalipour, A. (2010). Route optimization for roaming heterogeneous multi-homed mobile networks. *International Conference on Signal Processing and Communication Systems*, (pp. 1-7).

Murphy, S., Searles, M., Rambeau, C., & Murphy, L. (2004). *Evaluating the impact of network performance on video streaming quality for categorised video content.* In International Packet Video Workshop.

Murthy, C. S. R., & Manoj, B. S. (2004). *Ad hoc wireless networks, architecture and protocols*. New York, NY: Prentice Hall.

Murthy, S., & Garcia-Luna-Aceves, J. J. (1996). An efficient routing protocol for wireless networks. *Mobile Networks and Applications, 1*(2), 183–197. doi:10.1007/BF01193336

Mushtaq, M., & Ahmed, T. (2007). Hybrid overlay networks management for real-time multimedia streaming over P2P networks. *Proceedings of the 10th IFIP/IEEE International Conference on Management of Multimedia and Mobile Networks and Services: Real-Time Mobile Multimedia Services*, (pp. 1-13).

Mushtaq, M., & Ahmed, T. (2008). Smooth video delivery for SVC based media streaming over P2P networks. *5th IEEE Consumer Communications and Networking Conference,* (pp.447-451).

Mykoniati, E., Landa, R., Spirou, S., Clegg, R., Latif, L., Griffin, D., & Rio, M. (2008). Scalable peer-to-peer streaming for live entertainment content. *IEEE Communications Magazine, 46*(12), 40–46. doi:10.1109/MCOM.2008.4689206

MySpace. (n.d.). *Myspace home page*. Retrieved from www.myspace.com

Napster. (n.d.). *Napster home page*. Retrieved from http://free.napster.com/

Narita, N. (1993). Graphic scaling and validity of Japanese descriptive terms used in subjective-evaluation tests. *SMPTE Journal, 102*, 616–622. doi:10.5594/J03770

Narkhede, N. S., & Kant, N. (2009). The emerging H.264/AVC advanced video coding standard and its applications. *Proceedings of International Conference on Advances in Computing, Communication and Control*, (pp. 300-309).

Natsheh, E., & Wan, T. C. (2008). Adaptive and fuzzy approaches for nodes affinity management in wireless ad-hoc networks. *Mobile Information Systems, 4*(4), 273–298.

Nguyen, A. T., Eliassen, F., & Welzl, M. (2011). Quality-aware membership management for layered peer-to-peer streaming. *IEEE Consumer Communications and Networking Conference,* (pp. 720-724).

Nguyen, A. T., Li, B., & Eliassen, F. (2010). Chameleon: Adaptive Peer-to-peer streaming with network coding. *Proceedings of the IEEE INFOCOM,* (pp.1-9).

Nguyen, A. T., Li, B., & Eliassen, F. (2010). Quality- and context-aware neighbor selection for layered peer-to-peer streaming. *IEEE International Conference on, Communications*, (pp.1-6).

Ni, S.-Y., Tseng, Y.-C., Chen, Y.-S., & Sheu, J.-P. (1999). The broadcast storm problem in a mobile ad hoc network. *Proceedings of the Fifth Annual ACM/IEEE International Conference on Mobile Computing and Networking (MOBICOM 99).*

Nikander, P., & Tarkoma, S. (2010). *Data in context: The role of context in the RTFM network architecture.* Ideas for a paper within the PSIRP project.

Niraula, N. B., Kanchanasut, K., & Laouiti, A. (2009). Peer-to-peer live video streaming over mobile ad hoc network. *Proceedings of the 2009 International Conference on Wireless Communications and Mobile Computing: Connecting the World Wirelessly,* (pp. 1045-1050).

Nokia. (2004). *Quality of experience (QoE) of mobile services: Can it be measured and improved?* Whitepaper.

Nokia. (2005). *Finnish mobile TV: Pilot results*. Retrieved from http://www.mobiletv.nokia.com/download_counter.php?file=/onAir/finland/files/RI_Press.pdf

NORM. (n.d.). Retrieved from http://cs.itd.nrl.navy.mil/work/norm/

Nunes, R. P., & Cruz, R. S. (2010). Scalable video coding distribution in peer—to-peer architecture. *10a Conferencia sobre Redes de Computadores Conference*, (pp. 95-100).

Ogier, R., Lewis, M., & Templin, F. (2004). *Topology dissemination based on reverse-path forwarding (TBRPF)*. Retrieved March 8, 2011, from http://www.ietf.org/rfc/rfc3684.txt

Ogier, R., Templin, F., & Lewis, M. (2004). *Topology dissemination based on reverse-path forwarding (TBRPF)*. Work in Progress. Retrieved March 8, 2011, from http://www.faqs.org/rfcs/rfc3684.html

O'Hara, B., & Petrick, A. (2004). *IEEE 802.11 handbook: A designer's companion*. Chichester, UK: Wiley & Sons.

Oishi, J., Asukura, K., & Watanabe, T. (2006). A communication model for inter-vehicle communication simulation systems based on properties of urban area. *International Journal of Computer Science and Network Security*, *6*, 213–219.

Olista. (2007). *Live trials by Olista with European mobile operators demonstrate common barriers for mobile data services*. Press Release 120207-1.

Oliveira, L. B., Siqueira, I. Q., & Loureiro, A. A. (2003). Evaluation of ad-hoc routing protocols under a peer-to-peer application. *Proceedings of IEEE Wireless Communications and Networking*, (pp. 16-20).

Oram, A. (2001). *Peer-to-peer: Harnessing the benefits of a disruptive technology*. Sebastopol, CA: O'Reilly Media.

Oredope, A., Liotta, A., Morphett, J., & Roper, I. (2008b). P2P-SIP in highly volatile networks. *IEEE International Conference on Next Generation Mobile Applications, Services and Technologies* (pp. 76-82).

Oredope, A., & Liotta, A. (2008a). Service provisioning in the IP multimedia subsystem . In Mahbubur, R. S. (Ed.), *Multimedia technologies: Concepts, methodologies, tools, and applications* (pp. 491–500). Hershey, PA: Information Science Reference. doi:10.4018/978-1-59904-953-3.ch036

Ostermann, J., Bormans, J., List, P., Marpe, D., Narroschke, M., & Pereira, F. (2004). Video coding with H. 264/AVC: Tools, performance and complexity. *IEEE Circuits and Systems Magazine*, *4*(1), 7–28. doi:10.1109/MCAS.2004.1286980

Ozbek, N., & Tumnali, T. (2005). A survey on the H.264/AVC standard. *Turk Journal of Electrical Engineering*, *13*, 287–302.

Ozbilgin, T., & Sunay, M. O. (2008). Scalable video streaming in wireless peer-to-peer networks. *IEEE 16th Signal Processing, Communication and Applications Conference*, (pp. 1-4).

Padmanabhan, V. N., Wang, H. J., Chou, P. A., & Sripanidkulchai, K. (2002). Distributing streaming media content using cooperative networking. *Proceedings of the 12th International Workshop on Network and Operating Systems Support for Digital Audio and Video*.

Padmanabhan, V., Wang, H., & Chou, P. (2003). Resilient peer-to-peer streaming. *Proceedings of 11th IEEE International Conference on Network Protocols*, (pp. 16- 27).

Paila, T., Luby, M., Lehtonen, R., Roca, V., & Walsh, R. (2004). FLUTE: File delivery over unidirectional transport. *IETF, RFC 3926*.

Pallis, G., & Vakali, A. (2006). Content delivery networks. *Communications of the ACM*, *49*(1), 101–106. doi:10.1145/1107458.1107462

Pantos, R., & May, W. (2011). *HTTP live streaming*. IETF Draft. Retrieved from http://tools.ietf.org/html/draft-pantos-http-live-streaming-06

Papadopouli, H. S. M. (2002). Effects of power conservation, wireless coverage and cooperation on data dissemination among mobile devices. *Proceedings of ACM Symposium on Mobile Ad Hoc Networking and Computing*, (pp. 117-127).

Park, V. D., & Corson, M. S. (1997). A highly adaptive distributed routing algorithm for mobile wireless networks. *Proceedings - IEEE INFOCOM*, 1405–1413.

Patrick, A. S., Singer, J., Corrie, B., Nöel, S., Khatib, K., Emond, B., et al. (2004). A QoE sensitive architecture for advanced collaborative environments. In the *First International Conference on Quality of Service in Heterogeneous Wired/Wireless Networks* (pp. 319-322).

PeerCast. (n.d.). *PeerCast home page*. Retrieved from http://www.peercast.org/

Pei, G., Gerla, M., & Chen, T.-W. (2000). Fisheye state routing: A routing scheme for ad hoc wireless networks. *Proceedings of IEEE International Conference on Communications*, (pp. 70-74).

Pei, G., Gerla, M., & Chen, T.-W. (2002). *Fisheye state routing: A routing scheme for ad hoc wireless networks.* IETF internet draft.

Peltotalo, J., Harju, J., Saukko, M., Vaatamoinen, L., Bouazizi, I., Curcio, I. D. D., & van Gassel, J. (2009). A real-time peer-to-peer streaming system for mobile networking environment. *Proceedings of the IEEE INFOCOM Workshops,* (pp.1-7).

Peng, Z., Cheng, G., Yang, Z., & Chen, J. (2009). A scalable peer-to-peer video-on-demand system with asynchronous transfer. *Fourth International Conference on Communications and Networking in China,* (pp. 1-6).

Perkins, C. (2002). *IP mobility support for IPv4.* IETF RFC 3344.

Perkins, C. E., & Bhagwat, P. (1994). Highly dynamic destination-sequenced distance-vector routing (DSDV) for mobile computers. *Proceedings of ACM SIGCOMM's Conference on Communications Architectures, Protocols and Applications*, (pp. 234 – 244).

Perkins, C. E., & Royer, E. M. (1999). Ad-hoc on-demand distance vector routing (AODV). *Proceedings of the 2nd IEEE Workshop on Mobile Computing Systems and Applications*, (pp. 90-100).

Perkins, C. E., Royer, E. M., & Das, S. R. (2003). *Ad hoc on demand distance vector (AODV) routing.* IETF Internet Draft, 2003.

Perkins, C. E., Royer, E. M., Das, S. R., & Marina, M. K. (2001). Performance comparison of two on-demand routing protocols for ad hoc networks. *IEEE Personal Communication, 8*(1), 16-28.

Perkins, C., Belding-Royer, E., & Das, S. (2003). *Ad hoc on-demand distance vector (AODV) routing.* Retrieved March 8, 2011, from http://www.ietf.org/rfc/rfc3561.txt

Perkins, C. (2000). *Ad hoc networking.* Boston, MA: Addison Wesley.

Perkins, C. E., Royer, E. M., & Das, S. R. (2003). *Ad hoc on demand distance vector (AODV) routing*. IETF Internet Draft.

Picconi, F., & Massoulie, L. (2008). Is there a future for mesh-based live video streaming? *Eighth International Conference on Peer-to-Peer Computing,* (pp. 289-298).

Pirzada, A., Portmann, M., & Indulska, J. (2006). Performance comparison of multi-path AODV and DSR protocols in hybrid mesh networks. *14th IEEE International Conference on Networks*, (pp. 1-6).

Pirzada, A., Portmann, M., & Indulska, J. (2007). Hybrid mesh ad-hoc on-demand distance vector. *Proceeding of the Thirtieth Australasian Conference on Computer Science*, (pp. 49-58).

Poikselkä, M., & Mayer, G. (2009). *The IMS: IP multimedia concepts and services*. Chichester, UK: John Wiley & Sons Inc.

Postel, J. (1980). *User datagram protocol.* RFC 768.

Pouwelse, J., Grabacki, M., Epema, D., & Sips, H. (2005). The BitTorrent P2P file-sharing system: Measurements and analysis. *Peer-to-Peer Systems, IV*, 205 216. doi:10.1007/11558989_19

PPLIVE. (n.d.). PPLIVE. Retrieved from www.pplive.com

Prim, R. (1957). Shortest connection networks and some generalizations. *The Bell System Technical Journal, 36*, 1389–1401.

PSIRP: Publish Subscribe Internet Routing Paradigm. (n.d.). Retrieved from http://www.psirp.org

Pucha, H., Das, S. M., & Hu, Y. C. (2004). Ekta: An efficient DHT substrate for distributed applications in mobile ad hoc networks. *Proceedings of the 6th IEEE Workshop on Mobile Computing Systems and Applications*, (pp. 163- 173).

PURSUIT: Pursuing a publish subscribe Internet. (n.d.). Retrieved from http://www.fp7-pursuit.eu/PursuitWeb

Qadri, N. N., & Liotta, A. (2008). A comparative analysis of routing protocols for MANETs. *Proceedings of IADIS International Conference on Wireless Applications and Computing* (pp. 149-154).

Qadri, N. N., Alhaisoni, M., & Liotta, A. (2008). Mesh based P2P streaming over MANET. *Proceedings of the 6th International Conference on Advances in Mobile Computing and Multimedia,* (pp. 29-34).

Qiu, X., Wu, C., Lin, X., & Lau, F. (2009). InstantLeap: Fast neighbor discovery in P2P VoD streaming. *Proceedings of the 18th International Workshop on Network and Operating Systems Support for Digital Audio and Video,* (pp. 19–24).

Qiu, D., & Srikant, R. (2004). Modeling and performance analysis of BitTorrent-like peer-to-peer networks. *ACM SIGCOMM Computer Communication Review, 34*(4), 367–378. doi:10.1145/1030194.1015508

Raatikainen, P. (2005). On developing networking technologies and pervasive services. *Wireless Personal Communications, 33,* 261–269. doi:10.1007/s11277-005-0571-4

Rainey, S., Petty, G., & Cutten, M. W. (2005). Content acquisition challenges . In *IP Video Implementation and Planning Guide* (pp. 43–61). United States Telecom Association.

Rajagopalan, S., & Chien-Chung, S. (2006). A cross-layer decentralized BitTorrent for mobile ad hoc networks. *Proceedings of the 3rd Annual International Conference on Mobile and Ubiquitous Systems - Workshops,* (pp. 1-10).

Ramasubramanian, V., Haas, Z. J., & Sirer, E. G. (2002). SHARP: A hybrid adaptive routing protocol for mobile ad hoc networks. *Proceedings of the 4th ACM International Symposium on Mobile Ad Hoc Networking & Computing,* (pp. 303-314).

Ramzan, N., & Izquierdo, E. (2011). Scalable and adaptable media coding techniques for future Internet . In Domingue, J. (Eds.), *The future internet* (pp. 381–389). Berlin, Germany: Springer Verlag. doi:10.1007/978-3-642-20898-0_27

Ramzan, N., Quacchio, E., Zgaljic, T., Asioli, S., Celetto, L., Izquierdo, E., & Rovati, F. (2011). Peer-to-peer streaming of scalable video in future internet applications. *IEEE Communications Magazine, 49*(3), 128–135. doi:10.1109/MCOM.2011.5723810

Rappaport, T. (2001). *Wireless communications: Principles and practice*. New York, NY: Prentice Hall.

Rappaport, T. S. (2002). *Wireless communications: Principles and practice*. Upper Saddle River, NJ: Prentice Hall.

Ratnasamy, S., Francis, P., Handley, M., Karp, R., & Schenker, S. (2001). A scalable content-addressable network. *ACM Conference on Applications, Technologies, Architectures, and Protocols for Computer Communications,* (pp. 161-172).

Reeves, B., & Nass, C. (1998). *The media equation: How people treat computers, television, and new media like real people and places*. Chicago, IL: University of Chicago Press.

Reibman, A., Jafarkhani, H., Wang, Y., & Orchard, M. (2001). Multiple description video using rate-distortion splitting. *Proceedings of the IEEE International Conference on Image Processing,* (pp. 978-981).

Rekhter, Y., & Li, T. (1995). *A border gateway protocol 4 (BGP-4).* IETF, RFC 1771 (Draft Standard), obsolete by RFC 4271.

Reza, R. (2006). Anyone can broadcast video over the internet. *Communications of the ACM, 49*(11), 55–57. doi:10.1145/1167838.1167863

Ripeanu, M., Foster, I., & Iamnitchi, A. (2002). Mapping the Gnutella network: Properties of large-scale peer-to-peer systems and implications for system design. *IEEE Internet Computing Journal, 6*(1), 50–57.

Robert, E. M. (1986). On judging quality by price: Price dependent expectations, not price dependent preferences. *Southern Economic Journal, 52*(3), 665–672. doi:10.2307/1059265

Rodolakis, G., Laouiti, A., & Merarihi, N. A. (2007). *Multicast overlay spanning tree protocol for ad hoc networks*. International Conferences on Wireless/Wired Internet Communications.

Rosen, E., Viswanathan, A., & Callon, R. (2001). *Multiprotocol label switching architecture. IETF, RFC 3031.* Proposed Standard.

Rothenberg, C. E., Jokela, P., Nikander, P., Särelä, M., & Ylitalo, J. (2009). Self-routing denial-of-service resistant capabilities using in-packet Bloom filters. *5th European Conference on Computer Network Defense (EC2ND),* (pp. 46-51).

Rothenberg, C. E., Verdi, F., & Magalhães, M. (2008). Towards a new generation of information-oriented internetworking architectures. *Proceedings of the 2008 ACM CoNEXT First Workshop on Re-architecting the Internet (Re-Arch08)*, (pp. 1-6).

Rothenberg, C. E., Macapuna, C., Magalhães, F., Verdi, F., & Wiesmaier, A. (2010). In-packet Bloom filters: Design and networking applications. *Elsevier Computer Networks*, *55*(5), 1364–1378.

Rowstron, A., & Druschel, P. (2001). Pastry: scalable, distributed object location and routing for large-scale peer-to-peer systems. *IFIP/ACM International Conference on Distributed Systems (Middleware)*, (pp. 329-350).

Sadka, A. H. (2002). *Compressed video communications*. Chichester, UK: John Wiley & Sons, Ltd. doi:10.1002/0470846712

Sadka, A. H. (2002). *Compressed video communications*. Chichester, UK: Wiley & Sons. doi:10.1002/0470846712

Sánchez, Y., Schierl, T., Hellge, C., & Wiegand, T. (2010). P2P group communication using scalable video coding. *17th IEEE International Conference on Image Processing*, (pp. 4445-4448).

Sandvine. (2011). *Global internet phenomena report - Spring 2011*. Retrieved from http://www.wired.com/images_blogs/epicenter/2011/05/SandvineGlobalInternetSpringReport2011.pdf

Santos, R., Rangel, V., Mendez, A., García-Ruiz, M. A., & Edwards-Block, A. (2010). Analyzing IEEE 802.11g and IEEE 802.16e technologies for single-hop inter-vehicular communications. In M. Watfa (Ed.), Advances in vehicular ad-hoc networks: Developments and challenges (pp. 120-149). Hershey, PA: IGI Global. Retrieved from http://www.igi-global.com/bookstore/chapter.aspx?titleid=43168

Santos, R., Edwards, A., Edwards, R., & Seed, N. (2005). Performance evaluation of routing protocols in vehicular ad-hoc networks. *International Journal of Ad Hoc and Ubiquitous Computing*, *1*(1/2), 80–91. doi:10.1504/IJAHUC.2005.008022

Särelä, M., Ott, J., & Ylitalo, J. (2010b). Fast inter-domain mobility with in-packet Bloom filters. *Proceedings of the 5th ACM International Workshop on Mobility in Evolving Internet Architecture (MobiArch10)*, (pp. 9-14).

Särelä, M., Rothenberg, C. E., Aura, T., Zahemszky, A., & Nikander, P. (2011). Forwarding anomalies in Bloom filter based multicast. *Proceedings of the IEEE INFOCOM*, (pp. 2399-2407).

Särelä, M., Rothenberg, C. E., Zahemszky, A., Nikander, P., & Ott, J. (2010a). *BloomCast: Security in Bloom filter based multicast*. 15th Nordic Conference in Secure IT Systems (Nordsec)

Satamoto, K., Aoyama, S., Asahara, S., Yamashita, K., & Okada, A. (2008). Lecture Notes in Computer Science: *Vol. 5068. Relationship between viewing distance and visual fatigue in relation to feeling of involvement* (pp. 232–239). Berlin, Germany: Springer.

Schaumann, J. (2002). Analysis of the zone routing protocol. Retrieved March 8, 2011, from http://www.netmeister.org/misc/zrp/zrp.pdf

Schierl, T., & Wiegand, T. (2004). H. 264/AVC rate adaptation for internet streaming. *14th International Packet Video Workshop*.

Schierl, T., Ganger, K., Hellge, C., Wiegand, T., & Stockhammer, T. (2006). SVC-based multisource streaming for robust video transmission in mobile ad hoc networks. *IEEE Wireless Communications*, *13*(5), 96–103. doi:10.1109/WC-M.2006.250365

Schollmeier, R., & Schollmeier, G. (2002). *Why peer-to-peer (P2P) does scale: An analysis of P2P traffic patterns*. Second International Conference on Peer-to-Peer Computing.

Schollmeier, R., Gruber, I., & Finkenzeller, M. (2002). Routing in mobile ad hoc and peer-to-peer networks, a comparison. *Revised Papers from the NETWORKING 2002 Workshops on Web Engineering and Peer-to-Peer Computing*, (pp. 172 – 186).

Schulzrinne, H., Casner, S., Frederick, R., & Jacobson, V. (2003). RTP: A transport protocol for real-time applications. *IETF*, RFC 3550.

Schulzrinne, H., Rao, A., & Lanphier, R. (1998). Real time streaming protocol (RTSP). *IETF, RFC 2326.*

Schwarz, H., Marpe, D., & Wiegand, T. (2007). Overview of the scalable video coding extension of the H. 264/AVC standard. *IEEE Transactions on Circuits and Systems for Video Technology*, *17*(1), 1103–1120. doi:10.1109/TCSVT.2007.905532

Scott, J., Crowcroft, J., Hui, P., & Diot, C. (2006). Haggle: A networking architecture designed around mobile users. *Proceedings of the 3rd Annual Conference on Wireless On-demand Netwrok Systems and Services*, (pp. 78-86).

Secker, A., & Taubman, D. (2003). Lifting-based invertible motion adaptive transform framework for highly scalable video compression. *IEEE Transactions on Image Processing*, *12*(12), 1530–1542. doi:10.1109/TIP.2003.819433

Sentinelli, A., Marfia, G., Gerla, M., Kleinrock, L., & Tewari, S. (2007). Will IPTV ride the peer-to-peer stream? *IEEE Communications Magazine*, *45*(6), 86–92. doi:10.1109/MCOM.2007.374424

Setton, E., Yoo, T., Zhu, X., Goldsmith, A., & Girod, B. (2005). Cross-layer design of ad hoc networks for real-time video streaming. *IEEE Wireless Communications Magazine*, *12*(4), 59–65. doi:10.1109/MWC.2005.1497859

Shah, P., & Pâris, J. F. (2007). Peer-to-peer multimedia streaming using BitTorrent. *IEEE International Performance, Computing, and Communications Conference*, (pp. 340-347).

Shen, H., Zhao, L., Li, Z., & Li, J. (2010). A DHT-aided chunk-driven overlay for scalable and efficient peer-to-peer live streaming. *39th International Conference on Parallel Processing*, (pp. 248-257).

Shibata, N., Mori, M., & Yasamuto, K. (2007). P2P video broadcast based on per-peer transcoding and its evaluation on PlanetLab. *Proceedings of the 19th IASTED International Conference on Parallel and Distributed Computing and Systems*.

Shima, K., Uo, Y., Ogashiwa, N., & Uda, S. (2006). Operational experiment of seamless handover of a mobile router using multiple care-of address registration. *Journal of Networks*, *3*(1), 23–30.

Shollmeier, R., Gruber, I., & Finkenzeller, M. (2002). *Routing in mobile ad hoc and peer-to-peer networks, a comparison.* International Workshop on Peer-to-Peer Computing.

Si, J., Zhuang, B., Cai, A., & Cheng, Y. (2009). Layered network coding and hierarchical network coding for peer-to-peer streaming. *Pacific-Asia Conference on Circuits, Communications and Systems,* (pp. 139-142).

Siller, M. (2006). *An agent-based platform to map quality of service to experience in active and conventional networks.* Ph.D Thesis, University of Essex, Colchester, U.K.

Skype. (n.d.). *Skype home page*. Retrieved from http://www.skype.com

Sodergard, C. (2003). *Mobile television - Technology and user experiences report on the mobile-TV project*. VTT Information Technology.

Soldani, D., Li, M., & Cuny, R. (2006). *QoS and QoE management in UMTS cellular systems*. Chichester, UK: Wiley and Sons. doi:10.1002/9780470034057

Son, N., & Jeong, S. (2008). An effective error concealment for H.264/AVC. *IEEE 8th International Conference on Computer and Information Technology Workshops*, (pp. 385-390).

Sopcast. (n.d.). *Sopcast home page*. Retrieved from www.sopcast.com

Stallings, W. (2000). *Wireless communications and networks*. New York, NY: Prentice Hall.

Steinmetz, R. (1996). Human perception of jitter and media synchronization. *IEEE Journal on Selected Areas in Communications*, *14*(1), 61–72. doi:10.1109/49.481694

Stepanov, I., Hähner, J., Becker, C., Tian, J., & Rothermel, K. (2003). A meta-model and framework for user mobility in mobile networks. *Proceedings of the 11th IEEE International Conference on Networks*, (pp. 231- 238).

Stevens, W. R. (1993). TCP/IP illustrated: *Vol. 1. The protocols*. Boston, MA: Addison-Wesley.

Stockhammer, T. (2011). Dynamic adaptive streaming over HTTP–design principles and standards. *Proceedings of the Second Annual ACM Conference on Multimedia Systems*, (pp. 133-144).

Stoica, I., Morris, R., Karger, D., Kaashoek, M. F., & Balakrishnan, H. (2001). Chord: A scalable peer-to-peer lookup service for internet applications. *Proceedings of ACM SIGCOM.*

Stoica, I., & Morris, R. (2003). Chord: A scalable peer-to-peer lookup service for Internet applications. *IEEE/ACM Transactions on Networking, 11*(1), 17–32. doi:10.1109/TNET.2002.808407

Stoica, I., Morris, R., Karger, D., Kaashoek, M. F., & Balakrishnan, H. (2001). Chord: A scalable peer-to-peer lookup service for internet applications. *ACM SIGCOMM Computer Communication Review, 31*(4), 149–160. doi:10.1145/964723.383071

Stoica, I., Morris, R., Nowell, D. L., Karger, D. R., Kaashoek, M. F., & Dabek, F. (2003). Chord: A scalable peer-to-peer lookup protocol for internet applications. *IEEE/ACM Transactions on Networking, 11*(1), 17–32. doi:10.1109/TNET.2002.808407

Stojmenovic, I. (2002). Position-based routing in ad hoc networks. *IEEE Communications Magazine, 40*(7), 128–134. doi:10.1109/MCOM.2002.1018018

Sugawara, M., Mitani, K., Kanazawa, M., Okano, F., & Nishida, Y. (2006). Future prospects of HDTV-Technical trends towards 1080p. *SMPTE Motion Imaging Journal, 115*(1), 10–15. doi:10.5594/J11496

Sullivan, G., & Wiegand, T. (2005). Video compression—From concepts to the H. 264/AVC standard. *Proceedings of the IEEE, 93*(1), 18–31. doi:10.1109/JPROC.2004.839617

Sun, T., Tamai, M., Yasumoto, K., & Shibata, N. (2005). MTcast: Robust and efficient P2P-based video delivery for heterogeneous users. *Proceedings of the 9th International Conference on Principles of Distributed Systems.*

Sun, X., Wu, F., Li, S., Gao, W., & Zhang, Y.-Q. (2002). Seamless switching of scalable video bitstreams for efficient streaming. *ISCAS, 2002*, 385–388.

Taal, J. R., Pouwelse, J. A., & Lagendijk, R. L. (2000). Scalable multiple description coding for video distribution in P2P networks. *Proceedings of the Picture Coding Symposium.*

Taal, J., Pouwelse, J., & Lagendijk, R. (2004). Scalable multiple description coding for video distribution in P2P networks. *24th Picture Coding Symposium.*

Takai, M., Martin, J., & Bagrodia, R. (2001) Effects of wireless physical layer modeling in mobile ad hoc networks. *Proceedings of the 2nd ACM International Symposium on Mobile Ad Hoc Networking & Computing*, (pp. 87-94).

Tanenbaum, A. (2003). *Computer networks.* Upper Saddle River, NJ: Prentice Hall.

Tang, Y., Luo, J. G., Zhang, Q., Zhang, M., & Yang, S. Q. (2007). Deploying P2P networks for large-scale live video-streaming service. *IEEE Communications Magazine, 45*(6), 100–106. doi:10.1109/MCOM.2007.374426

Tarkoma, S. (Ed.). (2008). *Conceptual architecture of PSIRP including subcomponent description.* Deliverable D2.2 PSIRP Project.

Tarkoma, S., Rothenberg, C. E., & Lagerspetz, E. (2012). (To appear). Theory and practice of Bloom filters for distributed systems. *IEEE Communications Survey and Tutorials.* doi:10.1109/SURV.2011.031611.00024

Taubman, D., Reji, M., Maestroni, D., & Tubaro, S. (2004). *SVC core experiment 1- Description of UNSW contribution, MPEG doc. m11441.*

Taylor, I., Downard, I., Adamson, B., & Macker, J. (2006). Agentj: Enabling Java NS-2 simulations for large scale distributed multimedia applications. *Second International Conference on Distributed Frameworks for Multimedia*, (pp. 1-7).

Teunissen, K. (1996). The validity of CCIR quality indicators along a graphical scale. *SMPTE Journal, 105*(1), 144–149. doi:10.5594/J04650

Thomos, N., Argyropoulos, S., Boulgouris, N., & Strintzis, M. (2005). Error-resilient transmission of H. 264/AVC streams using flexible macroblock ordering. *Second European Workshop on the Integration of Knowledge, Semantic, and Digital Media Techniques*, (pp. 183-189).

Tonguz, O., Viriyasitavat, W., & Bai, F. (2009). Modeling urban traffic: A cellular automata approach. *IEEE Communications Magazine, 47*(5), 142–150. doi:10.1109/MCOM.2009.4939290

Touch, J., & Pingali, V. (2008). The RNA metaprotocol. *Proceedings of the IEEE International Conference on Computer Commications*, (pp. 1-6).

Touch, J., Wang, Y., & Pingali, V. (2006). *A recursive network architecture*. ISI Technical Report 626.

Touch, J. (2002). Those pesky NATs (network address translator). *IEEE Internet Computing*, *6*(4), 96. doi:10.1109/MIC.2002.1020334

Treiber, M., Hennecke, A., & Helbing, D. (2000). Congested traffic states in empirical observations and microscopic simulations. *Physical Review E: Statistical Physics, Plasmas, Fluids, and Related Interdisciplinary Topics*, *62*, 1805–1824. doi:10.1103/PhysRevE.62.1805

Triantafillou, P., & Pitoura, T. (2004). Towards a unifying framework for complex query processing over structured peer-to-peer data networks. *First International Workshop on Databases, Information Systems, and Peer-to-Peer Computing*, (pp. 169-183).

Trossen, D. (Ed.). (2009). *Architecture definition, component descriptions, and requirements*. Deliverable D2.3 PSIRP Project.

Trung, H. D., Benjapolakul, W., & Duc, P. M. (2007). Performance evaluation and comparison of different ad hoc routing protocols. *Computer Communications*, *30*(11-12), 2478–2496. doi:10.1016/j.comcom.2007.04.007

UMTS Forum report 11. (2000). *Enabling UMTS third generation services and application*.

Vaidya, N. H. (2004). Mobile ad hoc networks: Routing, MAC and transport issues. *Proceedings of the IEEE International Conference on Computer Communication*, tutorial.

Van der Auwera, G., David, P. T., Reisslein, M., & Karam, L. J. (2008). Traffic and quality characterization of the H.264/AVC scalable video coding extension. *Advances in Multimedia*, *2*, 1–27. doi:10.1155/2008/164027

Varsa, V., Hannuksela, M., & Wang, Y. (2001). Non-normative error concealment algorithms. *ITU-T VCEG Doc: VCEG-N62*, Vol. 62.

VEOH. (n.d.). *VEOH home page*. Retrieved from www.veoh.com

VQEG Final Report from the Video Quality Experts Group on the Validation of Objective Models of Video Quality Assessment. Retrieved from http://www.vqeg.org

Wakikawa, R., Devarapalli, V., & Thubert, P. (2004). *Inter home agents protocol (HAHA)*. IETF Internet Draft, draft-wakikawa-mip6-nemo-haha-01.

Wakikawa, R., Ernst, T., & Nagami, K. (2006). *Multiple care-of addresses registration. IETF, draft-wakikawa-mobileip-multiplecoa-05, RFC 5648*. Retrieved from.

Waldman, M., Rubin, A., & Cranor, L. (2000). Publius: A robust, tamper-evident, censorship-resistant web publishing system. *9th Conference on USENIX Security Symposium*, (pp. 59-72).

Wallach, D. (2003). A survey of peer-to-peer security issues. *International Conference on Software Security*, (pp. 42-57).

Wang, J., Huang, C., & Li, J. (2008). On ISP-friendly rate allocation for peer-assisted VoD. *Proceeding of the 16th ACM International Conference on Multimedia*, (pp. 279-288).

Wang, D., & Liu, J. (2008). A dynamic skip list-based overlay for on-demand media streaming with VCR interactions. *IEEE Transactions on Parallel and Distributed Systems*, *19*(4), 503–514. doi:10.1109/TPDS.2007.70748

Wang, Y., Reibman, A. R., & Lin, S. (2005). Multiple description coding for video delivery. *Proceedings of the IEEE*, *93*(1), 57–70. doi:10.1109/JPROC.2004.839618

Watson, A., & Sasse, M. A. (1998). *Measuring perceived quality of speech and video in multimedia conferencing applications*. In ACM Multimedia Conference.

Watson, A. B. (1994). Image compression using the discrete cosine transform. *Mathematica Journal*, *4*(1), 81–88.

Weber, S., Andrews, J. G., & Jindal, N. (2007). The effect of fading, channel inversion, and threshold scheduling on ad hoc networks. *IEEE Transactions on Information Theory*, *53*(11), 4127–4149. doi:10.1109/TIT.2007.907482

Webster, A. A., Jones, C. T., Pinson, M. H., Voran, S. D., & Wolf, S. (1993). An objective video quality assessment system based on human perception . In *SPIE Human Vision*. Visual Processing and Digital Display. doi:10.1117/12.152700

Wellens, M., Petrova, M., Riihijarvi, J., & Mahonen, P. (2005). Building a better wireless mousetrap: Need for more realism in simulations. *Proceedings of the Second Annual Conference on Wireless On-demand Network Systems and Services*, (pp. 150-157).

Wenger, S. (2003). H.264/AVC over IP. *IEEE Transactions on Circuits and Systems for Video Technology*, *13*(7), 645–656. doi:10.1109/TCSVT.2003.814966

Wenger, S., Knorr, G., Ott, J., & Kossentini, F. (1998). Error resilience support in H.263. *IEEE Transactions on Circuits and Systems for Video Technology*, *8*(7), 867–877. doi:10.1109/76.735382

Wiegand, T., & Sullivan, G. (2003). *Final draft ITU-T recommendation and final draft international standard of joint video specification* (ITU-T Rec. H.264 ISO/IEC 14 496-10 AVC).

Wiegand, T., Sullivan, G., Reichel, J., Schwarz, H., & Wien, M. (2006). *Joint draft 8 of SVC amendment.* ISO/IEC JTC1/SC29/WG11 and ITU-T SG16 Q.6 9 (JVT-U201), 21st Meeting, Hangzhou, China, Oct. 2006.

Wiegand, T., Noblet, L., & Rovati, F. (2009). Scalable video coding for IPTV services. *IEEE Transactions on Broadcasting*, *55*(2), 527–538. doi:10.1109/TBC.2009.2020954

Wiegand, T., Sullivan, G. J., Bjontegaard, G., & Luthra, A. (2003). Overview of the H.264/AVC video coding standard. *IEEE Transactions on Circuits and Systems for Video Technology*, *13*(7), 560–576. doi:10.1109/TCSVT.2003.815165

Wien, M., Schwarz, H., & Oelbaum, T. (2007). Performance analysis of SVC. *IEEE Transactions on Circuits and Systems for Video Technology*, *17*(1), 1194–1203. doi:10.1109/TCSVT.2007.905530

Winkler, S., & Dufaux, F. (2003). *Video quality evaluation for mobile applications.* In SPIE Visual Communications and Image Processing.

Winkler, S., & Faller, C. (2005). Maximizing audiovisual quality at low bitrates. In *Proceedings of Workshop on Video Processing and Quality Metrics.*

Woo, S.-C. M., & Singh, S. (2001). Scalable routing protocol for ad hoc networks. *Wireless Networks*, *7*(5), 513–529. doi:10.1023/A:1016726711167

Wu, J.-C., Peng, K.-J., Lu, M.-T., Lin, C.-K., Cheng, Y.-H., & Huang, P. … Chen, H. H. (2006). *HotStreaming: Enabling scalable and quality IPTV services.* Retrieved April 11th from http://nslab.ee.ntu.edu.tw/publication/conf/hotstreaming-iptv06.pdf

Wu, C., Li, B., & Zhao, S. (2008). Exploring large-scale peer-to-peer live streaming topologies. *ACM Transactions on Multimedia Computing . Communications and Applications*, *4*(3), 1–23.

Wu, H. R., & Rao, K. R. (2005). *Digital video image quality and perceptual coding*. Boca Raton, FL: CRC Press.

Wu, J. (Ed.). (2006). *Handbook on theoretical and algorithmic aspects of sensor, ad hoc wireless, and peer-to-peer networks*. Boca Raton, FL: Auerbach Publications.

Xiao, X., Shi, Y., Gao, Y., & Zhang, Q. (2009). LayerP2P: A new data scheduling approach for layered streaming in heterogeneous networks. *Proceedings of the IEEE INFOCOM,* (pp.603-611).

Xie, H., Gao, L., Zhang, L., Zhang, Z., & Yang, M. (2009). An efficient caching mechanism for video-on-demand service over peer-to-peer network. *Eighth International Conference on Embedded Computing Scalable Computing and Communications,* (pp.251-256).

Xu, K., Gerla, M. M., & Bae, S. (2002). How effective is the IEEE 802.11 RTS/CTS handshake in ad hoc networks? *Proceedings of IEEE GLOBECOM* (pp. 72-76).

Xu, Q., Mak, T., Ko, J., & Sengupta, R. (2004). Vehicle-to-vehicle safety messaging in DSRC. *Proceedings of the 1st ACM International Workshop on Vehicular Ad Hoc Networks*, (pp. 19-28).

Xu, T., & Cai, Y. (2007). Streaming in MANET: Proactive link protection and receiver-oriented adaptation. *IEEE International Conference on Performance, Computing, and Communications.*

Xu, Z., Mahalingam, M., & Karlsson, M. (2003). Turning heterogeneity into an advantage in overlay routing. *Proceedings of the IEEE INFOCOM*, (pp. 1499-1509).

Yan, L. (2005). Can P2P benefit from MANET? Performance evaluation from users' perspective. *Proceedings of the International Conference on Mobile Sensor Networks*, (pp. 1026-1035).

Yan, L., Sere, K., & Zhou, X. (2004). Towards an integrated architecture for peer-to- peer and ad hoc overlay network applications. *Proceedings of the 10th IEEE International Workshop on Future Trends of Distributed Computing Systems*, (pp. 312–318).

Yang, B., & Garcia-Molina, H. (2002). *Efficient search in peer-to-peer networks*. 22nd International Conference on Distributed Computing Systems.

Yang, B., & Garcia-Molina, H. (2003). *Designing a super-peer network*. 19th International Conference on Data Engineering.

Yang, J., & Gong, S. (2009). A content-based layered multiple description coding scheme for robust video transmission over ad hoc networks. *Proceedings of the Second International Symposium on Electronic Commerce and Security*, (pp. 21-24).

Yang, X., & Vaidya, N. (2005). On physical carrier sensing in wireless ad hoc networks. *Proceedings - IEEE INFOCOM*, 2525–2535.

Yong, L., Guo, Y., & Liang, C. (2008). A survey on peer-to-peer video streaming systems. *Peer-to-Peer Networking and Applications*, *1*, 18–28. doi:10.1007/s12083-007-0006-y

Yoon, S., Mao, M., & Kalman, M. (2006). Rate-distortion optimized video streaming for scalable H.264. *IEEE International Conference on Multimedia and Expo*, (pp. 2157-2160).

Zahemszky, A., Császár, A., Nikander, P., & Rothenberg, C. E. (2009a). Exploring the pubsub routing & forwarding space. *IEEE International Conference on Communications, Workshop on the Network of the Future*, (pp. 1-6).

Zahemszky, A., & Arianfar, S. (2009b). *Fast reroute for stateless multicast* (pp. 1–6). Ultra Modern Telecommunications & Workshops.

Zambelli, A. (2009). *IIS smooth streaming technical overview*. Microsoft Corporation. Retrieved from http://users.atw.hu/dvb-crew/applications/documents/IIS_Smooth_Streaming_Technical_Overview.pdf

Zapater, M. N., & Bressan, N. (2007). *A proposed approach for quality of experience assurance for IPTV.* In First International Conference on Digital Society.

Zattoo. Zattoo home page from www.zattoo.com

Zeng, P., & Jiang, Y. (2010). Robust scalable video with network coding for peer-to-peer streaming. *International Conference on Multimedia Technology*, (pp.1-5).

Zeng, X., Bagrodia, R., & Gerla, M. (1998). GloMoSim: A library for parallel simulation of large-scale wireless networks. *Workshop on Parallel and Distributed Simulation,* (pp. 154-161).

Zhang, G., & Yuan, C. (2010). Self-adaptive peer-to-peer streaming for heterogeneous networks using scalable video coding. *12th IEEE International Conference on Communication Technology,* (pp.1390-1393).

Zhang, L., et al. (2010). *Named data networking (NDN) project.* PARC, Technical Report NDN-0001.

Zhang, Q. (2005). Video delivery over wireless multi-hop networks. *Proceedings of the International Symposium on Intelligent Signal Processing and Communication Systems*, (pp. 793-796).

Zhang, J., Liu, L., Ramaswamy, L., & Pu, C. (2008). PeerCast: Churn-resilient end system multicast on heterogeneous overlay networks. *Journal of Network and Computer Applications*, *31*(4), 821–850. doi:10.1016/j.jnca.2007.05.001

Zhao, B. Y., Kubiatowicz, J. D., & Joseph, A. D. (2001). *Tapestry: An infrastructure for fault-resilient wide-area location and routing*. Technical report UCB//CSD-01-1141, U.C. Berkeley.

Zhao, B., & Huang, L. (2004). Tapestry: A resilient global-scale overlay for service deployment. *IEEE Journal on Selected Areas in Communications*, *22*(1), 41–53. doi:10.1109/JSAC.2003.818784

Zheng, H., & Boyce, J. (2001). An improved UDP protocol for video transmission over Internet to wireless networks. *IEEE Transactions on Multimedia*, *3*(3), 356–365. doi:10.1109/6046.944478

Zhengye, L., Shen, Y., Panwar, S. S., Ross, K. W., & Wang, Y. (2007). P2P video live streaming with MDC: Providing incentives for redistribution. *Proceedings of the IEEE International Conference on Multimedia and Expo*, (pp. 48-51).

Zhou, M.-T., Harada, H., Kong, P.-Y., Ang, C.-W., Ge, Y., & Pathmasuntharama, J. S. (2009). A method to deliver AODV routing messages using WiMAX mesh MAC control messages in maritime wireless networks. *Proceedings of IEEE International Symposium on Personal, Indoor and Mobile Radio Communications*, (pp. 1537-1541). doi:10.1109/PIMRC.2009.5449888

Zigbee Alliance. (2003). *Zigbee specification*. ZigBee Document 053474r06, Version 1.

Zou, X., Ramamurthy, B., & Magliveras, S. (2002). Routing techniques in wireless ad hoc networks –Classification and comparison. *Proceedings of the Sixth World Multiconference on Systemics, Cybernetics and Informatics*, (pp. 1-6).

Zuo, D.-H., Du, X., & Yang, Z.-K. (2007). Hybrid search algorithms for P2P media streaming distribution in ad hoc networks. *Proceedings of the 7th International Conference on Computational Science*, (pp. 873-876).

Zygmunt, Y. G., & Marc, R. P. (2001). ZRP: A hybrid framework for routing in ad hoc networks . In Perkins, C. M. (Ed.), *Ad hoc networking* (pp. 221–253). Boston, MA: Addison-Wesley.

About the Contributors

Martin Fleury holds a first degree from Oxford University, UK and an additional Maths/Physics based Bachelor degree from the Open University, Milton Keynes, UK. He obtained an MSc in Astrophysics from QMW College, University of London, UK in 1990 and an MSc from the University of South-West England, Bristol in Parallel Computing Systems in 1991. He holds a PhD in Parallel Image-Processing Systems from the University of Essex, Colchester, UK. He is currently a Visiting Fellow at the University of Essex, UK, having previously worked as a Senior Lecturer at the University of Essex. Martin has authored or co-authored over two hundred and twenty five articles on topics such as document and image compression algorithms, performance prediction of parallel systems, software engineering, reconfigurable hardware, and vision systems. His current research interests are video communication over MANs, WLANs, PANs, BANs, MANETs, and VANETs. He is a guest editor of a number of journal special issues and a reviewer for a good number of journals in the IEEE Transactions series. He is also an external examiner for the Arab Open University in Kuwait and the Open University in the UK.

Nadia N. Qadri received her Master's of Engineering (Communication Systems and Networks) and Bachelors of Engineering (Computer Systems), from Mehran University of Engineering and Technology, Jamshoro, Pakistan in 2004 and 2002, respectively. From 2006 to 2010, she took a PhD at the University of Essex, UK, where she was awarded her Doctorate with the title P2P Video Streaming in Mobile and Vehicular Ad Hoc Networks. She has more than six years of teaching and research experience at renowned universities of Pakistan viz. Mehran University of Engineering & Technology, Fatima Jinnah Women's University and COMSATS Institute of Information Technology (CIIT), Pakistan. She is currently an Assistant Professor and In-charge of the Telecommunication Engineering Program at CIIT, Wah Campus in Pakistan. Her research interests include video streaming for mobile ad hoc networks and vehicular ad hoc networks, P2P streaming, wireless sensor networks. She is a reviewer of many International conferences and journals including IEEE transactions.

* * *

Florence Agboma is a Lecturer in computer networks, multimedia applications, and system design. She received her PhD in electronics communication systems in 2009 and her MSc degree in computer information and networks in 2005, both from the University of Essex. Her research interests include content delivery networks, 3D QoE modelling over current infrastructure and multimedia transmission – with published articles in these areas. She also has interests and is an active participant in amateur radio.

Majed Alhaisoni is Assistant Professor at University of Ha'il, College of Computer Science and Engineering, Saudi Arabia. He has completed his PhD in the school of Computer Science and Electronic Engineering at the University of Essex, in 2010. Dr Alhaisoni earned his BSc in Computer Science from Qassim University, Saudi Arabia (2005) and his MSc in Computer and information Networks from the University of Essex, UK (2007). His achievements include three best paper awards and two distinguished achievements awards. He has contributed to peer-review journals and international conferences with 20 articles. His research interests encompass heterogeneous wireless systems, peer-to-peer networking and applications, video coding, and distributed computing.

Raul C. Almeida Jr. graduated as an Electrical Engineer from the Universidade Federal de Pernambuco (UFPE), Recife, Brazil, in 1999. In 2001 and 2004 he received, respectively, the M.Sc. and Ph.D degrees in Electrical Engineering from Universidade Estadual de Campinas (UNICAMP), Campinas, SP, Brazil. In 2005 he was a post-doc researcher at UFPE and in 2006 he joined the University of Essex, UK as a Senior Research Officer, where he has participated in a number of projects. He has experience mainly in telecommunication systems, network optimization, analytical and numerical modeling, optical resource allocation and management, physical impairments, quality of service guaranties, among others. His current research interests lie in the areas of elastic optical networks and future Internet technologies.

Fabrizio Bertone holds an MSc in Multimedia and Software Engineering from Politecnico di Torino, Italy. His expertise is on adaptive video streaming and its application to quality of experience management. He carried out his final-year project at Eindhoven University of Technology, the Netherlands, where he developed a QoE-aware streaming client for MPEG Dynamic Adaptive Streaming over HTTP (DASH). His interests include network security, communication protocols, and Internet multimedia.

Laura Carrea holds a 5-years degree (integrated Master) in Physics from the University of Turin, Italy. From 1999 to 2004 she worked at the Faculty of Electrical Engineering and Information Technology at the Chemnitz University of Technology, Germany with a Fellowship in the frame of the European Marie Curie TMR Network "Radar Polarimetry: Theory and Application." Her research interests focused on the theoretical aspects of polarization and scattering for radar remote sensing. After that, she has been with the University of Essex with a Fellowship in the frame of the European Marie Curie RTN Network "AMPER" continuing her work on polarimetry. Since January 2009 she has been working towards a Ph.D. degree at the School of Computer Science and Electronic Engineering, University of Essex, UK on routing and forwarding aspects of information centric networks. Her research interests include future Internet architectures for content dissemination, probabilistic data structures, information centric routing, and forwarding functionality within heterogeneous technologies.

Homer H. Chen received the Ph.D. degree in Electrical and Computer Engineering from University of Illinois at Urbana-Champaign, Urbana. Since August 2003, he has been with the College of Electrical Engineering and Computer Science, National Taiwan University, where he is Irving T. Ho Chair Professor. Prior to that, he held various R&D management and engineering positions with US companies over a period of 17 years, including AT&T Bell Labs, Rockwell Science Center, iVast, and Digital Island. He was a US delegate of the ISO and ITU standards committees and contributed to the development of many new interactive multimedia technologies that are now part of the MPEG-4 and JPEG-2000 standards. His research interests lie in the broad area of multimedia processing and communications. Dr. Chen was an Associate Editor of *IEEE Transactions on Circuits and Systems for Video Technology* from 2004 to 2010. He served as Associate Editor for *IEEE Transactions on Image Processing* from 1992 to 1994, Guest Editor for *IEEE Transactions on Circuits and Systems for Video Technology* in 1999 and Associate Editorial for *Pattern Recognition* from 1989 to 1999.

Arthur Edwards received his Master's degree in Education from the University of Houston in 1985. He has been a Researcher-Professor at the University of Colima since 1985, where he has served in various capacities. He has been with the School of Telematics since 1998. His primary areas of research are Computer Assisted Language Learning (CALL), distance learning, collaborative learning, multimodal learning, and mobile learning. The primary focus of his research is presently in the area of mobile collaborative learning.

Dan Grois received the B.Sc. degree in Electrical and Computer Engineering from the Ben-Gurion University (BGU), Beer-Sheva, Israel, in 2002. He received the M.Sc. degree in the Communication Systems Engineering field at BGU, in 2006, and is currently a Ph.D. candidate in the Communication Systems Engineering Department at BGU. Dan has extensive work experience in the fields of communications engineering and electronics in leading companies, such as Motorola and IAI (Israel Aerospace Industries). Dan has a significant number of academic publications presented in peer-reviewed international conferences and published by IEEE and Elsevier scientific journals, Wiley publisher, etc. In addition, Dan is a referee of top-tier conferences and international journals, such as the *IEEE Trans. in Image Processing, Journal of Visual Communication and Image Representation*, Elsevier, *IEEE Sensors,* and *SPIE Optical Engineering*. Dan is a member of the ACM and a Senior Member of the IEEE. Dan's research interests include image and video coding and processing, video coding standards, region-of-interest scalability, computational complexity and bit-rate control, computer vision, and future multimedia applications and systems.

Javier Gálvez Guerrero is a Research and Development Engineer at the Ubiquitous Internet Technologies Unit (UITU) at i2CAT Foundation. He received his Telecommunication Engineering degree from the Technical University of Catalonia (UPC) in 2008 and his Master's degree in Telematics Engineering in 2011. During his professional career he has been involved in different research projects including the design and development of an IPTV platform, a remote healthcare telematic service and the design and implementation of a DTN-based architecture for an environment monitoring service for smart cities. His main areas of interest are multimedia streaming services and platforms, wireless access technologies, seamless network mobility, and QoS management in wireless networks.

Ofer Hadar received the B.Sc., the M.Sc. (*cum laude*), and the Ph.D. degrees from the Ben-Gurion University of the Negev, Israel, in 1990, 1992, and 1997, respectively, all in Electrical and Computer Engineering. From August 1996 to February 1997, he was with CREOL at Central Florida University, Orlando, FL, as a Research Visiting Scientist, working on the angular dependence of sampling MTF and over-sampling MTF. From October 1997 to March 1999, he was Post-Doctoral Fellow in the Department of Computer Science at the Technion-Israel Institute of Technology, Haifa. Currently he is a faculty member at the Communication Systems Engineering Department at Ben-Gurion University of the Negev. His research interests include: image and video compression, routing in ATM networks, flow control in ATM networks, packet video, and transmission of video over IP networks and video rate smoothing and multiplexing. Hadar also works as a consultant for several hi-tech companies such as, EnQuad Technologies Ltd in the area of MPEG-4, and Scopus in the area of video compression and transmission over satellite networks. Hadar is a member of the SPIE and a Senior Member of the IEEE.

Anis Laouiti has been an Associate Professor at Telecom SudParis since 2006. Before this, he did his PhD research work and worked as a research engineer within Hipercom team at Inria-Rocquencourt, where he participated to the OLSR routing protocol design (RFC 3626). He is interested in unicast routing protocols as well as multicast protocols for data dissemination in ad hoc wireless networks. A special focus of his is on wireless ad hoc networks, used for emergency situations, and VANETs, where data dissemination is of great importance. He is also interested in mobile sensor network deployment and redeployment techniques in order to offer wide area coverage. His research covers different aspects in wireless ad hoc and mesh networks including protocol design, performance evaluation, and implementation test-beds.

Ling Lin received his MSc in Computer Science with distinction from the University of Surrey. Since 2008, he has been a research engineer at Vodafone R&D, UK. His expertise is on peer-to-peer systems, service management, security, voice over IP and mobile streaming. He is currently working on mobile presence, office communication, and cloud computing.

Antonio Liotta is a full Professor at Eindhoven University of Technology (TU/e) since 2008 and leads a research team on autonomic networks. Before joining TU/e, he held various academic positions in the UK, working on pervasive networks and service management. He is currently studying cognitive systems in the context of optical, wireless, and sensor networks. He has recently authored the book "Networks for Pervasive Services: Six Ways to Upgrade the Internet" (Springer, 2011 http://bit.ly/pervasive-networks).

Vlado Menkovski is a PhD researcher at Eindhoven University of Technology (The Netherlands) and he is a member of the Autonomic Networks group since 2009. He holds an MSc degree from Carnegie Mellon University (USA) in Information Networking. His research focus is on intelligent systems in multimedia services and autonomic networks. Recent work includes: the application of machine learning to objective and subjective video quality estimation, and quality of experience optimization for video delivery services.

Sulata Mitra is an Associate Professor in the Department of Computer Science and Technology, Bengal Engineering and Science University, Shibpur, West Bengal, India. She has published many research papers including a book chapter on mobile computing. Her current research areas include ad-hoc network and mobile computing.

Nobal B. Niraula received the B.E. in Computer Engineering from Pulchowk Campus, Tribhuvan University, Nepal in 2005, the M.E. in Information and Communication Technology and the M.Sc. in communication networks and services in 2008 from Asian Institute of Technology, Thailand and Telecom SudParis, France, respectively. From 2008 to 2010, he was a research engineer at INRIA, Saclay, France where he worked in Semantic Web, database systems, and P2P networks. He has started his PhD at the University of Memphis, USA since August, 2010. His research interests are primarily in natural language processing, Semantic Web, data mining, machine learning, P2P, and ad hoc networks.

Victor Rangel received a B.Eng (Hons) degree in Computer Engineering of the Engineering Faculty of the National Autonomous University of Mexico (UNAM) in 1996, an M.Sc in Telematics from the University of Sheffield, U.K. in 1998, and his Ph.D. in performance analysis and traffic scheduling in cable networks in 2002, from the University of Sheffield. Since 2002, he has been with the School of Engineering, UNAM, where he is currently a Research-Professor in telecommunications networks. His research focuses on fixed, mesh and mobile broadband wireless access networks, QoS over IP, traffic shaping and scheduling.

Raúl Aquino Santos graduated from the University of Colima with a BE in Electrical Engineering, received his MS degree in Telecommunications from the Centre for Scientific Research and Higher Education in Ensenada, Mexico in 1990. He holds a PhD from the Department of Electrical and Electronic Engineering of the University of Sheffield, England, UK. Since 2005, he has been with the College of Telematics, at the University of Colima, where he is currently a Research-Professor in telecommunications networks. His current research interests include wireless and sensor networks.

Jason Yao is now a consultant to executives at Fujitsu Laboratories of America. Previously Jason held various R&D and management positions at ASUS Computer International, Foxlink International and Fujitsu Network Communications. His research interest includes digital signal processing, audio and video compression, internet traffic engineering, P2P networks, and multimedia streaming over the Internet. Jason has published over 30 peer-reviewed journal and conference papers in these fields, and co-invented two patents related to P2P networks. Jason holds a PhD in Electrical and Computer Engineering from University of California, Santa Barbara and an MBA degree from Santa Clara University. During 2004-2007, he was a Visiting Scholar at National Taiwan University where he conducted research and taught Computer Network/Entrepreneurship classes. Since 2008 he has been an Associate Professor at Silicon Valley University and taught Software Engineering, Software Quality Assurance, and Cloud Computing.

Index

E

F

G

H

I

J

K

L

M

N

O

P

Q

R

S

T

U

V

W

Z